Proceedings
of the
1st MINE VENTILATION SYMPOSIUM

March 29-31, 1982
The University of Alabama
University (Tuscaloosa), Alabama

Howard L. Hartman
Symposium Chairman and Proceedings Editor

Sponsored by
Underground Ventilation Committee of SME-AIME
joint committee of the Coal Division
and Mining and Exploration Division
Department of Mineral Engineering
University of Alabama

Published by
Society of Mining Engineers
of
American Institute of Mining, Metallurgical, and Petroleum Engineers, Inc.
New York, NY•1982

Printed in the United States of America
by BookCrafters, Inc., Chelsea, Michigan

Library of Congress Catalog Card Number 82-071996
ISBN 0-89520-298-0

FOREWORD

Late in March 1982, 203 registrants gathered in Tuscaloosa, Alabama, at The University of Alabama to inaugurate the first national conference on mine ventilation ever held in the United States. Ventilation engineers, government scientists, equipment manufacturers, and university faculty snd students made up the audience that heard 32 state-of-the-art technical papers.

It was a first-time event in a branch of mining as old as the industry itself. Agreement was unanimous that another five centuries ought not to pass before the next symposium was held! In fact, action was taken by the Underground Ventilation Committee of SME-AIME, sponsor of the meeting, that symposia be held every two years, rotated among interested universities. The 2nd Mine Ventilation Symposium was approved by the UVC for the University of Nevada-Reno for 1985.

Subject matter covered in this the 1st Symposium varied over the all-inclusive environmental field of mine ventilation and was presented in 7 sessions over 3 days. Nine papers dealt with computer applications and modeling, 6 with auxiliary ventilation, 4 with fans and ventilation shafts, 3 with mine fires, 3 with methane and explosive gas control, 3 with mine air cooling, and 3 with general topics. The striking number of papers devoted to or encompassing the use of the computer in modeling systems and solving problems as well as monitoring and controlling ventilation circuits demonstrated how pervasive an influence it has become in the mining industry and mine ventilation field.

A fourth day of the Symposium was devoted to highly successful field trips to the surface and underground operations at the deep coal mines of Jim Walter Resources, Inc. in Brookwood, AL.

The Underground Ventilation Committee, sponsor of the Symposium and newly formed by SME-AIME, is itself a creature of the surging interest in the maintenance of a safe and comfortable underground environment. A group of ventilation practicioners--both mining and tunneling--petitioned the Society of Mining Engineers for recognition in fall 1980 and was officially granted unit (technical) committee status a year later. Led by six recently appointed officers, UVC plans to program at both the AIME Annual Meeting and SME Fall Meeting as well as sponsor the mine ventilation symposium and encourage the publication of significant papers in the field.

For further particulars on the organization and operation of UVC, see the introductory paper in this volume, "New Recognition for an Old Technical Field."

Georgius Agricola would be astonished indeed by the state of mine ventilation in the late 20th century. A scientific field dedicated to the preservation of human life and comfort, it has come a long way since the black days of medieval mining. It is to the further advancement of underground ventilation technology that this series of symposiums and the activities of UVC are dedicated.

Howard L. Hartman
The University of Alabama

SYMPOSIUM COMMITTEE

James L. Banfield, Jr.
Chief, Ventilation
Mine Safety & Health Administration
4800 Forbes Avenue
Pittsburgh, PA 15213

Jack Barry
President, Bonanza Fans
1622 Browning
Irvine, CA 92714

Robert W. Dalzell
Chief, Pittsburgh Health Technology Center
Mine Safety & Health Administration
4800 Forbes Avenue
Pittsburgh, PA 15213

Charles J. Hager
Consultant
1807 Seneca Road
Vestavia Hills, AL 35216

Howard L. Hartman
Professor and Holder of Endowed Chair of Mining Engineering
Department of Mineral Engineering
The University of Alabama
P.O. Box 1468
University, AL 35486

Bruce R. Johnson
Senior Research Engineer
Magma Copper Company
Box 37
Superior, AZ 85273

John D. Kalasky
Manager of Mine Ventilation
Island Creek Coal Company
349 Redding Road
Lexington, KY 40502

Richard J. Kline
Senior Engineer
Foster-Miller Associates
350 Second Avenue
Waltham, MA 02154

Frank Laird
Manager, Environmental Control
Anaconda Copper Company
555 17th Street
Denver, CO 80217

Pierre Mousset-Jones
Professor, Mackay School of Mines
University of Nevada-Reno
Reno, NV 89557

R. V. Ramani
Professor, Department of Mining Engineering
The Pennsylvania State University
Mineral Sciences Building, Room 126
University Park, PA 16802

John W. Stevenson
General Manager of Ventilation
Jim Walter Resources, Inc.
Box C-79
Birmingham, AL 35283

Robert H. Stevenson
Vice President/Sales
Peabody ABC Company
188 Firwood Drive
Bridgeville, PA 15017

Woods G. Talman
Consultant
2624 Thorntree Drive
Pittsburgh, PA 15241

PROGRAM

TECHNICAL SESSIONS

Monday, March 29

Presiding: John W. Stevenson, Jim Walter Resources, Inc., Brookwood, AL

8:00 AM Registration — Continuing Education Center

9:00 Welcome
Richard L. Sanford, Head, Department of Mineral Engineering
Joab Thomas, President
The University of Alabama

SESSION I: FANS AND VENTILATION SHAFTS

SESSION II: AUXILIARY VENTILATION

TECHNICAL SESSIONS (Continued)

Tuesday, March 30

Presiding: Bruce R. Johnson, Magma Copper Co., Magma, AZ

SESSION III: COMPUTER APPLICATIONS AND MODELING

SESSION IV: MINE FIRES

TECHNICAL SESSIONS (Continued)

Wednesday, March 31

Presiding: Pierre Mousset-Jones, University of Nevada, Reno, NV

SESSION V: METHANE AND EXPLOSIVE GAS CONTROL

SESSION VI: MINE AIR COOLING

SESSION VII: GENERAL VENTILATION

FIELD TRIPS

3:15-6:30 Surface and Underground Installations
Jim Walter Resources, Inc., Brookwood, AL

Thursday, April 1

8:00-noon Surface Installations
Jim Walter Resources, Inc., Brookwood, AL

TABLE OF CONTENTS

Keynote Address

Howard L. Hartman
Symposium Chairman and
Professor of Mining Engineering and Holder of Endowed Chair
The University of Alabama
University, Alabama

NEW RECOGNITION FOR AN OLD TECHNICAL FIELD

Howard L. Hartman

Symposium Chairman and
Professor of Mining Engineering and Holder of Endowed Chair

The University of Alabama
University, Alabama 35486

For The University of Alabama and the Underground Ventilation Committee of SME-AIME, I welcome you most cordially to the 1st Mine Ventilation Symposium.

To use a time-honored phrase, this _is_ a historic event. Thirty two papers are to be presented by 55 authors in this, the first U.S. and North American conference devoted to underground ventilation.

Mine ventilation engineers have long lamented the absence of an appropriate forum in which matters related to the underground atmosphere could be argued. True, an international congress is now being held every four years, but few of us are fortunate enough to be able to attend when it is abroad. The Mine Ventilation Society of South Africa enrolls a number of Americans as members, and its journal is a welcome depository for some of our papers, but again, that organization is based overseas.

Following the II International Mine Ventilation Congress at the University of Nevada in 1979, a number of us met to consider the formation of an American forum. An organizing committee was formed to pursue the establishment of an "Underground Ventilation Organization," preferably affiliated with the Society of Mining Engineers of AIME and open as well to non-AIME members from other North American countries and to representatives of the tunneling and underground space communities.

We succeeded in our mission partially if not wholly. I chaired the UVO committee, and we explored a variety of alternatives, both within and without AIME. A separate society is attractive but economically unrealistic at this time (perhaps feasible in the future if a sponsoring "angel" steps forward to subsidize it). Linking up with SME-AIME carries obvious advantages, but unfortunately some constraints too. Although the negotiating group for UVO held out for autonomy, none is possible under the AIME constitution (short of creating yet another society). Finally, we agreed on committee status, with the group to be jointly based in the Coal Division and Mining and Exploration Division. In November 1981, the _Underground Ventilation Committee_ was born.

Our structure is simple (the Committee's statement of "Purpose and Organization" and a list of officers are appended to these remarks). Anyone may affiliate with the Committee and attend our meetings (e.g., we meet tomorrow during lunch), although voting and holding office are limited to SME members from either of the two divisions. As you'll note, our purpose is "to foster engineering interest and technological progress in the ventilation of mines, tunnels, and other underground openings." Programmatically, our plans are to conduct joint Coal and M&E technical sessions at both AIME and SME meetings as well as to sponsor symposia every other year on university campuses on a rotating basis. Session papers and symposium proceedings will be published by SME.

If at some future time, the formation of an independent mine ventilation society in North America seems viable, the Underground Ventilation Committee of SME-AIME could well provide the organizational nucleus.

In addition to commemorating the establishment of the UVC and the initiation of these university symposia, we recognize another singular event at this meeting: the publication of two major text and reference books on mine ventilation in the U.S. and the planned release of a new edition of the South African work on gold mine ventilation. When can ventilation buffs recall it so good?--three new books to choose from in the space of a few months! Dr. Christopher Hall of the University of Idaho is the author of _Mine Ventilation Engineering_, published last fall by SME; and I and 24 others are coauthors of the new edition of _Mine Ventilation and Air Conditioning_, to be released by John Wiley & Sons next month. My unbiased (?) judgment of both U.S. books is that they contribute substantially to the text and reference literature in our field of mine ventilation.

As the magazine ad would remind us, "You've come a long way, baby." It seems only yesterday that Georgius Agricola, the first literary mining engineer, was lamenting the deplorable state of the mine atmosphere during the Middle Ages and the short, grim life of underground miners. That some of them survived, we suspect, is more a tribute to the stamina of the miners than to the genius of the mine ventilation engineers of those days!

Times _have_ changed. They changed substantially with the coming of the Industrial Revolution and the arrival of the machine age. Interestingly, however, the drastic changes occurring in mine ventilation and environmental control date from the years of the Post-Industrial Era, the years of our lifetime.

In Chapter 1, the new edition of _Mine Ventilation and Air Conditioning_ singles out four monumental events of the past two decades which account for--among other things--the ascendancy of atmospheric environmental control in mines:

1. The invention of the high-speed, electronic digital computer, permitting solutions to ventilation-system problems heretofore unsolvable, and allowing monitoring and remote control to become realities.

2. The development of the systems approach, which optimizes complex industrial operations, permitting men, materials, and methods to be coordinated in the most efficient way.

3. The enactment of tough new federal legislation, embodying strict regulations to improve the safety of mining operations.

4. The advent of socioengineering, the applying of technology with full consideration of the social, political, economic, and environmental consequences as well as the technical benefits.

What the full impact of these developments will be on mine ventilation is still unfolding. Clearly, the environment in the next decade can occupy no less influential a role in mining engineering than it did in the last. It is one of the purposes of the 1st Mine Ventilation Symposium to begin the assessment of that role and its definition for the future.

I look forward to being in attendance at the 10th Symposium in the year 2001, wherever it is held. For a long time, I've held the conviction that underground disasters caused by contamination of the atmosphere--fires and explosions, which cruelly snuff out many lives in an instant--<u>can</u> be eliminated from underground mining. An accomplishment of that magnitude is equivalent to finding a cure for cancer. But it is possible, and I believe it <u>will</u> occur through dramatic improvements in the way we ventilate mines. It <u>will</u> occur because the people in this room <u>can will</u> it to occur.

I here go on record today as predicting that atmospheric mine disasters <u>will</u> be eliminated by the time our 10th Symposium convenes. Whether it is the 9th, 10th, or 11th is not the point. Fires and explosions <u>can</u> be eliminated from mines. It is encumbent upon this assembled body of mine ventilation experts to see that they <u>are</u>.

The University, the Committee, and I welcome you to a historic event, the 1st Mine Ventilation Symposium.

\- - -

APPENDIX A

<u>Purpose and Organization</u>, Underground Ventilation Committee, Joint Coal Division - Mining & Exploration Division, Society of Mining Engineers, AIME.

<u>Name</u>: The Underground Ventilation Committee (UVC) is a unit or technical committee of AME-AIME that functions jointly across the Coal Division and Mining & Exploration Division.

<u>Purpose</u>: The purpose of the UVC is to foster engineering interest and technological progress in the ventilation of mines, tunnels, and other underground openings. Underground ventilation is here defined as the control of the quality, quantity, and temperature-humidity of the atmosphere of a designated underground space within specified limits. The UVC accomplishes its purpose by conducting technical sessions at SME and AIME meetings, sponsoring periodic symposia in cooperation with host universities, and soliciting papers for publication in <u>Mining Engineering</u> and SME transactions and proceedings. It also offers an affiliation home for SME members engaged in the practice of underground ventilation. In these ways, UVC seeks to encourage research, education, publications, and technology transfer in the field of underground ventilation.

<u>Membership</u>: Membership in UVC is open to any member of SME-AIME whose divisional preference is Coal or Mining & Exploration.

<u>Officers</u>: To satisfy the bylaws of both divisions, UVC shall have six officers, whose titles and duties are as follows:

1. <u>Chairman</u>: Administers the committee, presides at meetings, maintains the committee rolls, represents the committee in Society affairs, conducts elections. The Chairman and Vice Chairman for Programs represent UVC on the Executive Committees of their respective divisions and are responsible for communications liaison.

2. <u>Vice Chairman for Programs</u>: Serves as Chairman-Elect, assists the Chairman in administrative matters, chairs technical sessions sponsored by UVC, represents the committee on division program committees, contributes information to the divisions for the Annual Review issue of <u>Mining Engineering</u>.

3. <u>Vice Chairman for Program Planning</u>: Plans and solicits papers for technical sessions and symposia, assisted by the two Program Planners.

4. <u>Vice Chairman for Publications</u>: Solicits, processes, and reviews papers submitted for publication, represents the committee on division publication committees.

5. <u>Program Planner for Coal</u>: Assists the Vice Chairman for Program Planning, representing underground coal mine ventilation.

6. <u>Program Planner for Mining & Exploration</u>: Assists the Vice Chairman for Program Planning, representing underground noncoal mine ventilation.

TERMS OF OFFICE: The Chairman and three Vice Chairmen shall serve one-year terms. The two Program Planners shall serve staggered two-year terms.

LINE OF SUCCESSION: Persons appointed to either the fifth or sixth offices shall normally progress through the three Vice-Chairman offices to Chairman.

DIVISIONAL INTEREST: Since UVC is a joint committee, the six officers shall be chosen in alternating years to stagger representation from the two divisions; that is, three officers shall hold first preference for the Coal Division and three for the Mining & Exploration Division.

SELECTION PROCEDURE: All SME-AIME members who have requested membership in UVC may vote for officers. Normally, only the person nominated for either the fifth or sixth offices in UVC requires election in any given year, the other officers progressing through the various offices to Chairman. Persons nominated for office must be confirmed and appointed jointly by the two Division Chairmen, according to division procedures. The six current officers comprise a nominating committee to choose a candidate(s) for the coming year for the office of Program Planner (Coal or M&E).

Adopted February 15, 1982 by UVC.

APPENDIX B

Current (Founding) Officers, Underground Ventilation Committee, Joint Coal Division - Mining & Exploration Division, Society of Mining Engineers, AIME.

Chairman	Howard L. Hartman The University of Alabama	Coal	1983
Vice Chairman-Programs	Pierre Mousset-Jones University of Nevada-Reno	M&E	1984
Vice Chairman-Program Planning	John W. Stevenson Jim Walter Resources, Inc.	Coal	1985
Vice Chairman-Publications	Bruce R. Johnson Magma Copper Company	M&E	1986
Program Planner-Coal	Madan M. Singh Engineers International, Inc.	Coal	1987
Program Planner-M&E	Y. J. Wang West Virginia University	M&E	1988

1
Fans and Ventilation Shafts

Chapter 1

SHAFT SINKING USING THE V-MOLE - DESCRIPTION OF THE TMCI OPERATION IN ALABAMA

Klaus-Peter M. Hanke

Project Engineer
Thyssen Schachtbau GmbH
Tuscaloosa, Alabama

INTRODUCTION

In early 1979 Jim Walter Resources, Inc. (JWR) of Brookwood, Alabama approached TMCI Construction, Inc. (TMCI) to make a proposal on a program that involved the sinking of up to 10 ventilation shafts of approximately 6.7 m (22 ft) diameter and ranging in depth from 500 to 700 m (1650 to 2300 ft) for the JWR coal mines in Alabama. At this time TMCI was already constructing the first spiral underground bunker (capacity 2000 tons) in North America for the JWR organization at their No. 4 mine in Alabama.

The TMCI proposal was based on the use of the most modern large diameter shaft boring machine rather than sinking the shafts using the conventional drill-blast-muck technique. The proposal was made based on the experiences by the parent company, Thyssen Schachtbau, which has been using this type of machine in Germany for shaft boring since 1971.

As a result of the TMCI proposal JWR issued a purchase order to TMCI for the construction of four 6.7 m (22 ft) diameter, concrete lined, unfurnished ventilation shafts ranging in depth from 500 to 700 m (1650 to 2300 ft). An order was thus placed with WIRTH Machinen- und Bohrgeraete Fabrik GmbH, in Germany for the manufacture of a Model 650/850 E/Sch "Schachtbohrmaschine" (Vertical Shaft Borer = V-mole), which arrived on site in Alabama in early 1981.

The first V-mole GSB 450/500 was introduced in Germany in 1971 and was capable of enlarging in one step a pre-drilled 1.2 m (4 ft) pilot hole to 4.5 - 5.0 m (14.7 - 16.4 ft). This machine has sunk 9 staple shafts and deepened one surface shaft for a total of 2360 m (7740 ft) of shafts. On the last shaft boring operation in 1978 the machine was converted as an experiment to drill without a pilot hole using a hydraulic pumping system to remove the cutting debris.

A second generation machine, the SB VI 500/650, was introduced in 1977 for enlarging the pilot hole to a range of 5.0 - 6.5 m (16.4 - 21.3 ft) diameter. This machine is still in operation and has already drilled well over 2000 m (6500 ft) of shaft.

The third generation of V-mole, the SB VII 650/850, for diameters from 6.5 to 8.5 m (21.3 to 27.9 ft) was commissioned in May 1980 and has been used for two surface shaft deepenings totalling 606 m (1990 ft) with another scheduled for 1982.

The main advantages favouring the use of such V-moles were identified as:

1) A reduction in manpower to the crew required in a conventional shaft sinking operation.

2) A considerable reduction in time to complete a shaft compared to conventional techniques.

3) The use of the V-mole eliminates many of the hazards encountered in conventional sinking.

Based on the successful performance of the first three V-moles in Germany, Thyssen Schachtbau decided to employ this principle abroad. In 1980 a second machine of the third generation was built and is now operated by TMCI Construction, Inc. in Alabama.

The first shaft was completed at the end of 1981 and this paper describes the method of operation including some unique aspects not attempted on prior V-mole operations and some of the statistics arising out of the experiences during the first shaft boring operation.

THE NO. 7 MINE FAN SHAFT SITE

Jim Walter Resources, Inc. was formed in 1970 to exploit the coal field in Alabama on the southern tip of the Appalachian coal field. The coal reserves amount to around 650 million tons of mainly good quality coking coal of which about 350 million tons are to be extracted over the next 30 years. Shaft sinking and preparatory work began in 1972, and at present 6 mines are producing around 5.4 million t.p.a. Annual production is to expand to 10 million t.p.a. as soon as possible, and the ventilation shafts to be sunk by TMCI play a vital role towards attaining this goal.

The first shaft site is located at the No.7 mine, near Brookwood, Alabama. The actual location of the shaft relative to the production shafts is shown on the mine plan (Fig. 1), which also shows the room-and-pillar extraction system used at present. The mine plan further shows the conveyor route used for the muck removal.

The geological survey showed that the strata consisted of horizontal layers of mainly sandstone, sandy shale and shale interspersed with several coal seams. The seam being extracted at the No. 7 mine is a combined seam made up of the Blue Creek and Mary Lee seams at a depth of 513 m (1682 ft) and having an average seam thickness of about 2 m (6 ft).

At the beginning of September 1980 the surface site preparation and pre-grouting work was completed by JWR, and TMCI was able to commence with Stage I of the shaft sinking program - the drilling of the pilot hole.

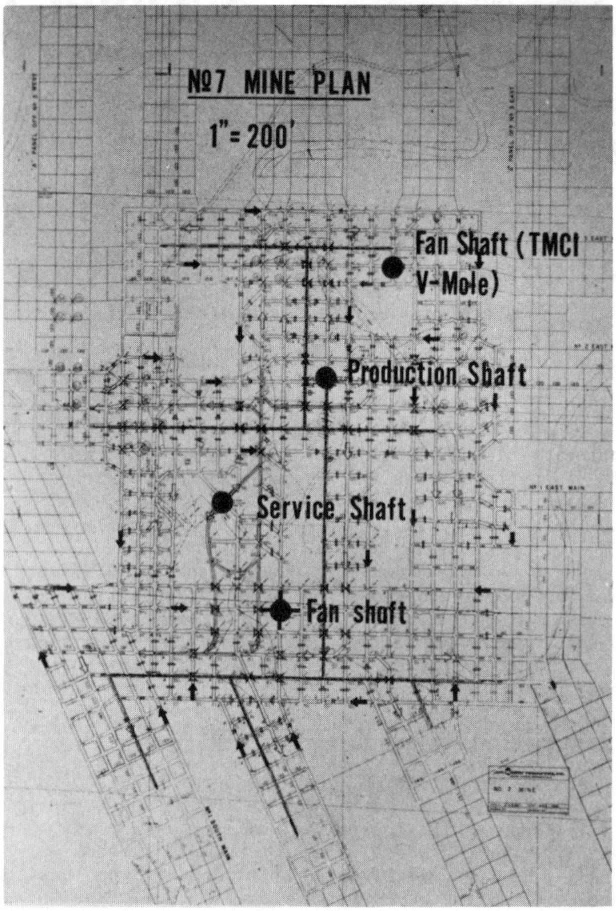

FIGURE 1. No.7 Mine Plan, showing shaft locations and conveyors

FIGURE 2. Robbins Raise Borer for Pilot Hole Drilling

FIGURE 3. 1.5 m (5 ft) Reaming Head

STAGE I - PILOT HOLE DRILLING

One prerequisite for all V-moles of the type described in this paper is a pilot hole connecting the top of the shaft to the prospective bottom. This pilot hole serves solely for the removal of the cutting debris produced by the V-mole. The debris falls to the bottom of the pilot hole and is removed there continuously during drilling using some appropriate means. A further prerequisite is thus at least a tunnel connecting the bottom of the pilot hole to the rest of the working mine or another shaft through which the debris can be removed. The drilling of the pilot hole was carried out by TMCI, and for this a Robbins raise borer, Type 7R-SPDC with corresponding 10 in. diameter drill pipe was acquired (Fig.2). The first operation was the drilling of a 12 ¼ in. target hole, and this commenced at the beginning of September 1980 and the bit holed through at 513 m (1682 ft) on November 14, 1980. Although machine malfunctions, a bit change and operator inexperience meant that only 1/3 of the time was used for actual drilling, giving a net drilling rate of 1.3 m/hr (4 fph), the operation was considered a success since of primary importance to the shaft boring machine operation is not the drilling rate but the accuracy. In the case of this particular hole the deviation was only 0.45 m (18 in.) over 513 m (1682 ft) which is a high degree of accuracy indeed. Accuracy is of importance for the V-mole operation as the manufacturer stipulates that the maximum deviation should not exceed one-half of the pilot hole diameter. In this case the eventual pilot hole diameter was 1.5 m (5 ft), and thus the deviation was well within the allowed 0.75 m (2½ ft).

After holing through with the 12 ¼ in. target hole, the bit was replaced with a 1.5 m (5 ft) diameter Robbins reaming head carrying 8 button cutters (Fig.3). During the back reaming, advance rates up to 94.5 m/day (310 fpd) were achieved with an average advance of 3.4 m/hr (11.4 fph). The reaming operation was completed on 19 December, 1980 and after the demobilization of the raise drill, work on the various parts of Stage II - the shaft drilling - commenced in early January 1981.

STAGE II - SHAFT DRILLING WITH THE V-MOLE

The first part of this section of the paper deals with the mobilization of the surface equipment, hoisting installations, V-mole assembly, headframe and concreting workdeck installation and erection of the underground muck removal system. Part 2 describes the actual operation of shaft sinking and lining and the last part deals briefly with the demobilization of all equipment.

Foreshaft and V-mole Assembly

For the assembly of the V-mole, an assembly shaft or foreshaft of at least 9 m (30 ft) depth is required since the gripper section of the V-mole has to push against the shaft wall for anchorage during drilling. As soon as the raise drill had been removed, the shaft collar was constructed and the first 9 m (30 ft) sunk conventionally and lined with concrete. The floor of the foreshaft was also covered with a layer of concrete to give a level surface for the V-mole assembly support struts to rest on. The diameter of the foreshaft was about 7.15 m (23.5 ft), which was about 0.15 m (6 in.) larger than the diameter of the cutter head of the V-mole to allow easier assembly.

Assembly of the V-mole began in early February 1981 as soon as the foreshaft was completed. All V-moles have been designed with underground transport systems in mind and could therefore be dismantled into relatively small pieces. These had been packed into crates and containers in the factory in Germany and began arriving on site in January and February 1981 (Fig. 4).

FIGURE 4. V-mole Parts arriving on site

Installation of the V-mole in the foreshaft began with the cutter head main bearing assembly followed by the inner kelly extension and main gear box plus the 6 electric motors and 6 planetary gear boxes (Figs. 5 and 6).This assembly is the first of two basic sections of the machine and is made up of the

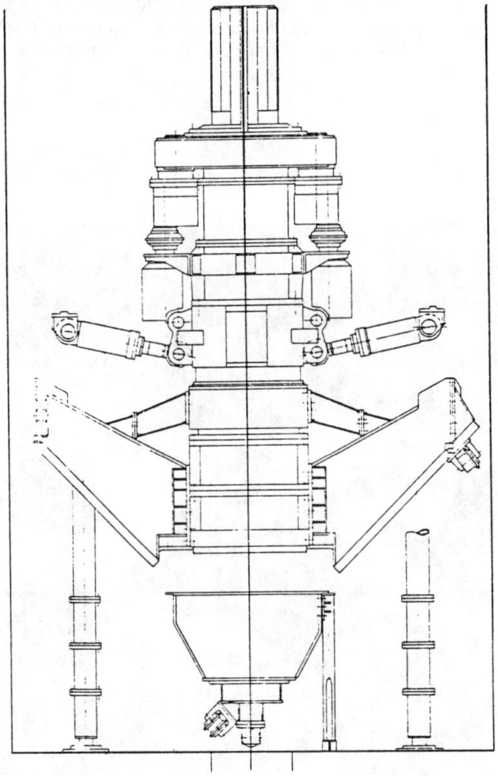

FIGURE 5. Inner Kelly Section with Cutter Head plus Drive Assembly

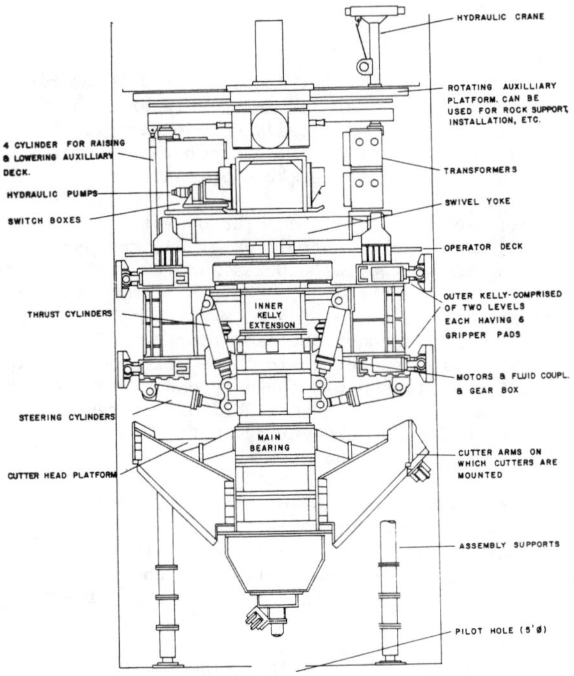

FIGURE 6. General Arrangement of V-mole SB VII

drive mechanism (motors and gear boxes) and the rotating cutter head. This section moves downward as drilling progresses. The second section of the machine is made up of the gripper units, operator deck with the control cabin and the hydraulic and electric installations and at the very top the auxiliary work platform. This second section of the V-mole remains stationary during drilling and provided the anchorage for the thrust and steering forces on the cutter head. The gripper ring is made up of six basic units each having two gripper pads (Fig. 7). These

FIGURE 7. Gripper Units

units are bolted together to form the gripper ring or outer kelly having two levels of gripper pads with six pads placed symmetrically on each level. The gripper pads are forced out against the shaft wall during drilling, locking the upper part of the machine in place. On top of the outer kelly is the operator deck (Fig. 6). The operator cabin with all the V-mole controls (Fig. 8), the hydraulic tank and pumps, two swith boxes and two transformers are situated on this deck. The auxiliary work platform is at the top of the machine. This deck can be rotated through 360° and four 2 m (6.5 ft) long cylinders connect it to the operator deck below (Figs. 6 and 11) and via these it can be raised and lowered as required. A small hydraulic crane is also provided as an aid during cutter changes. The total height of the machine is 13.4 m (44 ft).

The two basic sections of the V-mole described above are connected by 12 cylinders. Six cylinders are mounted near vertical and connect the top of the gripper ring to the inner kelly just above the cutter head (Fig. 6). These cylinders provide the thrust down on the cutter head during drilling. The remaining six cylinders are mounted near horizontally and connect the bottom of the gripper ring to the inner kelly (Fig.6). These steering cylinders can

FIGURE 8. Operator Control Cabin with Laser Display for Steering

FIGURE 9. Steering Yoke (Gelenkrahmen)

be individually controlled to allow continuous direction control during drilling. The top of the inner kelly square extension is mounted inside a yoke fixed to the outer kelly and this allows the cutter head to be steered on line even if the gripper ring is slightly off-set from centre (Fig. 9.). Figs. 10 and 11 show the set-up on completion of the V-mole

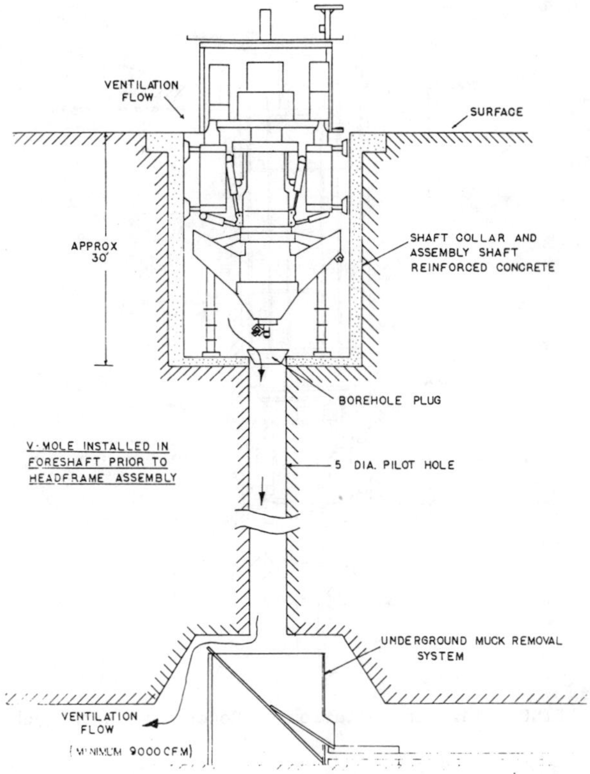

FIGURE 10. Completion of V-mole Assembly

FIGURE 11. Commisioning V-mole

assembly prior to commencement of drilling and commisioning of the V-mole. Table 1 gives the technical data of the machine. The diameter chosen for these shafts is 7.0 m (23 ft) and the cutter head, consisting of a nose cone (angle 55°) with six arms radiating from centre (angle 45°), as shown in Fig. 12, carries a total of 58 single disc cutters, 6 double disc cutters plus 8 smaller centre cutters. Power and water for cooling the motors plus dust suppression were supplied via hoses and cables stored on surface and lowered as needed.

TECHNICAL DATA:	
Machine Type	SB-VII 650/850 E/Sch
Boring Diameter	6500–8500 mm (21.3–27.9 ft)
Cutter Head	
Driving Power	6 x 122kW (6 x 160hp)
Revolution Number	6.5m dia. 5rpm
	7.0m dia. 4.7rpm
	7.5m dia. 4.4rpm
	8.0m dia. 4.1rpm
	8.5m dia. 3.8rpm
Max. Torque approx.	1260 1350 1440 1530 1670kNm
	414 444 474 503 549 tons-ft
Hydraulic System	2 x approx. 30kW (2 x 40hp)
Boring Pressure Thrust	up to 833 tons
Boring Stroke	800 mm (31 in.)
Gripping Power	1716 tons
Hydraulic Pressure	150–200 kPa/cm^2 (2100–2800 psi)
Total installed Power	Approx. 800kW (1072hp)
Current	three phase 60 Hz
High Tension Voltage	4160 V fed to machine
Motor Voltage	550 V
Transformers	2 x 630 kVA
Total Machine Weight	240 tons

TABLE 1. Technical Data of Shaft Borer

Assembly of the V-mole was completed in three weeks and after a further week for commisioning of the electrics and hydraulics, drilling was tried for the first time in the beginning of March 1981 (Fig. 13). Drilling was stopped at about 26 m (85 ft) to allow installation of the concreting work deck and assembly of the headframe (Fig. 14). The work deck and bottom section of the headframe

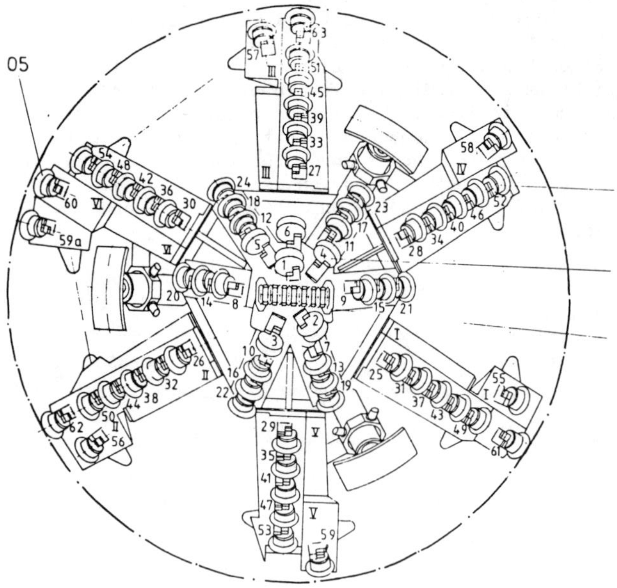

FIGURE 12. The Cutter Head

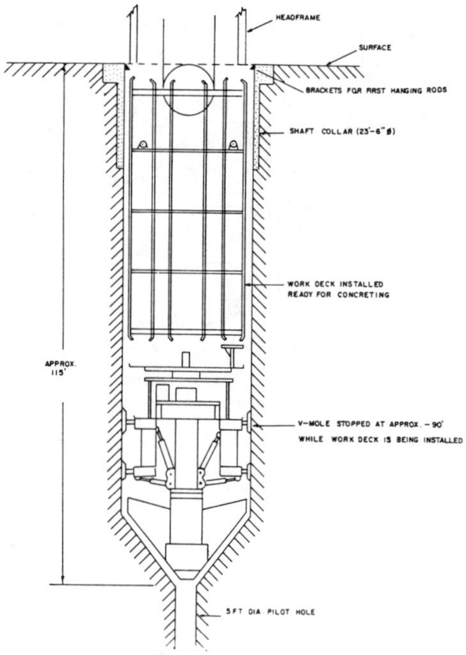

FIGURE 14. Installation of Concreting Work Deck

FIGURE 13. V-mole During Start-up

FIGURE 15. Main Hoist During Assembly

were installed by the end of March when work was halted for 3 months due to the UMWA strike.

Installation of Surface Equipment

During the foreshaft excavation and V-mole assembly, the assembly of the main hoist and service winches also took place.

The main hoist (Fig. 14) was a Fullerton, Hodgart and Barclay 3.65 m (12 ft) diameter, double drum hoist. The drive was an 1100 hp, 4160 V motor, which through a reduction gearing achieved a rope speed of maximum 8 m/s (1600 fpm).

The work deck for concreting was raised and lowered by two independent, electrically synchronized hoists. Each hoist was powered by a 20 hp motor through a reduction gearing to give a maximum rope speed of 0.10 m/s (20 fpm). The ropes were in double fall in the shaft, around a sheave on the work deck, the socket being anchored in the headframe.

The services winch used for lowering the power cable for the V-mole together with the compressed air, water and telephone lines was a small 20 hp electric winch. At approximately 10 m (33 ft) intervals,the services were clamped to the rope and lowered as required.

For emergencies, a diesel driven Timberland hoist was also installed. Fig. 16 shows the auxiliary winches.

FIGURE 16. The Auxiliary Winches

The headframe was a standard unit of 7.9 m (26 ft) by 3 m (10 ft) centres with an overall height of 32.2 m (106 ft) including sheaves.

The collar steel was secured independently onto the collar concrete and has two sets of double doors for the manriding and concrete buckets. An opening for the services and emergency cage was also included.

Other surface equipment included the batching/ mixing plant where the concrete for the lining was prepared and filled into concrete buckets. Fig. 17 shows the surface layout.

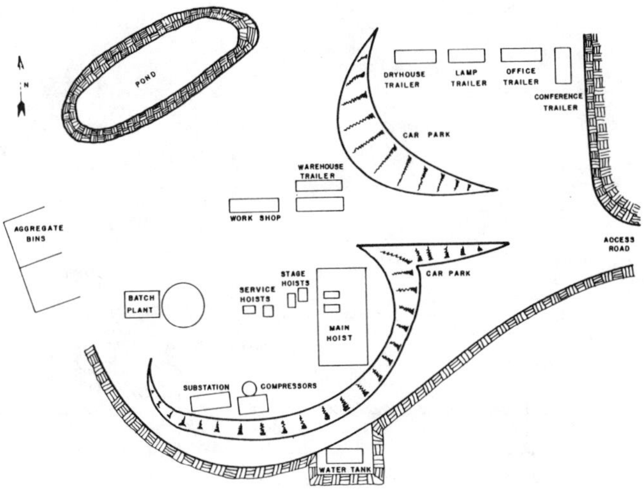

FIGURE 17. Surface Layout for No. 7 Fan Shaft

FIGURE 18. Muck Removal Chute

Installation of the Underground Muck Removal System

For removal of the cutting debris a simple rectangular box was designed with an I-beam structure and timber walls. Rails were used to deflect the cuttings through the chute onto the conveyor belts and out on the mine belts. Figs. 18 and 19 show the box and underground arrangement. In the past, in Germany, the muck was generally removed using a slusher. However, the simple box system worked very effectively and provided a continuous muck removal system. On the next shaft at the JWR No. 4 mine due to start drilling early April 1982, a Radmark stowing system will be installed in addition to the above set-up and the debris will be stowed in specially excavated headings thus making the shaft boring operation relatively independent of the mine, except in cases of stower malfunction.

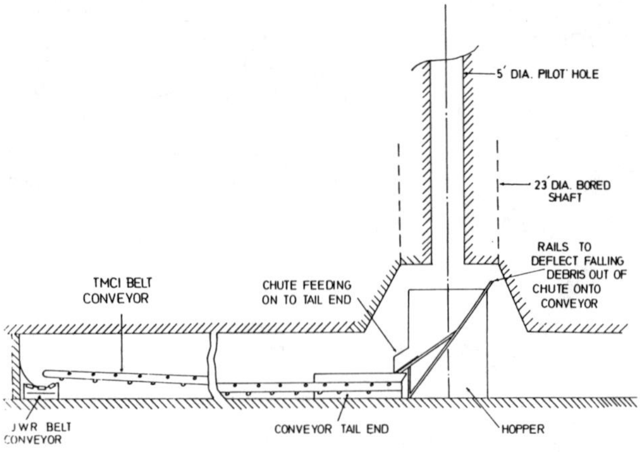

FIGURE 19. Muck Removal Underground

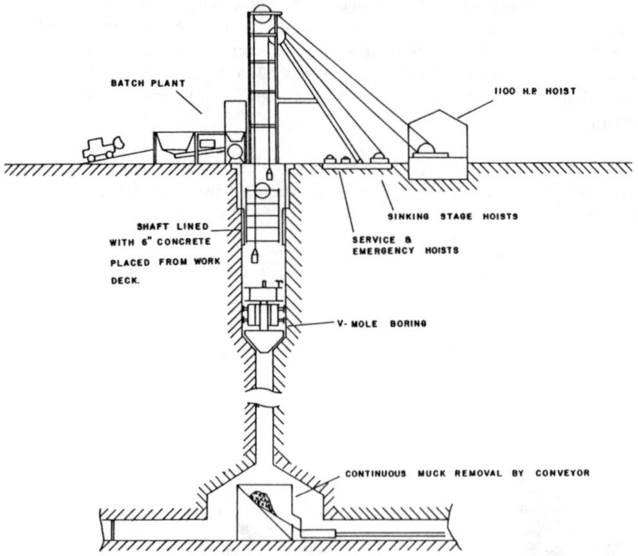

FIGURE 20. Set-up During Sinking

Sequence rollers were fitted on all conveyors, and to ensure boring stopped when a belt stopped, a visual and audio signal was relayed to the operator on the V-mole. Drilling only restarted on the verbal signal from the belt attendant.

Drilling and Lining

Work resumed at the end of June 1981 after the end of the UMWA strike, and once the work deck was installed completely and headframe assembly and roping up of hoists, etc. was complete, drilling and lining began at the start of August 1981 (Figs. 20 and 21).

Drilling - Drilling with the V-mole was a cyclic process of drilling one stroke length of the thrust cylinders and then resetting of the machine. During drilling, the gripper ring was securely clamped against the shaft wall by forcing out the gripper pads, and the rotating cutter head was thrust down into the rock face by the six thrust cylinders with a force up to 350 tons. Guidance of the machine was achieved by means of a self-aligning laser mounted on the shaft axis at the surface. The laser beam was picked up by a pattern of photo-electric cells in a receiver mounted centrally in the cutter head. At the operator's control desk was a monitor consisting of a set of LED's in exactly the same pattern as the photo-cells of the receiver. This showed the operator exactly where the cutter head was in relation to the shaft centre. By independent operation of the steering cylinders, steering of the V-mole to within fractions of an inch was possible.

On completion of one stroke length of 0.75 m (30in.), the drive motors were stopped and the cutter head pushed against the rock face. The gripper pads were withdrawn, the whole machine weight resting on the

FIGURE 21. Surface Layout

cutters, and the thrust cylinders retracted. This lowered the gripper ring plus the decks by one stroke length. The grippers were then pushed out again and drilling could be continued. The drilling time for one stroke was approximately 20-30 min. in the harder rocks with a resetting time of about 5-10 min. Operation of the V-mole required only one operator who was also responsible for maintenance, lubrication and cutter inspection.

Lining - As the V-mole drilled down, the lining operation followed directly behind. On previous shaft borings, the lining was placed from the auxiliary deck on the V-mole and consisted either of a temporary lining of rock bolts and mesh or a mesh and steel ring lining. Any concrete was placed from a work deck which was installed after the drilling and dismantling of the V-mole was finished. To avoid two operations in the JWR shafts, it was decided to suspend a separate work deck above the V-mole in the shaft and follow down simultaneously with the lining.

The lining, consisting of 0.15 m (6 in.) of concrete, was placed from a 5-deck work deck or sinking stage. The overall height of this work deck was 14 m (45 ft) with approximately 3 m (10 ft) between deck elevations corresponding to the working elevations during concreting. The deck was suspended by two ropes in double fall which also acted as guide ropes for the two buckets. Four compressed air winches were used to handle the formwork .

Concrete was mixed on surface and loaded into a concrete bucket which was lowered to the work deck by the main hoist and the concrete was poured through a rubber hose securely clamped to the bottom of the bucket. The height of the steel formwork was 6 m (20 ft) consisting of 8 rings, each 0.75 m (30 in.) high, allowing shorter lifts than 6 m should the need arise. Formwork was stripped only after the concrete had set sufficiently to be stable. An accelerator was added to speed setting times.

The formwork was suspended by eight 25.4 mm (1 in.) diameter hanging rods. A 6 m (20 ft) lift of concrete was carried out in the following sequence:

1) Lower work deck to new position 6 m below previous location

2) Detach kerb ring from hanging rods

3) Lower kerb and first 0.75 m (30 in.) form ring to new position

4) Place 8 new hanging rods and line and level kerb ring

5) Place 5cm (2 in.) diameter hose in annulus between kerb and shaft wall

6) Pour 1 bucket of concrete

7) Strike and lower form rings 2,3 and 4

8) Pour concrete

9) Strike and lower rings 5, 6, 7 and 8

10) Pour concrete

Each 6 m lift of concrete took about 20 m^3 (27 cu yd) of concrete and once a routine was established took about 6-7 hrs to place. Fig. 22 shows the set-up for concreting.

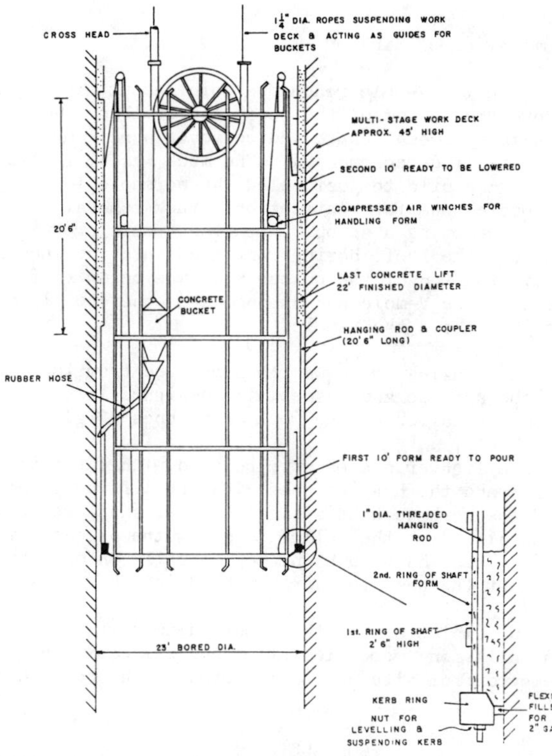

FIGURE 22. Concreting Arrangements

FIGURE 23. Lifting Frame for V-mole Dismantling, showing special shaft cover. Piece being loaded is main bearing assembly weighing approximately 35 tons.

Completion of Sinking

When the V-mole reached a point about 5 m (16 ft) above the bottom of the borehole. drilling halted. The muck removal chute and conveyor were removed and drilling recommenced. The muck was then allowed to form a pile to just below the borehole leaving adequate room for ventilation. Muck removal was done using a scoop. When the V-mole broke through the roof, boring continued into the muck pile until the top of the cutter head was below roof level. The V-mole was stopped and concrete placed to just above the V-mole.

The sinking stage was then removed from the shaft at the surface and a specially designed lifting frame installed on the work deck ropes (Fig. 23).

The lighter pieces were hoisted with the main hoist and the heavier ones with the lifting frame and the work deck hoists. A heavier shaft cover was also placed on the collar to allow the pieces coming out of the shaft to be placed directly onto a truck without the need for further handling (Fig. 23).

On completion of the V-mole dismantling, surface installations were dismantled and all equipment removed from site and transported to the next shaft.

CONCLUSION

The shaft was handed over to JWR in early March 1982 and, discounting the UMWA strike of three months, the total sinking time including the pilot hole drilling took 15 months. During this time several remarkable feats were achieved and some aspects of the operation tried, and successfully completed, for the first time. These can be summarized as follows:

1) Raise boring

- the 12 ¼ in. pilot hole deviation was only 0.45 m (18 in.) over a total length of 513 m (1682 ft)
- 94.5 m (310 ft) of 1.5 m (5 ft) diameter raise were completed in 24 hrs.

2) Shaft boring

- 32.25 m (106 ft) of 7 m (23 ft) diameter excavation cut in 24 hrs.
- 18 m (60 ft) of concrete shaft lining completed in 24 hrs.
- discounting commissioning and holing through an average of 9 m (30 ft) of shaft was completed per day
- at 513 m (1682 ft) this shaft was the deepest ever bored with a V-mole and it was the first shaft started from surface
- for the first time simultaneous drilling and concrete lining was carried out
- a continuous muck disposal system feeding directly onto a conveyor was installed for the first time
- the total manpower requirement for a two-shift operation was 25 men
- no time lost due to injuries during the entire project
- it was the first time the V-mole had been used outside Germany

The above list and a large number of impressed visitors to the site confirm the applicability and suitability of this type of machine for the task described in this paper and represents a proven step towards the desired mechanical method of safe and fast shaft sinking.

The next logical step of a blind shaft borer has already been tried both in Germany and the USA with limited success, and a further trial with a specially built machine is scheduled towards the end of 1982 in Germany, but until this method is refined the V-mole will remain one of the fastest and safest sinking methods.

Chapter 2

PNEUMATIC TRANSPORT IN CONJUNCTION WITH SHAFT CONSTRUCTION

William R. Eby

Eby Constructors
22005 84th Avenue South
Kent, Washington

Cheri Eby

Eby Constructors
22005 84th Avenue South
Kent, Washington

ABSTRACT

This paper describes the state-of-the-art of pneumatic transport systems that are capable of conveying waste cuttings generated by raise drills during shaft construction projects. Additionaly other examples of applications pertaining to pneumatic transport systems qualified to convey abrasive and nonabrasive materials (ie., gravel, shale, coal, ore, and limestone) will be presented.

INTRODUCTION

Pneumatic transport systems operate on one basic principle, that being: Material is transported through a pipeline by pressurized air and discharged at high velocity.

The employment of pneumatic transport systems is not unfamiliar in many industries such as:

> The pulp and paper industry utilizes this method of transport to move wood chips.
>
> Gunite systems employ air as the means of transporting select concrete substances as well as sand and other fine aggregates at tonnage rates from 5 to 20 tons-per-hour.
>
> Pneumatic stowing systems have been used in the European mining industry since the 1920's when the method was developed for stowing either select material or waste material into stopes and thus provide for proper roof support and ground stabilization.
>
> During the last decade the mining/ construction industry in the United States has increased its use of pneumatic transport systems for stope backfilling, mine sealing, and removing raise drill cuttings.

SYSTEM DESCRIPTION

The pneumatic transport system described here is primarily composed of off-the-shelf components. Proper size selection of the components is achieved once knowledge is acquired on:

1. Material characteristics
2. Distance of conveyance pipeline to include bends, inclines, and declines.
3. Site conditions
4. Desired quantity of material to be transported per hour.

Currently these types of pneumatic systems are being successfully used to convey three-inch-minus particle size material in tonnage ranging from 50 to 100 tons per hour.

The following componentry comprises this type of transport system:

1. Positive displacement, high volume-low pressure air blower, operating normally no higher than 18 psig.
2. Horse power to drive the blower may be obtained either by the use of a diesel engine or electric motor.
3. Rotary air lock feeder which meters the material into the air stream. The air lock feeder may be driven either hydraulically or electrically.
4. Control console.
5. Material conveyance pipeline, elbows, and discharge nozzle.
6. Infeed device (ie., slusher and/or conveyor) to transport material up to the air lock feeder.
7. Grizzly to prevent oversized material from entering the system.
8. Crusher unit, if required, to process oversized material.

PROJECTS

Frontier-Kemper Constructors of Evansville, Indiana, initiated the method, in the United States, of pneumatically hoisting raise drill cuttings from underground to surface.

The opportunity to prove this method developed while they were under contract to construct six raise-drilled shafts and seven conventionally sunk shafts during the Chicago Tunnel and Reservoir Project.

One of the raise drill sites was selected for a test site for the pneumatic system. Specific data as to the quantities of material hoisted was not recorded ; however, it was reported that the transport system ran "flawlessly".

Based on the results of the first application, Frontier-Kemper Constructors was able to enter into a contract where pneumatic muck removal was mandatory. This shaft construction project was performed for the Island Creek Coal Company at their Providence #1 Mine, in Providence, Kentucky.

The dimensions of the shaft were 10 feet in diameter, 325 feet deep.

Prior to the raise operation, three bore holes were drilled and cased from the surface to the mine level.

1. One bore hole four inches in diameter for communications and and underground power lines.

2. Two bore holes each eight inches in diameter: One for the air supply to the air lock feeder and one for the material conveyance pipeline.

The blower assembly consisted of a diesel engine (8V71-T Detroit) and a positive displacement blower (Gardner Denver 11 CDL 23, 4400 cfm @15 psig) were located on surface.

The blower assembly supplied the required air down through the first eight inch bore hole to the skid-mounted air lock feeder assembly that was located underground near the area where the cuttings were being deposited from the raise drill cutter head.

A slusher bucket collected the cuttings and carried them up a ramp to where a grizzly was located. At this point the material (three-inch-minus in particle size) fell through the grizzly onto an infeed conveyor and was delivered to the air lock feeder. The air lock feeder metered the material into the air stream that conveyed it horizontally approximately 50 feet to and through a 90^{0} elbow and hoisted it vertically 325 feet to the surface. Upon reaching the surface, the material passed through a second 90^{0} elbow and continued horizontally through 160 feet of pipeline and was deposited at the muck pile.

Water was added at the discharge end of the pipe to suppress dust.

The shaft was reamed at a rate of 1.65 feet per hour average. Tonnage rates averaged 35 tons per hour with surges of 50 tons per hour. During the entire raise operation, the pneumatic system kept pace with the raise drill. This marked the first time in history that a pneumatic system removed and disposed of the muck continuously while a raise drill was in operation.

If storage room is available underground, the material conveyance pipeline may be directed to that area and the drill cuttings deposited there. Futhermore the discharge end of the pipeline can be moved in a sweeping motion horizontally and vertically as the material is discharged, thereby allowing complete filling of an underground stope. Also, the need for a second bore hole to hold the vertical material conveyance pipeline is eliminated.

ADVANTAGES

The cost factor is always a prevalent worry in a major undertaking such as the construction of a new shaft, and muck disposal a problem that must be dealt with.

The incorporation of a pneumatic system to remove and dispose muck either on surface or in worked out areas underground provides the following benefits:

1. Eliminates the need for the mine owner to draw personnel and equipment away from the mining of profitable material and placing them on waste removal projects.

2. Since air is used as the conveying energy and not a fluid, a dewatering process is not required.

3. Eliminates raise drill cuttings from being transported from the mine level to and through the preperation plant.

4. By utilizing pneumatic backfilling and achieving a compacted fill in the underground worked out areas, mine ventilation requirements can be reduced.

OTHER ADVANTAGES OF PNEUMATIC STOWING

Mine refuse taken from surface to underground and pneumatically stowed into worked out areas reduces surface storage and aids in preventing future subsidence.

OTHER APPLICATIONS/CONCEPTS

Pneumatic transport systems may very well be utilized in other applications for large diameter shaft construction projects along with current mining equipment.

Examples of these applications and concepts are:

Shaft reamers could readily accept the incorporation of the transport system to work simultaneous with them. Once the pilot hole has been bored and the reamer placed on line, the transport system could be located underground near the pilot hole and would remove the material as the reamer descended down the shaft.

In shaft construction projects where the slashing method is used, the same theory would apply as it does with the reamer application. That being, as the slashing proceedure is being performed, the cuttings would be directed to and through a pilot hole down to the shaft bottom where the transport system would provide the option of either stowing the material underground or hoisting it to surface. The incorporation of a crusher would provide the means to properly size the material for the transport system to handle.

One concept that has shown promising potential is the combining of a vacuum pick up assembly with a blind shaft borer. During a test project conducted in Seattle, Washington, a vacuum pick up test fixture, designed to work with a Robbins 24-1-SB-184 bore machine, did indeed display that a vacuum technique would pick up muck from the cutter head.

The pick up assembly hoisted material from the back of the simulated cutter head up vertically 56 feet to where the machine carousel and skip station would be located. At this point it was planned that the material would be conventionaly hoisted to surface. Conceptually, material could be vacuumed up to the carousel and then hoisted by a pneumatic system up to the surface.

CONCLUSIONS

The demand for increased production of energy and minerals is rapidly approaching, and in order to meet this demand an increase in mining/construction activities will undoubtedly be noted in the near future.

The increase of mining and construction projects promotes the need for new shafts to be constructed, which inturn expands the need for modern day muck disposal methods.

The promises that pneumatic transport systems have provided so far indeed promote their use in other areas to be investigated and intergrated into other mining/construction projects.

Further research and developemnt is warranted in order to continue the expansion of the methods of material handling provided by pneumatic transport systems.

ACKNOWLEDGEMENTS

The authors wish to express their appreciation to the following companies and persons kind enough to allow the descriptions and slides of their projects to be presented here:

Frontier-Kemper Constructors: Mr. Dyke Howell

The Robbins Company: Mr. Jim Dawson

REFERENCES

Nash, W. R., 1982 "Raise Boring Shafts In Both Mining And Civil Applications" University of Wisconsin-Milwaukee, Frontier-Kemper Constructors.

Chapter 3

MAIN MINE FAN REVERSE PERFORMANCE CHARACTERISTICS

by

Michael F. Dunn and Francis S. Kendorski
Engineers International, Inc.
Downers Grove, IL

Raphael D. Schilz
Joy Manufacturing Company
New Philadelphia, OH

Edward D. Thimons
U. S. Bureau of Mines
Pittsburgh, PA

ABSTRACT

During a mine fire or other emergency, it may be desirable to reverse the air flow in order to provide an escapeway or to control a fire. Also, in colder areas, the air flow may be reversed to prevent ice build-ups. When reversing main mine fans, the operator usually does not know what operating characteristics of flow and pressure to expect. Laboratory testing of vane axial main mine fans has been completed in order to establish forward and reverse performance characteristics under controlled conditions, with the results presented here. Preliminary tests under field conditions have also been conducted. The data obtained suggest that reverse performance characteristics are dependent upon the blade angle and the hub to tip ratio. There is also evidence that reverse performance can be predicted for a family of blade angles in a given hub to tip ratio. In general, the air quantity delivered by the fan is 30% to 60% less in reverse than when operating in the normal forward mode.

INTRODUCTION

Ventilation of underground mines is an essential part of any mining operation. The ventilation air provides oxygen to persons and equipment, dilution of particulate and gaseous pollutants, and cooling (or heating). Generally, a shaft or adit is the mine air intake so that the miners can come and go freely in fresh air. However, in colder climates, the ice buildup can be detrimental, and a means for pre-heating the incoming air may not be available. Therefore, many mines will reverse their air flow on weekends or other shutdown times so that the shafts and adits are exhausting and consequently carry air heated by the mining equipment operation or the natural rock heat.

During a mine fire or other emergency, it may be desirable to reverse the air flow to provide an escapeway or to isolate a fire.

Some mines are not able to regularly reverse their air flow for environmental or other reasons, such as:

- Radon contaminated air - in uranium and thorium mines (and in some metalliferous mines), the exhaust air is contaminated with radioactive radon and radon daughters and miners are not permitted to move and work in such return air.
- Dusty mines - in bulk mining methods such as block caving, the air becomes extremely dusty and is generally exhausted through the cave area or through specially provided shafts.

Two methods exist for reversing the mine ventilation: an arrangement of ducts and doors to reverse the air path as it leaves the fan, and to electrically reverse the fan itself. However, fan manufacturers measure and report fan performance data only for the forward or normal operating mode, and not the reverse. Therefore, when a mine electrically reverses its vane axial main mine fan, it does not know what operating characteristics of flow and pressure to expect. Because of fan design, the efficiency is less in the reverse mode, but by how much?

The purpose of this study, funded by the Bureau of Mines, was to establish main mine fan reverse characteristics under laboratory and field conditions. To carry out this work, Engineers International, Inc., arranged with Joy Manufacturing Company of New Philadelphia, OH, to build and test two common main mine fans each of 2.4 m (96-in.)-diameter but with different hub to tip ratios.

Further work still will determine reverse performance characteristics of vane axial fans at operating mines around the country, to supplement the one field test reported here.

The laboratory testing developed forward and reverse performance data under reasonably controlled conditions. From these laboratory data, reverse fan characteristic curves can be established. Also, with these data, comparisons to the calculated performance, which is generally scaled from smaller laboratory tested fans by fan law relationships, can be made.

VANE AXIAL FANS

Mines often use vane axial main fans because of flexibility in installation and performance. However, these fans perform very inefficiently when reversed, their output being reduced substantially.

Vane axial main mine fans first developed in the 1930's are the most common fans used today. In order to cause air flow, these fans, like others, create a pressure difference in an airway. Vane axial fans accelerate the air through the fan, which results in both a pressure and velocity increase. Fans must supply enough pressure to overcome system pressure head losses and deliver the versatility of adjustment to a wide range of performance requirements. These fans basically consist of a cylindrical, housed impeller. The impeller has air-foil blades causing air flow axially through the fan, that is, parallel to the rotation axis.

Figure 1 illustrates a typical vane axial fan. The inlet bell and outlet cone reduce the pressure head against which the fan operates by reducing the shock losses at inlet and discharge areas of the fan. The main parts are the blades, rotor or hub, diffuser tail, straightening vanes, and drive shaft.

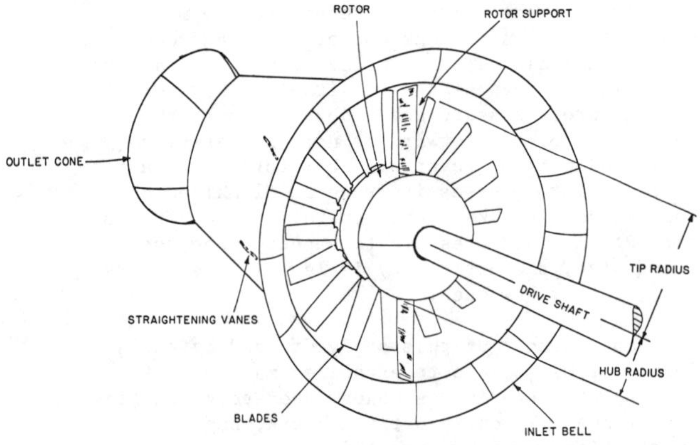

Figure 1 - Vane Axial Fan Parts

Problems involved with reversed main fans go beyond the limitations of the fan itself. After reversal of the main mine fans, the previously established air flow movement in the mine must be overcome and could take some time depending upon the circumstances. Any natural ventilation effect present can either aid the fan or add increased resistance depending on conditions. Thus, the field testing of main mine fans involves many other factors that influence performance.

Full fan evaluation is best done under both laboratory and field conditions. Data obtained from field fan tests can be compared for reliability with laboratory data. The hub to tip ratio which will be referred to is simply the ratio of the hub radius to that of the tip radius. The vane axial fans tested have adjustable pitch blades which can be set at various angles. A detailed illustration of the blade angle setting is shown in Figure 2.

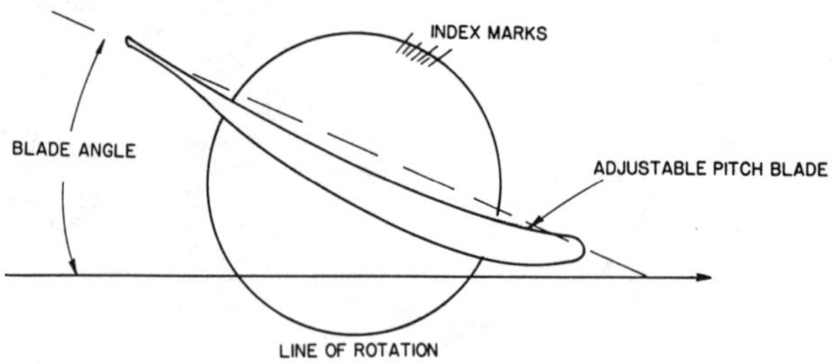

Figure 2 - Blade Angle Setting

LABORATORY TESTING

Comparing fan performance is done through characteristic curves which show pressure head, quantity, horsepower, and efficiency operating variables. The most significant information is the pressure head and air quantity relationship. Such curves are plotted for each different fan at a constant speed and standard air density. Fan laws (the mathematical relationship that adjusts data obtained under different conditions) are then used for comparison purposes. For field tests in particular, these relationships are used for pressure, fan diameter, air density, and fan rpm. These curves can reflect the fan performance at any number of different blade positions.

The Air Moving and Conditioning Association, Inc. (AMCA) has developed standards for testing fans which detail testing equipment and facilities. These standards assure uniformity of fan ratings, and AMCA Standards dictate that every care be taken to obtain as accurate results as possible, and that full documentation be carried out. When the testing apparatus is in working order, the fans must reach operating equilibrium before measurements are taken. All procedures are carefully recorded.

Figures 3, 4, and 5 show the fan test arrangement at the Joy Manufacturing Company's facilities. The fans were tested at 720 rpm for four different blade angles in both the forward and reverse

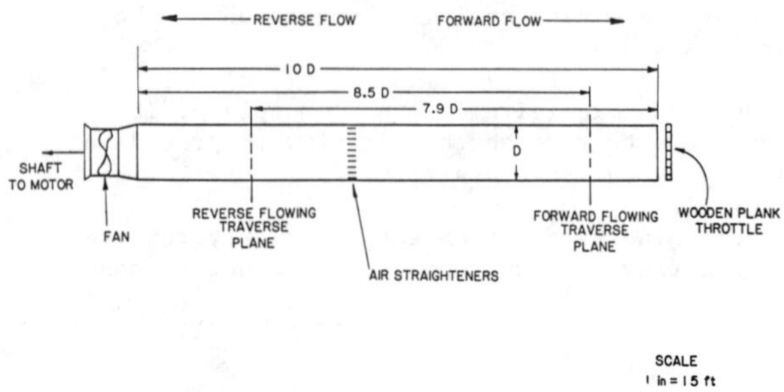

Figure 3 - Laboratory Fan Testing Arrangement

Figure 4 - Pitot Tube Measurements

Figure 5 - Lab Testing Facility

direction. A minimum of six performance determinations were taken at each blade angle which ranged from 10 to 42 degrees. For each determination, pressure readings from pitot tube, power input, rpm, wet and dry bulb temperature, and barometric pressure were recorded. The two fans tested were Joy models M96-43 and M96-58.

LABORATORY RESULTS

A typical performance curve for one of the fans tested is presented in Figure 6, showing forward and reverse performance, percentage of forward flow when in reverse, and input horsepower. Reverse fan summary curves of percentage of forward flow of air volume in the forward mode for each blade angle, are presented in Figures 7 and 8.

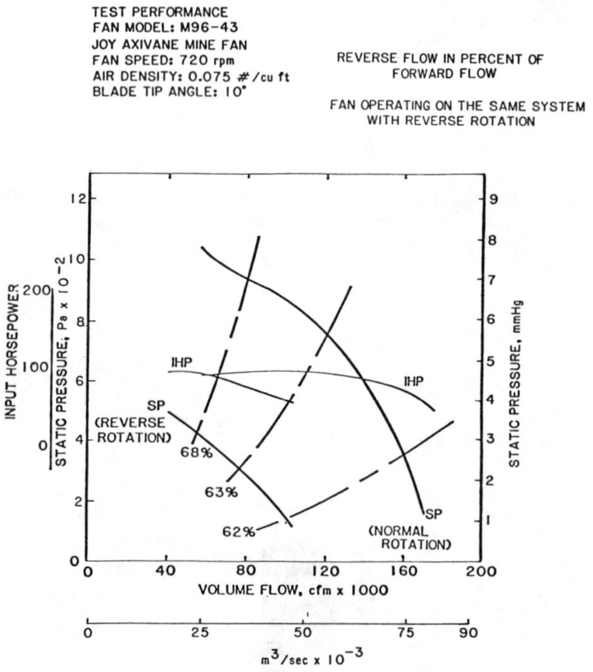

Figure 6 - Fan Performance With 10° Blade Angle.

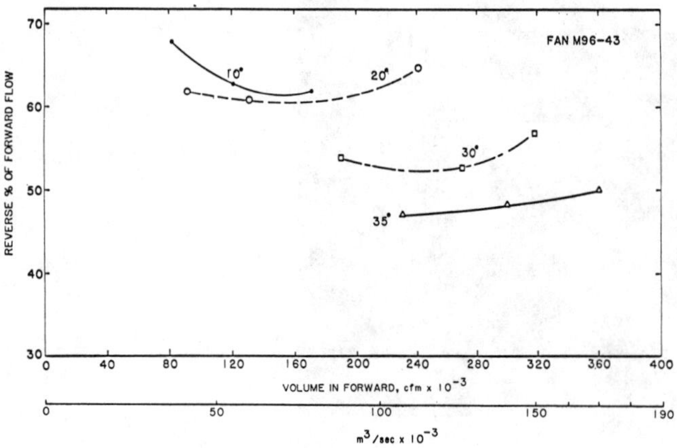

Figure 7 - M96-43 Reverse Performance Summary At Various Blade Angles.

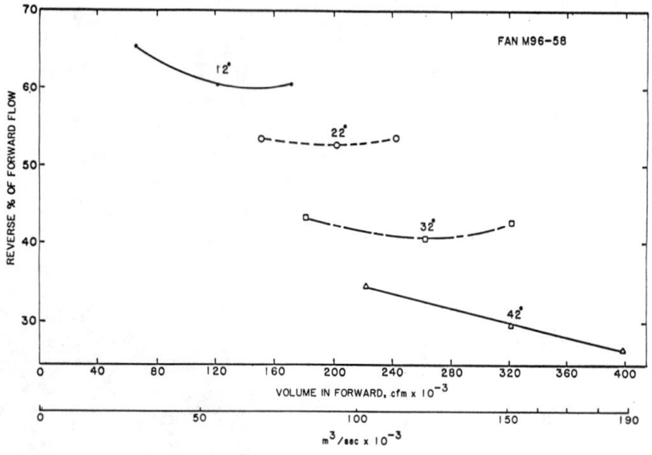

Figure 8 - M96-58 Reverse Performance Summary At Various Blade Angles.

To establish accurately the percent of flow in reverse (as compared to forward performance), it was necessary to first establish the forward performance during the test. The previously calculated data were found to be reasonably accurate when compared to the test results. Nevertheless, the curves presented are derived from the actual tests and not taken from calculated fan curves. Table 1 is the result of a comparison between calculated and lab test data for the M96-58 fan at the tested blade angles for forward flow. The difference between corresponding calculated and test data points is consistently less than 10%, often less than 5%, demonstrating the validity of the fan laws.

TABLE 1 - Calculated and Lab Test Data Comparison

Blade Angle	Calculated	Lab Test	Error
12°	49m³/s (104,000 cfm)	50m³/s (105,000 cfm)	-0.95%
12°	61m³/s (130,000 cfm)	66m³/s (140,972 cfm)	-7.78%
22°	87m³/s (185,000 cfm)	93m³/s (196,388 cfm)	-5.80%
22°	100m³/s (212,000 cfm)	109m³/s (230,416 cfm)	-8.00%
32°	127m³/s (270,000 cfm)	128m³/s (270,278 cfm)	-0.10%
32°	142m³/s (300,000 cfm)	149m³/s (315,000 cfm)	-4.76%
42°	146m³/s (310,000 cfm)	142m³/s (301,388 cfm)	+2.86%
42°	172m³/s (365,000 cfm)	170m³/s (359,722 cfm)	+1.47%

Data analysis reveals that reverse performance, P_r, decreases as the blade angle, B, increases:

$$P_r \alpha \frac{1}{B}$$

This generalization must be qualified by realizing that the relationship may be true only for the fan hub to tip ratios tested.

Analysis is carried further by plotting the average percent of forward performance, from curves in Figures 7 and 8, against the blade angle, as shown in Figure 9. The result shows a well defined slope for each tested hub to tip ratio. All data points fall into place with the exception of one point representing a blade angle of 10°.

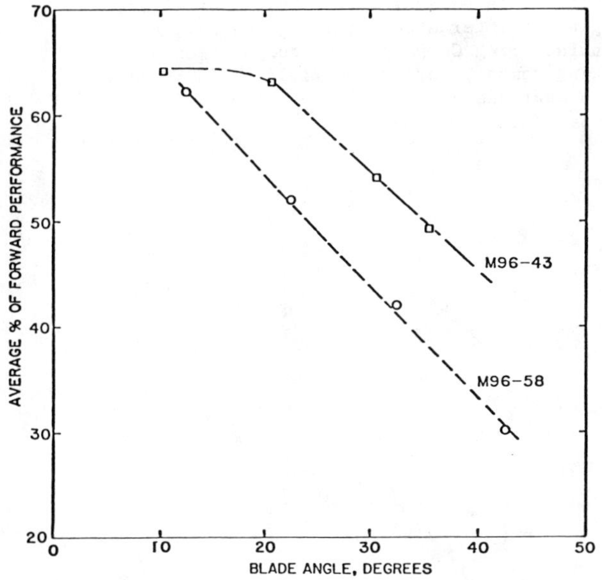

Figure 9 - Fan Average Performance vs Blade Angle.

Adopting these curves to fans of other hub to tip ratios is premature, although the order of data suggests that a series of points for a given hub to tip ratio will have their own slope for blade angles greater than 10°.

The data show, as expected, that fan pressure, P_s, changes as the square of the volume, Q:

$$P_s \alpha Q^2$$

Consequently, if the volume is reduced to 60% when the fan operates in the reverse direction, the pressure available from the fan will be 36%. This relationship has been proven with previous studies of smaller fans.

The laboratory testing provided useful data for generalizations. Field testing will be necessary to see if any generalizations can be made over a wider range of hub to tip ratios. It is clear that reverse flow is not a constant percentage of forward flow but depends upon blade angle and hub to tip ratio. The pressure volume relationship also seems to be applicable; that is, fan laws governing fan performance are valid for reverse flow. It appears that any hub to tip ratio has its own definable characteristic curve.

FIELD TESTING

The initial field test was performed on a 2.1 m (81-in)-diameter vertically mounted vane axial fan exhausting air from a room and pillar limestone mine. Auxiliary ventilation included two 28 m^3/sec (60,000 cfm) centrifugal fans operating at low pressure. These fans, consequently, had no effect on the main fan's performance, and no air doors or regulators were in use in the mine to influence the measurements.

Field testing of main mine fans followed as much as practicable the laboratory testing procedures with the same kinds of instruments used. Prior to testing, information that could directly influence the fan test was gathered at the mine site, and the overall mine ventilation scheme was studied in detail. Pertinent data included:

- Location of shafts and adits,
- Location and layout of intake and return airways,
- Distribution of seals and stoppings,
- Position of main mine fan,
- Position of auxiliary fans and their capacities,
- Location and type of ventilation doors, and
- Prior mine experience of fan reversal.

The fan installation was studied for:

- Type of fan - blowing or exhausting,
- Lay of the fan - vertical or horizontal,
- Fan ducting,
- Accessibility,
- Location of access holes for pressure survey purposes, and
- Location of guide vanes.

Other details were obtained from fan manufacturers, mine personnel, and information plates mounted on the fan assembly, including:

- Fan performance curves;
- Dimensions of the component parts of the entire fan; installation such as blade tip diameter, hub diameter: duct diameter and length, blade angle, and bell angles; and
- Drive motor specifications.

The field test was a two-step operation -- first with the fan's normal exhausting mode and then reversing, which was achieved by interchanging the motor's electrical connections. Sufficient time was allowed after reversal for air flow through the mine to reach a steady state.

The operations performed for each fan rotation were identical and consisted of determining the fan flow rates and pressure heads developed, by means of a pitot tube traverse, and, measurements were taken to determine air density, and amperes required.

FIELD RESULTS

Fan pressure values were corrected to standard density, 1.2 kg/m^3 (0.075 lb per cu ft), and corrections were made to adjust the affects of the evase and outlet cone in forward and reverse operation. After the fan operating points were identified, they were plotted on the fan manufacturer's performance curve, Figure 10. The measured

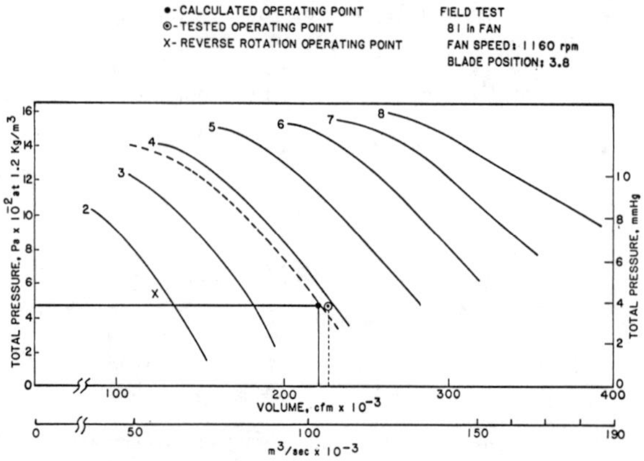

Figure 10 - Mine Fan Performance

operating point in the normal exhausting mode was very close to the calculated. The operating point in the blowing, or reverse mode, was at considerably less volume and at slightly higher total pressure. This is due to the increase in static pressure caused by conversion of velocity pressure, which is greater than the energy loss from expansion of air into the mine. Table 2 compares many of the important performance parameters of the mine fan in forward and reverse operation.

TABLE 2 - Performance Comparison

Parameter	Forward	Reverse
Volume, m^3/sec (cfm)	105 (226,000)	58 (125,000)
Pt, mmHg (in. W.G.)	-3.70 (-1.98)	4.38 (+2.34)
Pv, mmHg (in. W.G.)	3.24 (1.73)	1.81 (0.97)
Ps, mmHg (in. W.G.)	-6.94 (-3.71)	2.56 (+1.37)
Ampheres	151	127
Motor Output H.P.	122.6	103.1
Motor Efficiency	81%	62%

ACKNOWLEDGEMENTS

The support of the Bureau of Mines in carrying out this work under contract J0308044 is gratefully acknowledged, as is permission to publish this paper from the Bureau of Mines, Engineers International, Inc., and Joy Manufacturing Company. The cooperation of the mining company during the field tests is very much appreciated.

Chapter 4

SILENCERS FOR SECTIONAL FANS

Earl Kennedy

Supervisor Dust and Noise Control
Northern West Virginia Region
Consolidation Coal Company
Morgantown, West Virginia

INTRODUCTION

The Northern West Virginia Region of Consolidation Coal Company consists of ten underground mines. The Pittsburgh seam of coal is being mined. This seam is approximately seven feet in thickness in our area. The mining equipment consists of twelve longwall systems and a combination of Joy and Jeffrey ripper type miners. There are a few boring type machines still in use, but these are rapidly being phased out.

The auxiliary ventilation fans being utilized vary in type from single stage carriage mounted 20 horsepower Joys to 40 and 50 horsepower two-stage Joys, Buffalo Forge, and Peabody brands. The particular fans that we have primarily been working with on noise reduction are the two-stage 40 and 50 horsepower types.

HISTORY OF THE SILENCERS EVOLVEMENT

In 1977 the Loveridge Mine No. 22 began phasing into operation the Joy two-stage 50 horsepower fans. We immediately became aware of the increased noise level and began working on the problem. The noise level of a 50 horsepower two-stage fan is approximately 120 decibels (equivalent to a small aircraft engine) at the loudest point, the exhaust end. The level is approximately 104 decibels ten feet inby the fan (intake side).

The earliest experiment in 1977 was the application of Insul-coustic, or insulated quilt, barrier material manufactured by Birma Corporation of Centerville, New Jersey. The material was doubled and wrapped completely around the body of the fan. This early testing resulted in a substantial reduction of the noise level by six decibels ten feet inby the fan and by seventeen decibels at the exhaust end of the fan.

In 1978 an EAR ISODAMP C-1002 type of material was wrapped completely around the body of the fan. The reduction in the noise level was much less than experiment No. 1. The reduction was four decibels at the exhaust end of the fan, from the 119 to 115 level. The EAR material was left intact,and a section of the quilt barrier material from the earlier experiment was wrapped around the exhaust end adapter of the fan, reduction was noted,and the muffler idea originated.

I became involved in the testing for the first time in late 1978. My project, which is actually Test No. 3, involved the construction of a muffler using a section of 24" round ABC Rigi Duct ventilation tube five feet long. The tube was lined with two inch thick Geocoustic Foamglass made by Pittsburgh Corning Corporation. The foamglass was glued to the inside of the tube with contact cement. An expanded aluminum screen was placed over the inside surface of the foamglass to hold it in place. A significant reduction in the noise level was observed. The exhaust was reduced from 119 to 101 decibels (18 decibels) and the intake side (10 feet inby the fan) was reduced from 104 decibels to 98 decibels (6 decibels). The muffler was heavy, bulky, and it soon deteriorated. New and different material was immediately purchased.

Experiment No. 4 was conducted in early 1979. Experiment No. 4 was simply the previous experiment with a different material replacing the foamglass lining. The material used was cousticomposite 50-5-100T purchased from Ferro Composites in Norwalk, Connecticut. There was no significant difference in the noise reduction level comparison of the foamglass and Cousticompsite material. The major improvement was that the new material proved durable and flexible enough for further use as a covering material for the entire fan body, as was done early in 1977 with the insulated quilt material. Having worked with the noise reduction testing, more ideas evolved,and we have realized that further reductions can be obtained. (1) Fans can be checked (curtained) off or enclosed with strips of the noise damping material. (2) The mufflers can be attached to both the intake and exhaust ends. (3) It is also practical to maintain fans at least two tubes or more back in the crosscuts away from the mining area. Regardless of the method, or combination of methods used in the mining industry, one must keep in mind that location constantly changes. Therefore, the noise reduction procedures must be designed to require minimum handling by employees. A slide presentation will now be shown detailing the proper procedure for attenuating an auxiliary fan.

CONCLUSION

We can effectively reduce fan noise levels. The complete system of covering the body of the fan, installing the muffler, and work habits such as setting the fan at least two (2) tubes back in the crosscut is very effective. Also, it is to be noted that no reduction in the ventilation capabilities of our fans have been encountered during conduct of this noise reduction program. There are presently some (sound attenuators) being marketed. Ours are rapidly and easily assembled. The cost per muffler is approximately $125.00 (not including labor). In concluding, we have made the face area less noisy, with a minimum amount of time and cost involvement. The improvement of the work area can only be an asset in this situation. Employees may remain in the area for longer periods of time due to this reduction in the noise level. More can still be accomplished. Manufacturers could redesign fans for noise reduction and install noise control equipment during the manufacturing process. Indeed auxiliary fans are not the only contributor to noise in the mining process, but they are the constant contributor.

2
Auxiliary Ventilation

COAL MINE FACE VENTILATION SYSTEMS: NEW CONCEPTS AND UNDERGROUND RESULTS

Chapter 5

Edward F. Divers

U. S. Department of the Interior
Bureau of Mines, Pittsburgh Research Center
Dust Control and Ventilation Group
Pittsburgh, Pennsylvania

ABSTRACT

Face ventilation is generally the most critical area in the coal mine ventilation system. Various studies have repeatedly shown that good ventilation is by far the most effective and least costly method of controlling dust and gas problems created during face operations.

The Bureau of Mines has recently developed and assisted in the development of several new concepts in face ventilation systems, and in methods to accurately evaluate these systems underground.

A new type of scrubber and ventilation system design built into the auger support frame of a continuous miner was developed. The objective was to improve face ventilation and reduce respirable dust concentrations. With this arrangement, one scrubber is mounted into each side of the boom. Laboratory and underground tests of various scrubber discharge arrangements showed good to excellent dust and methane control.

Combined face ventilation systems, using both blowing and exhaust curtain, potentially offer dust and gas control and deeper cutting. Bureau tests at a mine using a combined face ventilation system and remote-control-operated continuous miners showed excellent dust and methane control.

The advancement of line curtain systems beyond permanent supports without exposing workmen to roof fall hazards can be accomplished only with specially designed extendable devices. Numerous concepts for line curtain extension have been developed and tested. For the past several years, the Bureau has been testing various types of extensible face ventilation systems. These extensible systems can generally be classified as floor-, roof-, machine-, and rib-mounted systems. In the past 6 years, approximately 40 systems have been evaluated for feasibility. From these, several have been selected as practical or innovative and are presented herein.

A technique developed by the Bureau to accurately and safely evaluate the effectiveness of face ventilation is also presented.

INTRODUCTION

Until recently, the only way of measuring face ventilation effectiveness was to measure methane at the face and in the return airway. If no methane was present, face ventilation could not be measured.

The Bureau of Mines has developed a simple and quick method for measuring ventilation at the working face that eliminates the need of measuring methane; it is called the face ventilation measurement (FVM) method (1). For sections using continuous miners a small volume of SF_6 tracer gas is released directly behind the cutters on the off-curtain side of the entry at the start of the mining cycle. At the same time, gas samples are taken periodically in the immediate return airway. These samples are analyzed for SF_6 content. A curve is then plotted of SF_6 concentration versus time. The area under this curve represents the volume of SF_6 that passed through the return airway during the test period. The area within each time increment is divided by the total area under the curve to give the fraction of SF_6 removed with each progressive time period. With this information, another curve is plotted showing the percentage of SF_6 removed with respect to time. This curve gives a graphic picture of how well the face is being ventilated. A well-ventilated face will show a high percentage of SF_6 removal in a short time period. For example, we generally consider a face ventilation system to be very effective when it will remove about 90 pct of the SF_6 within 1 min.

As another example, in full-scale mine entry model tests, at 1 min about 90 pct of the SF_6 is removed (Figure 1) when the curtain setback (curtain-to-face distance) is 10 ft or less. Studies have shown that face ventilation declines as the curtain-to-face distance is increased (2). The Mine Safety and Health Administration (MSHA) requires that the curtain-to-face distance be kept within 10 ft (3). Figure 2 also shows that as curtain setback is increased, the percentage of SF_6 removed in 1 min decreases. This is a clear indication that the FVM gives an accurate indication of how well the face is being ventilated.

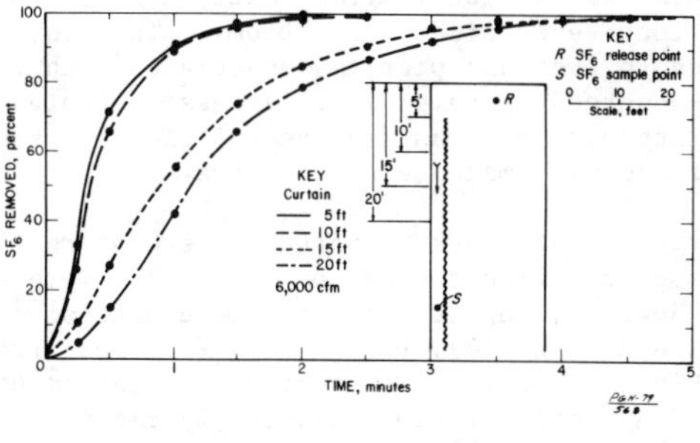

Figure 1. Percent SF_6 removed versus time at different curtain distances.

Data from these laboratory tests were also used to see if box penetration had any major effect on face ventilation. Each test used about the same air volume and curtain setback; only the box depth was changed

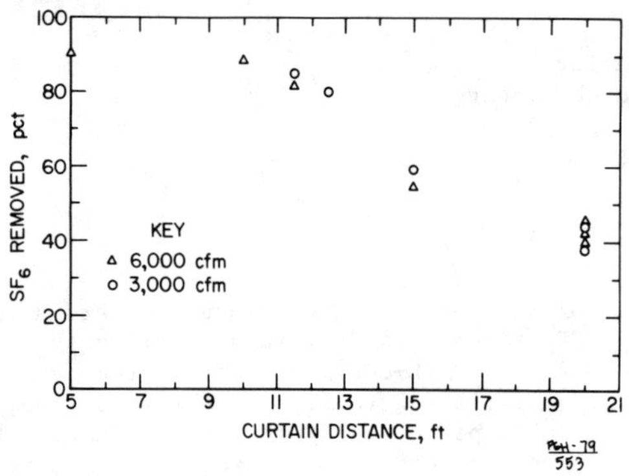

Figure 2. Percent SF_6 removed in 1 min with respect to curtain distance.

significantly, which had little effect on face ventilation.

THE TWIN SCRUBBER ARRANGEMENT

The Bureau of Mines and Jeffrey Mining Machinery Division, Dresser Industries, Inc., developed a new type of scrubber and ventilation system design built into the auger support frame of a Jeffrey 120-L continuous miner. The objective was to improve face ventilation and reduce respirable dust concentrations. With this twin-scrubber arrangement, one scrubber is mounted into each side of the boom. The system was developed to overcome problems associated with earlier machine-mounted systems; i.e., excessive height, inflexibility with various face ventilation systems, and interference with machine maintenance.

Early in this work, reports of methane recirculation associated with the use of scrubbers were evaluated. However, reports of other more extensive studies by the Bureau (4), by industry (5), and from the United Kingdom (6-7) were questioning the importance of avoiding recirculation caused by machine-mounted air movers such as diffuser fans and scrubbers. For example, Kissell and Bielicki (4) state that recirculation appears to become dangerous only when recirculated air is substituted for fresh air.

These reports were all considered in an effort to design a scrubber system with minimum recirculation. Scrubbers have the distinct advantage of being the only method for removing over 90 pct of the respirable dust from the face ventilating airstream, blowing or exhaust, thus allowing downstream operations such as roof bolting. The twin-scrubber concept resulted from these considerations.

TWIN SCRUBBER DESCRIPTION

The system delivers 5,000 cfm with one scrubber mounted on each side of the continuous miner's auger support frame. Each scrubber typically consists of a 2,500-cfm centrifugal fan, direct-driven at 4,000 rpm by a hydraulic motor, a flooded fibrous-bed panel approximately 14 in by 16 in by 1/4 in thick, water sprays (2 on each scrubber delivering 2.5 gpm at 150 psig), a wave-blade-type water droplet eliminator, and associated duct work.

Air intakes are located within 6 ft of the coal face and are protected against clogging during the mining cycle. The general arrangement of the scrubber system is shown in Figure 3. The clean air discharges

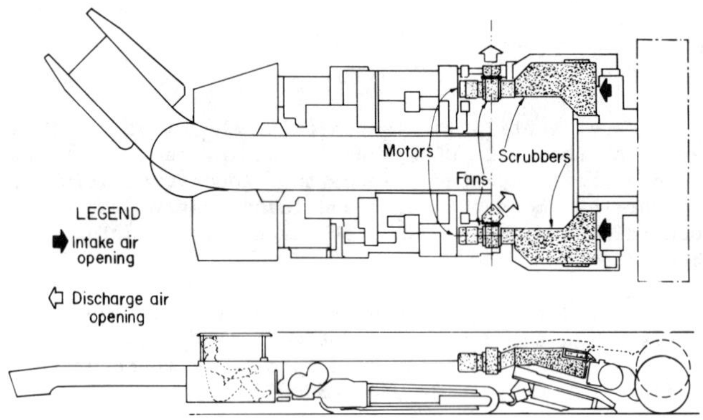

Fig. 3- Plan view and side elevation of scrubber layout.

Figure 3. Plan view and side elevation of scrubber layout.

to the entry through short lengths of adjustable duct work. The adjustment depends on the type of face ventilation system used on the section. The slurry water from the panel and droplet eliminator is pumped onto the coal on the machine conveyor. The two scrubber exhausts can also be tied together by the crossover duct and exhausted at the left rear corner of the continuous miner.

SURFACE TEST

Experience has shown that surface tests can expedite and reduce costs of systems developed for underground use. Consequently, preliminary tests were conducted at an existing MSHA facility at Bruceton, PA, that can simulate underground coal mine face methane emissions (8).

The facility is an L-shaped concrete block structure that simulates an underground coal mine entry. The test section was 20 ft wide and 3-1/2 ft high. Ventilation was supplied by an axial-flow mine fan. Exhaust curtain was located on the left-hand side. The last 25 ft of the curtain was built on a movable framework, which made it possible to vary the curtain distance from the face as needed.

Methane gas was released evenly along the face through perforated release panels in increments of 10 cfm until a maximum of 2.0 pct was detected at any sensor location, or until the next increment would have caused a methane concentration greater than 2.0 pct at any one of the monitors. When 2.0 pct was reached at any point, the gas release was stopped and the gallery was cleared of gas. From the above data, the flow rate of gas was calculated that would produce a peak of 1 pct methane. When necessary, the air quantity was changed and the line curtain moved for the next test. Gas liberation rates were measured using glass tube rotometers connected to the gas release panels. Concentrations of methane gas were detected in the model entry by 12 strategically placed methane monitors and recorded on strip charts in a remote instrument control room. The position of the monitors (igure 4) was determined from previous MSHA experience of underground face ventilation surveys.

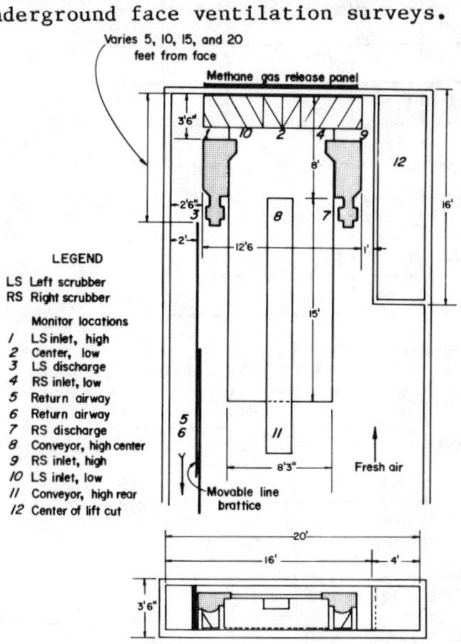

Figure 4. Location of methane monitors in model entry.

A plywood full-scale model of a continuous miner with a rotating cutting element and operating water sprays was placed on the left (curtain) side of the room, and a wall was built on the right side to simulate a cut 16 ft deep by 16 ft wide. The twin-scrubber system was mounted on the auger support frame of the continuous miner. Various scrubber discharge arrangements were tested (9) to determine which would provide the best face ventilation.

A successful scrubber configuration would yield equal or greater methane dilution than a baseline condition. This baseline condition with scrubber off was exhaust curtain at 10 ft from the face, 1.0 pct methane, and three exhaust airflows of 3,000, 6,000 and 9,000 cfm. Methane emissions at these three airflow conditions were normalized, and the baseline conditions were assigned an improvement factor value of 1.00 to facilitate comparison with other tests. The baseline was compared with the same airflows and methane concentration, four curtain-to-face distances (5, 10, 15, and 20 ft), and 27 scrubber exhaust positions, all with the scrubbers on. Fortunately, it was not necessary to test all combinations, as unfavorable trends could be readily spotted and subsequent similar tests stopped.

Configuration 1 (Table 1) differed from the other configurations in that additional inlet openings were present on the outer edge of each scrubber. Configuration 25 used expandable vent tubing to bring the exhaust of the left scrubber behind the curtain. Extension ducts were used on configurations 4-14 to cross the machine from right to left.

Smoke flares were used on specific tests to visually determine if recirculation would occur. Recirculation was visible evidence of smoke moving outby the end of the line curtain.

A completely successful scrubber configuration would have a methane dilution improvement factor greater than 1.00 for all 12 combinations of entry airflow and curtain distance. The determination of this factor followed rules previously set by MSHA for the surface facility tests. The relationship between these numbers presented in table 1 is linear; that is, 2.00 indicates twice the methane dilution capacity as 1.00. As an example, with 3,000 cfm of main ventilation and a 10-ft curtain distance, the base configuration without scrubbers could dilute 9 cfm of methane to 1.00 pct as first detected at methane monitor position 2. In comparison, configuration 24 with 3,000 cfm of main ventilation could dilute to 1.00 pct 25 cfm of methane at a 5-ft curtain distance, 19 cfm of methane at both 10- and 15-ft curtain distances, and 20 cfm of methane at a 20-ft curtain distance, all first detected at the comparable methane monitor position 10. The increased methane dilution capacity improvement factors are 2.80, 2.13, and 2.27, respectively. Each is a multiple of the 9-cfm dilution capacity found in the baseline condition.

TEST RESULTS

The "configuration" column on the left in table 1 is presented to rapidly show the exhaust (clean) air discharge configuration from each scrubber when viewed from the top; that is, in plan view. Only one successful configuration was needed for future underground applications.

Of the 27 different scrubber configurations tested, three were completely successful. Configurations 16, 17, and 24 exceeded all criteria of airflow and curtain distance. Note that 17 was a single-scrubber configuration. Six other configurations (11, 12, 13, 18, 23, and 25--Table 1) were partially successful because they met some airflow criteria but failed others.

The data in table 1 indicate that the success criteria become more difficult to meet as the face ventilation flow rate increases. While all configurations were not tested at 3,000 cfm, a total of 11 tests were successful (11, 12, 13, 13-1/2, 13-3/4, 16, 17, 18, 23, 24, and 25). This group was reduced to five at 6,000 cfm (11, 13, 16, 17, and 24). Only three were left at 9,000 cfm (16, 17, and 24). This may be due to the fixed volumetric flow rate from the scrubber having proportionally less effect when compared with increasingly effective face ventilation.

Recirculation was observed in configuration 16, but not for configurations 17 and 24. These tests show that recirculation tends to occur when the scrubbers are moving more air than the main ventilation system. Scrubber fan speeds of 4,000 rpm are moving 2,500 cfm each. Recirculation was observed on only three tests (13, 16, and 18), all with 3,000 cfm of main ventilation. When scrubber fan speeds were reduced in configuration 13 to 2,700 rpm and 1,800 rpm, recirculation disappeared. It is quite possible that more of the configurations would have produced recirculation if they had been tested at 3,000 cfm.

TABLE 1. - Methane dilution improvement factor

Test No.	Configuration[1] Fan	Duct	3,000 cfm at brattice distance of-- 5 ft	10 ft	15 ft	20 ft	6,000 cfm at brattice distance of-- 5 ft	10 ft	15 ft	20 ft	9,000 cfm at brattice distance of-- 5 ft	10 ft	15 ft	20 ft
Base			1.63	1	0.63	0.57	1.84	1	0.88	0.72	5.25	1	0.90	0.85
1							1.76	0.88	.84	.84				
2							1.68	.72	.68	.72				
3		•					1.68	.56	.60	.64				
4							1.92	.76	.88	.80				
5							1.52	.68	1.12	1.00				
6							1.12	.84	.64	.32				
7							1.92	.76	.72	.64				
8							.60	.68	.80	.44				
9							.88	.56	.52	.48				
10							1.32	.76	.60	.68				
11			1.67	1.77	1.43	1.50	1.76	1.24	1.12	1.08	3.50	0.95	0.85	0.90
12			2.03	1.60	1.13	1.33	1.52	1.52	.80	.96	3.25	1.25	.70	1.15
13			1.70	1.63	1.60	1.73	2.12	1.16	1.20	1.36	3.45	.80	1.00	1.00
13-3/4			3.10	1.47	1.33	1.50	2.40	.76	.80	1.00	3.15	.65	.80	.90
13-1/2			3.07	1.10	1.17	1.23	1.72	.56	.76	.68	4.25	.60	.50	.55
14	•						1.96	.72	.56	.60				
15							1.92	1.16	.40	.84				
16			3.50	3.43	3.87	4.00	2.64	2.20	1.88	2.12	4.05	3.10	2.05	2.35
17	•		2.83	2.80	2.63	2.63	2.40	2.00	1.68	1.72	2.50	1.85	1.75	1.75
18		•	2.27	1.43	1.50	2.00	2.48	1.56	.92	1.44	3.85	1.50	.80	1.15
19							2.96	.68	.92	.84				
20							2.28	.80	.52	.40				
21	•						2.48	.92	.52	.44				
22							3.00	1.60	.80	.84				
23			2.53	1.30	1.33	1.47	2.32	1.24	.96	1.04				
24			2.80	2.13	2.13	2.27	2.24	1.68	1.40	1.40	3.10	1.70	1.15	1.35
25		•		1.20	1.33	1.27		.76	.68	.76				
26								.72	.48	.48				
27		•						.52	.48	.36				

[1]Configuration shows a plan view of the position of the scrubber exhaust with the face at the top and the exhaust brattice to the left. Long horizontal bar extending from right scrubber denotes a crossover duct. A dot indicates scrubber not operating.

RESULTS OF IN-MINE METHANE DILUTION TEST

These series of tests measured the face ventilation effectiveness of the twin scrubber system. Tests were conducted at Eastover Mining Co.'s, Highsplint Mine, 001 Section. This mine is located in Harlan County, Ky, in the Harlan seam. The 001 section seam ranges from 4 to 5 ft in thickness, including a 1-ft band of rock located near the top of the seam. The twin scrubber was installed in Highsplint in mid-May 1981. After strengthening the system to withstand roof falls, the testing program started.

SF_6 was again chosen as the tracer gas because it is safe, odorless, chemically and thermally stable, and detectable at low concentrations, mixes well with the ventilating air, and does not occur normally in the underground environment. Since each test takes only a few min one can rapidly complete sufficient tests for high statistical confidence.

The SF_6 was released directly behind the cutting head on the off-curtain side, near a bank of water sprays. As the SF_6 was released, bottle samples were taken at specific time intervals for several min in the immediate return airway. The samples were analyzed, and a curve drawn of SF_6 removed versus time. Although the tests were conducted on a non-production shift, the continuous miner was positioned at the face to simulate the beginning of a box cut, and a shuttle car was parked behind the miner. Prior to and during each test, the sprays were turned on and the cutting head was rotated.

For these tests it was neccessary to connect both scrubber exhausts to a common flexible duct extending along the left fender and discharging opposite the operator. This had to be done to avoid changing the mine ventilation plan.

To determine the effect of the twin-scrubber unit on face ventilation, the curtain-to-face distance was varied between 10 and 40 ft. Identical tests were run with the scrubbers turned off for comparison. As expected, without the scrubber, face ventilation deteriorated as the curtain-to-face distance was increased beyond 10 ft. With the scrubbers operating, recirculation occurred when the line curtain was set inby the scrubber's exhaust, located approximately 23 ft from the face. This was due to the exhaust air striking the line curtain rather than being directed behind it. As the line curtain approached the scrubber's exhaust, at curtain-to-face distances of 25 and 33 ft, face ventilation was significantly improved, showing excellent effectiveness. However, when the curtain setback was increased to 40 ft, face ventilation fell off slightly, due to some recirculation. These tests are graphically shown in Figure 5, where time required to

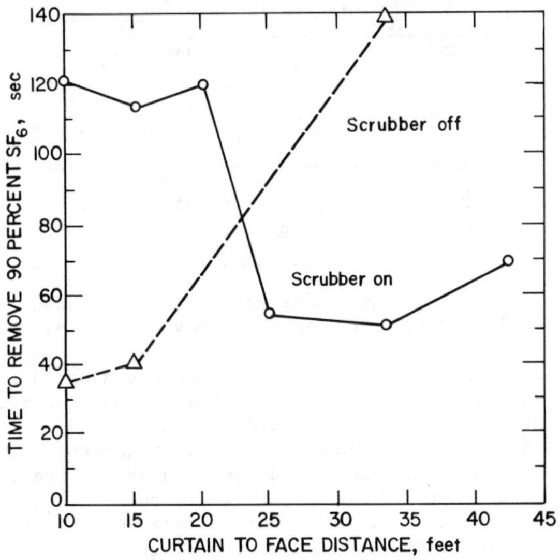

Figure 5. Time to remove 90 pct SF_6 with respect to curtain-to-face distance.

remove 90 pct of the SF_6 is plotted against the various curtain-to-face distances tested. These test results clearly show that scrubber exhaust must be directed behind the curtain when exhaust face ventilation systems are used. Fortunately the twin-scrubber system exhaust can be located far enough inby to satisfy even 10-ft curtain-to-face distances. The next series of underground tests will evaluate such a system.

EVALUATION OF A COMBINED FACE VENTILATION SYSTEM

There are three basic face ventilation systems: Blowing, exhaust, and combined.

Blowing systems can be very effective in sweeping gas from the face. They can also increase production by allowing deeper cuts. Unfortunately, blowing systems usually create severe dust problems by allowing the dust created by the continuous miner to roll back beyond the machine operator. The standard procedure for alleviating this problem is to place a mechanical dust collector (scrubber) on the continuous miner. Scrubbers are expensive and create additional maintenance problems, but increased production can justify their use.

Exhaust face ventilation can also offer good control when the line curtain is maintained reasonably close to the face. Maintaining this curtain is troublesome, and severe safety problems result from personnel extending curtain beyond permanently supported roof. These problems can also limit the depth of cut, thus lowering production.

Where conditions permit, combined face ventilation systems, using both blowing and exhaust curtain, potentially offer dust and gas control and deeper cutting. Required conditions are wide entries, 20 ft or more, to allow room for both curtains and reasonably good roof to allow deep cuts.

Solar Fuel Co., a division of Gulf+Western Industries, Inc., is using a combined face ventilation system that appears to offer high production with excellent dust and gas control. An underground test program was arranged by the Bureau of Mines to evaluate the effectiveness of the combined face ventilation system for both respirable dust and gas control. Since the mine selected (Solar #5A mine) did not have a significant gas problem, Bureau personnel would again use the tracer gas technique to determine face ventilation effectiveness.

MINE TEST SITE DESCRIPTION

Solar Fuel operates two drift mines in the upper Freeport "E" seam, just north of Somerset, Pa. Seam height is approximately 42 in. All mines utilize flight conveyors and radio remote control with Jeffrey 101 Continuous Miners. Typical development is three 20-ft-wide entries on 50-ft centers. Face crews normally consist of five men: A remote control miner operator, roof bolter operator and helper, bridge carrier operator, and brattice man-carrier. Typical shift production is 350 tons.

For our tests the blowing curtain was on the left and just behind the continuous miner. Spray water flowrate was 15 gpm and 70 psig at the nozzle. The exhaust curtain was on the right and extended inby 10 ft past the blowing curtain. Primary airflow at the end of the exhaust curtain was 4500 cfm. Both curtains were hung from roof bolt plates with wood wedges, and no special effort was made to eliminate curtain sag between roof bolts. Indeed, leakage from the blowing to the exhaust side of the entry allowed the remote operator to remain in fresh intake air. With deep penetration into the entry, about half of the air leaked to the exhaust side. Although the volume of leakage was approximately 4500 cfm, no measurable air movement was detected between the blowing and exhaust curtain. Since the continuous miner was cutting into a 14-in. rock band, large quantities of visible dust were being generated. This dust cloud seldom rolled back beyond the inby end of the exhaust curtain, and it was very easy for the remote control operator to stay out of the visible dust and still remain within 20 ft or so of the miner.

RESPIRABLE DUST CONTROL EVALUATION

For this respirable dust control evaluation, we relied on standard 8-hr gravimetric (personal) samplers, identical to those commonly used underground. A cluster of three was placed in the immediate return to evaluate total dust generated at the face. An identical cluster of three was placed at the usual cab position on the continuous miner, and two were connected to the radio remote control unit. Samples were taken for three consecutive daylight shifts. Production including rock was 350, 295, and 285 tons, respectively. During this period, 12 place changes were made. The remote operator for these tests positioned himself between 10 and 20 ft from the miner. Subsequent analysis of our gravimetric samples showed a 97-pct reduction (Table 2) in respirable dust concentrations at the remote position compared to the cab position on the miner. Dust reductions of this magnitude are rare. Also note that the 3.21-mg/m^3 concentration at usual cab location is not in compliance with Federal regulations.

TABLE 2. Solar Fuel Test 8-hr Gravimetric Sampling Results.

Date	10/26/81	10/27/81	10/28/81	
	Respirable Dust Concentration, mg/m^3			
Cab Position on Machine	3.27	2.70	1.84	
	4.12	3.59	2.20	
	5.62	3.71	1.80	
	(4.34)	(3.33)	(1.95)	(3.21)
Remote	.15	.10	.08	
	.13	.08	.09	
	(.14)	(.09)	(.08)	(0.11)

NOTE: Mean values appear in parentheses.

Comparison of Remote Operator to Cab Position on Machine

Ratio: 3.21/.11 = 29/1
Dust Reduction: 97 pct

FACE VENTILATION GAS CONTROL EFFECTIVENESS

Determining the gas dispersion ability of coal mine face ventilation systems can present problems, especially in mines like Solar Fuel's #5A mine, which is non-gassy. Fortunately, the SF_6 technique developed by the Bureau of Mines was proposed and used. A small quantity, 25 cu cm, of sulfur hexafluoride tracer gas (SF_6) was released immediately behind the machine cutter head. Simultaneous 10-cu-cm grab samples were taken at specific time intervals behind the exhaust curtain, just outby the continuous miner and at the remote control operator's position (Figure 6).

For these tests, the SF_6 was released through 1/4-in.I.D. polyvinyl chloride tubing strung from a position just behind the cutting head to a position just outby the right fender. A tee was placed in line with one end to the SF_6 syringe and the other to a hand-operated tire pump with the handle extended. Prior to this, ventilating air volumes, curtain setbacks, and face configurations are recorded. The total volume of SF_6 is injected, and the pump handle is depressed several times to insure complete dispersion. This is done with a pre-arranged signal to those taking the grab samples. For these tests, six grab samples were taken at each location at 5-second intervals, followed by six additional samples at longer time intervals.

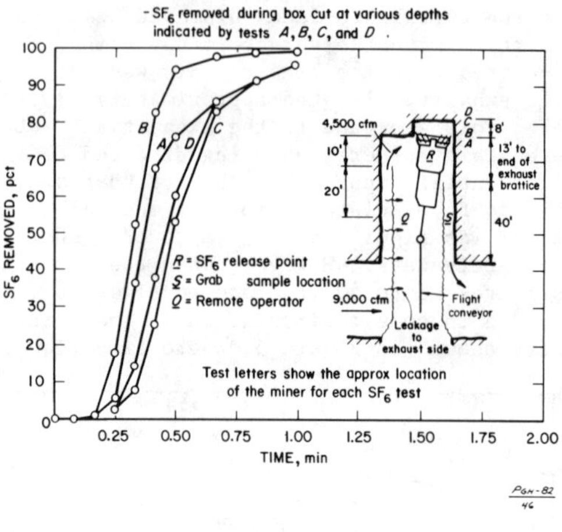

Figure 6. SF_6 Removed during box cut at various depth indicated by tests A, B, C, & D.

These air samples were analyzed for SF_6 content and used to plot graphs showing SF_6 removed vs time.

The first series of SF_6 tests were conducted with the miner making an 8-ft-deep box cut on the right (exhaust) side; four tests were conducted, and no effort was made to influence the remote control operator's standard cutting procedures. In gassy mines, it is assumed that the operator would box cut on the blowing side to better disperse face methane emissions. Even with the box cut opposite the blowing side, our test (Figure 7) showed good potential methane dispersion. SF_6 samples showed a rapid rise and

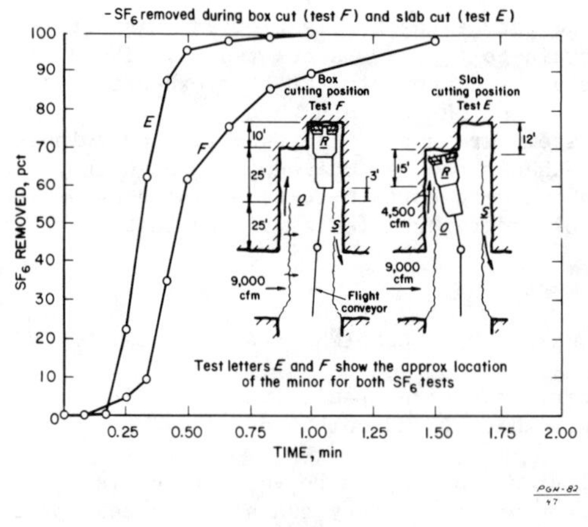

Figure 7. Percent SF_6 removed during box cut (Test F) and scab cut (Test E).

decay at the return location behind the exhaust curtain approximately 10 ft outby the continuous miner. This immediate sharp rise and decay is typical for good face ventilation.

The face ventilation measurement (FVM) conducted

by a previous underground study (1) showed that with conventional sprays, a 1-ft box cut, a 9.5-ft curtain distance, and 3600 cfm of air flow behind the curtain, 70 pct of the SF_6 was removed in approximately 1-min. In this study, 90 pct of SF_6 was removed in 1-min or less under all conditions of testing from A to F, indicating a very high degree of effectiveness for the double-brattice system.

These tests show that the combined face ventilation and remote control system used by Solar Fuel can offer excellent dust and gas control.

New Extensible Face Ventilation Systems

The study of extensible systems for control of dust and methane generated at the face has been and continues to be an important area of USBM investigation. These efforts have led to the development of new spray fans, diffuser fan layouts, and new extensible systems. Stringent dust and methane control requirements demand improved face ventilation techniques to assure maximum safety, if possible without adversely effecting production.

With ripper-type continuous miners, it is now relatively common with blowing systems to advance 20 ft or more before backing out for roof bolting. This advance distance can still permit the operator to remain under supported roof. The need to maintain the ventilation system within 10 ft of the furthest point of advance, as required by the Coal Mine Health and Safety Act of 1969, often makes extension of exhaust ventilation systems both difficult and dangerous.

With borer miners, unsupported roof presents fewer problems because of more stable roof geometries, but narrow entries, and space limitations on and around the bulky machines make effective face ventilation extremely difficult. The results can be inadequate face ventilation.

Line curtain or brattice cloth systems have historically been the principal coal mine face ventilation technique. These line curtains, when properly installed and maintained, can provide excellent ventilation to the operating face.

Presently, the advancement of the line curtain systems beyond permanent supports without exposing workmen to roof fall hazards can be accomplished only with specially designed extendable devices. Numerous concepts for line curtain extension have been developed and tested. Most have failed owing to the harsh working face environment and because they place additional burdens on face personnel.

The greatest problem now evident is the development of a practical means to safely advance these systems as the miner advances beyond the permanent roof supports. This problem will be magnified when remote control systems become more universal and maximum penetration into solid coal is increased from the present 20 ft to 40 ft or more. An examination of the state-of-the-art of the line brattice, tube, and other ventilation systems in light of projected mining equipment improvements certainly supports the need for new extensible concepts to keep pace with these improvements.

For the past several years, the Bureau has been testing various types of extensible face ventilation system in full-scale lab facilities. This lab work can reduce the extent of more expensive underground tests. The extensible systems tested can generally be classified as floor-, roof-, machine-, and rib-mounted systems. In the past 6 years, approximately 40 systems have been evaluated for feasibility. From these, about half were selected as practical or innovative. The chief advantages of these systems are low cost and high reliability. Primary disadvantage is that they require added effort from face personnel. Many of these systems are based on various types of cantilevered bars that support the extended brattice. Bureau tests have indicated that cantilevered systems are limited to approximately 20-ft maximum extensions.

Recent work by Foster-Miller Associates (FMA) under a U.S. Bureau of Mines contract has determined leak rates vs effectiveness of various (especially cantilevered) systems. Full-scale lab tests at FMA show air leakage over the top of extensible systems to be very significant. For example, a 10 ft-long extension that deflected 1 ft from the roof and 10-ft from the face in a 4-ft-high seam would bypass 70 pct of a 3,000-cfm primary airstream and 60 pct of a 9,000-cfm primary airstream. Although most of the air is lost to this triangular gap, these tests also show that almost any curtain extension, even with these high leak rates, will still have a pronounced beneficial effect on face methane dispersion. These same tests showed methane dispersion improved by a factor of 3 times with these "leaky" extensions. While line curtain leakage rates, of fairly large magnitude may be acceptable for dust control, intense methane liberation rates require much tighter face ventilation systems.

Six novel face ventilations systems have been selected from those examined to act as an idea springboard; most have not been tried underground. They have been selected because of their uniqueness, safety, simplicity, low cost, and versatility for various seam heights and mining conditions, and are presented herein.

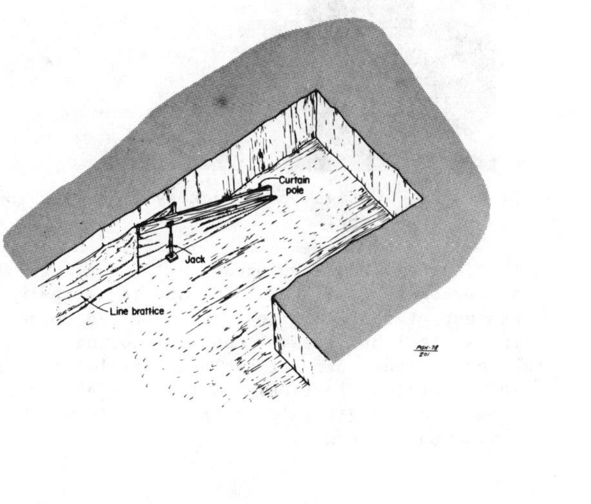

1. Figure 8. Kiski Curtain

This system, utilizing a cantilevered 2 x 4 with attached curtain, represents the utmost in simplicity and is used in some Pennsylvania mines. Maximum extension is about 10 ft.

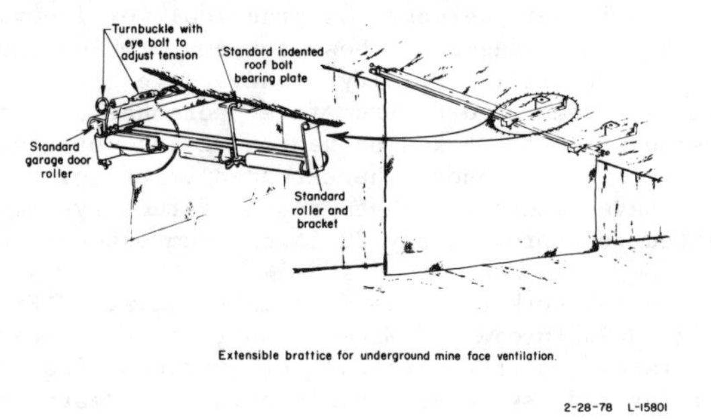

2. Figure 9. Garage door track system--Construction detail.

Figure 10. Garage door track system--View looking inby.

This system was developed by the Bureau. It is supported by two roof bolts and is largely independent of roof bolt location. It can be advanced in a straight line without dismounting, and utilizes a taut off-set cable to minimize vertical and lateral deflection. All components are made from standard garage door track fittings. Maximum extension is about 10 ft.

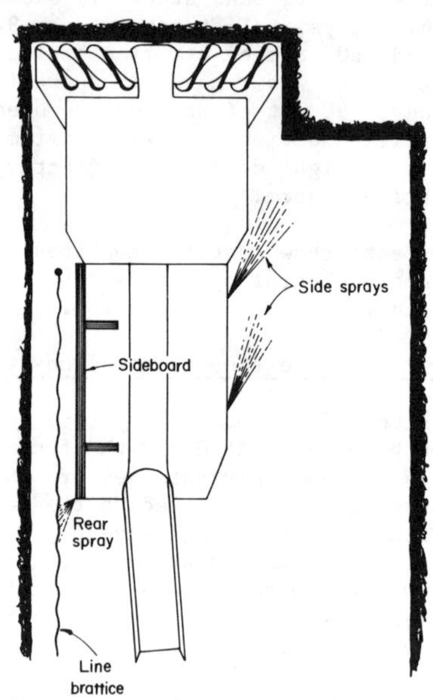

3. Figure 11. Sideboard with sprays.

This system utilizes a vertical sideboard mounted on the left fender of a continuous miner. A bank of water sprays opposite the miner operator helps prevent the ventilating air stream from short-circuiting between the sideboard and line curtain.

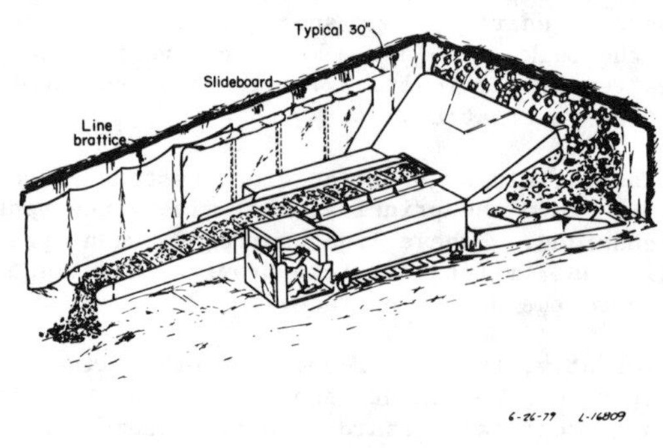

4. Figure 12. Slideboard Device

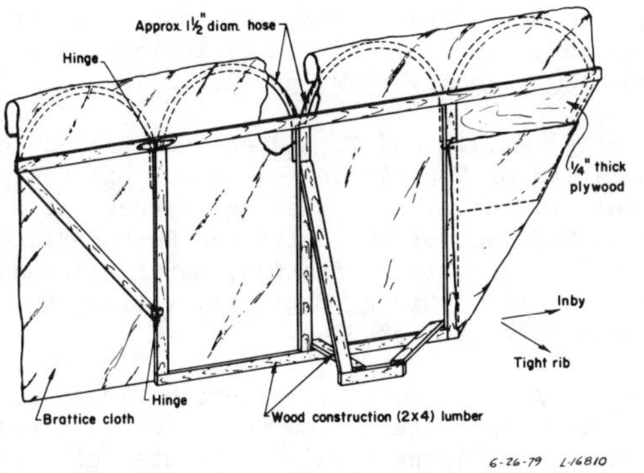

Figure 13. Slideboard Device

This floor-supported device is made of a wood frame, hose, plywood, and curtain. The helper pushes the device inby as required. Arc sections of hydraulic hose and overlapping curtain help to seal along the roof.

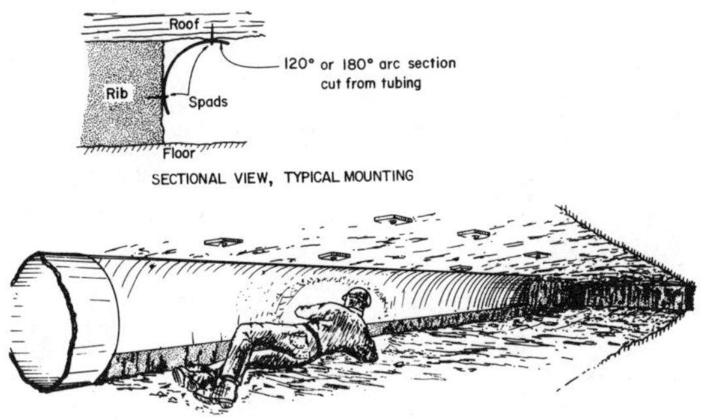

5. Figure 14. Coanda Device For Blowing Systems

This device capitalizes on the coanda, or wall effect, whereby a jet of air tends to hug a smooth surface and even travel an outside corner. It might be useful in low coal or where boring machines are still used. Use of the arc sections, as shown, will allow increases in the throw of a jet of air by 2 or 3 times.

Figure 15. Extensible brattice support system - Chariot Device

Figure 16. Extensible brattice support system - Half-moon slider.

Figure 17. Extensible brattice support system - Roof Jack supported.

6. Extensible Curtain Support System

This system represents a structures approach for cantilevered extensible systems. The typical maximum 10-ft extension of other cantilevered systems can be doubled by the use of long plywood gussets with a hinged angular support arm. The entire assembly is supported by three different methods: a dual-wheel "chariot" (Figure 15); a half-moon sliding arrangement (Figure 16); and a hinged support point from fixed roof jack (Figure 17). All are designed primarily for seam heights from 5 to 8 ft. Notice that the deflection of the outby angled arm "seats" the top arm against the roof.

REFERENCES

1. Vinson, R. P., F. Kissell, J. C. LaScola, and E. D. Thimons. Face Ventilation Measurement With Sulfur Hexafluoride (SF_6) BuMines RI, 8473, 1980, 16 pp.

2. Luxner, J. V. Face Ventilation in Underground Bituminous Coal Mines, BuMines RI 7223, 1969, 16 pp.

3. U.S. Code of Federal Regulations. Title 30, Mineral Resources; Chapter 1--Mine Safety and Health Administration, Department of Labor; Subchapter O--Coal Mine Health and Safety; Part 75, Mandatory Safety Standards--Underground Coal Mines. Subpart D--Ventilation. Revised annually.

4. Kissell, F. N., and R. J. Bielicki. Methane Buildup Hazards Caused by Dust Scrubber Recirculation at Coal Mine Working Faces, A Preliminary Estimate. BuMines RI 8015, 1975, 25 pp.

5. Hudak, M. J., Jr. Use of Dust Scrubbers on Ripper Miners. Min. Cong. J., v. 59, December 1973, pp. 47-50.

6. Leach, S. J., and A. Slack. Recirculation of Mine Ventilation Systems. Pres. at a Colloquium on Firedamp Measurement and Control, arranged by The Southern Counties Institute of Mining Engineers and held at MRE, Isle worth, Middlesex, England, May 3, 1968; Mining Engineer, v. 128, No. 100, January 1969, pp. 227-234.

7. Pickering, A. J., and R. Aldred. Controlled Recirculation of Ventilation--A Means of Dust Control in Face Advance Headings. Pres. at a General Meeting of the Notts and Derbys Branch of IMinE, Mines Rescue Station, Mansfield, England, Apr. 7, 1976; Mining Engineer, v. 136, No. 190, March 1977, pp. 329-345.

8. Haney, R. A. Preliminary Report on Improved Face Ventilation Systems Low Coal Tests. Pres. at 69th Mine Inspectors' Institute of America, Columbus, Oh., June 17-20, 1979, 13 pp.; available for consultation at Mine Safety and Health Administration, Pittsburgh Technical Support Center, Pittsburgh, Pa.

9. Divers, E., J. LaScola, and G. Hundman. New Twin Scrubber Installation for Continuous Mining Machines. BuMines TPR 112, Feb. 1981, 10 pp.

Chapter 6

IMPROVED COAL FACE VENTILATION THROUGH USE OF DUST SCRUBBER SYSTEMS LEADS TO GREATER PRODUCTION EFFICIENCY

A. D. S. Gillies

Assistant Professor of Mining Engineering
University of Missouri-Rolla
Rolla, Missouri

INTRODUCTION

One of the most difficult places to ventilate in any mine is the working face area. With an ever-changing configuration, continual modification and extension to the ventilation system is necessary to maintain an adequate supply of fresh air and to avoid mixing and recirculation in the face area. In coal mines, methane emissions and high dust levels as coal is broken and removed from the face area can produce unhealthy and dangerous conditions if ventilation is not adequately monitored.

Primary airflow to the coal face is normally achieved through use of ventilation tubing and auxiliary fan, or through the hanging of brattice curtain. Systems in which air is directed behind the brattice curtain or down the tubing to the face operate in the blowing or forcing mode, while in exhausting configurations, intake air moves down the entry and is returned along the tubing or curtain. Blowing ventilation is able to sweep the face with a strong air current and reduce the chance of a build up of pockets of high methane concentration in the immediate face area. This advantage over an exhausting system is, however, outweighed to a certain extent in that the blowing air current on sweeping the face picks up dust which is then passed over machine operators working in the heading. Where dust counts with this system are at unacceptable levels, the exhausting configuration may be used where face air is immediately removed from the working environment. Studies by Luxner (1969) and others demonstrate, however, that the ability of the intake air to ventilate the immediate face area is reduced with this configuration.

Current Federal and State Government regulatory requirements place constraints on face ventilation design. Federal regulations under the Mine Safety and Health Act state,

> Paragraph 75.301-1.."A minimum quantity of 3000 cubic feet a minute of air shall reach each working face from which coal is being cut, mined or loaded...."
>
> Paragraph 75.301-4(a)..."except in working places using a blowing system as the primary means of face ventilation...the minimum mean entry air velocity shall be 60 feet a minute in all working places..."

Further to these ventilation requirements are those that pertain to the presence of respirable dust and methane in the mine atmosphere. These determine that the average concentration of respirable dust over a shift length in the working place is to be less than 2 mg/m^3 of air, while the methane level as return air leaves the section is to be less than a prescribed level which varies with the situation, but does not exceed a maximum of 2 per cent by volume.

Mines which have difficulty meeting these requirements simultaneously are forced into searching for new means of improving the face air flow. Mining machine mounted tubing and exhaust fan systems which intake contaminated and dust laden air at the front of the machine and exhaust through a rear port have capability of improving face ventilation. Some units of this type incorporate a machine mounted dust scrubbing system to clean air before discharge. Auxiliary face ventilation systems will change the air flow pattern in the entry working area. Examination of their interaction on the primary face ventilation will allow optimization of design to increase air quantity reaching the face, allow improved sweeping of the newly exposed coal by the air current, and ensure recirculation of dust or gas laden air is kept to a minimum.

THE PEABODY DUST SCRUBBER SYSTEM

To alleviate problems from high respirable dust levels present in the working face area, a number of dust collection and scrubbing units have been developed and incorporated in the design of continuous miner machines. In conjunction with the Joy Manufacturing Company, the Peabody Coal Company, in 1978, placed into operation a machine mounted dust scrubber unit. In this system, intake ports located under the continuous miner cutting head suck dust laden air through duct work to a flooded bed dust filtering collector set before a fan adjacent to the rear discharge port. The layout and operative characteristics of the system have been described by Chironis (1977), Campbell, and others (1978) and Campbell (1979). The system layout is shown in figure 1.

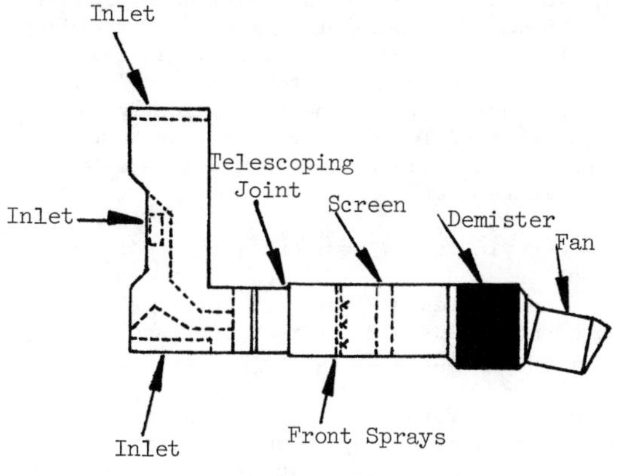

Figure 1. Flooded Bed Scrubber System for Continuous Mining Machine

The Peabody Scrubber System is powered by an 18.6 kW axial fan which is capable of moving 3.5 m^3/s of air at 3.75 kPa of static pressure. Dust laden air enters the system through three intake ports located back from the miner milling head and about 2.0 m from the face. Air passes through ductwork fingers which extend across the top of the machine to a common duct mounted on the left side of the mining machine. The scrubbing units, sprays and demister section are installed within the ductwork and so placed that cleaned air passes directly to the fan and exhausts to the mine entry through a rear port. Deflector vanes may be placed over this exhaust port to direct the high velocity air stream towards the entry left side rib and away from the shuttle car operator.

Campbell reports that in tests undertaken at a mine with a severe dust control problem, the Camp 11 Mine in Western Kentucky, the Peabody scrubbing system was able to achieve significant reductions in dust levels to below the federal standards. An example of recorded levels taken after ten shifts of operation indicates the improvements achieved, and is set down in table 1.

Table 1. Comparison of Dust Concentration Levels at Different Points in Face Area (after Campbell and others, 1978)

	Dust Concentration mg/m^3		
	Continuous Miner	Shuttle Car	Return
With Scrubber	1.5	1.5	1.5
Without Scrubber	4.9	9.2	15.9

A high capacity scrubber system on a continuous miner unit changes the air flow pattern within the face area and increases air velocity at the coal cutting point where air is sucked into intake ports. Improved ventilation at the point where methane gas or dust problems are generated by the breaking coal means that, with appropriate operator control and roof conditions, new mining geometries can be implemented.

As described by Wellman (1981), the Peabody Coal Company has been making use of the scrubber unit to implement a "deep cut mining" design in some of their mines. Through use of radio remote controls which allow the operator to position himself behind the machine and maintain adequate visual control, deep cycle cuts of 10.5 m have been achieved on production shifts. The main purpose for promotion of the deep cut concept is increased productivity. This is achieved by decreasing the continuous miner's tram time as the miner stays in the same place for as much as 50 per cent longer. Calculations based on actual underground time studies for 6 m cuts by the company indicate that

(a) a 9 m cut yields a 4 per cent increase in production,
(b) a 12 m cut yields a 7 per cent increase,
(c) a 15 m cut yields a 9 per cent increase, and
(d) a 18 m cut yields a 10 per cent increase.

Looking to the future, if the shuttle car change out time can be eliminated by use of a continuous coal haulage system, further significant increases in mining productivity can be achieved. In figures 2 and 3, increases that can be achieved have been set down in graphs based on data collected under mine conditions from a 5 day, 15 unit shift study.

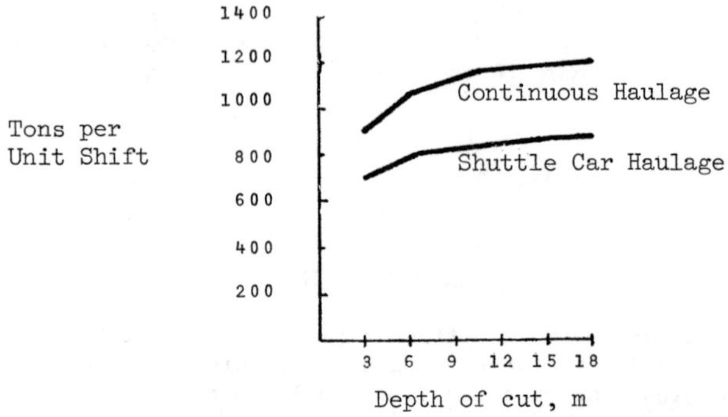

Figure 2. Calculated Effects of Deep Cut Mining and Continuous Face Haulage on Continuous Miner Productivity (after Wellman, 1981).

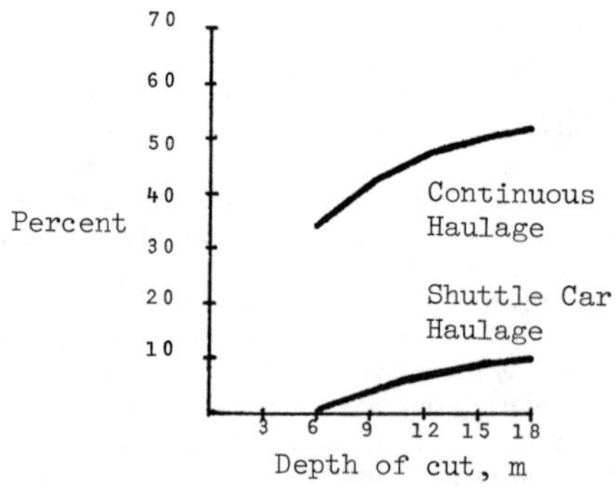

Figure 3. Calculated Percentage Increase in Continuous Miner Productivity with Deep Cut Mining and Continuous Face Haulage (after Wellman, 1981)

The dust scrubbing system has ability to improve face ventilation conditions. A study program has been undertaken to examine the extent to which the geometry of the mining cycle can be improved through use of the system while maintaining adequate air flow conditions. Mining cycles of different cut depth and which, if implemented, would lead to an improvement in production efficiency have been examined.

THE DETERMINATION OF FACE AIRFLOW CONDITIONS

Operation of a dust scrubber system changes the air flow pattern in the working area and may have some effect on the quantity of air being delivered to the face. Analysis of these changes and studies on the effectiveness of any ventilation system can be undertaken by

(a) underground tests at a working face,
(b) construction of full-scale mock-up on the surface, or
(c) construction of a scale model.

Practical determination of the ability of the system to adequately ventilate the face demands a testing method which will allow realistic determination of conditions under all mining sequences and configurations that the working area may take as entry advance occurs. A comprehensive examination of the system based solely on underground testing would be very expensive in terms of

(a) manpower deployment, and
(b) loss of production while lengthy testing is underway.

Further, detailed examination of airflow pathways in a confined area will not produce realistic and repeatable results if personnel undertaking tests, through the physical presence of their bodies or test instruments, distort air currents. For these reasons, it was decided to undertake the examination of face flow patterns using test results gathered from both the underground environment and simulation exercises using physical models. Model testing has the advantage of allowing

(a) a comprehensive set of test sequences to be set up and airflow patterns determined in a short time at relatively small cost,
(b) studies to be undertaken in a safe surface location, ventilation personnel do not have to spend protracted periods in newly driven entry areas and loss of production is avoided.
(c) distortion of airflow by bodies or instruments to be avoided by use of specially designed small sized probes operated from outside the model.

While use of a full-scale mock up would both avoid the disadvantages of underground testing and any scaling factor problem inherent in the use of models, it was considered that construction costs of all test configurations necessary to fully interpret flow patterns would not be justified.

The analysis of face airflow patterns in the laboratory has been undertaken using a model scaled to 1 in 10 of real dimensions. To ensure that results from test sequences using the model will realistically simulate underground conditions, model design and construction has been undertaken only after careful consideration of the physical laws appropriate to their usage. Further, an integral component of the analysis program is a series of underground tests taken as calibration checks of the model and to verify results at key points of the test sequences.

Murphy (1950), Sales and Hinsley (1952), Tomlin (1952), Mishra (1971), Ruglen and Wilson (1978), Mishra (1979), Shuttleworth (1963), Kissell and Bielicki (1974) and, Breslin and Strazisar (1976) have examined the laws of similitude necessary to the use of models in airflow studies. Gillies (1982) sets down details as to model design and the observance of the laws of similitude necessary for model usage. In figure 4, the model facility used for testing can be seen.

Figure 4. Model at scale of 1 in 10 showing continuous miner having excavated box cut by 6 m. Ventilation is delivered to face area by brattice line. Small flags show air pattern direction.

In order to establish airflow patterns in the face area for mining sequences which are or could be practiced under safe conditions with a scrubber in use, the following base conditions were established in model tests for comparison purposes. Dimensions are factored to underground, full scale size.

(a) Brattice rib distance was set at 0.6 m (10 per cent of entry width) and it was assumed that leakage of air over, under or through the line curtain was negligible.
(b) Primary entry ventilation was delivered by blowing brattice. However, a few comparison tests using exhausting line brattice were also undertaken.
(c) Primary ventilation delivery to the face area, as established at the end of the brattice line was 4 m^3/s. This level was established as being slightly in excess of the average 3.5 m^3/s delivery of the scrubber fan output and so direct recirculation should be avoided. This brattice line delivery quantity has been maintained on some of the early underground mine faces developed by the Peabody Company using the scrubber system.
(d) The flow pattern in the face area was established by taking velocity and direction readings across a grid pattern of mesh size 0.6 m square. Velocity readings were taken with a hot-wire anemometer, each reading being taken at the seam mid-height.
(e) Air flow patterns were further illustrated by additions of colored smoke to the model. To visually record the flow pattern within the face area colored smoke was added to the air stream. White smoke was added to the entry intake air, while orange smoke was injected through small 6 mm diameter holes placed at mid - seam height in front of the mining machine to simulate contaminants or airborne dust or methane gas liberated as mining progresses. As seen through the model transparent plexiglass top, a clear picture was formed of intake air mixing and removing contaminants, zones of face recirculation, vortex eddies formed in the return air in the entry and areas of stagnant face air where white intake and orange contaminated air paths were not mixing. Using high speed color slide film photography a record was made of flow in the model with one or both smoke additives present. Smoke sources were navigation flares and fire smoke simulator generators with a source life of about 2 minutes. Further, a television camera and videotape recording equipment were used to form a dynamic record of smoke patterns. This system provides a powerful tool for recording patterns and particularly, for examining the occurrence of stable eddy vortices and zones of recirculation. With ability to record moving flow patterns over time, television allows close study of air currents in the face area. Pathways taken by contaminated air travelling from the face were readily identified and a definitive answer formulated to the question as to whether these currents have a possibility of recirculating to active mining areas.

Prototype underground testing was undertaken in off-shift producing coal entries which were uniform in dimension and as nearly as possible approached model simulation dimensions of a 2 m seam height and 6 m entry width. With a continuous mining machine, and at times a shuttle car in the entry, both a machine operator and assistant were present to move these machines and operate the scrubber unit. Details on the procedures adopted are described by Gillies (1982).

SATISFACTORY FACE AREA VENTILATION

Before the results of a study on air flow patterns in the face area of a producing heading can be analyzed, some interpretative guidelines are necessary to assist in defining what level of face ventilation is adequate or safe under a particular set of circumstances. Normally the supply of air to the working area will be sufficient to maintain oxygen levels and respiration; so it is to the removal of methane and the limitation of respirable dust counts that guidelines must be directed.

It is essential that air is moving at a point very close to the face so as to pick up and dilute any gas liberated from the breaking coal and to stop lighter-than-air gases accumulating as roof layers. Further, air movement is needed to break up a dust cloud that may form around the cutting head and to disperse this dust so that men in the entry are not breathing an atmosphere containing unacceptable levels.

A second essential feature is that some fresh intake air must be reaching the newly exposed coal face. While a certain level of recirculation of air currents in the face area is acceptable, (the level of allowable recirculation being determined by the methane liberation rate and either the amount of dust produced as coal is broken or the efficiency of a scrubbing dust removal system to take this out of the environment), some purging of the air in the immediate face zone must occur to ensure that dangerous levels of contaminants are not reached.

For the purposes of this study, interpretative guidelines as to the acceptability of a face ventilation pattern have been established. An acceptable configuration meets the following requirements.

(a) Air velocity in the face zone must be consistently perceptible, or on average at or above 0.3 m/s. This level was adopted on the basis of recommendation made by Kingery and others (1969) in a study on face ventilation conditions necessary to control respirable coal mine dust.
(b) A significant quantity of air reaching the immediate face zone must be fresh intake air, and recirculation of air containing high gas or dust levels must be at a low level.

Conditions for these requirements have been adopted on the approach that they are conservatively based and appropriate for arduous mining situations. Air in the immediate face zone should not be stagnant (with a velocity at or below 0.1 m/s) and a significant flow stream should pass through the area. While there may be small pockets or areas shadowed by equipment or coal pillars in which flow velocity is marginal, a steady and perceptible movement should be observable through most of the working entry. Further, air flow patterns in which contaminated air is not removed from the zone, but rather recirculates, must be avoided. Recirculating patterns of air are, in themselves, not necessarily unacceptable. Many safe mining patterns allow some recirculation, with new fresh intake air continually moving into the system, purging the zone of contaminants and not allowing unsafe conditions to develop. Dangerous situations will arise where the level of contaminants in recirculated air is at a point where addition of intake air does not decrease readings below the safe prescribed point. The degree of acceptable recirculation flow will vary from mine to mine. Any minor air vortices or eddies occurring back from the face area and down the entry have been ignored in the evaluation as their presence does not directly contribute to the recirculation of air to the immediate zone where coal mining is taking place.

RESULTS OF SOME TESTS

The number of configurations of possible coal mining sequences which could be tested in a study of this type is very large. During testing, many entry face geometries have been set up for the determination of air flow patterns. Measurement results given emphasize both ventilation flow developed on faces where adequate air quality is maintained without use of auxiliary systems such as a dust scrubber unit, and air flow to faces in mining configurations which extend the ventilation system and require use of auxiliary systems.

The Peabody Company scrubber system was introduced to allow satisfactory respirable dust levels to be maintained in face areas ventilated by blowing systems. Generalized results of tests are set down for analysis with reference to the face area configuration established by the brattice curtain to face distance. In a previous study, Gillies (1982), face area flow patterns established during the steps necessary to the completion of various proposed mining cycles were discussed. Results discussed below emphasize air current behaviour at different brattice face distances, examine the formation of zones of turbulent eddy formation or recirculation, and predict equations appropriate for the calculation of methane concentration levels in the immediate face area.

Conditions with Face-Brattice Distance Short

On a continuous miner section operating under this system, mining advance occurs from the last roof support (the point at which brattice curtain is attached) for a cut of up to 6 m depth. In many cases the face is advanced in two halves; coal on the brattice curtain side is mined, followed by the removal of the remaining pillar on the other side.

Test results from model and underground measurements with the dust scrubber system both operating and not operating demonstrate that adequate air flow reaches the face. Further, adequate air flow is present without any assistance from the dust scrubber unit. Operation of the scrubber slightly increases air flow reaching the face and maintains or increases an already adequate ventilating flow throughout the entry. In figure 5, flow patterns can be seen for face geometries in which the mining machine has advanced a cut distance of 6 m on both sides of the entry; examples with both the scrubber unit not in operation, and operating can be seen.

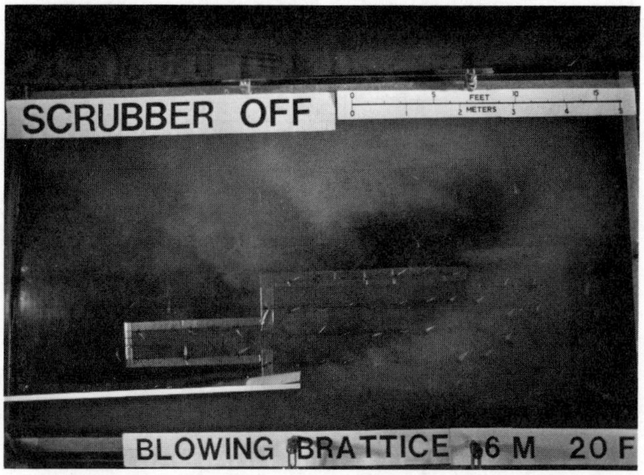

Figure 5. Airflow patterns with blowing brattice set at 6 m. Intake air stream is carrying white smoke.

The flow of fresh air entering the face area is controlled by the line brattice distance, and, in the case when the scrubber unit is operating, the scrubber fan. A simple theoretical analysis of flow in the entry can be established based on a similar approach to that put forward by Kissell and Bielicki (1974). A zone AB is defined in the immediate face area. As set down in figure 6, the quantity of fresh air passing along the line brattice is Q, the methane gas seeping into the zone from the coal is V, the quantity flow rate through the scrubber unit is R, and additional air flow out of the zone is Z.

An airflow balance of movement into and out of the zone establishes that

$$Q + V = Z + R \tag{1}$$

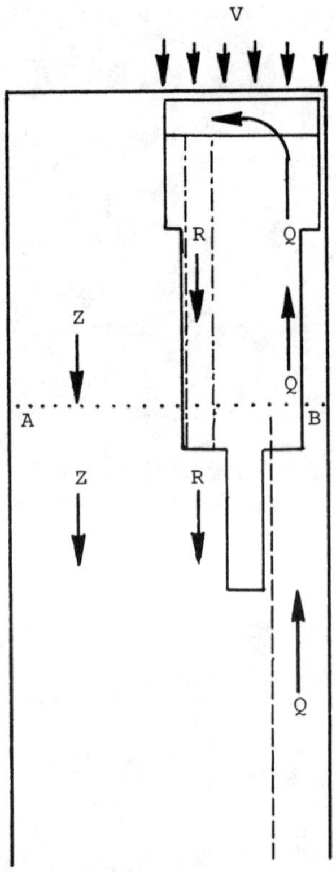

Face - Brattice Distance Short

___ Blowing Line Brattice
_._. Machine Dust Scrubber
. . . . Face Area Boundary

Figure 6. Analysis of principal airflow currents in face area with scrubber unit operating

and a methane balance, assuming steady-state conditions with fresh air containing no methane is (C is methane concentration)

$$V = CZ + CR \qquad (2)$$

Now in (1) $Z = Q + V - R$

Then $V = C(Q+V-R) + CR$

From this $C = \frac{V}{Q+V}$, But $Q >> V$

Therefore $$C \simeq \frac{V}{Q} \qquad (3)$$

The concentration of methane in the face area defined by line AB is equal to the methane intake divided by the fresh air intake, and is independent of the air quantity passing through the scrubber. Where the face brattice line distance is short, conditions of turbulent eddy formation in the entry or recirculation to the face zone will be minor. As the brattice line distance back from the face is extended, these conditions may become more significant.

Conditions with Face-Brattice Distance less than 9 m

Examining conditions in which face ventilation has been extended, flow patterns in which the entry has been cut to a point 9 m in advance of the brattice curtain have been established. In figure 7, photographs of model test results with both the scrubber unit not in operation and operating can be seen.

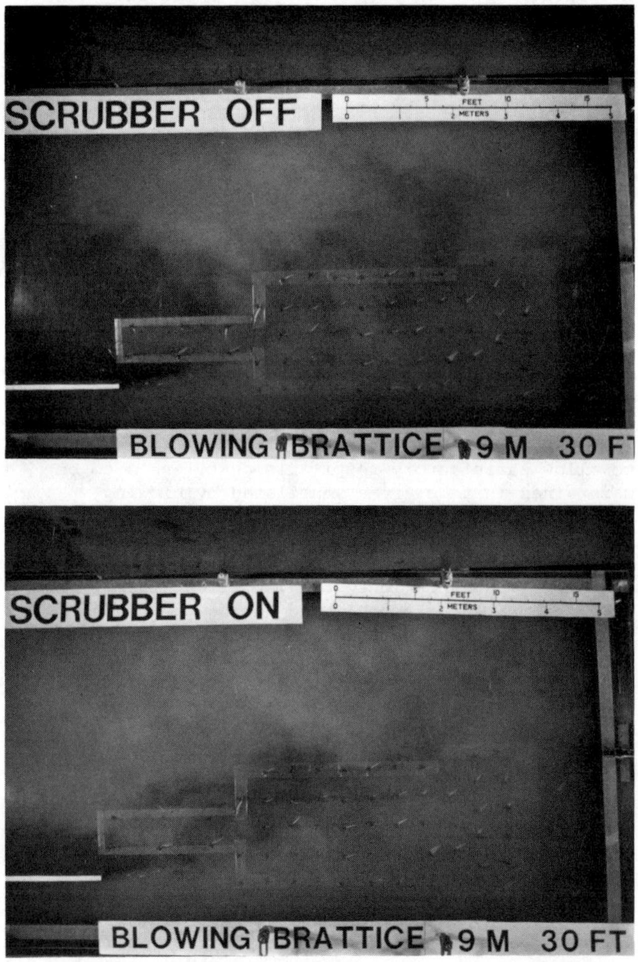

Figure 7. Face airflow patterns with blowing brattice set at 9 m. Intake air stream is carrying white smoke

With this face geometry some intake air is carried to the face without the assistance of a scrubber unit, but much is deflected behind and across the top of the mining machine and does not enter the immediate face zone. With the scrubber in operation, air is pulled to the front of the miner, a larger proportion passes through the unit itself, while some continues across the face and returns down the entry. The high velocity air stream exhausting from the scrubber unit enters entry air in advance of the end of the line brattice. Some of this discharge on slowing down will enter turbulent eddy zones and be carried by intake air back to the face. The start of this recirculation pattern with the scrubber operating is visible in the photograph in figure 7 above the miner chain conveyor boom.

Making use of the flow pattern symbols established in figure 6, a theoretical analysis of this ventilation system is presented in figure 8.

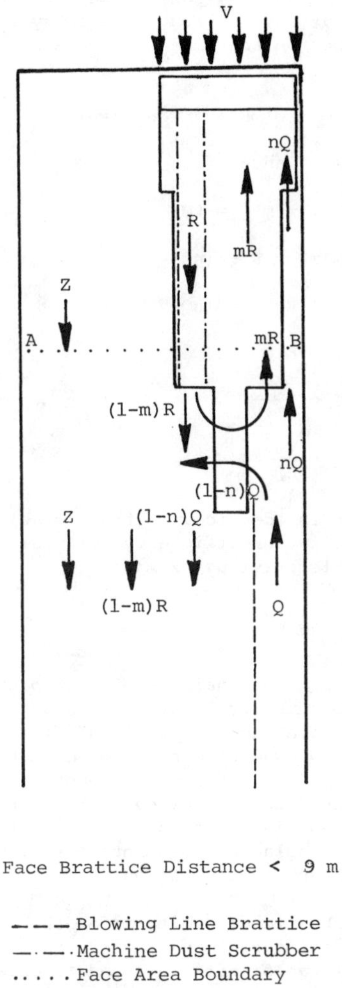

Figure 8. Analysis of principal airflow currents in face area with scrubber unit operating

The flow patterns establish that the intake air divides to two streams; component nQ enters the zone in front of AB while the rest, (1-n)Q, returns down the entry without sweeping the face. Further, a component of the scrubber discharge, mR, recirculates to the face zone, while (1-m)R enters the mine return immediately. An airflow balance on the immediate face zone delineated by AB establishes that

$$nQ + mR + V = R + Z \quad (4)$$

Further, a methane balance gives

$$V + mRC = ZC + RC \quad (5)$$

Now in (4) $Z = nQ + mR + V - R$

Therefore $V + mRC = nQC + mRC + VC - RC + RC$

Then $V = nCQ + CV$

And $C = \frac{V}{nQ+V}$

$$\simeq \frac{V}{nQ} \quad (V << nQ) \quad (6)$$

Also, if we establish C_r as the methane concentration level in return air, "face ventilation efficiency" is calculated from C_r divided by C.

Now $C_r = \frac{V}{Q}$ (7)

Therefore $\frac{C_r}{C} = n$ (8)

As was seen from figure 6, the concentration of methane in the zone is equal to the methane intake divided by the fresh air intake into the zone. The quantity R has dropped out of the equation, indicating that recirculation has no direct effect on concentration of methane build up. Of course, it must be realized that recirculation airflow will, to some extent, displace fresh air flow to the zone.

Conditions with Face-Brattice Distance Advanced Beyond 9 m, and less that 21 m

The mining of deeper cuts which extend the Face-Brattice Distance leads to the development of changed entry flow patterns. In figure 9, photographs of test results with line brattice set back 15 m from the face and with both the scrubber unit not in operation, and operating are shown.

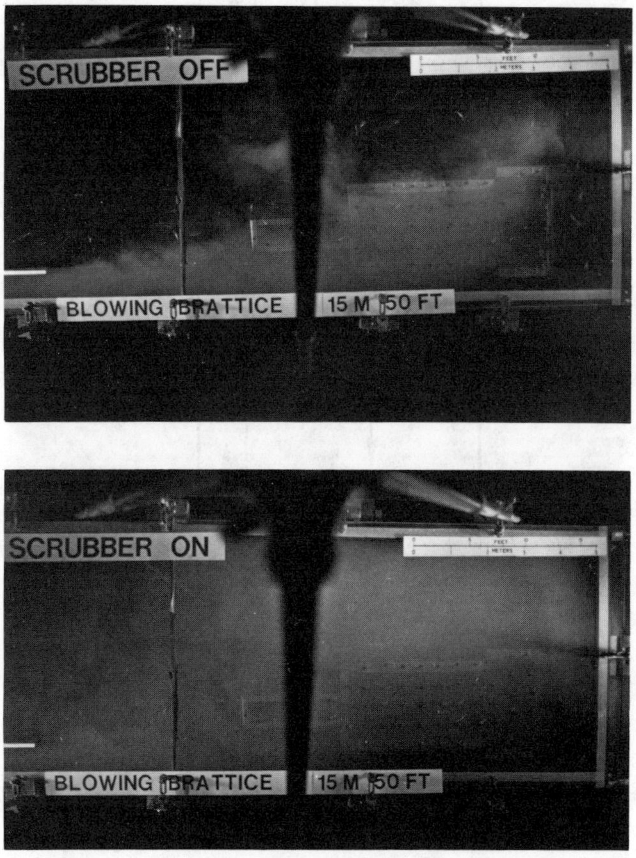

Figure 9. Face airflow patterns with blowing brattice set at 15 m. Intake air stream is carrying white smoke (the camera tripod frame shadows the center of the photographs)

With the scrubber unit not operating, little or no air reaches the immediate face area; some passes in a stream which is deflected by the bulk of the mining unit to the return, while a component passes across the machine, though having lost momentum, little of this air reaches the cutting point. With the scrubber fan on, air is brought directly to the face. Close analysis of patterns has shown that there is significant recirculation of air from the scrubber exhaust back into the intake air path.

A theoretical analysis of the system is presented in figure 10.

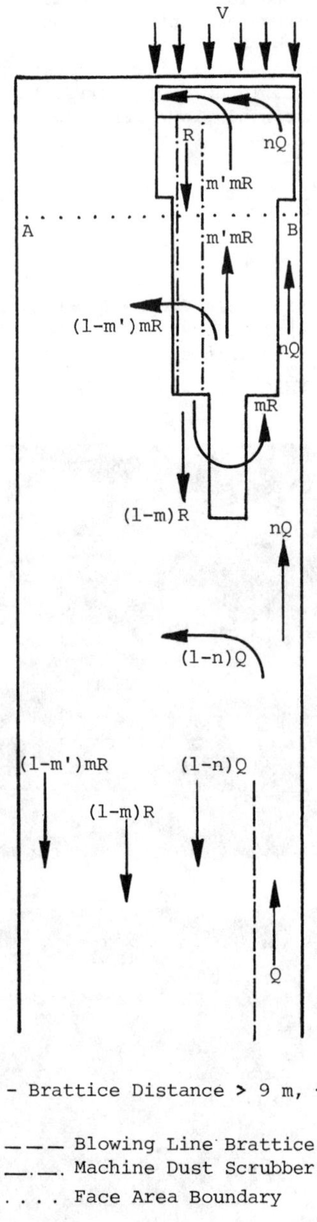

Figure 10. Analysis of principal airflow currents in face area with scrubber unit operating

Examination of the flow system demonstrates that only air flowing to the scrubber intake ports passes across the coal cutting face and into the face zone AB. Intake air, Q, divides to two principal currents as does the scrubber discharge, R. Further, scrubber recirculating air, mR, is seen to follow two passageways. Some, m'mR, recirculates to the face while the rest remains in an eddy zone above the mining machine and eventually passes through the entry return. An airflow balance on the immediate face zone establishes that

$$nQ + m'mR + V = R \quad (Z=0) \tag{9}$$

and a methane balance gives

$$V + m'mRC = RC \tag{10}$$

Now in (9) $m'mR = R - (nQ+V)$

Therefore $V + RC - C\,(nQ+V) = RC$

Then $V = nCQ + CV$

And $C = \frac{V}{nQ+V}$

$$\simeq \frac{V}{nQ} \quad (V<<nQ) \tag{11}$$

This equation is identical to that established in (6) above. "Ventilation efficiency" will also remain the same as in this previous example:

$$\frac{C_r}{C} = n$$

As was seen with the analysis of flow patterns in figure 8, the quantity R has dropped out of the above equations, indicating that recirculation has no direct effect on the concentration of methane within the face zone. However, as the quantity of air brought to the face is limited to the scrubber capacity, R, any increase in the recirculation component, m'mR displaces fresh intake air, nQ, and so increases methane concentration.

It should be recognized that for the particular mine system configurations studied, the intake air component, nQ, was significant, and in excess of recirculation quantity m'mR for brattice line face distances of 12 m and 15 m. Considerations such as the mining system practiced, entry width, seam thickness and machine height will affect conditions in a particular mine.

Conditions with Face-Brattice Distance Beyond 21 m

The studies have indicated that the mining of cuts which extend the brattice line face distance beyond 21 m will lead to excessive recirculation conditions in the operation of the Peabody scrubber system in the adopted mining situation. In figure 11, photographs of test results with line brattice set back 24 m from the face and with both the scrubber unit not in operation, and operating are shown.

In the situation where the scrubber unit is not operating, intake air has little or no influence in the immediate face zone; it advances no more than 9 m or 12 m into the opened entry. With the scrubber operating, two sequences are shown.

(a) The first (upper) photograph was taken a very short time after the scrubber was switched on. Some intake air can be seen being sucked forward; however the scrubber discharge interferes with this flow and a zone of turbulence is set up

Figure 11. Face airflow patterns with blowing brattice set at 15 m. Top photograph shows conditions with scrubber off. Bottom two show flow with scrubber on. Intake air stream is carrying white smoke (the camera tripod frame shadows the center of the photographs).

behind the machine. Discharge from the scrubber slows down as it expands in the entry until it meets intake air travelling into the entry. These opposing currents meet, much of the scrubber exhaust is diverted back into the face area and a recirculation condition is initiated.

(b) The second (lower) was taken some time later. Due to the build up of smoke in the entry, some flow patterns are unclear, however, the highly turbulent recirculation condition above the miner can be seen.

A theoretical analysis of the system is presented in figure 12.

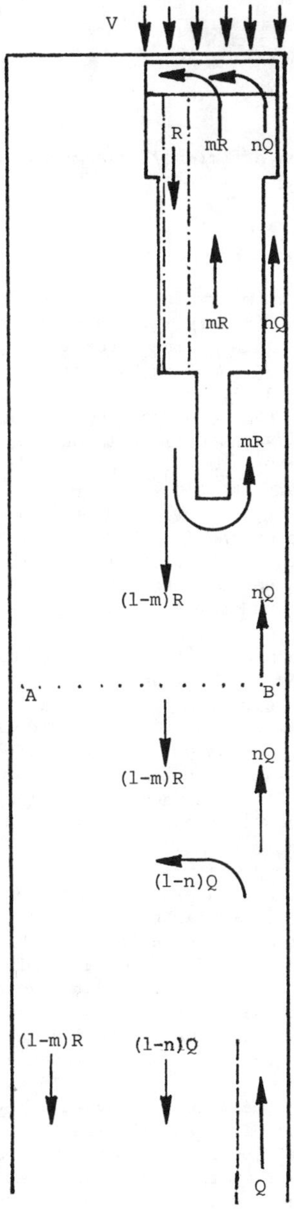

Face Brattice Distance > 21 m

— — — Blowing Line Brattice

—·—· Machine Dust Scrubber

· · · · Face Area Boundary

Figure 12. Analysis of principal airflow currents in face area with scrubber unit operating

An airflow balance on the zone delineated by AB establishes that

$$nQ + V = (1-m)R \tag{12}$$

and a methane balance gives

$$V = (1-m)RC \tag{13}$$

Therefore $C = \frac{V}{(1-m)R}$

Now in (12) $(1-m)R = nQ + V$

Then $C = \frac{V}{nQ+V}$

$$C = \frac{V}{nQ} \quad (V<<nQ) \tag{14}$$

and "ventilation efficiency" can be calculated as before

$$\frac{C_r}{C} = n$$

As in the face geometry examined, the scrubber unit output does not directly influence methane concentration level. This mining situation differs in that intake quantity nQ reaching the immediate face area is small compared with the recirculation current mR. Due to the fact that the opposing air currents Q and R meet at line AB and deflect large components, (1-n)Q and (1-m)R of air in opposing directions, an "air curtain" effect is seen to be developed at this point in the entry. Flow across this interface is low; quantities nQ and (1-m)R will in magnitude be approximately equal (V is small in absolute value). The quantity nQ can be considered as replacement air entering an independent air cell.

To allow the machine to advance to mine deeper cuts without formation of an "air curtain" which deflects intake air from the face area, some means of keeping currents Q and R apart, and in parallel but opposite direction pathways,is needed. Wide diameter tubing attached to the scrubber exhausting port would allow an extended advance and solve this problem, although physical support of this auxiliary system may be cumbersome.

Results from the material balances on the proposed face zones have indicated that the scrubber output or the amount of recirculation from a machine mounted unit is not the major factor in determining the average methane concentration in the zone. The average level is determined by the methane flow rate and the amount of fresh air entering the area. Recirculation becomes a problem when the recirculated air is substituted for fresh air. In situations where the mining machine has advanced beyond the normal flow path of intake air (further than a distance of 9 m depending on the entry configuration), a movement of air in the immediate face zone will be solely determined by the scrubber fan operation. Under these conditions, there is limited flow within the face area and any recirculation will replace fresh air flow.

APPLICATION OF TEST RESULTS

As reported earlier, dust scrubber systems have been developed to the point where, with proper maintenance, respirable dust levels can be kept to satisfactory levels under difficult mining conditions. It is for the dilution of methane gas contaminants that additional airflow in the immediate face area is necessary. Primary face ventilation may be adequately supplemented by the scrubber fan systems.

To ascertain, by example, how far deep mining cuts may be advanced with satisfactory ventilation, typical methane emission rates need to be incorporated in the equations developed earlier. Previous studies by others such as Haney (1979), Kissell (1979), and Luxner (1969) have used rates appropriate to some Eastern seams of about 0.005 m^3/s, to 0.015 m^3/s in tests on face ventilation systems. Methane emission rates may increase to much higher levels of 0.05 m^3/s or more in seams at great depth. For comparison purposes an average methane concentration of 1.0 per cent in face air is considered satisfactory. In table 2, maximum seam emission levels which may be adequately diluted are set down against cut advance distance past the line brattice. These have been calculated against standard conditions of intake airflow behind brattice (Q) of 4 m^3/s, scrubber output flow (R) of 3.5 m^3/s, and the study entry dimensions and mining system described earlier. For these calculations, the equation

$$C = \frac{V}{nQ} \quad (3,\ 6,\ 11,\ 14)$$

derived earlier as appropriate for all study conditions has been used. The factor for the proportion of intake air reaching the face zone (n) has been estimated from test results. This factor will vary under different mining situations and should be estimated from underground observation, or appropriate model tests, for individual mine applications.

Table 2. Maximum Seam Methane Emission Level for Cut Advance Distances by a Mining Machine Incorporating a Dust Scrubber Unit

Cut Advance Distance m	Intake Air Factor n*	Face Methane Allowable m^3/s
3	1.0	0.040
6	0.8	0.032
9	0.7	0.028
12	0.6	0.024
15	0.5	0.020
18	0.4	0.016
21	0.2	0.008
24	0.1	0.004

* n factor established from underground and model test results

From these results appropriate to the prescribed standard conditions, a deep cut to 18 m would be possible before ventilation capacity for typical Eastern seam emission rates described above may prove inadequate. Deep seams with high emission rates would restrict the advance achievable.

With the typical rate set down above, a theoretical maximum machine advance distance would be less than 3 m. In practice the intake air quantity would have to be increased significantly here. In fact, mines with methane emission rates of this level do require much higher air quantities than those used in the study to maintain satisfactory condition on even short cut advances.

While the above test results cannot be generalized directly to individual mine situations, a method has been described for determining how specific cases can be analyzed and mining systems improved to increase production efficiency.

CONCLUSIONS

Through the use of mine ventilation modeling, a most valuable tool is available for improving the efficiency of mine production through demonstration of how available air can be put to work advantageously. Model tests allow accurate examination of flow on a detailed basis without the problems that underground tests cause in interrupting production. Techniques which are available, such as the use of smoke as a color indicator,allow a clear view of air flow which cannot be seen by other methods.

The use of a dust scrubber system does allow improvement to the underground face ventilation system and the adoption of new mining configurations that can be tied to efficient production cycles. The correlation of specific ventilation requirement of individual mines to a full understanding of how the scrubber system can assist air flow at the face of the production entry should lead to the maintenance of a safe and healthy working environment practices.

ACKNOWLEDGEMENTS

Financial support which has enabled the research program to be completed, and from which the results discussed in this paper were gathered, has been provided under the Missouri State Mining and Mineral Resources Research Institute Program of the Office of Surface Mining. Financial assistance, co operation with underground visits and on - site measurements, and advice on interpretation of results has been provided by the Peabody Coal Company. The contribution of Mr. Jerry Tien and Mr. Donald Richard of the Peabody Coal Company are expressly noted. The support of these parties is most appreciated and hereby acknowledged. Finally, thanks is extended to the University of Missouri, the Faculty, Staff and Students for the many and various efforts that have contributed to the completion of the project.

REFERENCES

1. Breslin, J. A. and Stazisar, A. J., 1976. "Dust-Control Studies Using Scale Models of Coal Mine Entries and Mining Machines," Bureau of Mines RI 8191.
2. Campbell, J. A. L. and others, 1978, also Campbell, J. A. L., 1979. "Peabody Resolves Respirable Dust Emissions at Camp II" American Mining Congress Mining Show, October 1978, and Mining Congress Journal, March 1979.
3. Chironis, N. P., 1977. "Peabody's Mines Find Better Way to Reduce Dust," Coal Age, June 1977.
4. Cook, N. G. W. (Chairman), 1980. "Measurement and Control of Respirable Dust in Mines," Report of the Committee of Measurement and Control of Respirable Dust, National Research Council.
5. Gillies, A. D. S., 1982. "Studies in Improvements to Coal Face Ventilation with Mining Machine Mounted Dust Scrubber Systems," A.I.M.E. Annual Meeting, Preprint 82-24.
6. Haney, R. A., 1979. "Preliminary Report on Improved Face Ventilation Systems, Low Coal Tests," Mine Inspectors Institute of America.
7. Kingery, D. S., Doyle, H. N., Harris, E. J., Jacobson, M., Peluso, R. G., Shutack, J. B., and Schlick, D. P., 1969. "Studies on the Control of Respirable Coal Mine Dust by Ventilation," USBM Technical Progress Report-19.
8. Kissell, F. N. and Bielicki, R. J., 1974. "Ventilation Eddy Zones at a Model Coal Mine Working Face." Bureau of Mines RI 7991.
9. Kissell, F. N., 1979. "Improved Face Ventilation By Spray Jet Systems," Second Annual Mining Institute, University of Alabama.
10. Luxner, J. V., 1969. "Face Ventilation in Underground Bituminous Coal Mines," Bureau of Mines RI 7223.
11. Mishra, G. B., 1971. "Model Studies on the Resistance of Airways Supported with Timber Sets." Trans. A.I.M.E., Vol. 250.
12. Mishra, G. B., 1979. "Resistance of Fan Drifts, A Model Study." Proc. Second International Mine Ventilation Congress, A.I.M.E.
13. Mundell, R. L., 1976. "Blowing Versus Exhaust Face Ventilation for Respirable Dust Control on Continuous Mining Sections." Mine Enforcement and Safety Administration IR 1059.
14. Murphy, G., 1950. "Similitude in Engineering" The Ronald Press Company, New York.
15. Ruglen, N., and Wilson, P. H., 1978. "Aerodynamic Studies of Shaft/Airway Intersection Losses and Mine Cage Resistance," Annual Conference, Australasian Institute of Mining and Metallurgy.
16. Sales, T. J. R. and Hinsley, F. B., 1952. "The Use of Models in the Study of Airflow in Mines." Trans. Institution of Mining Engineers, vol. 111.
17. Shuttleworth, S. E. H., 1963. "Ventilation at the Face of a Heading, Studies in the Laboratory and Underground." International Journal Rock Mechanics and Mining Sciences, vol. 1.
18. Tomlin, N., 1952. "The Investigation of Fan Drift Design by Means of Models." Trans. Institution of Mining Engineers, vol. 113.
19. Wellman, T. S., 1981, "Deep Cut Remote Control Mining," Mining Congress Journal, October 1981.

Chapter 7

FACE VENTILATION SYSTEMS PERFORMANCE IN LOW-HEIGHT COAL SEAMS

Robert A. Haney, Stephen J. Gigliotti and James L. Banfield, Jr.

United States Department of Labor
Mine Safety and Health Administration
Pittsburgh Health Technology Center
Pittsburgh, Pennsylvania

ABSTRACT

A series of investigations were made by the Mine Safety and Health Administration to evaluate the performance of various face ventilation systems as part of a program to reduce frictional ignitions of methane and to develop face ventilation systems that could be used with the line brattice more than 10 feet from the face. Tests were conducted in a full-scale laboratory model of a face area. Systems tested included unassisted blowing and exhaust line brattice, diffuser fan and spray fan systems. Systems were evaluated at air quantities ranging from 1.4 to 7.0 m^3/s (3,000 to 15,000 cfm), and brattice-to-face distances from 1.5 to 9 meters (5 to 30 feet). Tests were designed so that systems could be evaluated for the location of the high-point methane concentration in the face area, methane dilution capacity, and face ventilation index. The methane dilution capacity is the total amount of methane a system can dilute to 1.0 volume-percent. The face ventilation index is the efficiency at which the available face air volume is being used.

INTRODUCTION

One of the primary functions of a coal mine ventilation system is to supply a sufficient air volume to the working faces to dilute and carry away methane and respirable dust generated during mining activities. The face ventilation system must be compatible with the mining conditions so that a safe and healthful work environment can be provided.

As mining conditions and priorities change, new types of face ventilation systems are introduced. These systems may be installed in very gassy mines in an effort to reduce frictional ignitions of methane or in mines where it is feasible to maintain line brattice more than 3 meters (10 feet) from the face, thus reducing the need to set temporary roof supports. Whenever a new face ventilation system is introduced, it is prudent to evaluate the performance of that system. To assist in this evaluation, a full-scale laboratory model of an underground face area was constructed. The use of laboratory tests provides a uniform method of evaluating face ventilation systems under controlled conditions. Additionally, the hazards that may be associated with field testing a novel face ventilation system can be eliminated by conducting tests under controlled conditions.

To assist in the evaluation of the conditions that exist during mining and of the ability of a face ventilation system to control these conditions, the Ventilation Division, of MSHA's Pittsburgh Health Technology Center developed several quantitative indices to describe the ventilation constraints in the face area. The use of these indices has provided a more uniform approach to evaluating the effectiveness of a face ventilation system. These indices include the peak face-area methane liberation rate, the methane dilution capacity and the face ventilation index. The peak face-area methane liberation rate is the maximum rate at which methane is released during mining. The methane dilution capacity is the total methane flow rate that a face ventilation system can dilute so the face methane concentration does not exceed 1.0 volume-percent, and the face ventilation index is the efficiency at which the available face air volume is used to dilute methane. By comparing obtained values for each of these indices, an evaluation can be made to determine the suitability of various face ventilation systems to a particular mining situation.

FACE VENTILATION APPRAISAL AND TESTING FACILITY

The Face Ventilation Appraisal and Testing Facility is an "L" - shaped, concrete block structure with a steel and gunite roof and a reinforced concrete floor.

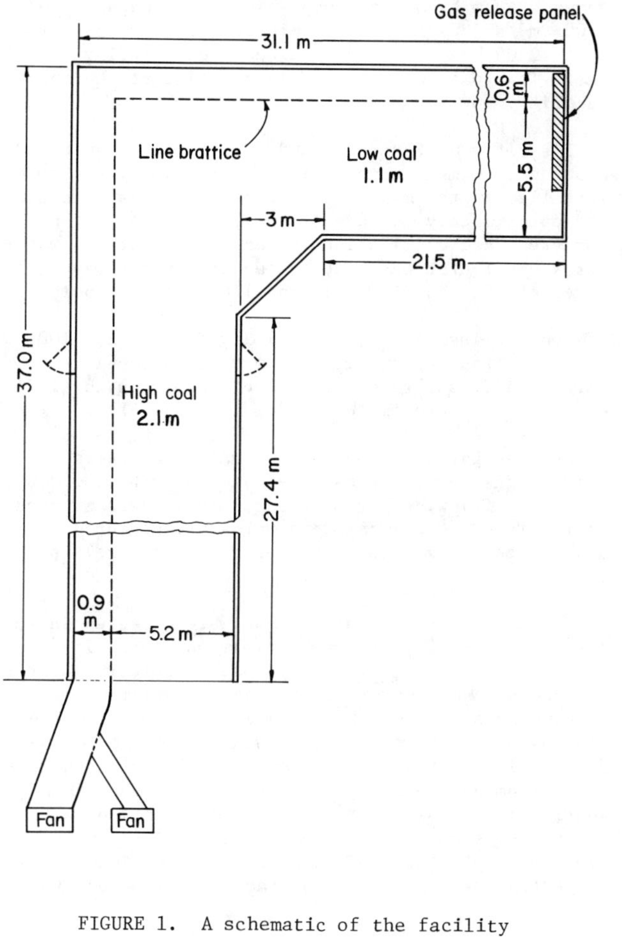

FIGURE 1. A schematic of the facility

The facility simulates an underground face with a 6.1 meter (20 ft) wide entry and mining height capacity of 2.1 or 1.1 meters (7 feet or 3½ feet). This series of tests was performed in the low-coal section. Figure 1 is a schematic of the facility.

Airflow is induced in the test facility by a permissible Series 1000 Joy Axivane Fan Model 23¼ - 17½ - 3450 with a 30 kW (40 hp) motor. A second, identical fan is available either for parallel fan operation when air quantities in excess of 4.2 m^3/s (9000 cfm) are desired or to induce airflow for blowing face ventilation systems.

Airflow is delivered to the face of the test facility by a nylon reinforced line brattice installed on the left side of the facility 0.6 meters from the wall. The last 9 meters (30 feet) of line brattice were built on a moveable framework for easy adjustment to or from the face.

A wooden, full-scale model continuous miner with rotating drum and operating water sprays was placed at the face to simulate an actual mining place. The model mining machine can be placed at various positions in the face area to simulate various phases in the face advance process. This set of tests was conducted with the model mining machine at the end of the sump cut position. Mining at the end of the sump cut is generally considered to be the most difficult position to ventilate.

Water is supplied to the continuous-miner water spray system by a Bean pump which has a capacity of 0.002 m^3/s (30 gallons per minute) at a pressure range up to 3400 kPa (500 psi). Excess water during the tests is bypassed into the floor drains of the test facility.

For diffuser fan tests and unassisted exhaust and blowing systems, a conventional water spray system was installed on the miner. Water pressure was maintained at approximately 500 kPa (75 psi). For spray-fan assisted face ventilation system tests, various water spray configurations were tested at water spray pressures of 700 kPa (100 psi) and 1400 kPa (200 psi).

Two permissible 4 kW (6 hp) Joy Model I 2RF 100, two-stage fans are available to test as machine-mounted diffuser fans, with each fan having an airflow capacity of approximately 1.2 m^3/s (2500 cfm).

Methane gas emissions are simulated by releasing natural gas evenly across the face through two, 3.6 meter (12 ft) wide perforated release panels mounted on the face. Gas liberation rates are measured using glass tube rotameters connected to the gas release panels.

Concentrations of natural gas were detected in the facility by a 12-point combustible-gas detection system, and transmitted to strip chart recorders in an instrument and control room. The calibration of each methane sensor is checked each day prior to the beginning of any tests. The 12 sensor heads were installed in the test area at critical locations determined from previous experience in underground face ventilation surveys. Methane sensor locations were checked by sampling throughout the face area with hand-held detectors to insure that the maximum methane concentration present in the face area was detected.

TEST PROCEDURE

Prior to testing any improved face ventilation system, it was necessary to run baseline tests on unassisted exhaust and blowing face ventilation systems. These tests would serve as a comparison to indicate the amount of improvement obtained from the modified system. To check the baseline performance of a system, the air quantity at the inby end was adjusted to one of the various test air quantities, nominally 1.4, 2.8, 4.2, 5.6 or 7.0 m^3/s (3000, 6000, 9000, 12,000 or 15,000 cfm); the line brattice was set at 1.5 meters from the face; the water sprays were turned on and the bits were rotated.

Natural gas was released into the face area at rates which were progressively increased by increments of 0.005 or 0.010 m^3/s (10 or 20 cfm) until a maximum of 2.0 volume-percent was detected at any sensor location, or until the next 0.005-m^3/s increment would have caused a methane concentration greater than 2.0 volume-percent at one of the monitors. At each gas flow rate, the system was allowed to come to equilibrium and the methane concentration at each methane sensor location was recorded. After 2.0 volume-percent was reached at any one point, the gas release was stopped, the gallery was cleared of gas, the brattice was moved, and, if necessary, the air quantity was changed for the next test. Brattice distances of 1.5, 3.0, 4.5, and 6.0 meters were tested with the unassisted exhaust face ventilation system. Brattice distances of 1.5, 3.0, 4.5, 6.0, 7.5, and 9.0 meters were tested with the unassisted blowing face ventilation system. Following the completion of baseline testing, various improved face ventilation systems were tested at air quantities ranging from 1.4 to 7.0 m^3/s and brattice distances ranging from 1.5 to 9 meters.

DATA ANALYSIS

The methane concentrations from each of the methane sensor locations, the air quantity at the inby end of the line brattice and the gas release rate for each system at each brattice-to-face distance were coded for computer analysis. A computer program was developed that would give the location of the highest methane concentration and calculate the methane dilution capacity and the face ventilation index for a system.

The location of the highest methane concentration is particularly important during an optimization process. For exhaust systems that perform well, the highest methane concentration usually occurred in the brattice-side corner of the face. This indicates the airflow is being properly coursed to and across the face. When the high-point concentration occurs in the off-brattice side corner on an exhaust face ventilation system, good air distribution is not being obtained. For an unassisted blowing face ventilation system the high-point methane concentration occurred near the face on the off-brattice side of the entry. The location of the highest methane concentration is also useful in determining the positioning of the machine-mounted methane monitor.

The methane dilution capacity is the total amount of methane a system can dilute so that the high-point methane concentration is 1.0 volume-percent. The methane dilution capacity is determined from the test data by placing a least-squares regression through the maximum methane concentration versus the methane release rate data at a specific air quantity and brattice-to-face distance, and solving the regression for the release rate that would give 1.0 volume-percent maximum methane concentration.

The methane dilution capacity cannot exceed 1 percent of the air volume at the inby end of the line brattice. When the intake methane concentration is 0.00 volume-percent, the methane dilution capacity is also the face-area methane liberation that a system can dilute to the 1.0 volume-percent. However, for field applications where the intake methane concentration may not be 0.00 volume-percent, the peak face-area methane liberation rate that a system can dilute to 1.0 volume-percent can be found by multiplying the methane dilution capacity by the difference between 1.0 volume-percent and the intake methane concentration. Therefore, a system with a methane dilution capacity of 0.025 m^3/s and an intake methane concentration of 0.04 volume-percent can dilute a face liberation of 0.024 m^3/s. (0.024 m^3/s = 0.025 m^3/s x (1.00 - 0.04).) When the intake concentration is low, the difference between the methane dilution capacity and the maximum liberation rate that a system can dilute is negligible. However, with higher intake methane concentrations, it is important to note the difference between these two quantities.

For comparison of field and laboratory result, the methane dilution capacity for field tests can be determined by dividing the peak face-area methane liberation rate by the maximum methane concentration in the face area less the intake methane concentration,

$$MD = \frac{L}{(C_{MAX} - C_i)} \qquad (1)$$

where:

MD = methane dilution capacity, m^3/s (cfm)

L = peak face-area methane liberation rate, m^3/s (cfm)

C_{MAX} = maximum detected methane concentration, percent CH_4

C_i = intake methane concentration, percent CH_4

For field tests, the peak face-area methane liberation rate can be determined by multiplying the peak face-return methane concentration during mining less the intake methane concentration by the face-return air quantity, and dividing by 100:

$$L = \frac{(C_R - C_i)Q_R}{100} \qquad (2)$$

where:

L = peak face-area methane liberation rate, (m^3/s (cfm)

C_R = peak face-return methane concentration, percent CH_4

C_i = intake methane concentration, percent CH_4

Q_R = face-return air quantity, m^3/s (cfm)

Also calculated for the laboratory tests is a value called the face ventilation index. This value indicates the portion of the available air volume actually used to dilute the face area methane. For the laboratory studies, the face ventilation index is found by dividing the methane dilution capacity by 1 percent of the air volume at the end of the line brattice,

$$FVI = \frac{MD}{.01 \times Q_B} \times 100\% \qquad (3)$$

where:

FVI = face ventilation index, percent

MD = methane dilution capacity, m^3/s (cfm)

Q_B = air quantity at inby end of line brattice, m^3/s (cfm)

If all of the air at the end of the line brattice were used to dilute the methane released from the face, the methane dilution capacity would be 1.0 percent of the air quantity and the face ventilation index would be 100 percent.

The face ventilation index is indicative of the limitations of a face ventilation system and gives an indication of where system improvements can be made. If improvements are necessary for systems with low face ventilation indices, the face air quantity should be increased.

TEST RESULTS

Unassisted Exhaust and Blowing Line Brattice Systems

Results of laboratory tests on unassisted exhaust and blowing line brattice systems are shown in Table 1. By comparing the methane dilution capacity results to the peak face-area methane liberation rate at a mine, it can be determined if the system is adequate for the existing conditions. The methane dilution capacity results show that for a given air quantity, the performance of the exhaust face ventilation system deteriorates as the brattice-to-face distance increases. For a constant air quantity, the performance of a blowing face ventilation system fluctuates but does not show any great deterioration as the brattice-to-face distance increases from 3 to 9 meters. For either an exhaust or blowing face ventilation system, an increase in air quantity at a given brattice-to-face distance generally resulted in an increase in the methane dilution capacity.

For the unassisted exhaust system, the full benefit of increasing the face air quantity was generally realized for all brattice distances. This is illustrated by the fact that there is no rapid decrease in the face ventilation index for a given brattice-to-face distance as the air quantity increases.

For the unassisted blowing face ventilation system, the full benefit of an increase in face air quantity was not realized at brattice-to-face distances greater than 3 meters. This is indicated by a decrease in the face ventilation index at a given brattice-to-face distance as the face air quantity increases.

Diffuser Fan Assisted Exhaust Line Brattice System

Historically, machine-mounted diffuser fans have been considered the primary means of improving a face

TABLE 1 - LABORATORY RESULTS OF METHANE DILUTION CAPACITY AND FACE VENTILATION INDEX FOR UNASSISTED EXHAUST AND BLOWING FACE VENTILATION SYSTEMS

System	Air Quantity m^3/s (cfm)	Methane Dilution Capacity, m^3/s (cfm) Brattice Distance, m, (ft)						Face Ventilation Index % Brattice Distance, m, (ft)					
		1.5 (5)	3 (10)	4.5 (15)	6 (20)	7.5 (25)	9 (30)	1.5 (5)	3 (10)	4.5 (15)	6 (20)	7.5 (25)	9 (30)
Baseline	1.4 (3,000)	.006 (13)	.004 (8)	.003 (7)	.002 (5)			44	25	22	17		
Exhaust Brattice	2.8 (6,000)	.022 (46)	.006 (13)	.005 (11)	.004 (8)			76	22	18	14		
	4.3 (9,100)	.043 (91)	.010 (22)	.007 (15)	.006 (12)			100	25	16	14		
	5.7 (12,000)	.057 (120)	.012 (26)	.009 (19)	.008 (16)			100	22	16	13		
	6.9 (14,700)	.067 (143)	.017 (35)	.008 (18)	.008 (17)			97	24	12	12		
Baseline	1.5 (3,100)	.006 (13)	.006 (13)	.007 (15)	.009 (19)	.007 (15)	.006 (12)	42	43	49	61	47	38
Blowing Brattice	2.9 (6,100)	.008 (16)	.009 (19)	.010 (22)	.009 (20)	.010 (21)	.009 (19)	26	31	35	32	35	30
	4.3 (9,200)	.020 (42)	.014 (30)	.009 (20)	.009 (19)	.008 (18)	.008 (17)	45	33	22	20	20	18
	5.8 (12,300)	.033 (70)	.022 (46)	.012 (26)	.011 (23)	.009 (19)	.010 (21)	57	37	21	19	15	17

ventilation system. While combination blowing and exhaust systems, and more recently, spray fan (water spray assisted) systems have been used to increase the capacity of a face ventilation system, diffuser fans are still used where high face-area methane liberation rates exist.

Airflow for the diffuser fan systems tested was induced by Joy Model I 2RF100 two-stage fans. Two fans, each with a capacity of approximately 1.2 m^3/s, were available for the tests. The duct work used to direct the diffuser fan discharge had a cross section of 0.15 meters by 0.30 meters. The velocity at the diffuser fan discharge was approximately 26 m/s for systems with a single discharge and 13 m/s for systems with dual discharges from a single diffuser fan. For both single and dual fiffuser fan tests, the diffuser fan inlet was maintained at approximately 5.8 meters from the face.

In order to determine which diffuser fan configurations should be tested at the various air quantities and face distances, a series of preliminary tests were conducted. The preliminary testing consisted of evaluating a diffuser fan configuration at a nominal airflow of 2.8 m^3/s or 5.7 m^3/s at a 3-meter brattice-to-face distance and with a 0.005-m^3/s methane release rate. Once the high-point methane concentration was determined, the approximate dilution capacity and face ventilation index were calculated from equations (1) and (3), respectively.

Preliminary testing was done on 23 single diffuser fan systems (1.2 m^3/s airflow) and 27 dual diffuser fan systems (2.4 m^3/s). Results of the preliminary tests on single and dual diffuser fan systems indicate that the addition of a diffuser fan does not guarantee an improvement in the face ventilation system. Systems with fan discharges too close to the face or discharges directed toward the brattice-side corner of the face generally resulted in little if any improvement to the face ventilation system. Care must be taken to insure that the inlet and discharge of a diffuser fan system are properly designed and installed. For diffuser fan assisted face ventilation systems that performed well, the high-point methane concentration occurred near the face on the brattice side of the entry.

From the preliminary test result, seven systems were selected for final testing: four single diffuser fan systems and three dual diffuser fan systems. None of the diffuser fan systems selected for final testing were symmetrical and each of these systems discharged air toward the off-brattice corner of the face. Figure 2 is a schematic of the optimum diffuser fan system tested for either a 1.2-m^3/s or 2.4-m^3/s

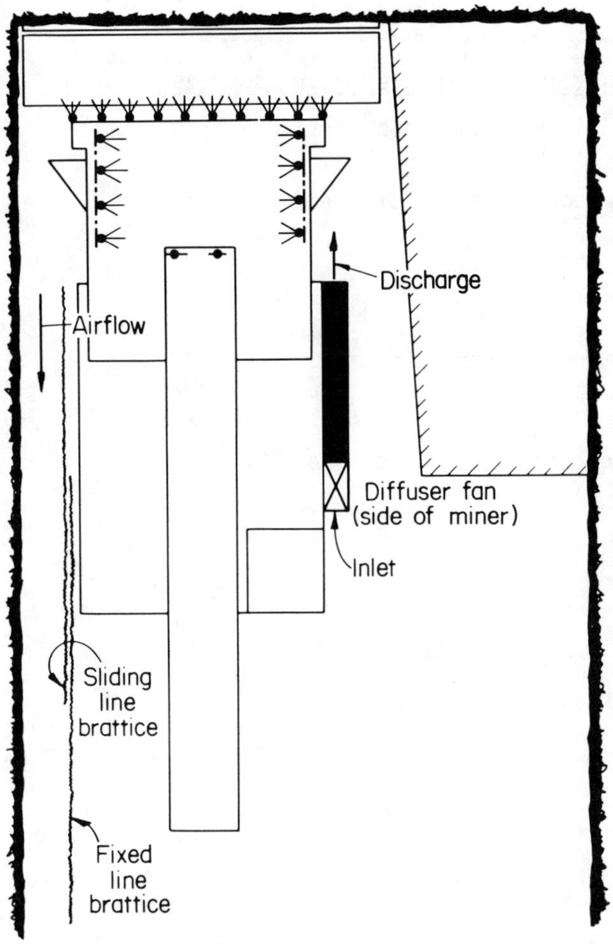

FIGURE 2. A schematic of the optimum diffuser fan system

diffuser fan system. The diffuser fan system is mounted on the side of the mining machine with the discharges 3.4 meters from the face. Results of tests made to determine the methane dilution capacity and face ventilation index for the optimum 1.2-m^3/s and 2.4-m^3/s diffuser fan systems are shown in Table 2.

At a given brattice-to-face distance, an increase in the face air quantity generally resulted in an increase in the methane dilution capacity and a decrease in the face ventilation index. At a given face air quantity, both the methane dilution capacity and face ventilation index generally decreased as the brattice-to-face distance increased.

Test results indicate that through the use of diffuser fans, the performance of a face ventilation system can be improved through a more effective use of the available face air quantity. With properly designed diffuser fan systems, this improvement can be such that the performance of the diffuser-fan assisted face ventilation system with the brattice at 9 meters from the face provides methane control equivalent to an unassisted exhaust line-brattice set at 3

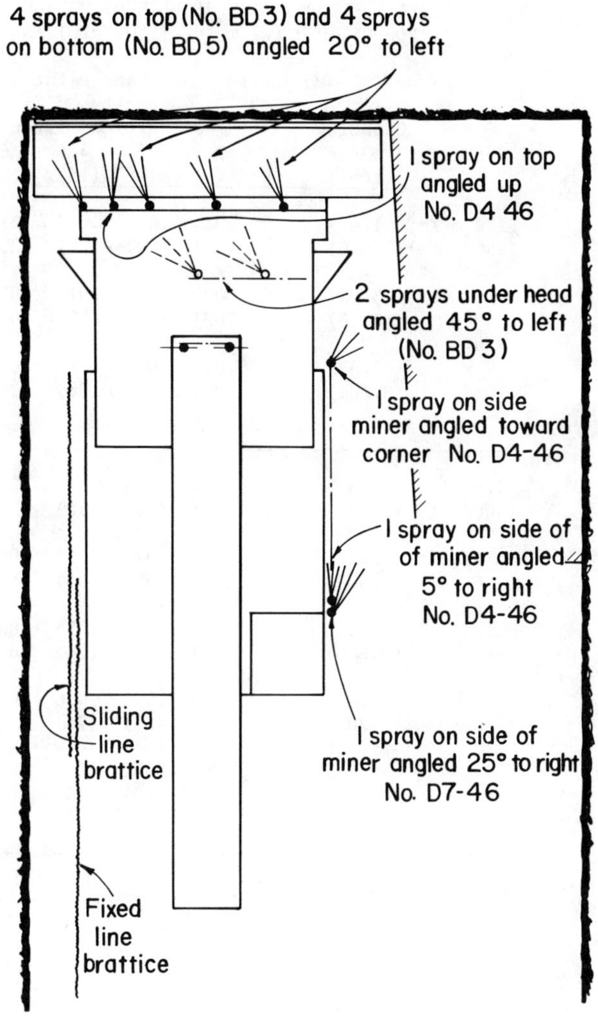

FIGURE 3. A schematic of the spray fan system

meters from the face.

For the systems selected for final testing, the recirculation of airflow in the face which occurred when the diffuser-fan air quantity exceeded the face air quantity did not create any additional hazard with respect to methane buildup. The final systems tested with diffuser-fan air quantities in excess of the line brattice air quantity performed better for methane control than the unassisted exhaust line-brattice system. However, because recirculation may have an adverse effect on the respirable dust concentrations at the miner operator's location, it may be necessary to limit the diffuser fan airflow during underground mining operations to approximately one-half the air quantity behind the line brattice and/or maintain the line brattice inby the machine operator.

Water-spray Assisted Face Ventilation Systems

Recent developments in the area of face ventilation systems have incorporated the use of specially oriented water sprays to assist and improve the face ventilation system. One particular system that has received much attention is the "spray fan" system developed by Foster-Miller Associates under U.S. Bureau of Mines contract.

TABLE 2 - LABORATORY RESULTS OF METHANE DILUTION CAPACITY AND FACE VENTILATION INDEX FOR OPTIMUM 1.2 m^3/s, (2500 cfm) AND 2.4 m^3/s, (5000 cfm) DIFFUSER FAN SYSTEMS

System	Air Quantity m^3/s (cfm)	Methane Dilution Capacity, m^3/s (cfm) Brattice Distance, m, (ft) 1.5 (5)	3 (10)	4.5 (15)	6 (20)	7.5 (25)	9 (30)	Face Ventilation Index % Brattice Distance, m, (ft) 1.5 (5)	3 (10)	4.5 (15)	6 (20)	7.5 (25)	9 (30)
1.2 m^3/s (2500 cfm)	1.4 (3,000)	.014 (30)	.014 (30)	.013 (27)	.012 (25)	.008 (17)	.007 (15)	100	100	91	83	56	50
Diffuser	3.0 (6,300)	.030 (63)	.017 (36)	.018 (39)	.016 (34)	.014 (30)	.011 (24)	100	57	62	53	47	38
Fan	4.3 (9,200)	.043 (92)	.022 (47)	.018 (38)	.017 (37)	.017 (35)	.015 (32)	100	51	42	40	39	34
2.4 m^3/s (5000 cfm)	1.4 (3,000)	.014 (30)	.014 (30)	.014 (30)	.014 (30)	.014 (30)	.014 (30)	100	100	100	100	100	100
Diffuser	2.7 (5,800)	.027 (58)	.027 (58)	.027 (58)	.027 (58)	.027 (58)	.027 (58)	100	100	100	100	100	100
Fan	4.3 (9,200)	.043 (92)	.035 (74)	.035 (74)	.033 (69)	.031 (66)	.028 (59)	100	80	80	75	72	64
	5.3 (11,200)	.053 (112)	.041 (87)	.036 (76)	.034 (72)	.036 (76)	.028 (59)	100	73	64	60	64	49
	6.9 (14,700)	.069 (147)	.052 (110)	.034 (73)	.032 (67)	.034 (71)	.025 (54)	100	75	49	45	48	36

TABLE 3 - LABORATORY RESULTS OF METHANE DILUTION CAPACITY AND FACE VENTILATION INDEX FOR OPTIMUM SPRAY FAN ASSISTED FACE VENTILATION SYSTEMS

System	Air Quantity m^3/s (cfm)	Methane Dilution Capacity, m^3/s (cfm) Brattice Distance, m, (ft) 1.5 (5)	3 (10)	4.5 (15)	6 (20)	7.5 (25)	9 (30)	Face Ventilation Index % Brattice Distance, m, (ft) 1.5 (5)	3 (10)	4.5 (15)	6 (20)	7.5 (25)	9 (30)
Spray	1.4 (3,000)	.008 (18)	.008 (17)	.008 (17)					59	58	56		
Fan	2.8 (5,900)	.011 (23)	.009 (19)	.008 (17)					39	32	29		
700 kPa (100 psi)	4.1 (8,700)	.015 (32)	.009 (19)	.009 (19)					37	22	22		
Spray	1.4 (3,000)	.011 (23)	.010 (22)	.010 (22)					75	73	73		
Fan	2.8 (5,900)	.013 (27)	.011 (24)	.010 (22)					45	40	38		
1400 kPa (200 psi)	4.2 (9,000)	.017 (37)	.011 (24)	.010 (22)					41	27	25		

The spray fan system incorporates a series of water sprays mounted along the side of the mining machine with angled sprays across the front of the machine. The side sprays are used to cascade air to the front of the machine. The front sprays are angled toward the brattice to direct the air across the face. Figure 3 is a schematic of the spray fan system tested in the low-coal section of the test facility.

In addition to testing the complete spray fan system, tests were conducted on various face ventilation systems where only the three sprays indicated on the side of the miner in Figure 3 were added to a conventional water spray system. This modification could be easily made to a mining-machine water-spray system. Tests were run on various side spray systems using hollow cone, full cone, and flat sprays with varying expansion angles at water spray pressures of 700 and 1400 kPa.

For the side spray tests, the line brattice was set at 3 meters, the air quantity was adjusted to approximately 2.8 m^3/s, and 0.005 m^3/s of methane was released into the test facility. From the high-point methane concentration data, the methane dilution capacity and face ventilation index were calculated from equations (1) and (3), respectively. For water-spray assisted face ventilation systems, the high-point methane concentration generally was near the face on the line-brattice side of the mining machine.

Each of the side spray systems showed an improvement over the unassisted exhaust face ventilation system (baseline). Systems at water spray pressures of 1400 kPa showed a 10 to 20 percent improvement over systems at 700 kPa. The side spray systems performed best (methane dilution capacity of 0.01 m^3/s at 1400 kPa) when hollow cone, narrow angle (0.70 radian expansion), or flat sprays were used. Because alignment is not critical with the hollow cone spray from a practical standpoint, hollow cone, narrow angle would be the most advantageous spray choice for side sprays.

In addition to the tests on side sprays, tests were run on a complete spray fan system. The complete spray fan system utilized narrow angle (0.70-radian expansion) hollow cone sprays as side sprays and wide angle (1.4-radian expansion) hollow cone sprays as front sprays. Results of the tests on spray fan systems at brattice-to-face distances of 3 to 6 meters for various air quantities are shown in Table 3 for water pressures of 700 and 1400 kPa.

The complete spray fan system improved the face ventilation system performance more than the addition of the side sprays alone. For a given brattice-to-face distance and air quantity, the 1400-kPa system showed an improvement over the 700-kPa system.

At face distances greater than 3 meters, an increase in air quantity resulted in little, if any, change in the methane dilution capacity. As a result, the face ventilation index decreased as the face air quantity increased. For brattice-to-face distances of up to 6 meters, the spray fan systems at 700 kPa or 1400 kPa were equivalent to the simple exhaust system at a 3-meter brattice-to-face distance.

SUMMARY

Figure 4 gives a comparison of the laboratory results of the tests of methane dilution capacity for various face ventilation systems at approximately

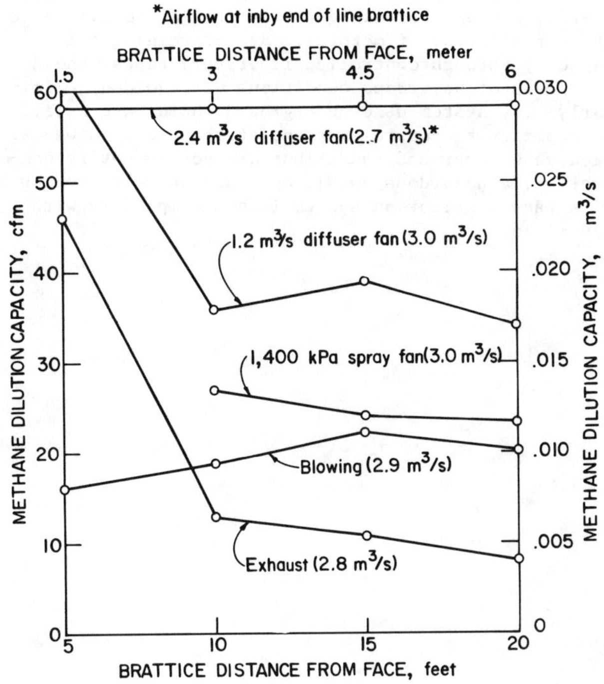

FIGURE 4. A comparison of laboratory testing results

2.8 m^3/s of airflow and brattice-to-face distances of 1.5 to 6 meters. By comparing the methane dilution capacities to the peak face-area methane liberation rate at a mine, one can determine if a system would be suitable for a specific application.

For either unassisted or improved exhaust face ventilation systems, the high-point methane concentration generally occurred near the face on the brattice side of the entry. For the unassisted blowing face ventilation system, the high-point methane concentration occurred near the face on the off-brattice side of the entry. These results should be used to determine the most suitable location for the machine-mounted methane monitor.

From the results of the laboratory tests, the methane dilution capacity of a system appears to be a more useful indication of a system's performance than the face ventilation index. The methane dilution capacity can be compared to the face-area methane liberation rate to determine if a face ventilation system would be adequate for the existing mining conditions. The face ventilation index gives the percent of available air that is actually being used for diluting methane in the face area, and can only be used to compare systems at the same face air quantity. This is illustrated by an example comparing the two face ventilation systems: if each system had a face air quantity of 1.4 m^3/s, then the system with the higher face ventilation index would be the better of the two face ventilation systems; however, comparing a 1.4-m^3/s system to a 2.8-m^3/s system with the same face ventilation index (50 percent), the 2.8-m^3/s system would have twice the methane dilution capacity of the 1.4-m^3/s system (0.014 m^3/s and 0.007 m^3/s, respectively).

All of the systems tested, regardless of brattice-to-face distance or air quantity, had a limited capacity for dilution of methane. By determining this capacity through controlled laboratory tests, the development of hazardous conditions was avoided. Similarly, any system used underground would have a limited capacity to control methane at the face. However, because underground conditions are not as easily controlled, a hazardous condition could quickly develop if a face ventilation system is not compatible with the mining activities.

Chapter 8

DUST CONTROL USING WET-TYPE DUST COLLECTORS

Bob J. Rawicki, President

Underground Surface Pollution Control, Inc.
2040 Regency Road, Suite F
Lexington, Kentucky

TYPES OF WET DUST COLLECTORS

Basically, there are two types of wet-type dust collectors. One is mechanical, incorporating pumps, motors, fans, sprays, filters, or flooded beds. These come in many forms, but their operating principle is basically the same. The screen or flooded bed is wetted by a series of sprays. Polluted air is drawn in through a fan; here the dust particles impinge on the screen and are flushed off into a settlement tank. The air then passes through spray eliminators to the atmosphere. A great deal of time and money has been spent developing this type of collector.

In laboratory tests where clean water at constant pressure, air at constant volume, and dust at low but constant mass to volume were applied, these collectors gave high percentage dust collection efficiencies. Theoretically and technically, they are very good. In practice, however, under rugged mining conditions where there are no constants, and dust load, particle size, water pressure, and volume of air vary with time, their efficiency varies dramatically.

Some of the most common problems are:

1. clogging of screens
2. blocking of nozzles
3. replacement of damaged pump stators

Any of the above result in loss of air flow and dust collection efficiency. Generally, there are so many mechanical moving parts, something is always breaking down. This type of collector is very expensive in parts and labor to maintain.

The second type of dust collector, which I personally developed, with the help of the British National Coal Board of Great Britain, has none of the above mentioned problems and it works on a completely different principle.

My collector, the Mark III Precipitaire Wet-Type Dust Collector, (an improved version of previously successful models) has no internal moving parts, no flooded beds, screens, spray nozzles, etc. It is highly efficient - 99.8% efficiency on total collected dust and 97% on respirable dust. The pressure drop through the collector is constant, hence, air flow remains constant. There is no maintenance expenditure, be it labor or parts. The only maintenance that there is, is desludging. This, of course, varies with the dust load. With a low dust load, desludging may only be required every 4 weeks, but this is a very simple operation that can be handled by unskilled men and it takes only a very short time. My collector is powered by a ventilation fan that is placed on the clean air side for a longer fan and motor life. Dust is extracted solely by water action.

The collector's operating principle is: Dust laden air is drawn in along ducting into a tapered scrubber section, shaped to create a self-induced curtain of water. This action washes the dust particles from the air and the collected dust settles into the bottom of the tank. Cleaned air is then exhausted via a series of spray eliminators where it is completely dried and on into the atmosphere.

These collectors have been sold in Great Britain, to the National Coal Board, and throughout Europe. They have been working very well for many years now.

GENERAL APPLICATIONS

Dust is very dangerous; silicosis and black lung kill and disable many workers each year. Many of the mine explosions can be attributed to coal dust. By the use and application of dust collectors, these deaths and disasters can be avoided. There really is no excuse.

However good and efficient a dust collector may be, it's positioning and installation arrangement is critical. The collector may remove 99% of dust supplied to it, but it takes the skill and experience of a dust control engineer to design and fit ducting, hoods, etc. to remove polluted air at source and deliver it to the collector. In some cases this is very simple; in the case of longwall machines, it is very difficult. In general, dust collectors should be used to collect dust at every point of dust generation.

I have noted a great reliance on dilution as a means of dust control. I personally think this is a dangerous and unreliable method. As mines become deeper and the use of larger and more productive equipment becomes more common, reliance on dilution will be impossible and dust collectors will be essential.

Some of the most typical arrangements are

1. To extract and collect dust created by Dosco roadheaders. A typical arrangement is shown here:

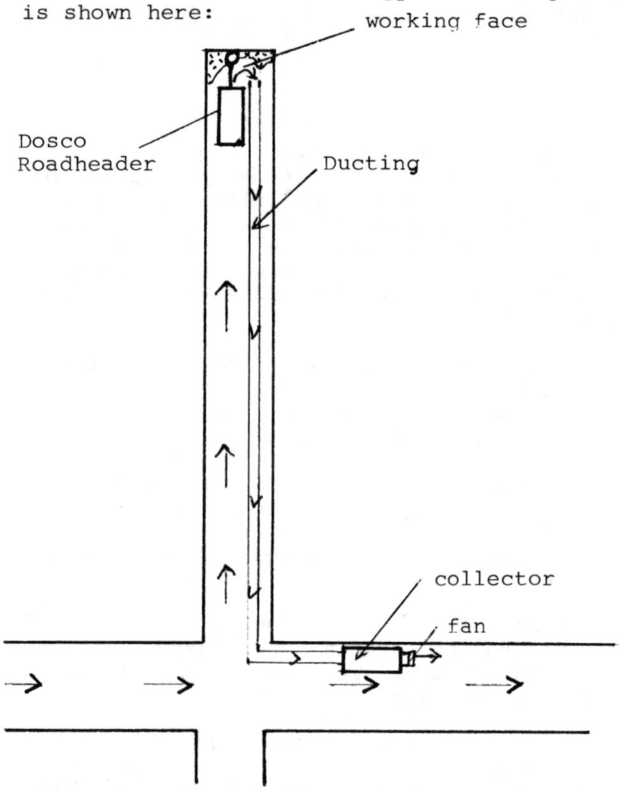

The collector can be put in any convenient location, then ducting is taken to the development or productive face. As the dust is produced, it is constantly drawn into the ducting and to the collector. The operator works in a healthy environment. The machine can work at full production speed and clean, dry air is returned into the mine.

2. Collectors coupled to fans may be placed anywhere within a mine to act as free standing filters, while at the same time aiding air flow and ventilation.

3. At transfer points they can collect the dust produced and prevent pollution of the ventilating air. The life of conveyors will be prolonged if they work in a dust-free atmosphere.

4. To clean and extract air from roadheaders or drum miners in development driveages and productive faces.

5. At the end of longwall face where the ripping is being taken in line with the face.

6. In controlled air recirculation systems at the end of a longwall face where air from the face is cleaned in the roadway before being directed into advance development.

7. To clean air from continuous miners, roof bolters, underground crushers, hoppers, etc.

On the surface, my dust collectors have been used to extract and clean air from: dryers, conveyor transfer points, crushers, screens, loading points, coal hoppers or silos, and coal preparation plants.

The Precipitaire can be hung from monorails, or mounted on rails and moved forward with other equipment. One I have recently supplied, in West Virginia, is positioned on a transfer point platform. This allows constant sludge discharge directly to the settling pit, eliminating manual periodic desludging.

There are no standard procedures. The collector has to be adapted to whatever conditions prevail.

In most mines, dust at transfer points, is suppressed by using water sprays. This method reduces the amount of dust, but only slightly. On the other hand, it causes sludge formation on the conveyor drive and surrounding equipment, which often results in friction and overheating. Equipment requires frequent cleaning and replacement.

Using the wet-type collector and extracting dry dust can prevent the aforementioned.

The working life of a Precipitaire is in the region of 20 years without the need for any replacement parts.

If a transfer point is moved to a new location, the Precipitaire can be moved and relocated with the other equipment.

By enclosing transfer points, less air needs filtering, therefore, one can reduce the size and the cost of the collector. For example: If a conveyor transfer point is enclosed partially, say 20-30%, 1000 to 1500 CFM per ft. of conveyor width will be required. If 80% is enclosed, 400 to 600 CFM per ft. of conveyor width will be sufficient. Therefore, less power is used, the collector and fan are cheaper.

The aim of a good dust engineer should always be not only to design an efficient system, in terms of dust collection and extraction, but also a cost effective system.

By the use of enclosures and the application of well-designed hoods, one can reduce the amount of air that needs to be filtered, and hence, one can substantially reduce both the power consumption and purchase cost of collectors and fans. Unfortunately, one can not always use hoods and enclosures , for instance, in development areas and operational areas.

For example: Where a road stone heading machine is in operation, the only way to extract the dust at source is to run ducting from the collector to the face. Here it is very important to keep the ducting around 8 feet from the face. Concentrations of dust have to be studied and the entry placed as close to the heaviest dust concentration as possible. The end of the ducting should be tapered out, to reduce entry velocity. This will increase dust capture considerably. The above principle apply to continuous miners as well.

One of the most productive machines available today is the longwall machine. It will produce the highest tonnage per man, but it also produces more dust than any other machine. To my knowledge no one has yet produced a method of extracting dust directly from the shearing drums where the dust is generated. I think this problem could be solved, but government pressure and finance would be required to get dust control engineers and machine manufacturers to co-operate.

SURFACE APPLICATIONS

Vibrating screens produce large quantities of dust. A lot of research has been done on hoods for vibrating screens, the problem being access to the screen for maintenance.

I know of one dust control company that allows 200 CFM per sq. ft. of screen area. If the screen is 100 sq. ft. this means they recommend 20,000 CFM for this screen. This is a lot of air for a small screen.

A well designed hood would only need 50 - 70 CFM per sq. ft. and still allow access for maintenance. I would like to add here that although hood design is very important, the positioning of extraction points is of equal importance.

Recently, a tipple plant was quoted $180,000.00 for the installation of dust extraction and collection equipment. They also stated that 40,000 CFM would be required in total. The collection equipment they quoted for was the conventional bag-type filter.

I designed and supplied this plant with dust extraction and collection equipment using only 16,000 CFM using the Precipitaire Wet Type Dust Collector for a price of $28,000.00. It has been working with a throughput of 450 tons per hour for some time now, and the plant is 100% dust free.

To obtain a consistent product from a dryer, the air flow must be constant. So, when choosing a dust collector for a dryer, a collector that provides constant volume with low maintenance is the only choice.

HOW THE SIX CLEANEST U. S. LONGWALLS STAY IN COMPLIANCE

Chapter 9

Charles D. Taylor, Industrial Hygienist
Robert A. Jankowski, Physical Scientist

U. S. Department of the Interior
Bureau of Mines, Pittsburgh Research Center
Dust Control and Ventilation Group
Pittsburgh, Pennsylvania

INTRODUCTION

The objective of this program was to conduct dust surveys at six longwall sections having double-drum shearers, that were regularly in compliance, and to identify the dust control techniques that were the most effective. As each of the mines selected had a record of regular compliance with the Federal Dust Standard, all sampling was "short-term" (often less than 30 minutes) and no attempt was made to relate the survey results to dust-sampling results obtained by the companies for compliance purposes. The sampling procedures (1) used during this survey were developed jointly by the Bureau of Mines and Foster-Miller Associates, Inc. In addition to monitoring dust levels, measurements were also taken throughout the shift to monitor airflow along the face and water usage on the shearer.

This program was a cooperative effort of the Bureau of Mines and the Mine Safety and Health Administration (MSHA). The following MSHA health specialists contributed to its successful completion: Paul Lenyo, CMS&H, Dist. 2; Ellis Mitchell, CMS&H, Dist. 3; Carl Jenkins, CMS&H, Dist. 4; David Stritzel, CMS&H, Dist. 8; and John Prelc, CMS&H, Dist. 9.

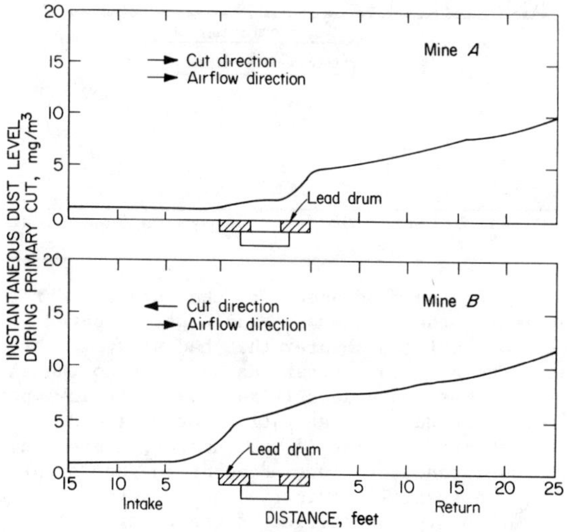

Figure 1. Dust level profile around shearer due to position of the lead drums, Mines A & B.

RESULTS

CUTTING SEQUENCE - WORK LOCATION

At all mines surveyed, an important factor in controlling dust exposure at the shearer is the design of the cutting sequence so as to allow the shearer operator to work most of the time on the intake-air side of the lead drum.

It is generally accepted that most of the dust generated by a double-drum shearer comes from the lead drum, which in the raised position normally takes a full sump. Because the position of the drum that is in the raised position determines the horizon control, the operator must be close enough to carefully monitor its operation. Therefore, his position relative to the lead drum has a significant effect on his dust exposure.

Dust profiles, such as those constructed from data collected at Mines A and B (Figure 1), show the large gradient in dust concentrations that can exist between the intake and return air side of the lead drum. When the primary cut is taken head to tail, the lead drum is downwind and dust levels can be maintained at a low level until air has passed the shearer (Mine A). Thus, shearer operators can stay in relatively uncontaminated air. When the primary cut is taken tail to head, the lead drum is upwind of the shearer controls, resulting in higher dust concentrations at the shearer operator's work location (Mine B). At the mines where this cutting sequence was used, it was company policy for the shearer operators to walk on the intake side of the lead drum, returning to the controls only to adjust the height of the lead drum.

DIRECTED WATER SPRAYS AND PASSIVE BARRIERS

It is not always possible to keep the shearer operator on the intake air side of the lead drum. Various techniques have been tried to keep the dust generated by the shearer away from the machine operators when they work adjacent to the shearer. At all mines visited, the external water spray nozzles were directed so that dust from the shearer was directed downwind and toward the face, and away from the shearer operators.

The effectiveness of using external water sprays to keep dusty air close to the face was noted during Bureau of Mines efforts to develop improved water spray systems (2). Laboratory and underground tests showed that water sprays located properly and directed downwind can create aiflow patterns that move dust generated by the shearer toward the face and away from the shearer operators. Sprays directed upwind, into the airflow, can actually move dust away from the face into the shearer operators' work area. Sprays pointed crosswind can also raise walkway dust levels under many circumstances.

The location and direction of the water spray nozzles at Mine C are illustrated in Figure 2. To demonstrate the effectiveness of the water spray system for

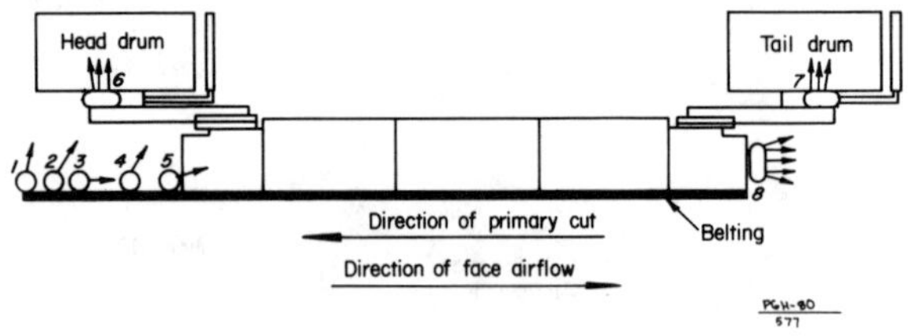

Figure 2. Location of Water Sprays on Shearer at Mine C.

controlling dust adjacent to the shearer, the dust profile for the shearer at Mine C was compared with a dust profile for a shearer that had sprays pointed upwind. The dust concentrations adjacent to the shearer at Mine C are much less (Figure 3). It is important to note that dust concentrations at Mine C do increase on the return-air side of the shearer, indicating that the lower concentrations adjacent to the shearer are due to a spatial redistribution of the dust cloud rather than to suppression of the dust. It is also important to supply water at adequate pressure to the spray nozzle. An average water pressure of 150 psig was supplied to the external nozzles at the mines visited. The higher water pressure increases the airflow over the machine and also helps to reduce nozzle plugging and improves average airborne dust capture.

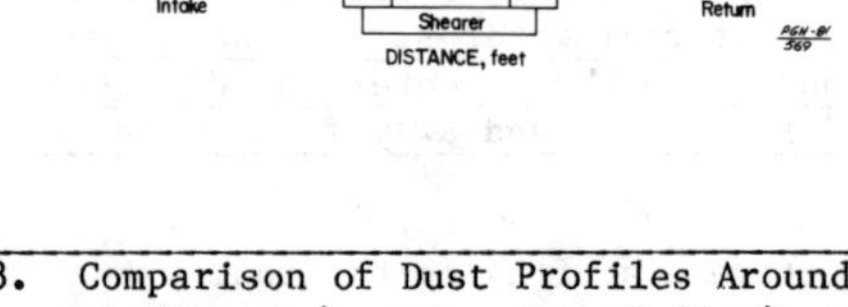

Figure 3. Comparison of Dust Profiles Around Shearer at Mine C (sprays with airflow), and a Mine with the sprays oriented against airflow.

In addition to the proper positioning of water sprays to achieve the desired airflow over the shearer, some type of passive barrier system was used at all mines surveyed. Figure 4 illustrates a typical passive barrier system.

The two basic parts of the passive barrier system are

1. The splitter arm mounted on the intake and return air sides of the shearer. The arm "splits" the intake airflow so that one part passes over the shearer, removing dust generated by cutting, and the other part remains relatively clean and passes over the shearer operators.

2. Belting mounted on top of the shearer. This belting, which should be as high as possible without obstructing the operator's view, forms a partial block between the air at the face and air passing over the operator.

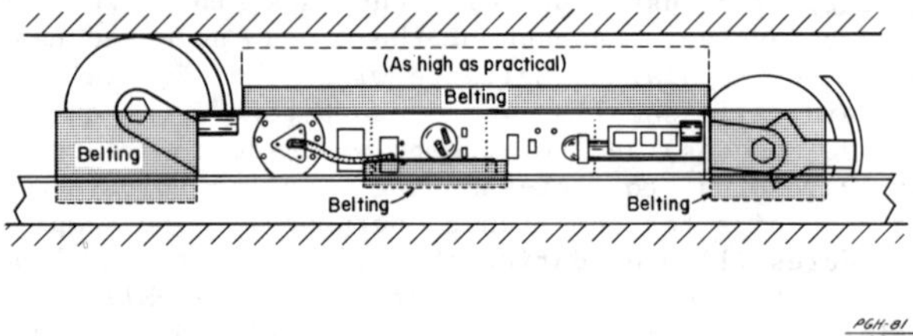

Figure 4. Typical locations for passive barriers mounted on the shearer body.

WATER QUANTITY AND PRESSURE

Although there are no specific guidelines for water usage on longwall shearer sections, previous foreign and Bureau of Mines studies indicated that increasing the quantity and pressure of water supplied through the cutting drums reduces dust generated by the shearer. Larger-than-normal water quantities and higher water pressures were used at most of the longwall sections visited (Table 1). The greater water quantities and pressures supplied particularly through the drum water sprays aid in dust knockdown and help to prevent dust suspension. However, possible adverse effects of additional water include soft bottom and increased moisture in the mined coal.

At Mine B, approximately 80 gpm of water is supplied to the shearer. This is about twice the average water quantity normally provided to a shearer. The average quantity of water supplied to the shearer on the six sections visited was 60 gpm. At these sections, a significant portion of the water was supplied through the drums and loaded out with the coal, thus reducing problems resulting from water accumulating on the bottom.

OTHER CONTROL TECHNIQUES

The dust control techniques described above are the most important controls that could be identified with the sampling procedures used. However, other dust control methods were also applied. At two of the mines surveyed, worker exposure to dust generated by support advance was reduced by moving the supports immediately on the return-air side of the shearer during the tail-to-head cut. This permitted all face workers

CLEANEST LONGWALL COMPOSITE

TABLE 1

	Shearer Make	Cut Direction (Primary)	Cutting Height, in.	Tram Speed, fpm	Average Face Vel, fpm	Water Pressure, psig	Water Flow, gpm	Production, tps
A.	Anderson Mavor 300	Head-to-Tail	53	12	200	90 (all)	26 total 3/(5.5 gpt)	700
B.	Eickhoff 150-2L	Tail-to-Head	66-78	16	280	140 (Drums) 165 (Ext)	80 total (8.3 gpt)	1,500
C.	Eickhoff 300	Tail-to-Head	92	12	255	60 (Drums) 125 (Ext)	85 total (8.7 gpt)	1,500
D.	Eickhoff 300	Head-to-Tail	84	17	260	60 (Drums) 160 (Ext)	55 total (4.6 gpt)	2,000
E.	Anderson Mavor 300	Bi-directional	1/60	17	2/200	30 (Drums) 75 (Ext)	40 total (5.3 gpt)	800
F.	Eickhoff 300	Figure Eight	66-78	14	315	300 (Drums) 275 (Ext)	60 total (7.1 gpt)	1,000
Average			72	15	250	115 (Drums) 150 (Ext)	60 total (6.6 gpt)	940

1/Tail drum ~ 47", Head drum ~ 60" - all other operations had symmetrically sized drums.

2/Airflow from Tail-to-Head - all other operations had airflow Head-to-Tail.

3/Reported as gallons per ton of coal mined, a more appropriate indicator of water usage. Calculated from extraction rate in tons per foot of cut, average tram speed, and water flow (gpm).

to remain on the intake-air side of the support movement. Several operational constraints exist with this method of advance. The support system must be of the immediate forward support (IFS) type; bottom conditions must be such that the support movers are able to keep up with the shearer; and the dust control procedures must be such as to permit support advance in an area where dust from the shearer is still confined to the face area.

It is interesting to note that although the specific effects were not determined, the majority of operations surveyed employed reduced drum speeds, ranging from 45 down to 37 rpm. One operation, Mine C, had drums specially designed for slow-speed deep cutting, and the bidirectional operation, Mine E, had the slowest drum speed, 37 rpm.

In general, much of the success in reducing dust concentrations at the mines surveyed is attributable to preventive maintenance, especially the regular cleaning and repair of the shearer water spray system and the replacement of worn bits.

CONCLUSIONS

Although most longwall mining operations that use double-drum shearers frequently have difficulty limiting dust exposures of face workers, several operations have been identified that maintain regular compliance with the mandatory dust standard. At the six cleanest operations surveyed, three major dust control techniques were recognized: (1) a cutting sequence such that the shearer operators work most of the time on the intake-air side of the lead drum; (2) a passive barrier and/or external water spray system design that directs the airborne dust produced by the shearer downwind and away from the operators; (3) large quantities of water through the sprays, particularly the drum sprays, to aid in dust knockdown and prevent dust suspension.

Application of these control techniques on other longwall operations will not necessarily guarantee compliance, but should help in meeting the Federal Dust Standard.

REFERENCES

1. Taylor, C. D., et al, "The Use of Instantaneous Samplers to Evaluate the Effectiveness of Respirable Dust Control in Underground Mines," presented at the International Symposium on Aerosols in the Mining and Industrial Work Environment, November 1-6, 1981, Minneapolis, MN.

2. Kissell, F. N., et al, "Reducing Dust at Longwall Shearers by Confining the Dust Cloud to the Face," BuMines TPR 111, 1981, 21 pp.

Chapter 10

MINE SEALANTS - THEIR USES & BENEFITS

M. E. Herrera

Vice-President, Research & Development
American Energy Products Corp.
8210 Industry Avenue
Pico Rivera, CA 90660

The use and effectiveness of sealants in underground mines is a recognized and well accepted practice by many people in the mining industry. Their use however in many cases is to correct, or attempt to correct,problems that might have been avoided with sealant use in the early stages of development, and continued thereafter.

With the passage of the Federal Coal Mine Health & Safety Act and each subsequent revision, ventilation as covered in the act can cause the increase of problems associated with air slacking of roof and ribs. The act requires that: "The volume and velocity of the current of air shall be sufficient to dilute, render harmless and to carry away flammable, explosive, noxious and harmful gasses, and dust, and smoke and explosive fumes." Standards are set for minimum quantities of air to reach the last open cross cut,etc.

To meet these standards mine companies find that they must install more or larger fans as the working face gets further away and the ventilation space gets larger. As these huge quantities of air move thru these ventilated areas, air slacking problems and subsequent deterioration of tunnel surfaces can begin. Deterioration starts with drying of the tunnel surfaces, continues to fine crazing, checking, and cracking, till finally pieces of roof and ribs begin to fall. This causes a constant problem since the air slacking can lead to the damaging cycle starting all over again on surfaces freshly exposed as pieces fall off or worse, to large pieces falling capable of causing serious injury to miners or damage to equipment. This also requires the constant clean up of the fallen debris and the frequent scaling of loose roof to avoid the possibility of injury. Bolting chain-link or wire mesh to the roof is only of temporary help since this only keeps the large pieces from falling to the ground and holding them on the mesh. The weight of so much fallen roof on the mesh with roof bolts partially exposed sometimes requires the removal of the mesh, bolting it even higher up. This of course again does nothing to stop the cycle of air slacking from continuing.

The problem can be further aggravated in the summer months when moisture in the warmer outside air is condensed on the cooler tunnel surface as it is drawn into the mine. When this occurs the shale absorbs the moisture and expands. As the outside temperature drops, the moisture content in the outside air also drops. This drier air now tends to dry the shale causing it to shrink and contract. This constant expansion and contraction due to moisture absorption, then drying and expansion/contraction due to thermal changes in the strata can cause much checking and cracking. As the ventilation air gets in these checks and cracks, the process of air slacking begins again.

Mine sealants can help keep these problems from starting and also help stop them from continuing once they've started, in several ways.

The first is by simply sealing the moisture in the strata and protecting the tunnel surface from the ventilating air flow with a layer of sealant.

The second is by thermally insulating the tunnel surface from temperature changes in the outside air.

The third is by filling and sealing from the ventilation air, cracks, gaps, and fissures that have already occurred on the tunnel surface.

The fourth and, in my opinion after seeing many mine applications, the most important if the tunnel surface is badly weathered, the ability to cement and knit together into a tough solid surface the "jigsaw puzzle" pattern of loose, broken, and crumbling pieces of roof and ribs to keep them intact.

What type of sealants do we need to accomplish these four points as easily and economically as possible and with long-lasting results?

With the exception of urethane foam which possibly may not be used much longer in underground mines except in fire and rescue operations, there are two classes of sealants with several types in each class. These are liquid sealants and cementitious sealants.

In the liquid sealant class some of the types are two-component epoxies, two-component water-containing epoxies, inorganic silicate solutions, and latex emulsions of acrylics, vinyls or neoprene, either clear or white in color.

In general the liquid sealant class will do a good job of protecting against weathering of freshly mined tunnels or existing tunnels that are not yet badly degraded. They do this as mentioned in the first example by sealing the moisture in the strata and by protecting the tunnel surface from the ventilating air flow. This in turn reduces the erosion of the tunnel surface and the amount of coal dust in the tunnel reducing the frequency of rock dusting. However, they cannot insulate the tunnel surface from temperature changes, fill, seal or bridge cracks or fissures, or cement the broken surface of badly weathered roof or rib together, or help smooth their surface.

Most liquid sealants are easy to apply and generally lower in cost than cementitious sealants, since they are applied much thinner using lower cost equipment. In cooperation with the Spokane Mining Research Center, we at American Energy Products have developed a liquid sealant which can be applied with a simple low-cost pressurized hand garden sprayer. This sealant has been applied in several mines during the mining cycle with very good results keeping air slacking from starting in the areas sprayed. A recent test was conducted in a local mine where the application was easily made with airless spray equipment powered by the hydraulic system of a mine power scoop.

Because their application is relatively thin, liquid sealants should be very flexible and slow drying so they can penetrate through minor amounts of rock dust if it is present and bond to the tunnel surface. It should also be possible to apply them during cold periods at temperatures down to 40°F and have them form a film. They should not present a fire hazard during or after their application or be toxic during use.

In the cementitious sealant class some of the types are Magnesia cements, Portland cements, Gypsum cements, and lightweight varieties of all three.

Cementitious sealants regardless of type should have the following properties in addition to the ones already mentioned for liquid sealants, especially when they are applied to badly weathered tunnel surfaces. Also, by virtue of their higher cost, they should offer more benefits than just sealing the moisture in the strata and protecting the tunnel surface from the ventilating air flow. These additional benefits are:

1. It should be possible to apply them to a minimum 1/2" thick to help protect the tunnel surface by insulating it from temperature changes in the outside air and reduce cracking due to thermal expansion and contraction.

2. They should easily fill and seal cracks and adhere strongly to the sides without shrinking or pulling away. This is important to help reduce gas seepage from the strata, also to smooth the tunnel surface and reduce friction of the ventilation air against otherwise rough, cracked, broken, and jagged surfaces. When friction is reduced, back-pressure against the fan is reduced and its efficiency and air flow are improved.

3. They should bond to the strata and to the surface of products used in mines, such as cement block, steel, wood, or burlap brattice, and if necessary, applied to a thickness of at least three inches to fill and seal gaps and holes around stoppings, bulkheads or pipes, without sagging or slumping.

4. They should be completely noncombustible to help fire-protect the surface they cover. Also, they should not give off any smoke or harmful gases during fire conditions.

5. They should be light in color to increase light reflectance and visibility.

6. They should allow slight water percolation to relieve hydrostatic pressure so the sealant doesn't lose its bond even with water pushing behind it, and still have very low air permeability to help reduce the seepage of gases in the strata.

7. They should harden quickly even under damp conditions or at temperatures down to 40°F within 12 hours so they can't be damaged during normal mining operations.

8. They should be safe to apply without requiring <u>excessive</u> precautions. They should not require spray pressures so high that the sealant could cause serious injury if the spray hit someone or that it erodes fragile or deteriorated strata over which it is sprayed.

Cementitious sealants are normally applied in one of two methods.

One method is to mix the dry powdered sealant with water into a "slurry" just prior to spraying it. The slurry is then pumped under pressure through hoses and mixed with pressurized air as it comes out of the nozzle in a spray form. There are many types of "slurry" spray equipment available, but the principle is the same in all. Slurry applications in general have less spray "rebound loss" than most "dry process" spray applications, but they have other disadvantages. These are 1.The time needed to mix the slurry smooth and lump free before spraying. 2.The need for a greater amount of water used for proper pumping and spraying than is needed for hydration of the cement portion. This limits the thickness that the slurry can be applied to without shrinking, cracking, slumping, and even falling off. 3.Some coal shales are very sensitive to the water in slurries and begin to swell and fall within a few minutes of their application. 4.Pieces of roof falling into the mixer are a too common cause of plugging the nozzle. When this occurs, either the entire batch has to be dumped or you endure the mess and the time it takes to clear the plugs. 5.The worse drawback, however, with the slurry systems, and one that has happened to me more times than I care to recall, is the possibility of a power failure or the need to quickly evacuate the spray area due to any number of reasons with slurry in the mixer pump and line. If you're lucky you may lose only a couple of lengths of hose. If not, you can expect a substantial repair bill to get the equipment clear of hardened sealant and back in operation at a later date.

The other method, the "dry process" spray, also has its drawbacks, but they are far outnumbered by the advantages.

In this method you simply pour the dry powdered sealant into a hopper where it is then blown through the hose. Just before it comes out the hose, it mixes with a spray of water in a special nozzle. This is the only place where water is introduced in the system. One big advantage with this is that the sealant can be applied just damp enough for cement hydration but dry enough to allow very thick applications, especially around gaps in bulkheads and stoppings. We have several times closed gaps around stoppings with our sealant that were big enough for a man to squeeze through.

The biggest drawback with dry process spray is that some of the equipment used requires large amounts of air at 80 to 100 PSI. This is not readily available in most coal mines.

Rebound spray loss is a big problem with the dry

process sealants that are mostly sand and Portland cement; Portland cement, sand and aggregate; or Portland cement, sand, and steel fibers. Since these materials are heavy they require high air pressure to carry them through and out the hose. The edges of the spray hit the surface with such force they bounce off since they are not sticky. Rebound losses of roof applications when using some of these sealants can be as much as 25 to 30%. This is offset to some extent because they are usually the lowest cost per bag sealants. Because they are heavy, consideration for cost must also be given to their coverage per bag in square feet and thickness. When this and rebound losses are figured in, their cost may be considerably more than originally thought; however they are very strong.

Static electricity is also a problem with most Portland cement sand mixes, and the potential for shock especially during wet conditions is very high when spraying these with dry process spray equipment.

Magnesia cements are generally stronger than Portland cement and have superior bonding properties. These sealants can be made without sand to have equal or greater strength at the same thickness as sand containing Portland cement mixes. Since they don't contain sand, they are lighter in weight. This means that a bag goes further. Also, because they are lighter in weight, they don't require more than 50 to 60 PSI spray pressures, so they don't hit the surface with much more force than a slurry; and because Magnesia cements are quite sticky and contain fiberglass, they cling to the surface and have little rebound loss. Also, static electricity does not seem to be a problem with Magnesia cement sealants and unlike other dry process sealants, they can be sprayed with slurry-type equipment if dry process equipment is not available or there is insufficient air to use it.

Unlike Portland cement sealants in slurry form, Magnesia cement slurries can be applied at freezing temperatures, and although they will take longer than normal to harden they won't freeze. This is because the water in the slurry becomes a salt solution that has a lower freezing point than water.

Our Magnesia cement sealant has been applied many times during freezing conditions in coal and uranium mines in Utah and Colorado with no apparent ill effects.

Magnesia cement sealants can be made to set extremely fast without using caustic or corrosive additives that irritate the skin. This is desirable when sealing damp areas or attempting to divert water or for cold weather applications.

The stable form of Magnesia cement has molecularly bound water attached to each crystal. This results in a less porous, non-shrinking sealant that prevents the drying and shrinking of the shale, sandstone, coal or other strata over which it is applied. This is especially important to help reduce radon gas emanations in uranium mines, in areas such as lunch rooms, supply stores, or repair shops that cannot be properly ventilated.

Dry process spray equipment is very small and portable and gives high production rates of as much as one 50-pound bag per minute (60 bags per hour) with a 3 man crew. The Reed Guncrete Sova Model III does an excellent job with our Magnesia cement mine sealant. This unit is about the size of a 30-gallon drum, weighs slightly more than 400 pounds, is on wheels, and very portable. Again, its only drawback is that it requires a lot of air, not as much as the larger units used in spraying the heavier sand containing products, but still a minimum of 325 CFM.

To overcome this drawback, we expect to have in the very near future a spray unit approximately the same size as the Reed, that to operate will be quick coupled to the hydraulics of a mine vehicle. This will solve the problem of air or electricity as a power source in mines that can't easily supply either. This unit will also have the advantage that dry process spray equipment has over the slurry units--that in case of power failure or if it becomes necessary to evacuate the spray area, you simply drop the hose and leave, not worrying about hardened lines, plugged pumps, or solid mixers.

Our most exciting recent development in the underground application of mine sealants, and one that the mining industry has long desired, is the handling of the sealant in bulk and conveying it pneumatically into and throughout the mine in the same manner as rock dust, and applying it with the MSA pressure duster to the roof, ribs, stoppings, etc. The benefits of this are obvious; these are no capital expenditure of new spray equipment to apply the sealant; no handling of full bags and accompanying breakage; no problem with having flammable empty paper bags in the mine, and the need of their prompt removal to the surface; putting an expensive piece of equipment, the pressure duster, to an alternate use, getting more out of it and its cost; etc. We have sprayed our sealant through 200 feet of hose with such ease with the pressure duster, that if necessary we feel we can add another 100 feet without problems.

In summary, sealants are being used extensively on stoppings and bulkheads of all types to make them airtight and reduce ventilation losses. They are used to reduce and control air slacking. They are stabilizing and strengthening the surface of deteriorated and caved areas. They are reducing the frequency of rock dusting. They help reduce fire hazards when applied greater than 1/2" thick, help reduce methane gas seepage and radon gas emanations, increase visibility, and more.

We must also recognize what sealants cannot do. Sealants do not hold the roof up or keep it from falling. They do not prevent rib failure, or structurally strengthen the tunnel in any way, although a combination of wire mesh and strong sealant will reinforce the surface. They do not stop running water, but can help divert it away from critical areas to keep them drier.

Regardless of what class sealants are used or what method is used to apply them, when applied over properly prepared and scaled tunnel surfaces, mine sealants help make underground mining safer, more comfortable, and less costly.

3
Computer Applications and Modeling

Chapter 11

COMPUTERIZED CONTROL AND WARNING SYSTEM FOR URANIUM MINES

C. T. Sheeran

Mining Engineer
U.S. Department of Interior
Bureau of Mines, Spokane Research Center
Spokane, Washington

and

J. C. Franklin

Supervisory Physical Scientist
U.S. Department of Interior
Bureau of Mines, Spokane Research Center
Spokane, Washington

ABSTRACT

A commercially available microprocessor-based system capable of monitoring 512 channels has been interfaced with monitors for radon, working level, air velocity, and fan power. The basic system utilizes both Z80 and 8080 microprocessors in a desktop central processing unit (CPU). CPU hardware includes keyboard, video display (CRT), and disk drive; a printer is used to keep permanent data records. Signals from all channels are transmitted to the computer in digital form where they are processed for alarm status.

Software developed for the system allows for audiovisual alarms in the event of low and high readings, rate change, change of state, or communication failure. Up to six channels can be continuously displayed on the CRT for current readings. Shift reports and trend logs may be generated to help the ventilation engineer determine average working levels and ventilation effectiveness. Additional software permits the operator to program command sequences which may be used to automatically restart fans after a power outage.

INTRODUCTION

In protecting the health and safety of underground personnel, the uranium mining industry must contend with the radioactive gas radon and its daughter products. Radon and radon daughters are decay products of uranium and although found in nearly all types of mines, their extreme concentrations in uranium mines pose more of a health hazard.

The radon daughters ^{218}Po and ^{214}Po emit alpha particle radiation which may induce certain forms of cancer. Exposure to these daughters in mine air has produced a high incidence of lung cancer among uranium miners. Present federal safety standards state that the maximum allowable concentration of radon daughters in a work environment is one working level (WL).[3] Total exposure per person is limited to 4 working level months (WLM)[4] for any calendar year.

[3] One working level is defined as any combination of short-lived daughters in 1 liter of air that will result in the ultimate emission of 1.3×10^5 MeV of alpha energy. This includes the radon daughters from polonium-218 through polonium-214.

[4] The working level month is defined as the working level exposure, times the hours exposed, divided by 173.

Most of the uranium mined in the United States is in sandstone formations which have a high porosity/ permeability compared to other rocks and may be highly fractured. Because of this, radon is released from the sandstone very readily and its emanation rate can be profoundly affected by barometric pressure changes.

Ventilation techniques such as blowing, exhausting, and push-pull are the primary controls used by the industry to reduce the radon daughter concentrations. As uranium mines have become larger and deeper, ventilation costs have increased sharply--ventilation now accounts for a major portion of production expenditures. Bates (1981) estimated that in order to comply with the 4 WLM standard, these costs ranged from \$4.68 to \$5.41 per ton of ore removed in 1980. This estimate included costs for ventilation shafts, fans, vent bags, and other associated equipment.

An interruption in mine power causing fan shutdowns can cause drastic changes in the radon/radon daughter concentrations. Sometimes shutdowns may go undetected for some time before fans can be restarted. Franklin (1978) and Musulin (a, b) have shown that fan shutdowns from 5- to 30-minutes duration can increase radon concentration from three to five times normal. Franklin (1981) reports that the mining activities of slushing and blasting may increase this concentration up to 40 and 400 percent, respectively.

The uranium industry presently uses batch sampling to determine personnel exposure and to detect areas where ventilation changes are needed to maintain desirable radon daughter levels. These samples are pulled at various intervals and are usually very accurate point-in-time measurements. However, since concentrations are continually changing, other methods of monitoring are needed to minimize miner exposure.

The Bureau of Mines has been developing instrumentation that can help in detecting excessive working levels. Droullard (1977) developed a continuous working level monitor that has been used for several years in research activities in both experimental and active mines. Shaw and Franklin (1982) took this continuous working level monitor and interfaced it to a microcomputer to develop an alarm system which has been successfully field tested. It was concluded from this field test that an alarm system with expanded capabilities would be more useful to the ventilation engineer. Therefore, work began on the new system described in this paper.

SYSTEM DESCRIPTION

The system investigated is a modified Senturion-200 manufactured by Conspec Controls, Ltd.[5]

[5]Use of brand names are for identification purposes only and do not imply endorsement by the U.S. Bureau of Mines.

Modifications were performed by Conspec in both hardware and software according to Bureau specifications, and the Bureau has made further hardware modifications to outfit the system for mine usage.

Hardware

The Senturion-200 is advertised as a "microcomputer-based real-time data acquisition, monitoring, and control system." Its basic hardware consists of a central processor, data/event printer, communication trunk, and accessors as illustrated in Fig. 1.

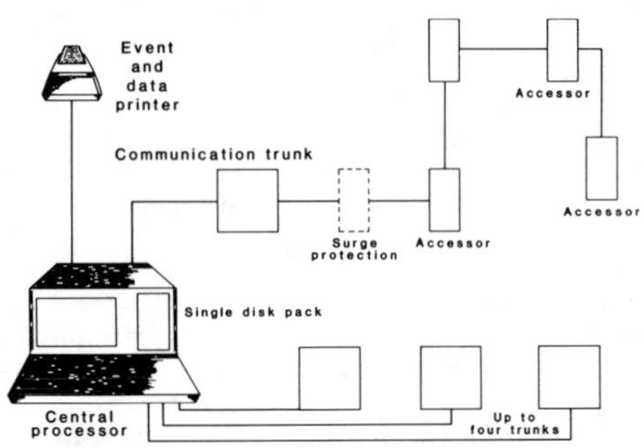

FIGURE 1. Monitoring system.

Central Processor. The central processor is a desktop unit which contains the main processor, video terminal, keyboard, disk drive, and accessor trunk drivers. The main processor is a Zilog-based microcomputer system which utilizes a Z80 central processing unit with both parallel and serial input/output ports. Memory consists of 64K of dynamic RAM (Random Access Memory) with an additional 7K of EPROM (Erasable Programmable Read-Only Memory). A floppy disk controller is used to interface to the system disk. Other components include DMA (Direct Memory Access) logic, an onboard PROM (Programmable Read-Only Memory) programmer, and a 4-channel counter-timer.

An Intel 8080 microprocessing unit is used to control the video terminal (CRT) and keyboard. The alphanumeric keyboard is laid out similar to an office typewriter with additional special function keys used to simplify operator commands when in the main program.

The system disk utilizes 8-inch single-sided floppies and provides 128K words of additional memory. Disks are used as non-volatile storage for programs and data.

Data/Event Printer. A 43 Teleprinter is used with the system for character-at-a-time receive-only operation. The system is programmed for automatically and manually requested printouts. Automatic printouts are obtained for all alarms and events with their corresponding time of occurrence. The printer can also be requested from the keyboard to print out point descriptions, shift reports, trend logs, and other system configurations and parameters.

Communication Trunk. The communication trunk is a line driver through which the system communicates with the accessors. The trunk is connected to the central processor via an RS-232 interface and to the accessors via a single 4-wire cable. The trunk uses two of these wires to provide 24v DC power to the accessors; the remaining two wires are used to transmit and receive digital signals. Trunk-to-accessor communication is carried on asynchronously at a 4800 baud rate.

Accessors. Accessors provide the interface between the system and the field devices to be monitored or controlled. Each accessor is actually a point-multiplexer with its own unique address. Addresses are selectable from 0 to 127 on dipswitches within the accessors.

Accessors are classified as either analog or binary. Analog accessors digitize a reading from a sensor, and binary accessors control or report status from contacts. The Spokane Research Center purchased both types of accessors for field testing in uranium mining situations. Analog accessors tested include the pulse-integrating and process types. Binary accessors tested include the supervised contact and two-state commandable types.

The pulse-integrating-type accessor accumulates pulses for a preset time period and has been interfaced to both radon and continuous working level monitors. Process-type accessors digitize analog output from a field device and have been tested with anemometers and temperature/humidity sensors. Supervised contact-type accessors sense contact position as well as condition of the line, and have been used to monitor power to detectors or the running status of a motor. Two-state commandable-type accessors control motors, lights, etc., and also contain a proof circuit which may be used to verify command execution.

Software

Senturion-200 software is made up of two parts: the operating system and application programs. The operating system used is CP/M (Control Program for Microprocessors). Applications programs, written for the Bureau by Conspec, accomplish data acquisition, monitoring, and control of the various points.

Operating System. CP/M consists of programs which execute user commands and access hardware resources such as the CRT and printer. These programs are stored on disk and loaded into RAM with a bootstrap loader stored in an EPROM. Once the operating system has been loaded, it responds to user commands to perform CP/M activities or to enter into the application programs.

Application Programs. The application software includes the scan, alarm processing, and sequencing programs. Together, these programs make up the main program referred to later in this paper.

SCAN. Once in the main program, the processor continuously sends messages to the accessors in a predetermined sequence. Messages consist of an address and a command, to which the appropriate accessor responds by returning its status. Status consists of digitized information which may represent an analog reading or the position of contacts as shown in Fig. 2.

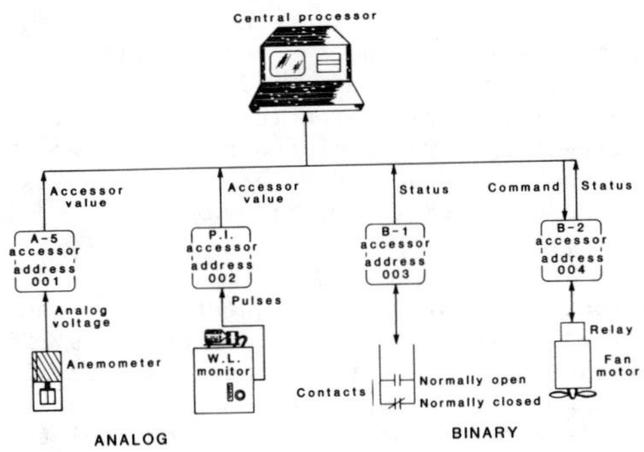

FIGURE 2. Communication between processor and accessors.

After requesting information from an accessor, the processor must wait several milliseconds for data to be returned. During this time it performs the remainder of its tasks, such as updating the CRT or performing calculations. If no data are returned from an accessor, as would be the case with broken communication lines, a communication-fail alarm would be generated. The processor would then continue with the scan starting with the next address. A flow diagram that summarizes the scan procedure is shown in Fig. 3. The scan only checks those accessors which the operator has enabled through the keyboard. Scan time per accessor is on the order of 10 milliseconds; total scan time depends upon the number of enabled accessors on-line.

ALARM PROCESSING. Once data is returned from an accessor, it is processed for alarms by the computer. For binary accessors, the new data are simply checked against the old data. If there is no change, the new data are ignored. In the event of a change, the processor decides whether to alarm or return-to-normal, and updates the printout and display accordingly.

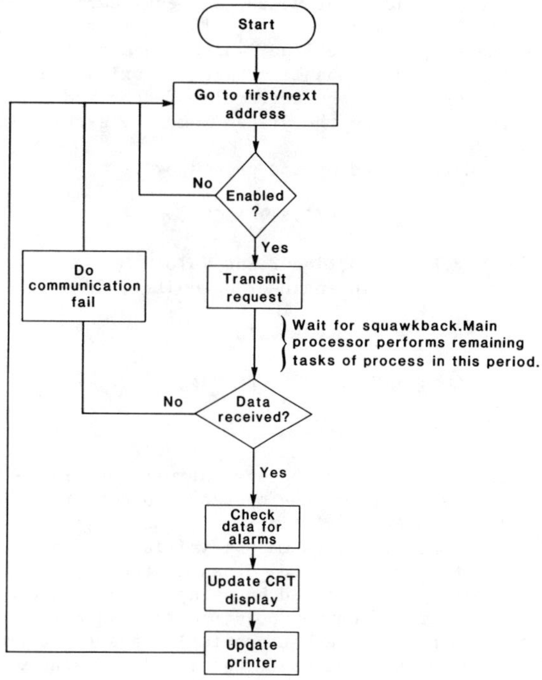

FIGURE 3. Basic scan procedure.

For analog accessors, processing is more complex due to the different types of alarm situations. Three different analog alarms, low, high, or rate, may occur with possible time delays. The processor checks the value received from the accessor to determine the status of alarms, and updates the "delay" to alarm counter accordingly.

The processor also performs a mathematical operation on input from analog accessors prior to printout or display on CRT. A conversion to chosen engineering units is made through the use of a correction factor and offset. This conversion assumes there is a linear relationship between analog device output and engineering units.

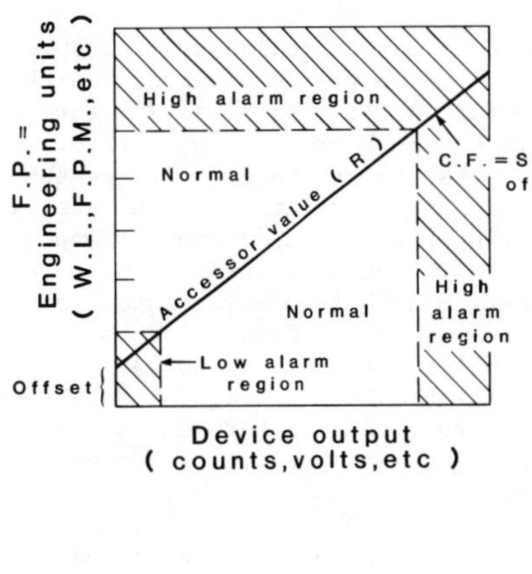

FIGURE 4. Representation of math performed by processor including decision regions.

Fig. 4 summarizes the relationship between the device output, accessor value, and engineering units, with normal and alarm regions. The correction represents the slope of the line, while the offset represents the intersection of the line on the Y axis. The operator determines the correction factor and offset and enters them into the system memory via the keyboard.

The equation used by the processor is

$$F.P. = R(C.F.) + OFS,$$

where

F.P. = Floating Point Value (in engineering units),

R = Accessor Value,

C.F. = Correction Factor,

and OFS = Offset. (1)

SEQUENCING. The sequencing software available with the system allows the operator to construct control sequences consisting of a sequence of controlled events. These sequences may be initiated either manually (through keyboard) or automatically. Auto-initiation is accomplished by using a keyboard command to set a "trigger" on any point in the system. The proper sequence is then automatically started after the trigger point changes to the specified state. Triggers can be set for alarm conditions (low, high, on, off, and rate) or return-to-normal conditions. Up to eight sequences may be in operation at the same time.

SYSTEM USE

After installation of underground and surface hardware, the system is ready for startup. Operating procedure consists of booting (loading from disk into RAM) the main program and entering the data base and access code. From this point on, the system automatically scans for alarms and monitors the status of all enabled points.

Startup is accomplished by powering-up the system and disk drive, inserting the floppy disk, and typing two keys (space bar, F) on the keyboard. This boots the CP/M into memory. A command typed in at this point will cause the application programs to be loaded from the disk. All operator commands from this point on consist of three-letter mnemonics.

The system data base contains all the specific information for accessors and sequences which is needed for operation. The data base is entered into the memory via the keyboard and consists of point numbers, addresses, descriptors, alarm status, analog parameters, and control sequences. The operator is aided in entering these data by prompts appearing on the CRT screen. The data entered may be saved on disk and read from disk into memory upon later startups. Saving the status on disk prevents it from being accidentally lost. Printouts may be obtained of the status for use in checking for data entry errors. Parameters or descriptors may be changed at any time by using commands to correct errors or to update information.

After data entry, the operator must enable each accessor through use of a special command. This causes each accessor to be entered into the scan and establishes communication between the accessor and the processor. This command may also be used to delete a point from the scan for maintenance or other purposes.

The access code is a feature which protects system status from change by unauthorized personnel. After starting the main program and reading in the status, no further input is accepted until the access code is entered. Two codes may be keyed in which correspond to different levels of access: operator or supervisor.

The operator code allows access to maintain system operation and permits CRT display manipulation, enabling/disabling points, commanding control points, temporarily changing analog parameters, and requesting certain print routines. The supervisor code allows a higher level of access which includes all operator capabilities. This code also permits entrance of the data base and of necessary information for automatic operation of shift reports and sequences.

The operator/supervisor may manipulate the system to change CRT display, command controlled points, alter status files, and request print routines. All incoming alarms and change-of-states are automatically displayed on the CRT and printed. A horn alerts the operator of new additions which may then be acknowledged and deleted. Up to six points may be selected for continuous status monitoring.

Print routines that can be requested from the keyboard include shift reports and trend logs. Both of these, once initiated, run continuously until they are stopped by the operator. Shift reports consist of averaged readings for each channel that are automatically printed up to five different times per 24-hour period. The trend log routine prints present readings for up to ten channels per time interval variable from 1 to 99 minutes.

URANIUM MINING APPLICATIONS

The computerized system investigated offers many advantages to the uranium mining industry, particularly in the use of its monitoring and control capabilities. The system requires only a 4-wire cable between the main processor and the accessors to provide 24v DC power and communication. This cable may be wired in parallel, series, or branched to allow for accessor installation over distances of up to 10,000 feet from the CPU. Up to 128 accessors may be used per line, and 4 lines can be supported by the system hardware/software for a total capacity of 512 accessors. These accessors can be interfaced to fans, pumps, radiation detectors, air doors, anemometers, etc.

Monitoring Capabilities

The system's monitoring capabilities may be used to keep records, initiate command sequences, and to assist in troubleshooting system hardware.

Records. Recordkeeping is automatic for all changes-of-state or alarm occurrences. The operator is notified of these occurrences at the console so that remedial action may be taken if necessary. This can be applied to continuous radiation monitoring as well as to keep track of air door positions or the operational status of fans and other motors. In continuous radiation monitoring, a rate alarm may be set to warn of rapidly changing conditions before critical levels are reached. Shift report printouts may be used in figuring the average working level exposure in monitored areas. Trend logs may be used to study radiation variation in connection with mining activity, barometric changes, or fan shutdowns. These studies may be used to improve ventilation effectiveness and thereby reduce ventilation costs.

Sequences. Command sequences can be constructed for either manual or automatic operation. This feature may be used to turn on, or off, a number of fans from the console in a predetermined sequence with time delays between steps. The status monitoring then gives feedback to inform the operator of command execution and if indeed the fans went to the commanded state. Anemometers interfaced to analog accessors could be placed in the drifts as further verification that the fans were on/off. By using an accessor to monitor mine power, the above fan sequence may be set to automatically start up fans after a power outage. In this case, the power monitor would serve as the sequence trigger under return-to-normal conditions. Other sequence applications include blowing warning horns prior to fan startup, turning on underground alarms for high radiation readings, or automatic startup of secondary pumps triggered by a water level monitor.

Troubleshooting. Troubleshooting system hardware (cables, accessors, and field devices) can be aided by using the system's monitoring capabilities. Certain types of alarms may be used to diagnose hardware problems and their locations. Basic hardware problems that will be detected by the computer include cut computer cable (to accessors), broken signal cable between accessor and field device, broken power cord to field device, and electronic malfunction in field device.

CONCLUSION

With the continually changing radon daughter concentrations in all underground uranium mines, minimizing worker exposure is difficult. The system described in this paper has the capability to continuously monitor critical situations such as fan operation and radon daughter concentrations and alert the ventilation engineer when excessive measurements are being obtained. It also has the capability to control ventilation, sound underground warnings, and automatically restart fans after power failures. Recordkeeping features of this system will help the ventilation engineer to control radiation hazards and to predict where future problems will occur.

REFERENCES

Bates, R. C., 1981, "Ventilation Cost Impact of Reduced Radon Daughter Working Levels," Proceedings, Radiation Hazards in Mining: Control Measurement, and Medical Aspects, Oct. 4-9, pp. 1066-1070.

Droullard, R. F. and R. H. Holub, 1977, "Continuous Working Level Measurement Using Alpha or Beta Detectors," Bureau of Mines RI8237, 14 pp.

Franklin, J. C., et al, 1978, "A Continuous Radon Survey in an Active Uranium Mine," Mining Engineer, Vol. 30, No. 6, June, pp. 647-649.

Franklin, J. C., 1981, "Control of Radiation Hazards in Underground Uranium Mines," Proceedings, Radiation Hazards in Mining: Control, Measurement, and Medical Aspects, Oct. 4-9, pp. 441-446.

Musulin, C. S., J. C. Franklin, and F. A. Roberts, a, "Effects of a Fan Shutdown on Radon Concentration in a Positive Pressure Ventilation Mine," Bureau of Mines RI to be released soon.

Musulin, C. S., J. C. Franklin, and V. W. Thomas, b, "Effects of Diurnal and Fan Shutdowns on Radon Concentration in an Experimental Uranium Mine," Bureau of Mines RI to be released soon.

Shaw, D. M. and J. C. Franklin, 1982, "Continuous Area Monitoring and Alarm System," Engineering and Mining Journal, accepted and scheduled for publication in May.

Chapter 12

SUNNYSIDE NO. 3 — A CASE STUDY IN VENTILATION PLANNING

Malcolm J. McPherson and Michael Hood

Department of Materials Science and Mineral Engineering
University of California, Berkeley

INTRODUCTION

Sunnyside Mines, owned and operated by the Kaiser Steel Corporation, are situated near the city of Price, Utah. The complex comprises three adjacent mines, named simply Nos. 1, 2 and 3, all connected underground. Two seams, the upper and lower Sunnyside have been worked. These dip at about 10 percent to the north-east. The surface cover is variable due to the mountainous nature of the topography. The Sunnyside upper seam varies from 5 1/2 ft (1.7m) to 9 ft (2.7m) in thickness whilst the lower seam remains at about 6ft (1.8m). The separation between the two seams has ranged from 7 to 45 ft over the mined area (2 to 14m). Longwall mining has been practiced at Sunnyside for over 20 years due to difficulties of roof control encountered when using the room and pillar system.

Number 3 mine is bounded on the north and south sides by mines Number 1 and 2 respectively. Whilst current production is concentrated into Number 1 mine, much of the future of the complex lies in the further development of deeper reserves in Number 3 mine. Workings in this latter mine were curtailed in 1978 due to difficulties in ventilation. Present developments are ventilated partially from the neighboring Number 2 mine where no workings are in progress.

The layout of Number 3 mine is illustrated on the schematic Figure 1. Trunk airways extend down dip from the surface at No. 2 Canyon and the Water Canyon for a distance of some 9,600 ft. (2930m). The area between the two sets of trunk airways has been worked extensively in both seams as have the corresponding reserves on either side in the connected adjacent mines. At the present time exhausting fans are sited at the top of a shallow shaft in No. 2 Canyon and an 8 ft (2.4m) diameter shaft sunk to a depth of 1013 ft (310m) closer to the current developments (Figure 1). The current airflow system, even with an additional 116,000 cfm ($55m^3/s$) entering from No. 2 Mine, is adequate only for the development work now in progress but will be unable to support new longwall faces further downdip.

The basic ventilation problem of this mine may be stated quite simply. In a situation where all intake and return airways pass through extensive old workings, a ventilation system design was required that would be effective, efficient and economic for the foreseeable future of the mine.

ORGANIZATION OF THE PLANNING PROCEDURE

The procedure followed during the study is illustrated on Figure 2. Initial ventilation surveys established the current state of the airflow system and provided the necessary data for setting up a Basic Network File in a computer store. The data in this file was a mathematical model of the ventilation system of the mine. The basic network was analysed by a ventilation network analysis program in order to correlate the measured and computed airflows and to establish the basic network as a true representation of the mine as it stood at the time of the surveys. The network model could then be extended to simulate the future development of the mine and alternative ventilation designs investigated.

The remaining sections of the paper outline the work involved in each of these main phases of the planning procedure.

VENTILATION SURVEYS

Conduct of Surveys

Two types of measurements were conducted simultaneously throughout the air-carrying routes of the mine:

(i) Airflow measurements were made by anemometer traverse or smoke tube at 221 selected stations. Anemometer traverses were repeated at each station until at least three gave results to within 5 per cent.

(ii) Pressure drop measurements were made across ventilation doors, regulators and, wherever possible, across stoppings. Additionally, frictional pressure drops were measured along airways where such pressure drops were significant (above 0.01 inches of water gauge or 2.5 Pa over a 100m distance). The trailing hose method was used to determine these frictional pressure drops. This involved laying out 100m of abrasive resistant plastic tubing (3 mm internal diameter) with a 4 ft. pitot-static tube facing into the airflow at either end and a low range pressure gauge connected into the line. The trailing hose method was preferred to the alternative barometer technique for this study because of (a) the relative ease of access between measuring points and (b) the greater accuracy within individual airways.

The anemometers used were Davis Biram Type A/2-3" (30 to 5,000 ft/min) and Airflow Developments AM-5000 digital (50 to 5,000 ft/min). The pressure gauges employed were Dwyer magnehelic instruments. These were preferred to liquid in glass manometers because of their portability and dependability under adverse mining conditions. A checklist of the equipment used in the survey is given in Appendix 1.

The instruments were calibrated before and after the surveys in the mine ventilation laboratory at the University of California, Berkeley. The survey occupied two teams, each of three men, for ten working days. The work consisted

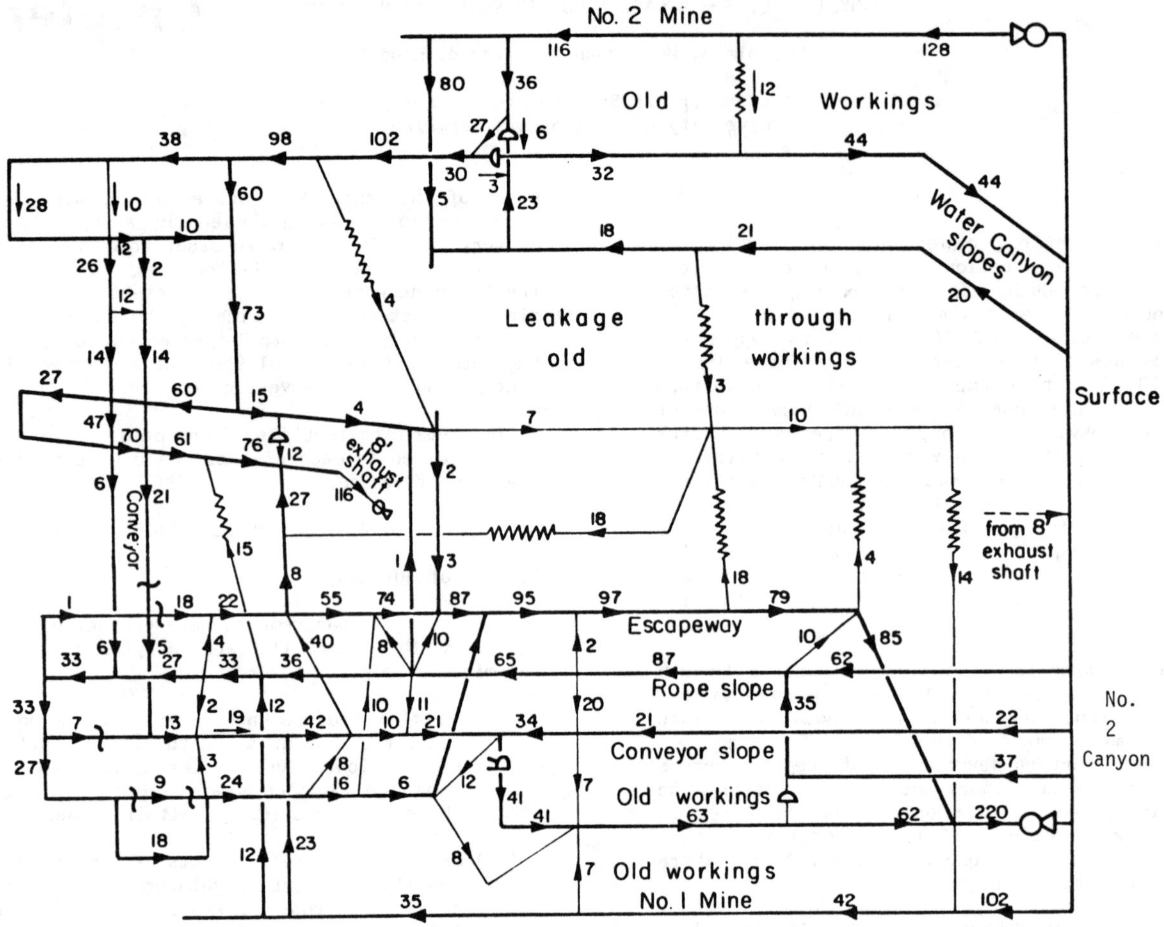

FIGURE 1. Network representation of Sunnyside No. 3 mine current ventilation system. Airflows are given in thousands of cubic feet per minute.

essentially of traversing around pre-determined routes measuring airflows and pressure differentials, and making copious notes on the geometry and type of lining of the underground excavations. A detailed and voluminous log of the survey was amassed.

Each evening the data were transcribed on to mine plans and checks made on the inherent compatability of the measurements. Such checks consisted essentially of (i) ensuring that the algebraic sum of airflows at each main junction was zero, and (ii) verifying that the algebraic sums of measured pressure drops around closed loops were each also zero, within observational accuracy. Any uncertainties were re-measured the following morning. During the evening discussions, the team leaders agreed the routes to be followed by the two groups on the following day. Sufficient overlap was allowed between the traverses to provide cross-checks and continuity.

Survey Results

The first step in the analysis of ventilation survey results was to draw up a network representation of the mine layout. This is shown on Figure 1. The network is not a scale plan of the mine but merely indicates the manner in which the main air-carrying routes are interconnected. Complete worked out areas are represented by leakage paths to which are allocated "equivalent resistances", calculated or estimated from the measured data. Airways and leakage paths that are effectively in parallel are represented as single lines and the airflows and resistances each compounded accordingly.

Surface connecting shafts (excepting an 8 ft diameter exhaust shaft) and slopes are shown joined to a single manifold representing the free surface atmosphere across the end of the network. The adjoining No. 2 and No. 1 mines are shown as manifolds on the top and bottom boundaries respectively.

Without wishing to pre-empt the detailed analyses which were to be conducted on the data, there were some observations that could be made immediately and without computer simulations. Of 380 thousand ft^3/min (kcfm) or $180m^3/s$ airflow passing through No. 3 mine, only 73 kcfm ($34m^3/s$) reached working areas. This gives a volumetric efficiency of 73/380 or 19 per cent, a very low value.

With a few exceptions, the quality of stoppings

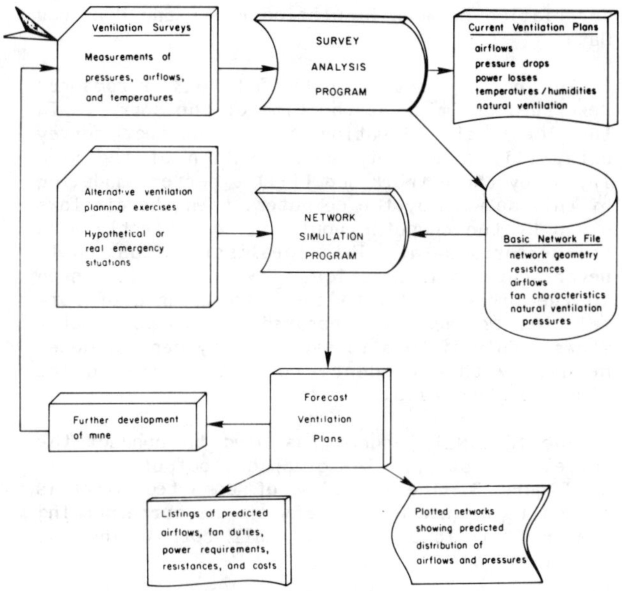

FIGURE 2. Airflow planning procedures for an established mine.

and ventilation doors is good at Sunnyside. Excessive leakage due to inadequate seals is not the problem. The difficulty lies rather with the poor utilization of existing airways and the proximity of main intakes and returns. In particular:

(i) The intakes down the rope slope and adjoining parallel airways are bounded on either side by returns or old workings connected to return airways.

(ii) The return air from the working areas is concentrated into two return routes. One of these, shown on the network as the Escapeway, carries an airflow varying between 79 kcfm and 97 kcfm (37 to $46 m^3/s$) along its length. This airway has deteriorated over the years to an extent that of the main exhaust fan pressure of over 4 inches of water gauge (1 kPa), more than 2 inches (0.5 kPa) is consumed in this one airway alone. The second main return for No. 3 mine leads to an 8 ft diameter shaft and is also in a poor state throughout its length.

(iii) In contrast to the heavily overloaded, poor quality and high resistance airways into which the return air is currently concentrated, there are three airways (combined into two lines on the network) emerging to the surface at Water Canyon. These airways are of large cross-section and hence of low resistance along most of their length but are not being utilized at all in the ventilation of the mine.

(iv) Brattice cloths are installed in headings to course the ventilating airstreams. This is common practice in U.S. coal mines. The introduction of such restrictions in the working areas maximises pressure differentials and hence leakage between intake and return airways.

The situation in No. 3 mine would appear to be typical of many coal mines. Long term ventilation planning had been neglected in favor of short term ameliorative remedies. This had resulted in an inefficient and increasingly ineffective system that now had to be completely revised if the mine was to achieve the planned level of production.

METHOD OF ANALYSIS

Choice of simulation method.

Of the several simulation programs available for the network analysis of mine ventilation two were chosen to analyze the Sunnyside ventilation system. The programs VNET and MINIVNET were selected for the following reasons:

(i) Some 500 network exercises were to be carried out during the study. Hence the program had to be efficient in computer time and storage, and capable of multi-exercise runs.

(ii) For the purposes of the case study there was no requirement for an investigation of the distribution of smoke or gases throughout the network. However, operating costs were required.

(iii) The basic VNET program has been in routine practical use for ventilation planning since 1965 and the investigators were confident of its accuracy and reliability.

(iv) The graphical input and output facilities of MINIVNET made it particularly suitable for investigating variations at each of the major planning phases.

The method of using both programs is quite straightforward. The mine is drawn up as a network of airways and a identifier number allocated to each junction. The geometry of the existing mine layout is fed into the computer as a series of branches given in terms of FROM and TO junctions together with the corresponding resistances and fan pressures. These data are used by the computer to form a mathematical model of the mine and produce an airflow distribution that can be checked against the measured airflows. Provided that a satisfactory correlation is achieved the model can be modified to simulate the future development of the mine and to investigate alternative means of ventilating the workings.

Data Preparation

The basic network of Sunnyside No. 3 consisted of 145 branches. A junction number was allocated to each junction.

The airflow through each branch of a mine ventilation network is governed by the approximation.

$$p = RQ^2 \text{ (Square Law)}$$

where p = frictional pressure drop

Q = airflow
and R = airway resistance

The resistance of every branch in the basic network was required. These were determined by four methods.

(i) For most of the airways, measured p and Q values were available from the surveys. Then

$$R = \frac{p}{Q^2}$$

(ii) For some branches, either the airflow, the pressure drop or both were unavailable due to: (a) values too low for meaningful measurement, or (b) areas inaccessible. Whenever possible, missing airflows were determined by difference from those airflows that could be measured. This was the only practicable way of estimating leakage that occurred through large areas of old workings. Again, wherever possible, missing pressure drops were found by summing the measured pressure drops around closed loops and utilizing the check measurements of pressure differentials made across doors.

(iii) For airways that were accessible for inspection but for which the airflow and/or pressure drop could not be measured or deduced, the resistance was calculated from

$$R = \frac{k \times \text{length} \times \text{perimeter}}{(\text{cross-sectional area})^3}$$

where k = friction factor dependent upon the roughness of the airway lining.

The k values employed were correlated against airways where reliable p and Q measurements could be made.

(iv) There remained a few leakage paths through abandoned workings where only the airflow could be deduced. The programs used allowed these paths to be entered as 'fixed quantity branches'. The computer then calculated the resistance required to give each of those fixed airflows.

The remaining data needed to initiate the analyses were the ventilating pressures. Fan characteristics curves for the mine fans were not immediately available. However, measurements were made of all the fan pressures relevant to the No. 3 mine and these were used in the establishment of the basic network on the simulation models. The fan pressures are shown (in milli-inches of water gauge) on Figure 3.

The airflows in the Water Canyon slopes were promoted by natural ventilating effects only. Hence the pressure drops summed around these slopes gave a direct measure of natural ventilating pressure. This came to 0.19 inches of water gauge (47Pa). As the temperatures in the return airways were stable and similar this value was used as a basis for allocating natural ventilating pressures in the other return slopes.

Correlation between Simulation Model and Measured Data

The primary set of data fed to the computer described the mine at the time of the survey. In the idealistic situation of (i) perfect survey data, (ii) a perfect representation of the mine layout by the network and (iii) a perfect modeling of this network by the computer, then the airflows given by the computer would agree perfectly with the measured data. This idealistic situation is never achieved in practice. However, a vital step in the procedure is to check the degree of correlation between the measured and computed airflows. Only if this is satisfactory can the model be used with confidence to investigate future planning strategies.

The MINIVNET program was used to conduct the correlation study. The graphical output is shown on Figure 3. The majority of computed airflows agreed to within 1 or 2 kcfm of the corresponding measured airflows. The overall correlation was calculated as

$$\frac{\text{Sum of actual errors}}{\text{Sum of measured airflows}}$$

taking into account all branches. This gave a value of 3.66 per cent. The maximum value normally acceptable in order to proceed with planning exercises is 10 percent. Hence, the Sunnyside correlation was good and the 'basic network' could be approved as a reliable basis for future planning.

DESIGN OF VENTILATION SYSTEM FOR FUTURE WORKINGS

Airflow Requirements and Projected Resistances

The future plans for No. 3 mine involve workings continuing to the north-east, that is, down slope from the current 16th level (See Figure 4). It is the intention that at any one time there should be one longwall face and two continuous miner districts in production. Experience at Sunnyside has shown that in order to maintain methane concentrations at an acceptable level, the outbye ends of continuous miner districts should be supplied with some 45 kcfm (21 m^3/s) of air. For the purposes of gas dispersal and dilution the air velocity on a longwall face at Sunnyside should be about 600 ft/min (3m/s). The face air velocity should not exceed 800 ft/min (4m/s) in order to avoid excessive dust pick-up. At an average face cross-section of 55 sq ft, the given airflow is

$$600 \times 55 = 33{,}000 \text{ cfm } (16\text{m}^3/\text{s})$$

The planning exercises aimed to provide each longwall face with an airflow in the range 30 through 35 kcfm (14 through 17 m^3/s).

The resistances of all airways in the basic network were determined from the survey data. For subsequent planning exercises these resistances were increased, where appropriate, to allow for roadway deterioration. The resistances of new airways were calculated on the basis of

(i) length from development plans

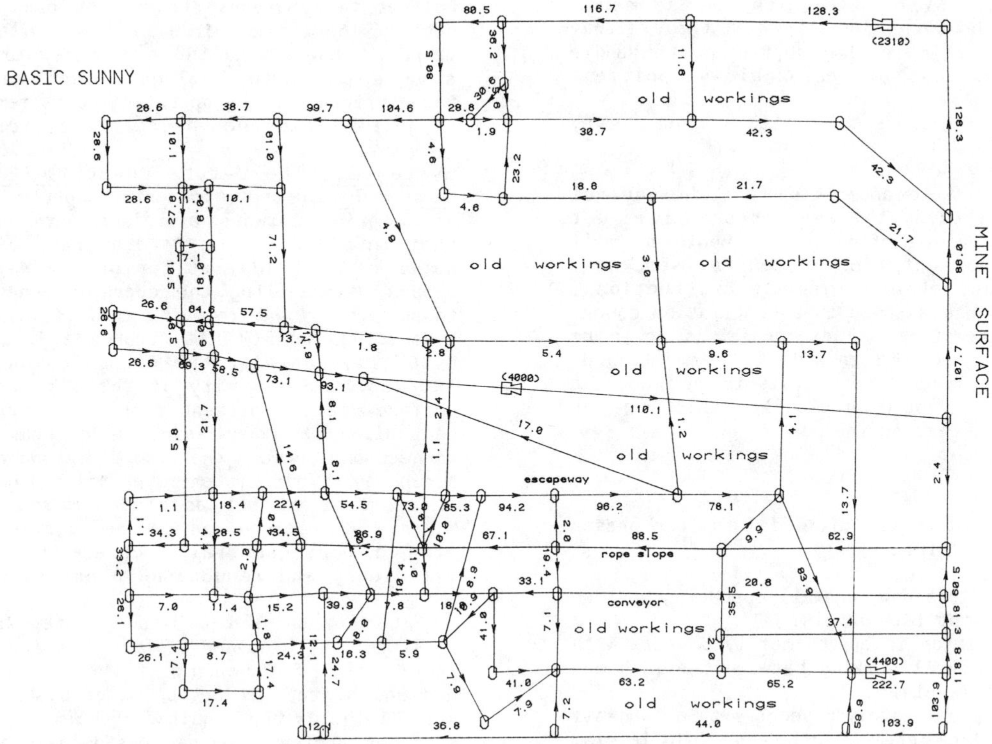

FIGURE 3. MINIVNET autoplot giving computed airflows for basic network.

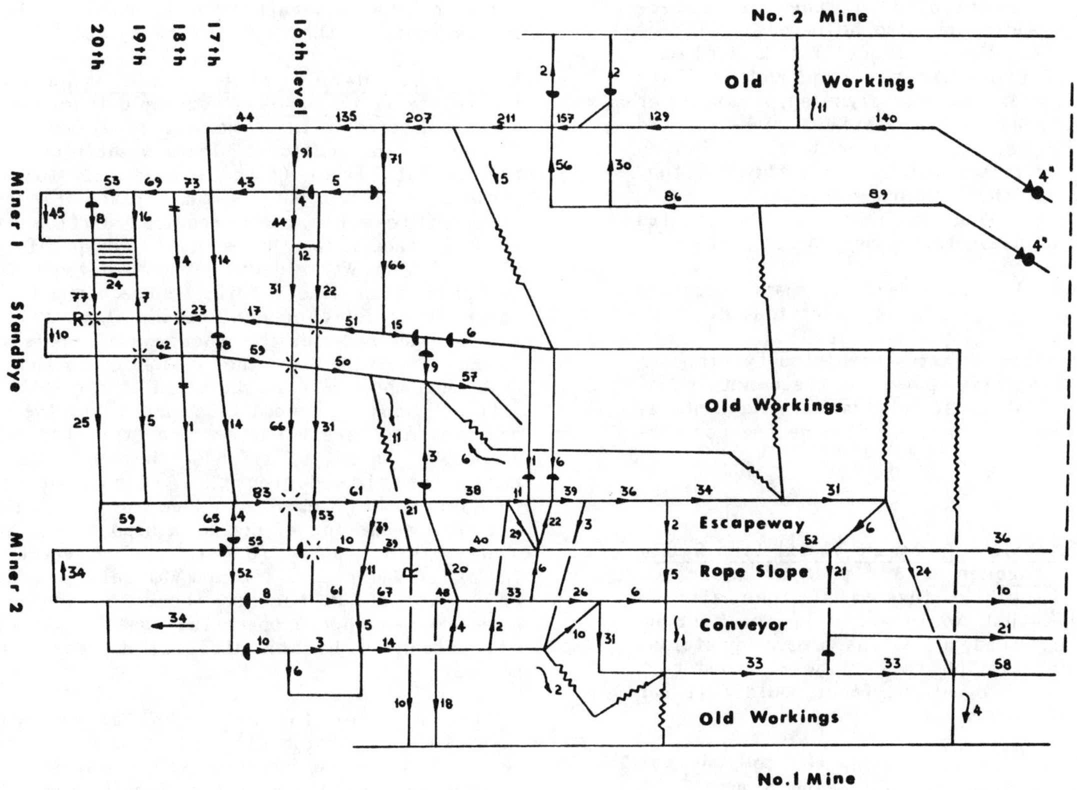

FIGURE 4. System B. U-tube with 4" w.g. forcing fans in Water Canyon. Op. Cost = $54,040/year. Vol. Eff. = 49 percent.

(ii) airway sizes from mine management. (Throughout the investigation these airway sizes were more than sufficient to handle the airflows at reasonable velocities)

(iii) friction factors correlated against survey data

Whilst due allowance was made for normal closure of airways, the exercises carried out assumed that main airflow routes would be well maintained and kept free of debris, stock materials or any other permanent obstructions. Similarly, it was assumed that ventilation doors, regulators, stoppings and air-crossings would be constructed to a good standard and be maintained, where relevant, thereafter. This is an important point since, as noted previously, this has not always been the case in the past.

Planning Phases

The investigation was divided into four phases with respect to future workings.

Phase 1: The first new longwall operating between 15th and 16th levels.

Phase 2: Situation in approximately 3 years with a longwall working between the 16th and 17th levels.

Phase 3: At approximately 8 years with a longwall working between 19th and 20th levels.

Phase 4: At approximately 16 years when the workings will have reached the 24th level.

It was important to avoid any large expense to facilitate the ventilation of the near future workings that would not also be necessary in the longer term. Any major changes of this kind must be capable of providing the required airflows throughout the sixteen year planning period. For this reason, Phase 3 (8 years) was investigated first, and in detail, in order to establish an economic and efficient ventilation system for that time. The other three phases were then studied in the light of the Phase 3 findings. The results are outlined below in that same order.

Within each planning phase, alternative designs of ventilation systems were investigated. The Exercise numbers given in the text refer to individual airflow patterns predicted by computer simulation. Exercise 1 is the basic network of the mine as it stood at the time of the ventilation survey and that was verified by the correlation study.

Phase 3: At 8 Years

System A. Extension of Present Ventilation System (Exercises 2 through 6) A number of exercises were carried out to investigate the airflow distributions that would occur if the current system were extended, using the present positions and duties of existing fans. These showed that the airflows on all working faces would fall far short of requirements.

At the time of the survey, the company was pursuing the possibility of sinking a new 16 ft diameter exhaust shaft between No.1 and No.3 mine to a depth of 1,000 feet at the 15th level. However the analyses showed that even this would not provide the required airflows in an economic manner. Connecting the new shaft further downslope at the 20th level gave some improvement in the airflows but the operating costs remained high (Figure 6) and the volumetric efficiency low.

System B. The U-tube Concept: (Exercises 7 through 10) In order to achieve better utilization of existing airways exercises were carried out that involved siting forcing fans of 4 inches water gauge (1kPa) at top of the Water Canyon slopes, dismantling the doors at the bottom of those slopes and removing both of the existing exhausting fans. This created a large inclined U-tube system with the Water Canyon slopes as intakes, the majority of the air returning to surface via the present rope slope and all adjacent airways. Doors were removed from all interconnections between the rope slope and the current Escapeway. This system makes No. 3 mine virtually independent of the adjoining No. 1 and No. 2 mines. Air leakage is reduced very considerably with a consequential increase in volumetric efficiency and reduction in costs (Figure 6).

Figure 4 shows the U-tube system in its simplest form with no new shafts. The longwall face, at 24 kcfm, falls short of the required 30 kcfm and one of the production miner districts, at 34 kcfm, is below the required 45 kcfm. However, the airflows achieved are a considerable improvement upon previous exercises and at much lower operating cost.

As shown on Figure 6, amendments to the airways such as driving a new return in parallel to the existing Escapeway failed to yield any significant improvements to the basic U-tube system.

System C. U-tube with an additional shaft (Exercises 11 through 16) Further U-tube exercises were conducted with a new 16 ft.diameter shaft in one of two locations. The new shaft positioned at the 15th level (Site 1) gave only minor improvements to the basic U-tube system (Figure 6). In order to achieve the required airflow on the longwall face it was necessary to equip this shaft with a 5 inch water gauge exhausting fan. One of the production miners fails to reach the required 45 kcfm. Furthermore, the relatively high fan pressure leads to increased leakage, low volumetric efficiency and a high operating cost. Only when the new shaft is sited at the 20th level (Site 2) do all production areas receive their required airflows. Furthermore, this is achieved without the need for a fan on the new shaft. For the first time during this investigation the volumetric efficiency exceeded 50 percent. Figure 6 shows the effect of equipping the Site 2 shaft with an exhausting fan. A 3 inch water gauge (0.75kPa) fan raises the longwall airflow from 32 to 38 kcfm, but further increases in the fan pressure (and hence operating costs) do not give significantly higher airflows in the working places.

System D. The Through-Ventilation System: (Exercises 17 through 21). When intake and return airways are connected via doors, stoppings, aircrossings or old workings then leakage of air will occur. In the case of an old mine that has been worked extensively then there will be a very large number of such leakage paths. This is the situation at Sunnyside. Even with the seals well built and maintained, the total leakage will remain high and the volumetric efficiency low. If such a system is extended as the mine develops then the situation can only deteriorate.

A system that is designed to minimize leakage and maintain a high operating efficiency and effectiveness in the conditions pertaining at Sunnyside must incorporate one or more of the following features:

(i) Pressure control to minimize the pressure differentials across leakage paths.

(ii) Physical separation of intakes and returns (as in the U-tube concept.).

(iii) A "through-ventilation" layout.

In its simplest form, a through-ventilation system involves surface connections at opposite ends of the area to be mined and arranging for the air to flow through the system in one general direction.

The method is particularly suitable where existing airways pass through old workings. Again, this is the situation at Sunnyside. Figure 5 quantifies a through-ventilation system for the Phase 3 network. As all exercises favored strongly Site 2 (20th level) for the location of the new shaft, this was the position chosen for the through-ventilation exercises. Effectively, the whole mine is ventilated by intake air except those relatively short airways returning from the workings to the new shaft connection. An exhausting fan is sited at the top of the new shaft, and there are no other main fans operating in No.3 mine.

The advantages of a through-ventilation system applied to this mine are illustrated on Figures 5 and 6, and include the following:

(a) A high volumetric efficiency; loss of air through leakage is small, and most of the airflow is used effectively in the production areas.

(b) The multiple intakes give a very low total resistance to airflow. Hence the system is amenable to upgrading, without major expenditure, should additional faces be required in the future. Additional exercises showed that an upgrading of the fan from 4 to 5 inches of water gauge (1 to 1.25 kPa) gives more than sufficient air for two longwalls operating simultaneously.

(c) The operating cost is low. Even with two longwalls working, the ventilating cost is only 65 per cent of the

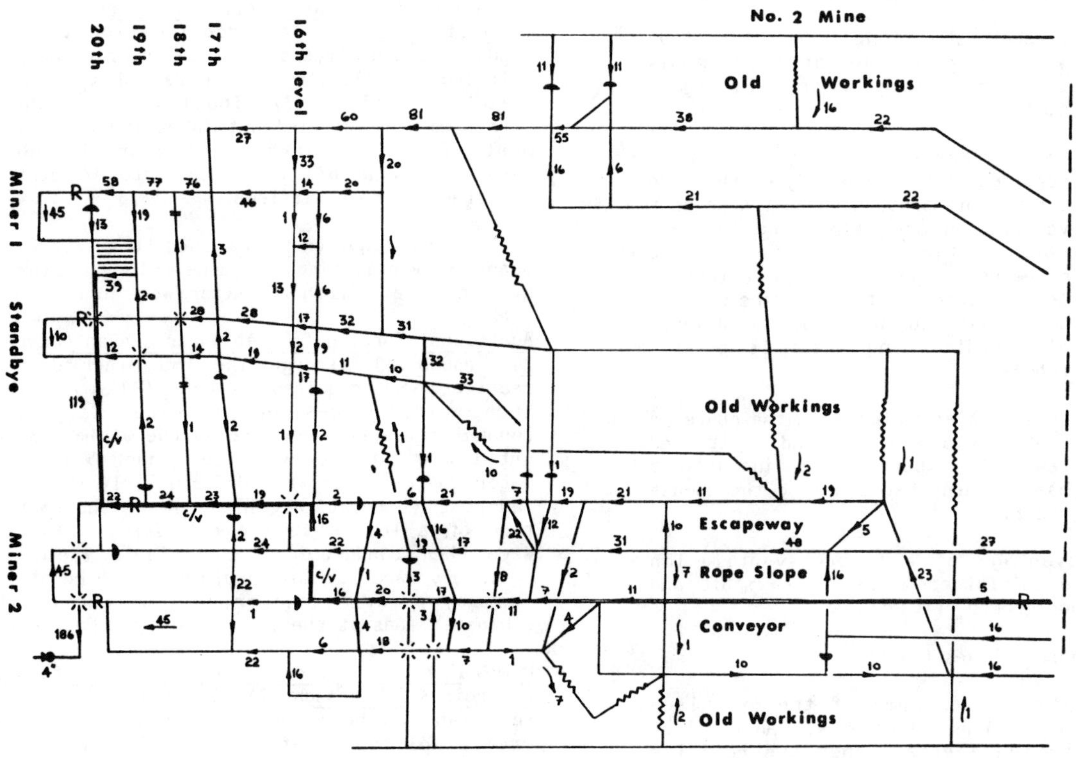

FIGURE 5. Phase 3. System D. Through-ventilation. New shaft on Site 2 with 4" exhausting fan. Op. Cost = $43,797/year. Volume efficiency = 75 percent.

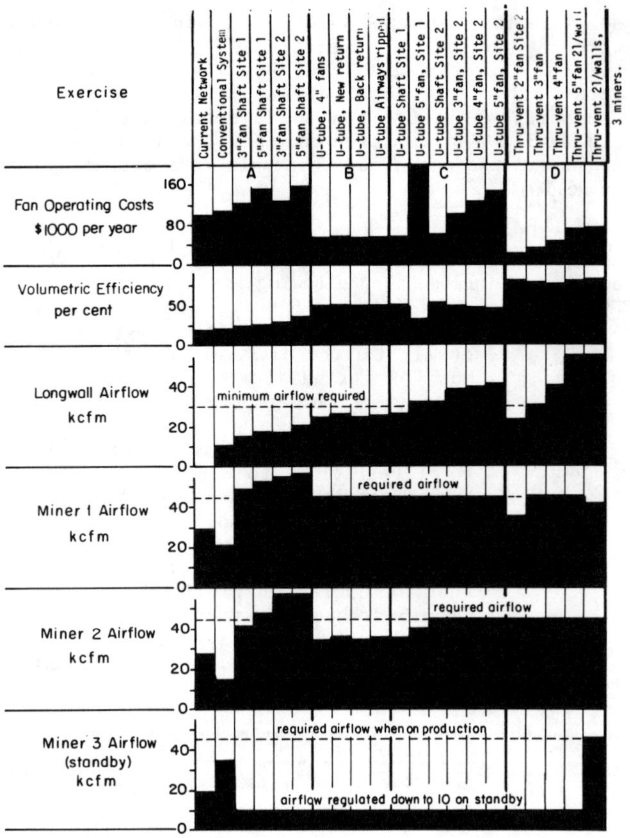

FIGURE 6. Summary of Phase 3 exercises

present situation. The comparison with some of the other systems investigated may be seen in Figure 6.

(d) There are relatively few doors required in the main body of the mine except for maintaining an independent ventilation path along the conveyor route. The consequence is that movement of personnel and materials is unimpeded. Furthermore the cost of building and maintaining doors, air-crossings and seals is greatly reduced.

(e) It is easier to maintain the conveyor route as a separate and controlled ventilation path exhausting into a main return that leads to the upcast shaft.

The disadvantage of a through-ventilation system is the initial cost of the new shaft and associated underground drivages.

Summary of Phase 3 Investigation

Figure 6 summarizes some of the exercises carried out on the Phase 3 network. An examination of this Figure establishes the relative merits of the four ventilation designs investigated. It is clear that an extension of the current system (System A), even with a new shaft, is unsatisfactory. The operating costs are high, the volumetric efficiency low and the required longwall airflow is not achieved.

Separating main intakes and returns by the U-tube principle (System B) improves the volumetric efficiency and lowers operating costs quite dramatically. However, the face airflows remain unsatisfactory. Combining the U-tube with a new shaft (System C) provides a viable design but costs escalate rapidly if additional airflows are required for extra mine production.

Moving to the "through-system" (System D) shows a large increase in efficiency and a further drop in operating costs. In this system, almost all of the mine is in intake air with the total airflow exhausting through a new shaft at the 20th level.

A 4 inch water gauge (1kPa) fan at the top of this shaft will provide adequate airflows in all working places. Additional airflows for enhanced production can be achieved by upgrading this fan and at modest increases in operating cost.

Phase 1: First New Longwall Face

At Phase 1 the network layout differs little from the current system except that a longwall face is operating between the 15th and 16th levels as shown on Figure 7. Two continuous miners developing down to the 17th level each require some 45 kcfm (21 m^3/s) of air. The Phase 1 investigation proceeded along four stages. Summaries of the results are shown on Figure 9.

Extension of Current Ventilation System. (Exercises 22 and 23): Modification to doors, regulators and air-crossings in order to adapt the existing ventilation system failed to give the required airflow distribution. A number of exercises were carried out involving the enlargement of poor airways and the driving of new returns. None of these succeeded in achieving satisfactory face airflows.

The investigation then turned to a search for a means of ventilating the mine prior to sinking a new shaft and without major work underground.

Additional forcing fans at Water Canyon (Exercises 24 through 26) Although the two miner districts receive their required 45 kcfm (21 m^3/s) with 3 inch water gauge fans at Water Canyon, the longwall face does not exceed the minimum 30 kcfm until the Water Canyon fans reach 5 inches of water gauge (1.25kPa). Unfortunately the operating cost is relatively high due to the continuation of a low volumetric efficiency. Exercises were then conducted to investigate whether the necessary face airflows could be achieved at lower cost by removing one or both of the existing exhauster fans at the 8 ft. shaft and No.2 Canyon.

Removal of Exhauster Fan from top of 8 ft shaft (Exercises 27 through 29). The effect of removing this fan was to reduce operating costs and increase volumetric efficiency (Figure 9). The airflows were only a little lower than stage B when all fans were operating. Again, the minimum face airflow for the longwall was not reached

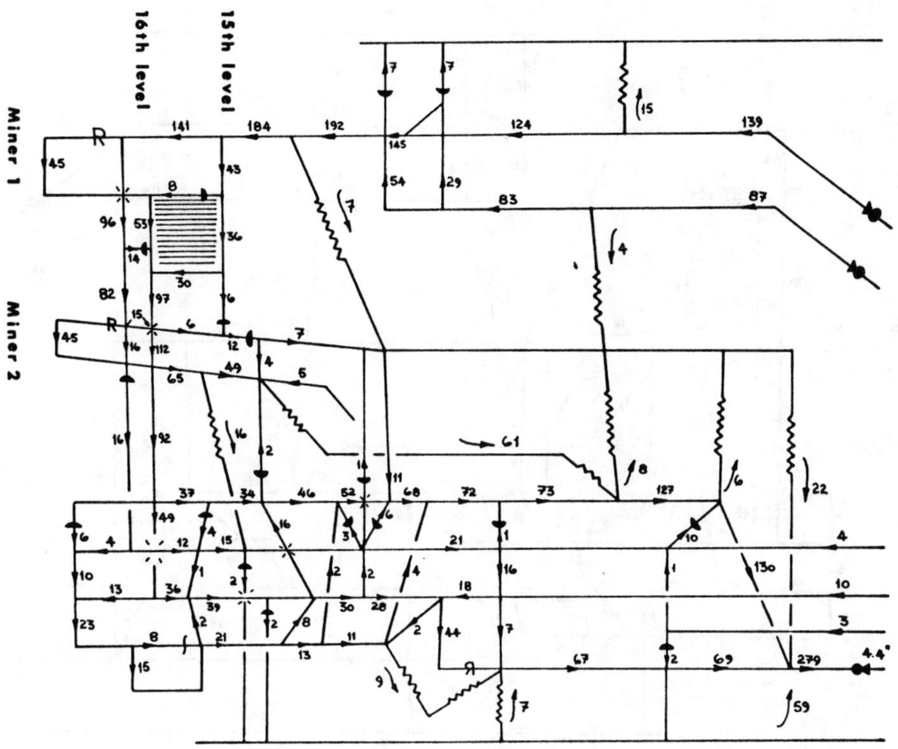

FIGURE 7. Phase 1. Water Canyon fans at 5" w.g. 8 ft. shaft fan switched off. Operating Cost = $138,750. Volume Efficiency = 39 percent.

until the Water Canyon fans delivered a pressure of 5 inches of water gauge (1.25kPa). The resulting distribution is shown on Figure 7.

Water Canyon fans only (Exercises 30 through 32). With the removal of both of the existing exhaust fans, the system then reverts to the U-tube concept. The elimination of adjacent intakes and returns results in an improvement in volumetric efficiency and a dramatic reduction in operating costs. (Figure 9). On the other hand, the face air flows are almost the same as stage C when the main exhausting fan was still operating. Figure 8 gives the air distribution with the Water Canyon fans again set at 5 inches of water gauge. (1.25kPa)

Summary of Phase 1 investigations

The Phase 1 exercises illustrate one of the real benefits of a quantified ventilation planning study. A result has become evident that would have been difficult to foresee without network analysis - the removal of two main fans reduces operating costs markedly but with little effect on face airflows. The explanation of this phenomenom can be found by studying Figures 7 and 8.

With forcing and exhausting fans at opposite extremities of the "U-tube" (Figure 7) the 8 ft shaft is virtually at a neutral pressure with respect to surface. Hence the flow through this shaft is very low. Effectively the shaft is no longer available to assist in the ventilation of the mine.

An examination of the airflow in the rope slope on Figure 7 reveals a similar state of affairs. The airflow in this large open airway is low compared to other systems. Indeed, at one point on the rope slope the airflow becomes zero. This arises from the pressure distribution given by the combinations and sites of the main fans. In simple terms, the Water Canyon fans are attempting to promote flow in a uniform direction around the U-tube. The No.2 Canyon exhausting fan, however, endeavors to maintain the rope slope and conveyor road as downcasting intakes. Hence, these airways are also virtually sterilized and, once again, all return air is concentrated into the high resistance escapeway.

With the removal of the exhauster fans (Figure 8) the Water Canyon fans are able to promote the U-tube ventilation using all existing airways. The system is much less constrained, the airflows better distributed and the pressure pattern designed to assist rather than oppose the ventilation of the mine. The resulting effect on costs and ventilating efficiency is shown on Figure 9.

Phase 2: At 3 Years

At Phase 2, a longwall face will be operating between the 16th and 17th levels with continuous miners developing down to the 18th level. A series of network exercises confirmed the Phase 1 findings that (i) an extension of the existing ventilation arrangements would not provide the necessary face airflows, and (ii) a U-tube system with 4 inch water gauge (1kPa) forcing fans at the

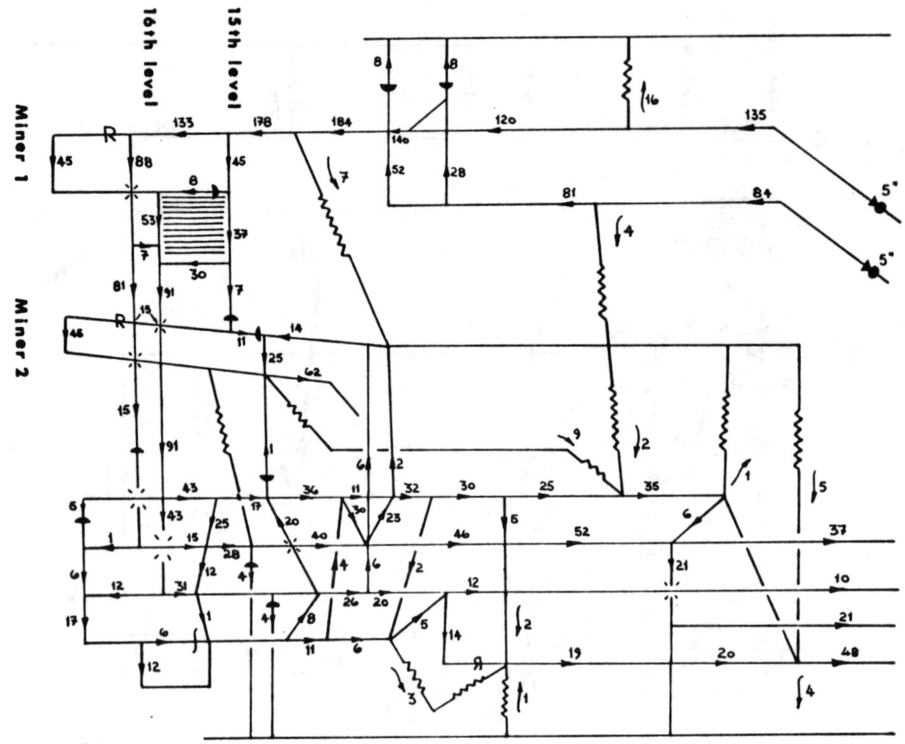

FIGURE 8. Phase 1. U-tube. Water Canyon fans at 5" w.g. No other fans. Operating Cost = $64,636/year. Volume Efficiency - 55 percent.

Water Canyon portals was more effective than alternative fan sites.

Progressing to a longwall face position between the 17th and 18th levels required the Water Canyon fans to be raised to 5 inches water gauge (1.25 kPa). It is during this period that the new shaft should be sunk and the necessary drivages constructed to connect at the 20th level.

Phase 4: At 16 Years

The Phase 4 network has the longwall face operating between the 23rd and 24th level, with continuous miners developing at the 25th level. The exercises were concentrated on the through-ventilation system.

A 4 inches water gauge (1kPa) fan sited at the top of the new shaft achieves the required minimum of 30 kcfm (14m^3/s) on the longwall and also satisfies the airflow requirements in the two miner districts. At a fan pressure of 5 inches water gauge (1.25kPa) the system is capable of satisfying the airflow requirements of two longwalls and two continuous miners all operating simultaneously. In this situation the ventilating operating cost is still some 36 per cent below the current level (present day values).

DISCUSSION OF PLANNING EXERCISES

This case study has concentrated on the strategies of ventilation planning in order to demonstrate the ease with which alternative designs may be investigated through an organized and computer based procedure.

For the purposes of comparison, the most economic ventilation system that gave satisfactory face airflows at each of the four phases is shown on Table 1. These selected exercises are all based on the simultaneous operation of one longwall and two continuous miner panels. The positions and duties of the main fans are also given for the purposes of fan specification. Any fans that are purchased must be capable of meeting the required specifications over the sixteen years.

Figure 10 further illustrates the corresponding variations of volumetric efficiency and operating cost over the four phases, and gives a comparison with the present day situation. The large improvement in efficiency and fall in operating cost on conversion to the U-tube system is clearly shown. Further improvements are achieved when the new shaft and the through-ventilation system are brought into operation. As the workings progress down slope beyond the site of the new shaft then leakage will again increase with a consequential reduction in volumetric efficiency and raising of operating costs. This is shown by the Phase 4 results on Figure 10 and further illustrates the advantage of siting the shaft connections down slope from the present workings.

The ventilation survey at Sunnyside provided the necessary raw data for the establishment of an accurate simulation of the mine. The correlation

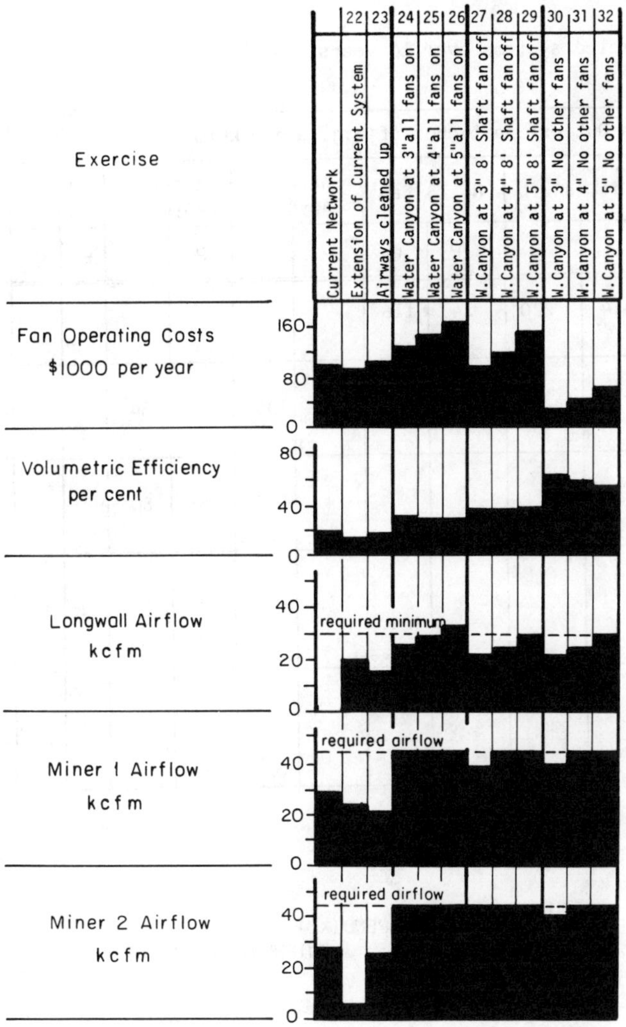

FIGURE 9. Summary of phase 1 exercises

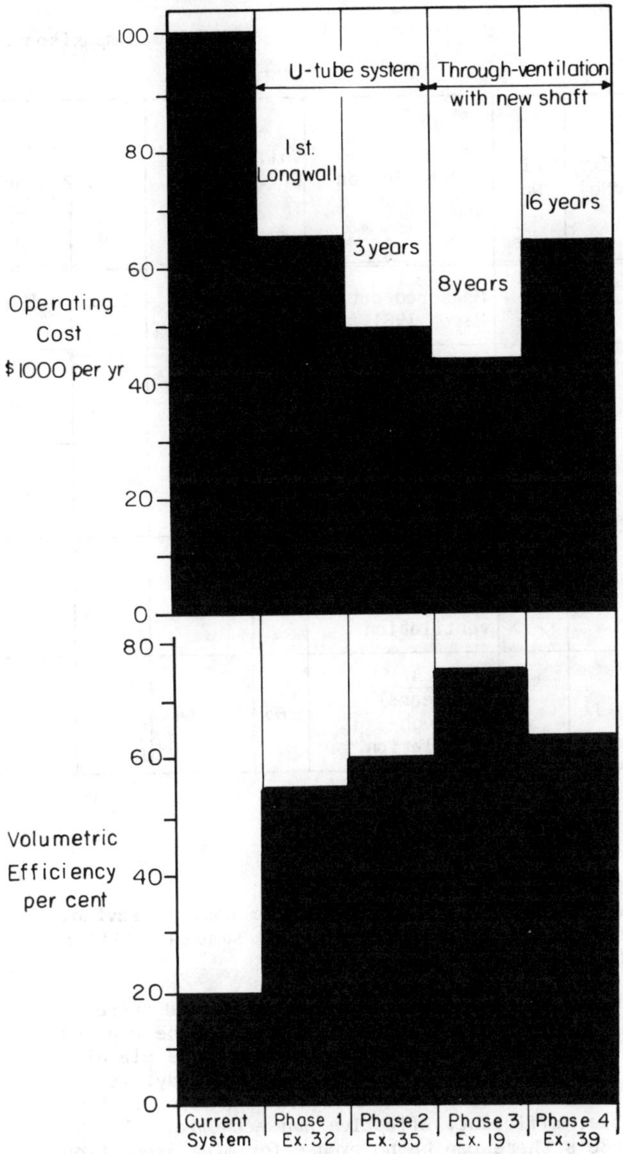

FIGURE 10. Comparison of operating costs and volumetric efficiency over 16 years.

exercise verified that the basic network established on the computer could be used with confidence for planning the future ventilation arrangements. The planning exercises carried out were directed towards providing sufficient airflows in the workings and with emphasis on the efficiency and economics of the ventilation designs investigated. These systems were limited to layouts that are in accord with current U.S. coal mining practice. Other techniques such as the use of booster fans were not investigated in detail. It is possible that the latter would enable the mine to work through the sixteen year planning period without the capital cost of a new shaft. This is the method that would most probably be adopted in some other countries, the network exercises being directed towards finding the best sites and duties of the booster fan(s).

During the course of the investigation, results arose that would have been difficult to foresee, even by the most experienced of ventilation planners, without computerized network analysis. This is a consequence of the complexity of the ventilation system of a large mine and the interaction of many variables. The effect on design methods is twofold. First, the ventilation planner is able to observe the influence of his chosen exercises on the total mine system and not only in localized parts of it; and second, trends observed in consecutive flow analyses suggest further exercises to improve the system. An organized planning procedure with computer assisted network analysis is therefore a most powerful aid to system design.

At the time of writing, work is in progress for a new fan installation at the Water Canyon portals and the establishment of a U-system of ventilation (Phase 1). The study showed that the site originally planned for the new shaft was inappropriate for the future workings and a new location for connecting the workings to the shaft has been established. Furthermore, it has been shown that by improved utilization of existing airways, the sinking of a new shaft may be postponed for

TABLE 1. Comparison of selected systems over 16 years.

Ex. No.	Fig. No.	Description	Vol. Eff. %	Oper. Cost k$/yr	Fan Duties: p inches of w.g. and Q kcfm											
					No. 2 Mine		8 ft. shaft		No. 3 fan		Water Canyon 1		Water Canyon 2		New shaft	
					P	Q	P	Q	P	Q	P	Q	P	Q	P	Q
1	1	Measured data: March 1981	19	101	2.31	128	4.4	220	4	116						
32	8	PHASE 1 1st new longwall U-tube system	55	65							5	135	5	84		
35		PHASE 2 (3 years) U-tube system	60	49							4	126	4	80		
19	5	PHASE 3 (8 years) Through-ventilation	75	44											4	186
39		PHASE 4 (16 years) Through-ventilation	63	64											5	219

several years. This involves a capital saving, estimated by the company, of some 4 million dollars.

No two underground mines are alike, and there is no single ventilation system that can be applied universally. However, by following the planning procedure demonstrated by this case study, ventilation schemes may be designed that are practicable, efficient, effective and economic. In the 1980's there can be no excuse for mine production being disrupted by ventilation problems except in the most abnormal of circumstances.

ACKNOWLEDGEMENTS

This case study formed a part of a larger project funded by the Department of Energy. The authors are also grateful to the Kaiser Steel Corporation for giving their permission to use Sunnyside Mines as the subject of this paper.

APPENDIX 1
VENTILATION SURVEY: EQUIPMENT CHECKLIST:

No.	Item
2	AM 5000 Anemometers
3	Davis Anemometers (2 medium and 1 high range)
4	4 ft. Pitot Tubes
2	100 m lengths of strong (pref. nylon) plastic tubing: 3 mm inside diameter
8	4 ft. lengths of connecting (rubber) tubing
1 box	connector clips
2	sharp penknives
1 box	aerosol paint tins
2	0.5 inch magnehelics and traverse cases
1	1 inch magnehelic
2	2 inch magnehelics
1	4 inch magnehelic
2	measuring tapes (20ft.)
1	measuring tape (100 ft.)
3	mechanical stop watches
3	clipboards
1 box	airflow booking sheets
1 box	pressure booking sheets
	notebooks
	pencils
1	30 ft. length spare flexible tubing
1	aneroid barometer
2	methanometers
2	whirling hygrometers and spare wick
2	programmable calculator
2	rucksacks

DEVELOPING INPUT DATA FOR COMPUTER SIMULATION OF MINE VENTILATION SYSTEMS FROM A PRESSURE-QUANTITY SURVEY

Chapter 13

George W. Luxbacher

Manager - Technical Data and Engineering Support
Island Creek Coal Company
Lexington, Kentucky

R. V. Ramani

Professor of Mining Engineering
Pennsylvania State University
University Park, Pennsylvania

ABSTRACT

Computer simulation of mine ventilation systems, after a slow start in the early sixties, is gaining increasing acceptance for planning, design, and evaluation purposes. Numerous programs are readily available from government, university, consulting, and industry sources. The computer adage "garbage in, garbage out" is as applicable in ventilation engineering as in any other application of computers, and has been responsible for the questionable results obtained in some instances.

The source of input data for a ventilation simulation is most often a pressure-quantity survey of the mine, or, in the case of a new mine, information derived from a survey of an existing mine which has conditions similar to those expected in the new mine. The crucial step is to generate simulation input data from the ventilation survey. Problems such as quantities that do not balance, pressure gradients opposite to measured air flow, differences in the pressure datum in a multi-shaft mine, modeling of leakage, and deviations in flow from the square law ($H=RQ^2$) are frequently encountered and must be resolved. The purpose of this paper is to discuss these aspects and to propose procedures which can lead to input data for valid simulations of mine ventilation systems.

INTRODUCTION

Computer input data for an operating or proposed mine is developed on the basis of a detailed ventilation survey which includes air quantity, air quality, pressure loss and fan performance measurements. Quantity measurements must be reviewed extensively and assumptions made about leakage quantities and airflow paths. Pressure measurements must be reviewed and developed into a pressure gradient for the mine system. Air quality measurements are used to determine methane sources and amounts. Finally, the fan test data is used to develop the correct (or operating) fan characteristic curve. Quantities and pressure losses for each branch in the system are used to calculate the branch resistances which are then utilized in the ventilation simulator to model the mine network.

Although computer oriented analysis of mine ventilation systems was introduced over two decades ago, it has yet to receive widespread acceptance in the United States. Central to this problem is the fact that, while many computer programs for network analysis have been developed, little or no attention has been paid to the development of computer input data. Consequently, program applications have been infrequent, or have produced less than satisfactory results. The elements in the development of the input data, as outlined above, should be carefully reviewed as they pertain to computer simulation (Luxbacher, 1977). The discussion in this paper is limited to pressure-quantity survey as the bulk of input data for computer simulation consists of resistances of airways and leakage characteristics of ventilation controls.

QUANTITY BALANCE

The quantity of air entering a mine based on field measurements will rarely equal the quantity learning the mine, even if allowances are made for methane or compressed air input to the system. The same statement can be made for airflow through any isolated part of the mine ventilation system such as individual working section or main aircourses leading to an area isolated from the rest of the mine. The reason for the apparent discrepancies is the accuracy of the indicated quantities which in turn is based on the accuracy of the individual velocity and area measurements. Even with measurements following the recommended guidelines, this error can only be minimized but not eliminated. As a result the ventilation system must be balanced so that, for any loop, the quantity in plus any methane or compressed air injection will equal the quantity out (Figure 1). The continuity of flow described by the balanced system must approximate, as closely as possible, the flow conditions that exist in the mine.

LEAKAGE CONSIDERATIONS

In most ventilation systems the intake aircourses parallel the return aircourses and are separated by stoppings. Air leakages through these stoppings are a function of the pressure differential across the stopping and several other parameters. Leakage in an existing circuit cannot be considered on the basis of an individual stopping or a group of stoppings, because of the tremendous variation in stopping conditions in the mine. Such factors as the condition of the stopping interface with the ribs, roof and bottom, the effect of ground pressure, care in initial construction, and any doors or holes which might exist all invalidate the ideal parameters in an existing circuit. Kharker et al, (1973) have shown that the majority of the system leakage occurs in the initial phase of the system, as had been suggested by

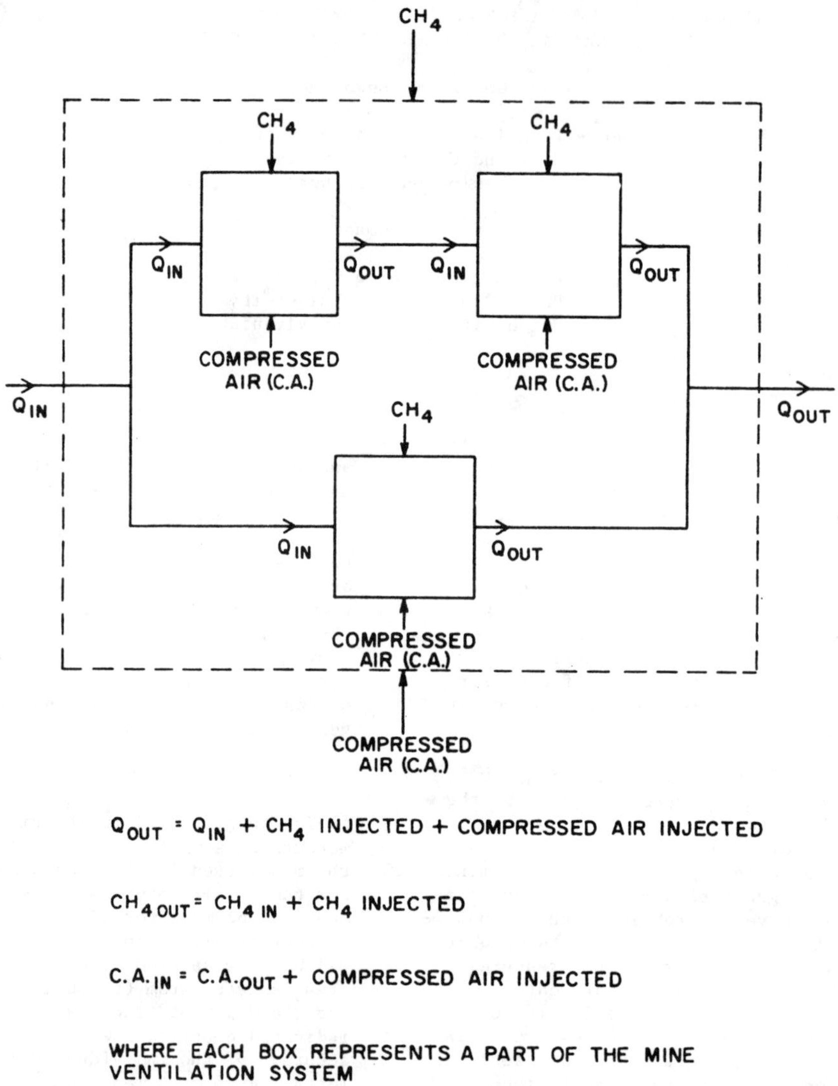

Figure 1. System Quantity Balance.

Mancha (1942). For example, in a panel the majority of the leakage occurs at the panel mouth where the pressure differential between the intake and return is high and very little leakage occurs at the face where it is low. Although this is easily understood, its application to simulate leakage conditions in computer models is difficult. In addition, the factors mentioned earlier could easily alter the assumptions; for instance a stopping in poor condition may easily leak much more air than a better stopping subjected to a much greater pressure differential. In general, the best treatment of leakage in an existing mine is to utilize a leakage branch at the midpoint of the parallel paths between which leakage occurs. Usually no more than 25 or 30 stoppings should be combined in this fashion; larger numbers of stoppings should be split into separate leakage paths. The number of leakage splits is obviously a function of the leakage, the total quantity flowing, and the length of the split. If desired, the number of pressure and quantity measurements can be increased during the ventilation survey to further locate major sources of leakage. Using this data the number of leakage branches can be increased, and a more accurate model of the system can be developed.

PRESSURE BALANCE

Pressures at each measurement point are calculated from altimeter survey data utilizing one of several well known procedures. Differential pressures are calculated between points and summed, so that the pressures from each survey period are relative to the base used during that period. The total pressure should be selected as the datum and defined as zero pressure at the base (Figure 2). The differential total pressure can then be summed to give the total pressure at each point relative to the base datum. The total head is recommended for three reasons. First it accounts for the conversion between velocity head and static head and hence reflects more accurately the headloss between points. Situations where a static head regain occurs due to an expansion in area can be properly handled when working with the total head. Secondly, on an exhaust system the rated fan static pressure is found by measuring the total

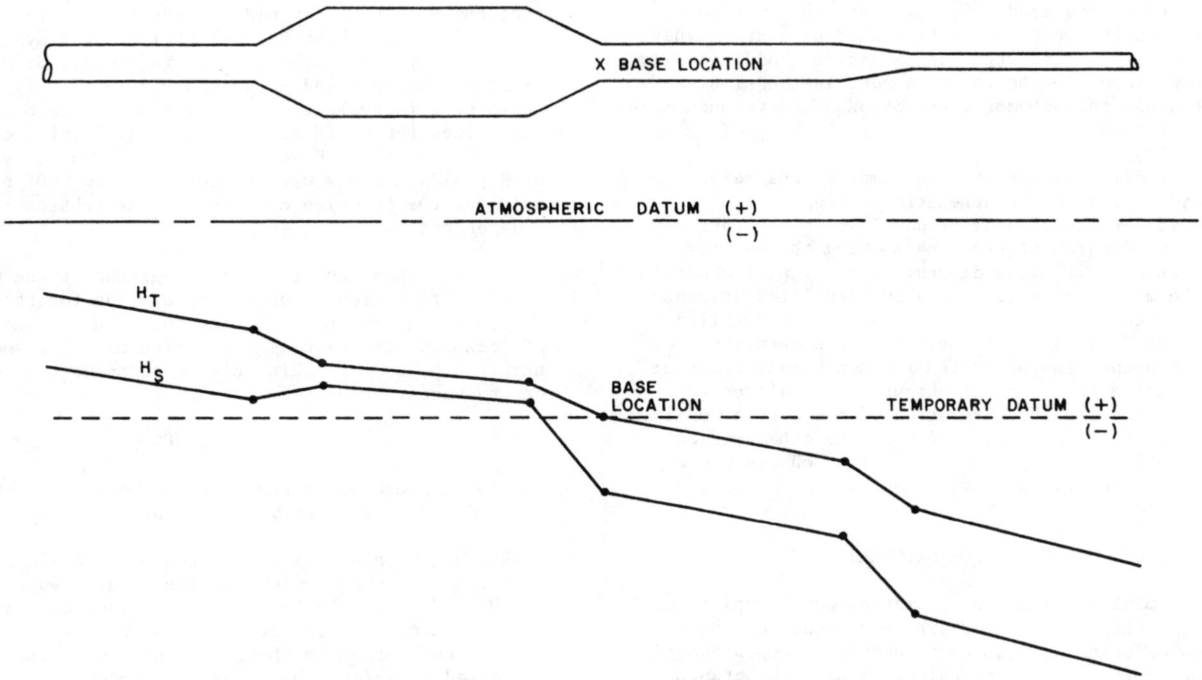

Figure 2. Base Station Location as a Temporary Datum.

head directly in front of the fan; by making all measurements to total head rather than static head the measurements are consistent with the fan rating. Thirdly, since at the entrance of a ventilation circuit the total head is equal to zero while the static head is equal to negative of the velocity head, the use of the total head allows the use of a zero datum.

Balancing of the pressures is much more simple compared to that of the quantities and requires only that the pressure gradient obtained from the survey indicate the observed direction of airflow. Of particular importance are closely spaced measurement points, and branches with extremely low airflow; under these conditions, headlosses will be very small and the accuracy of the pressure and quantity measurements plays a big role. Calculations may indicate airflow in the opposite direction and require adjustment.

In the multi-shaft mines that are being developed today, the pressure gradient is much more complex that the simple closed loop ventilation systems of small or newly developed mines. Interaction between the various shafts must be considered in the evaluation of the survey data. For instance, the pressure gradient for one mine studied by the authors was based on atmospheric pressure at the top of the intake shaft in the active portion of the mine since this was the primary area under study. As the pressure gradient was developed from the active portion of the mine through the inactive portion to the pit mouth it became obvious that pressures in the intake aircourses in the inactive portion of the mine were positive with respect to the atmospheric datum. This can be explained by the fact that the atmospheric pressure at the other shaft locations differed from that at the shaft selected for the atmospheric datum mainly due to the effect of elevation. The atmospheric datum happened to be the shaft with the highest elevation, hence lowest absolute atmospheric pressure. The other shafts were at lower surface elevations and as a result had higher absolute atmospheric pressures which caused pressures at these shafts to be positive with respect to the atmospheric datum. The observed pressure difference between shafts is also affected by air density and acceleration due to gravity in addition to elevation. However, elevation is the major factor, responsible for 90% of the observed difference. Additionally natural ventilation pressure (NVP) probably accounted for some of the observed pressure gains. This pressure offset, once recognized, was easily handled by a simple network modification for computer simulation.

SCHEMATIC DEVELOPMENT

After the pressures and quantities in the ventilation system have been balanced the transition can be made between the line diagram used up to that point and a schematic suitable for computer input. An example of the procedure to follow in this transition is given by Stefanko and Ramani (1972). Parallel flow branches can be combined to greatly simplify the schematic. Any complex circuit can be reduced to one or more branches between the surface and the fan; obviously over simplification of the ventilation system will result if too many branches are combined during schematic development. The main danger of oversimplification is the loss of resolution with regard to branch flows that are required for verification and planning purposes.

The final schematic is usually a tradeoff between simplified computer input and requirements for a user oriented output. Initially simplified computer input may be the overriding factor since a simplified system will reduce required computer time; this may be particularly important on time sharing systems where the cost of input/output is excessive. Additionally, as experience in the use of a ventilation simulator increases, the schematic can be expanded to much more complex cases.

User oriented output is ultimately the major consideration in the schematic development. The network is left as closely as possible in the form of the line diagram of the mine showing the various aircourses. The line diagram of the inactive portions of the mine can be considerably simplified whereas that of the active portion, considerably detailed (Figures 3 and 4). Evaluation of the quantity flow in any branch in the system is then simply a matter of checking the computer output. In addition to simplifying the interpretation of the output, maximum flexibility is given in updating the schematic as the ventilation layout or conditions change since each segment of the ventilation system can be easily identified.

RESISTANCE CALCULATION

The mine ventilation schematic, when complete with the quantity flowing through each branch and pressures existing at each node, provides the information required for resistance calculations. The branches in the system should be broken down into normal or leakage branches.

Normal branches are those branches in which the airflow follows the square law as defined by Atkinson in 1854.

$$H = R\,Q^2$$

where H is the headloss in the branch in inches water gage
Q is the quantity flowing in the branch of 10^5 ft^3/min
R is the branch resistance in inches water gage per 10^{10} ft^6/min^2

Many computer programs limit their treatment of the network to this type of flow although other flow conditions, such as leakage may exist.

UNCERTAINTY IN RESISTANCE CALCULATIONS

Although Atkinson's equation for turbulent flow has been used to determine the resistance of normal branches for the simulator, some controversy exists over its applicability to all branches in a mine network. The deviation from the square law of head loss ($H=RQ^2$) can be treated in one of two ways.

1. The relationship can be expressed as $H=RQ^n$, where n can take any value. Most ventilation simulators either already have this feature or can be easily modified to incorporate this feature.

2. The squared law can be retained ($H=RQ^2$); however the value of the coefficient of friction, hence resistance, is adjusted according to velocity as represented by the Reynolds number.

That Atkinson's formula may not describe flow conditions accurately was noted as early as 1926 by McElroy (1926). He reviewed all published data and the Bureau of Mines experimental work and concluded that the effect of Reynolds number on coefficient of friction was greatest for smooth lined airways in the range of Reynolds numbers found in mines. Sales and Hinsley (1951-52) and Brown and Hinsley (1957) have shown that for values of Reynolds number below 50,000 the values for coefficient of friction varied considerably with the Reynolds number. When the value of Reynolds numbers was greater than 100,000, the value of the friction coefficient was substantially independent.

Despite these uncertainties, because of the practical difficulties in obtaining a value for the quantity exponent or in correcting the value of the coefficient of friction as a function of the Reynolds number, the use of Atkinson's equation continues to be practiced.

LEAKAGE BRANCHES

Flow through stoppings, doors, checks and gob areas deviates from Atkinson's law.

The behavior of air flow through gob areas and stoppings is more complex and was studied as early as 1916 (Storrow, 1916-17) with the idea of reducing such leakages to eliminate the possibility of spontaneous combustion in the gob. This research attempted to define the conditions under which the square law holds true and at which point it should be replaced by the relationship H = RQ. Storrow's conclusions were that the linear relationship was approached as the material in the job became finer, the air current velocity was decreased and the length the air must travel was increased. More recently, Peascod and Keane (1955) and Low (1956) have also used a direct relationship between headloss and quantity in modeling leakage flows across a porous barrier represented by the gob on an advancing longwall face.

Pickering (1969) has conducted experiments on airflow in the waste area directly behind the longwall face and has found the flow exponential to be 1.78. Although this conflicts with the direct relationship (H=RQ) used by other researchers, the conditions of flow in the waste immediately behind the face differ from that in a total gob area. The waste behind the face is not as closely packed as the gobs, hence the flow tends to be more turbulent. Moreover, the flow in the waste is not continuous but rather varied in proportion with air in the face area (McPherson, 1969).

Leakage through stoppings, may be handled using an equation developed by Kawenski and Mitchell (1963). According to their experimental work, the quantity of air leaking through a stopping is expressed as follows:

$$Q = aP^n$$

where P = pressure differential across the stopping in in. of water
Q = quantity leaking in ft^3/min
a = leakage in cfm across the stopping in ft^3/min at a pressure differential one inch water guage for 100 ft^2 area of the stopping, and

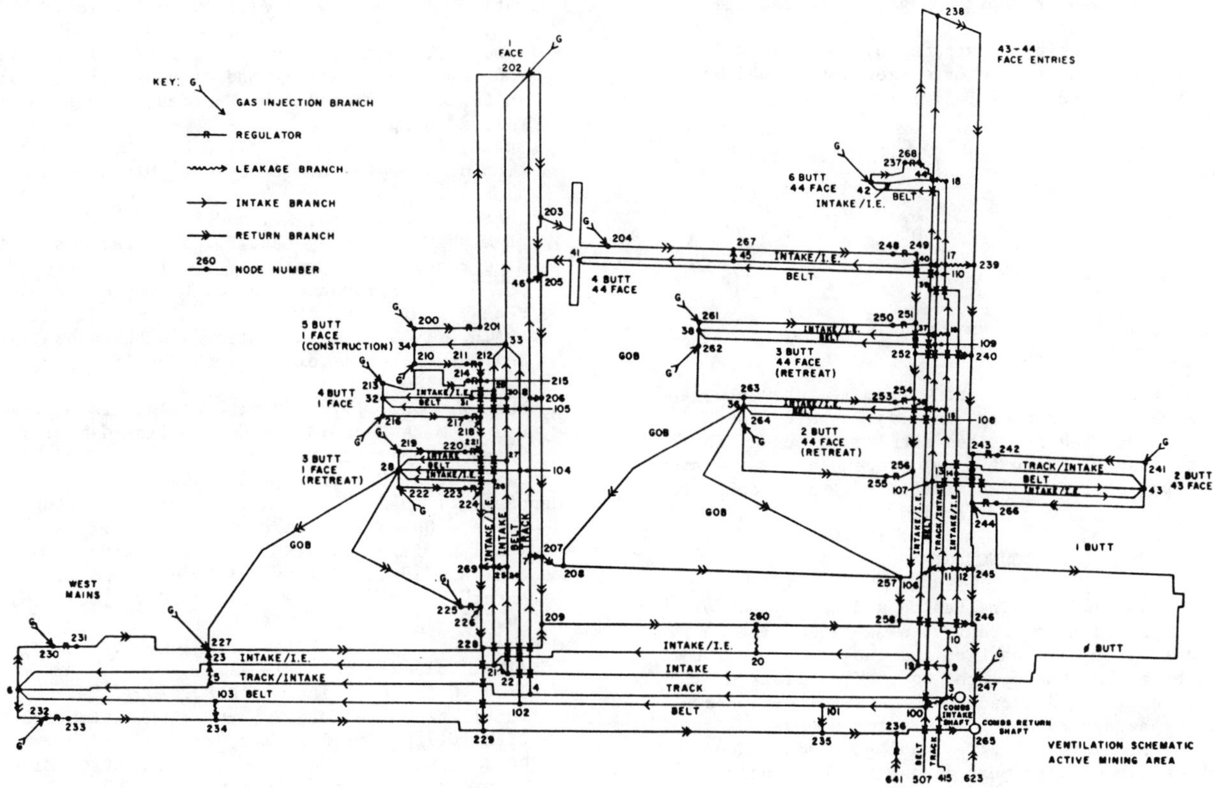

Figure 3. Ventilation Schematic for the Active Mining Area.

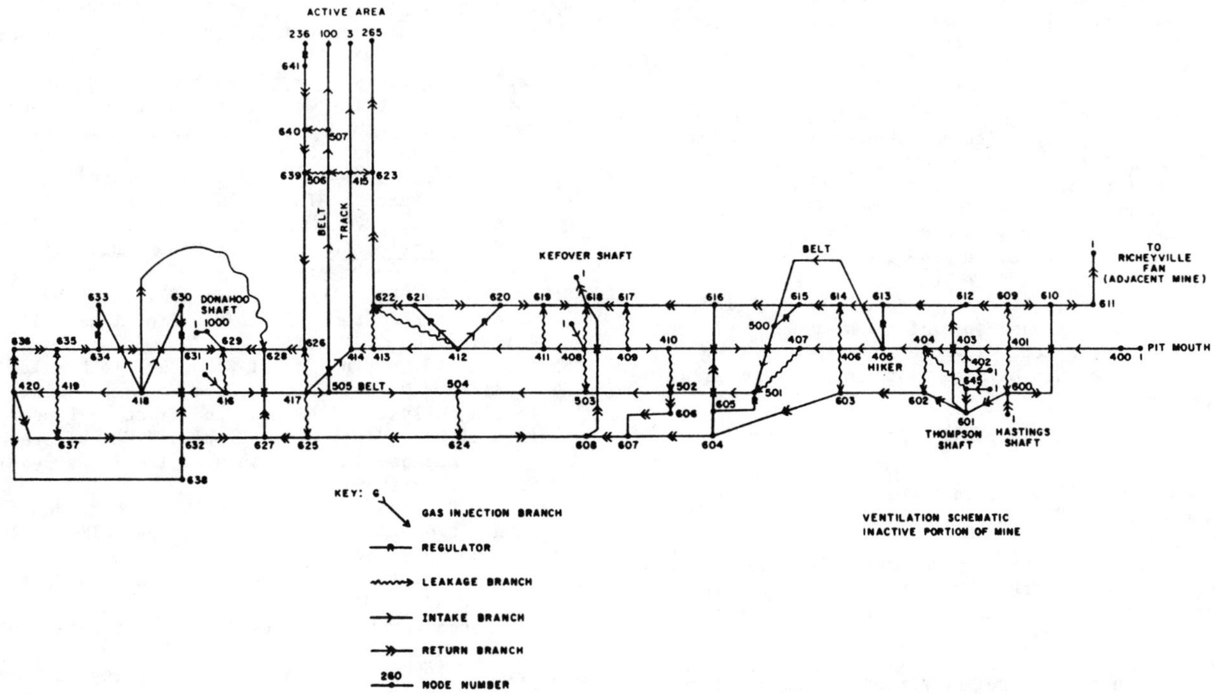

Figure 4. Ventilation Schematic for the Inactive Portion of the Mine.

n = exponent for the pressure, representing the logarithmic change of air leakage with pressure.

For a given stopping construction, the values for a and n are constants, and have been determined by Kawenski and Mitchell (1963). It is readily seen that

$$P = \left(\frac{1}{a}\right)^{\frac{1}{n}} \cdot Q^{\frac{1}{n}}$$

$$P = R\,Q^{c}$$

where $R = \left(\frac{1}{a}\right)^{\frac{1}{n}}$ and $c = \left(\frac{1}{n}\right)$.

Knowing the values of P and Q from measurements, and the value of "n" for the stopping, a value for R can be calculated from the ventilation survey for a leakage branch. Values for c, depending on the stopping type, vary from 0.91 (solid, drywall, concrete block stopping) to 3.33 (hollow core, drywall, block).

CONCLUSION

The Federal Coal Mine Health and Safety Act of 1969 initiated major changes in mine ventilation with regard to both air quantity and quality. Requirements have also been placed on the ventilation of beltlines, gob and face areas. Air velocities on trolley haulage roads have an upper bound. Escapeways, separate from trolley haulage roads and on an intake split of air, must be maintained continuous to the surface. These and other similar requirements have not only precipitated several modifications to existing ventilation systems but also have greatly increased the complexities of new mine ventilation planning. Since a mine ventilation system is a complex inter-dependent system, evaluations, modifications and extensions should be performed on a total system basis. Therefore, the use of computers in ventilation system analysis and design is a necessity.

Consideration must be given, however, to the development of input data with the thought that the validity of computer simulation results depends on the quality of the input data. Concepts such as all network branches obey Atkinson's law, and the use of fixed quantity branches to model leakage, affect the accuracy of simulation. However, depending on the availability of input data from survey results and suitable programs, these concepts can be modified. This paper has reviewed some of these aspects of the input data development with the hope of improving the state of the art of mine ventilation computer simulation.

REFERENCES

Atkinson, J.J., 1854, "On the Theory of the Ventilation of Mines," Trans. N. E. Inst., Vol. 3, pp. 118.

Briggs, H., 1931-32, "The Underground Leakage of Air, Investigation of the Porosity Coefficient," Transaction of the Institution of Mining Engineers, Vol. 82, pp. 111-121.

Brown, J.R., and Hinsley, F.B., 1957, "The Flow of Air in Bends and Straight Airways," Transactions of the Institution of Mining Engineers, Vol. 116, pp. 845-867.

Hodkinson, J.R., and Leach, S.J., 1959, "Some Observations on the Law of Airflow Resistance," Colliery Engineering, Vol. 36, pp. 526-628.

Holland, C.T., and Skewes, W.J., 1962, "Investigating Construction Materials and Methods for Stoppings in Coal Mine Ventilation Systems," Trans. AIME, Vol. 223, pp. 21-30.

Kawenski, E.M., and Mitchell, D.W., 1963, "An Evaluation of Stopping Construction," Mining Congress Journal, Vol. 49, pp. 57-61.

Kharkar, R., Stefanko, R., and Ramani, R.V., 1973, "Analysis of Leakage and Friction Factors in Coal Mine Ventilation Systems," Special Research Report No. 99, Coal Research Board, Commonwealth of Pennsylvania, pp. 73.

Low, A.H., 1956, "Leakage Effects with Unequal Intake and Return," Colliery Engineering, Vol. 33, pp. 178-180.

Luxbacher, G.W., 1977, "Practical Aspects of Computer Use in Mine Ventilation," Unpublished Master of Science Thesis in Mining Engineering, The Pennsylvania State University, pp. 456.

Mancha, R., 1942, "Effects of Underground Stopping Leakage Upon Mine Fan Performance," Trans. AIME (Coal Division), Vol. 149, pp. 178-182.

Mancha, R., 1946, "Surveys of Underground Mine Pressure," Transactions AIME (Coal Division), Vol. 169, pp. 106-118.

McElroy, G.E., 1927, "Mine-Air Flow," Transactions AIME, Vol. 74, pp. 297-311.

McElroy, G.E., and Kingery, D.S., 1957, "Making Ventilation-Pressure Surveys with Altimeters," USBM IC 7809, NTIS PB-222 086/1.

McPherson, M.J., 1969a, "A New Treatment of Mine Barometer Surveys," The Mining Engineer, Vol. 129, No. 109, pp. 23-34.

McPherson, M.J., 1969b, "Air Flows and Methane Emissions on Fully Caved, Mechanized Coal Faces in South Nottinghamshire," The Mining Engineer, Vol. 129, No. 110, p. 104.

Peascod, W., and Keane, A., 1955, "The Effect of Leakage on Mine Ventilation," Colliery Engineering, Vol. 32, pp. 207-211.

Pickering, A.J., 1945-46, "Air Flows and Methane Emissions on Fully Caved, Mechanized Coal Face in South Nottinghamshire," The Mining Engineer, Vol. 105, pp. 631-654.

Potts, E.L.J., 1945-46, "Ventilation Surveying Technique," Transaction AIME, Vol. 105, pp. 631-654.

Quilliam, J.H., 1974, "Pressure Surveys," Ventilation of South African Gold Mines, Mine Ventilation Society of South Africa.

Sales, T.J.R., and Hinsley F.B., 1951-52, "The Use of Models in the Study of Air Flow in Mines," TIME, Vol. 111, pp. 774-791.

Saran, A., 1975, "The Use of Computers in Simulation of Airflow Distribution in the Working Faces, and in the Analysis of Leakage and Friction Factors in Coal Mine Ventilation Systems," Unpublished Mining Engineering M.S. Thesis, The Pennsylvania State University.

Stefanko, R., and Ramani, R.V., 1972, "Mine Ventilation Network Analysis," Transactions AIME, Vol. 252, pp. 382-387.

Storrow, J.T., 1916-17, "The Flow of Air Through Small Coal and Other Broken Material: Report to the Doncaster Coal-Owners Committee," TIME, Vol. 55, pp. 313-323.

Teply, E., 1970, "Influence of Variability of Flow Resistance in Mine Airways on the Calculation of Ventilation," Mining and Metallurgy Quarterly, No. 4, pp. 31-38, (Translated from Slovenian), NTIS TT-70-56005/4.

Vujec, S., 1970, "Determination of Resistance in Mine Airways," Mining and Metallurgy Quarterly, No. 1, pp. 74-85, (Translated from Slovenian), NTIS TT 70-56005/1.

Williamson, J.N., 1932a, "How to Measure Pressure Losses in Ventilation," Coal Age, Vol. 37, No. 8, pp. 297-298.

Williamson, J.N., 1932b, "How to Make Air Survey and Calculate Results," Coal Age, Vol. 37, No. 9, pp. 331-333.

MODELING THE VENTILATION NETWORK FOR AN OLD COAL MINE

Chapter 14

Robert L. Grayson

Superintendent, Nemacolin Mine
Jones and Laughlin Steel Corporation
420 S. Washington Road
McMurray, PA 15317

R. Mike Mishra

Chief Mining Engineer
Emway Resources, Inc.
90 W. Chestnut Street
Washington, PA 15301

Y. J. Wang

Associate Professor of Mining Engineering
College of Mineral and Energy Resources
West Virginia University
213 White Hall
Morgantown, WV 26506

ABSTRACT

Prediction of ventilation needs in long range mining projections requires serious consideration when formulating those projections. Predicting the behavior of the ventilation system of an old, complex mine poses an especially provocative problem. Airflow around and through mined-out areas via bleeder systems, leakage flows through stoppings or other air control devices, and the use of multiple fans all add to the basic complexity of a ventilation system and make predictions of ventilation changes very difficult. Fortunately, the task of making ventilation decisions can be greatly assisted by ventilation surveying and computer simulation.

The Nemacolin Mine of Jones and Laughlin Steel Corporation, located in Southwestern Pennsylvania, is one such old mine which used the computerized approach successfully. A mine-wide pressure-quantity survey was used to obtain realistic values for airway resistance factors. These values, together with network topology, formed the basic input for computer simulation. Results from this simulation indicated that air quantities generated at critical airways by the model were generally within a couple of percent of those measured in the actual survey. It was then concluded that the model was acceptable for practical application, and it was used immediately for studying the effects on the ventilation system of changing the blade settings of fans. Additionally, the model was modified to incorporate projected mine development and used to predict future ventilation.

INTRODUCTION

The use of digital computers to solve mine ventilation networks began in the United States at Pennsylvania State University by Hartman and Trafton (1963). After successive expansion of program capabilities by Trafton and Hartman (1964), Wang and Hartman (1967), and Wang and Saperstein (1970), a program which is presently the most widely used ventilation simulator was presented by Luxbacher, Ramani, and Stefanko (1977). The version developed by Wang and Saperstein (1970) was the one used for this application. It contains provisions for multiple external fans, natural ventilation pressure, and fixed quantity branches.

Air leakages through stoppings involves complex parameters as reported by Holland and Skewes (1962), Kawenski and Mitchell (1963), and Kharkar, Stefanko, and Ramani (1974). A single leakage path was established to include generally no more than 25 to 30 stoppings. This is consistent with the findings of Kharkar, Stefanko, and Ramani (1974).

THE NEMACOLIN MINE

The Nemacolin Mine has been in operation since 1918. During 60-plus years of operation, a very complex system of airways has evolved, as depicted in Figure 1. The condition of the mine at the time of the ventilation survey is illustrated, showing the extent of mine workings and remaining reserves, predicted to last about 10 to 15 more years. Nearly 4.9 km^2 (1200 acres) remained to be mined, with 3.6 km^2 (900 acres) of these located in the last major, solid reserve.

The mine is equipped with four mechanical ventilating fans, ranging from 673 mm (26.5 in.) to 2743 mm (108 in.) in diameter. All basic fan data, along with fan curves from the factory, are maintained at the mine engineering office. Locations of the four exhaust fans are also indicated in Figure 1. Air is supplied at five locations. Numbers 2, 3, and 4 air shafts are all dual compartment shafts, while air ingresses additionally down the slope and skip shaft at the portal near the Monongahela River.

Referencing Figure 1 again, at the time of the survey, the 8 Road area was entirely in a development stage, while the "C" Mains area, except for the 10 Right section at the back end of the mains, was strictly on pillar extraction. To further complicate the existing ventilation system, one section along the main haulage and one section in the 8 Road area were splitting old pillars which were previously abandoned. Specifically, three sections were on

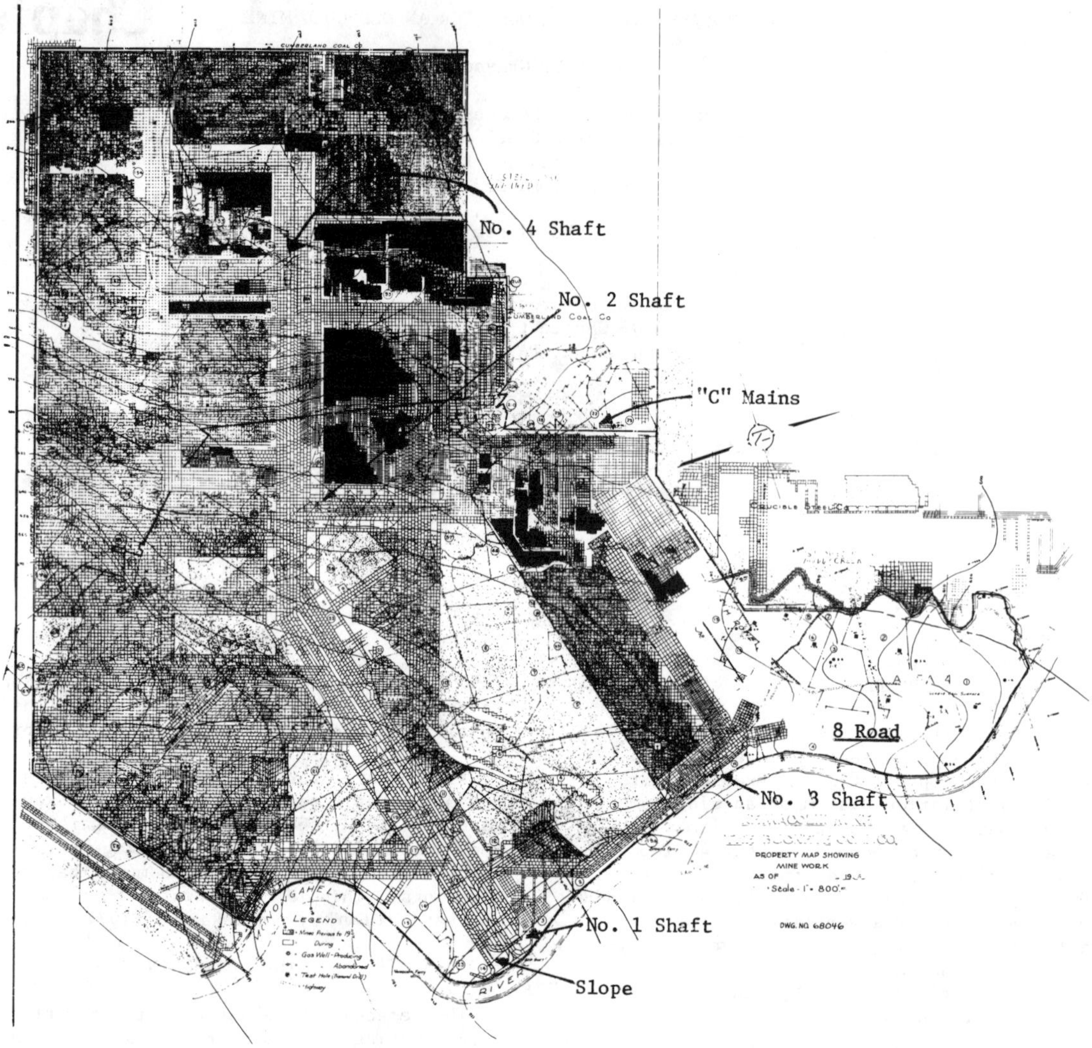

Figure 1. The Nemacolin Mine

development, two sections were splitting pillars, and two sections were on retreat.

As previously mentioned, the 8 Road area of the mine was on the eve of expansion. During formulation of mining projections for this expansion, the ventilation model of the mine was used to predict effects of the differing mining projections and to determine the maximum number of sections which could be operated in this area without the addition of another air shaft. Indeed, a real need existed for just such a model, and during miners' vacation work began to obtain one.

PLANNING THE PRESSURE-QUANTITY SURVEY

Prior to the beginning of miners' vacation at the mine, a meeting was scheduled to fully plan the approach for surveying and modeling the ventilation system. Locations for air quantity and pressure measurements were spotted on 1 to 1200 scale (100 feet equals one inch) mine maps for dissemination to proposed members of survey teams. Nearly 150 air quantity survey points and 106 pressure survey points were selected. Quantity survey points were chosen such that quantities entering and leaving distinct airways or sets of airways, both before and after junctions or splits, would be known. Accurate air readings were desired in order to pinpoint leakages. Two anemometer readings, or more if a definite correlation was not obtained, were planned at each measurement point. Air reading locations were marked with tags inside the mine, indicating each survey location number and the area of that location as measured by the engineering department. In this way, members of survey teams would have no doubt about measuring quantities at proper locations. Pressure survey stations were selected in accordance with suggestions contained in Bureau of Mines Information Circular 7809 (McElroy and Kingery, 1957) at "splits, junctions, and at major changes in the physical characteristics along main airways, at which pressure and quantity data are desirable." The method of survey chosen was to use a base station and multiple roving altimeters. Members of teams were to be briefed upon survey procedures just before deployment, as well as in advance of the survey date. The briefing warned members of possible sources of error and explained the way in which certain problems could be avoided. Requiring special care were the need to watch for altimeter creep and the effect of air velocities exceeding 2 m/s (400 ft/min). A survey date was established, but it was contingent upon weather conditions. Three survey teams, consisting of two members each, were selected so that the entire pressure survey could be completed in one day. The quantity survey could then be completed the following day. An additional person was planned at a base station in the lamphouse on the surface. He would take altimeter readings at 5-minute intervals, along with wet- and dry-bulb temperature readings every 15 minutes. Equipment gathered for the survey included four Wallace and Tiernan altimeters, three low-velocity velometers, two high-velocity anemometers, three recently calibrated anemometers, three 15 m (50 ft) cloth tapes, three magnehelic pressure-difference measuring devices, and three watches with second hands. Sling psychrometers were provided with each altimeter. Distilled water was obtained and placed in containers for use in the survey during wet-bulb temperature measurements. Separate, predesigned forms for recording measured data were provided for use by each team.

THE SURVEY

The day planned for the survey was forecast to remain calm, except for possible late afternoon thundershowers. A decision was made early to deploy on the pressure survey. One person was assigned to monitor the base station and to keep a watch on meteorological developments, noting significant transitions such as increased wind velocity and gathering storm clouds. Each of the three teams was equipped as described and began the survey by 9:00 A.M. The first team surveyed from the slope to 7 South section and the number 2 air shaft area. The second team took the span from number 1 air shaft to the sections near number 3 air shaft. The third team surveyed from number 4 air shaft to the sections in "C" Mains. With continuous base readings being taken, the various roving altimeters were read from station to station along their assigned traverse. Prior to entering the mine, watches were synchronized and altimeters were calibrated to read identically. At each survey point, two altimeter readings were taken with a short time lapse of about 1 minute allowed between them (unless conditions of creep warranted a longer wait), dry-bulb and wet-bulb temperature readings were taken, and the time, along with all other data, was recorded on pre-prepared forms. The elevation of each point was taken from an accurate mine map provided, with adjustments for height due to fallen material noted. Velocities of 1 m/s (200 ft/min) or greater were recorded for possible adjustments of velocity pressure effects. Accomplishment of the pressure survey eventuated smoothly until 1:30 P.M., when weather conditions began to fluctuate. This could have been the source of error in one team's problem of not obtaining a good traverse close, which eventually led to a resurvey of an area for better correlation. All teams completed their traverse and reported back to the base station by 4:30 P.M., at which time one member took a quick survey outside from one surface location to another. Thus, the entire pressure survey was done in a single day as planned; however, as mentioned, one route had to be resurveyed for a better tie.

The quantity survey was also accomplished in one day. Quickness in the survey was facilitated by having all survey stations marked beforehand. The mine was again divided into thirds, roughly along the same routes that were assigned in the pressure survey. Problems commonly encountered in quantity surveys conducted during production time were avoided by scheduling the survey during vacation. As many air readings as necessary to obtain accurate correlation were taken at each station in an effort to pinpoint leakage along each air course and to keep all results within an acceptable error. Intakes, returns, flows over gob areas, and bleeder entry air readings were taken as planned. Areas for mine openings were traversed, and the difference between readings at a station was kept to less than 5% (additional readings were taken as required to obtain them). The air readings which were thusly obtained were then averaged.

DEVELOPMENT OF THE VENTILATION MODEL

Once all of the field data were obtained, compilation of results followed. Each traverse was required to close to within less than 4% of the greatest cumulative water gage calculated along the traverse. Initial results gave closures of 0.1%, 3.8%, and 4.0% difference, but the 4.0% path was resurveyed to 0.3% error because it was in the critical 8 Road expansion area.

After the ventilation results were calculated and balanced, they were placed on a previously developed schematical representation of the mine. Figure 2 is the form by which the results were forwarded to corporate headquarters for computer simulation. The cumulative water gage was given at each node, and the quantity for each entry or set of entries in a branch between nodes was given. Leakage paths and quantities associated therewith were added in order to balance survey results. Flows across mined-out areas were also shown in the manner of leakages.

A modified network was developed for use in connection with the computer. Airway resistances were given in lieu of cumulative water gages at nodes, and junctions were numbered in preparation for computer input.

Solution 1 was the result of inputting, from the modified network, the resistance for each airway (except for regulators), the air quantity at each regulator, and the pressure for each fan. Upon comparing critical quantities obtained in solution 1 with those measured, a favorable correlation existed. Table 1 summarizes these results as columns (1) and (2) are compared. The percentage of deviation between columns (2) and (1) is seen to be less than 5% in all cases, but generally the difference is less than 1%. The one outstanding difference at 4.93% occurs at number 3 fan in the return, but this was expected because of quantities which could not be balanced at the faces of two sections in that area due to the complexity of the system.

RESULTS FROM USE OF THE MODEL

In view of the results obtained from simulation, it was determined that the model developed accurately represented the ventilation system of Nemacolin Mine, at least to within the accuracy of the actual pressure-quantity survey itself. The first application of the model was in the prediction of changes in the ventilation system resulting from proposed resetting of blade positions at fans number 2 and 4 in an attempt to conserve power. It was believed by management that adequate ventilation could be maintained at working faces and yet lower the blade

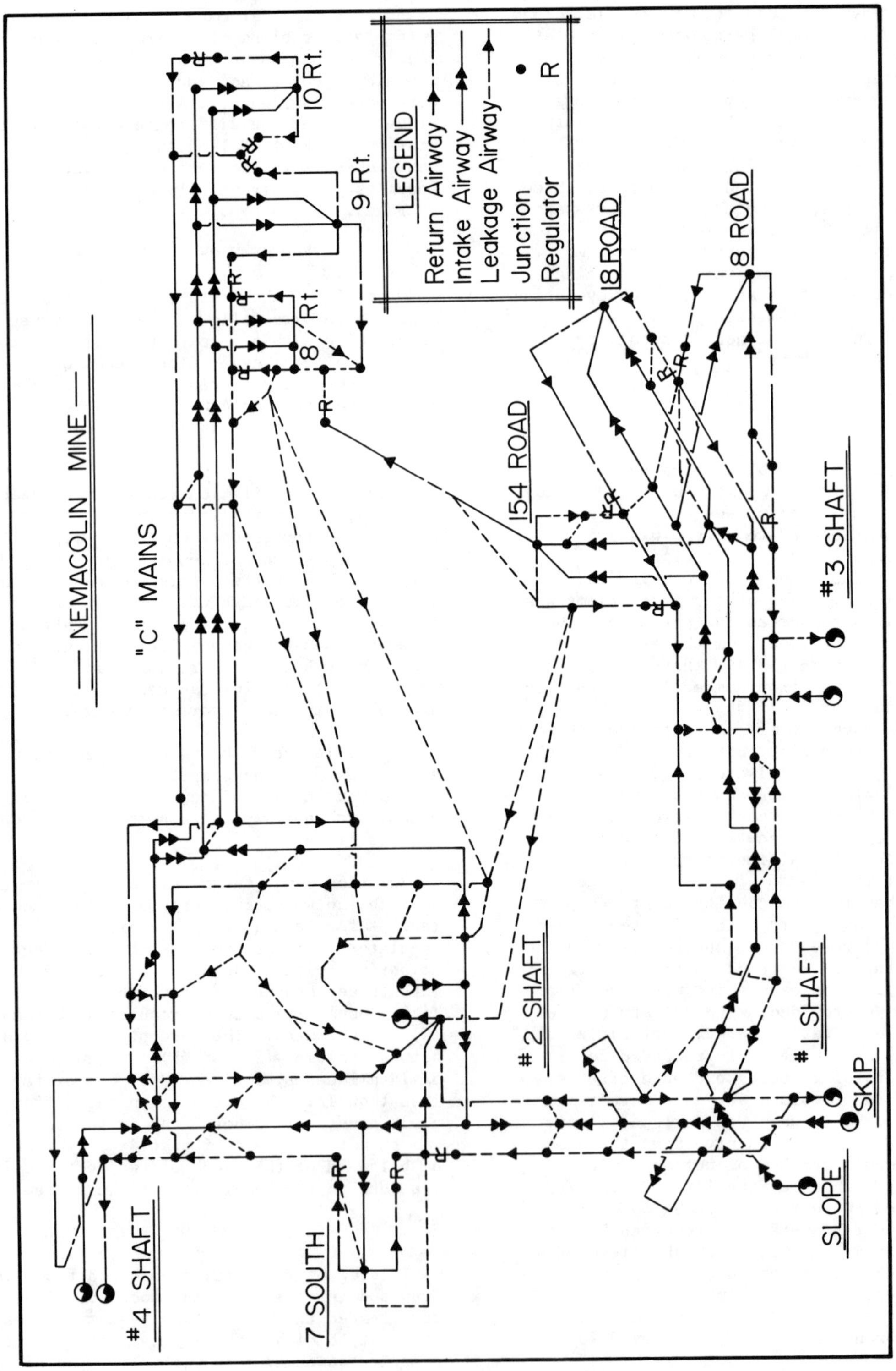

Figure 2. Schematical Representation of the Nemacolin Ventilation System

Table 1. Air Quantities, m^3/s (CFM), at Shafts and Working Faces

	MEASURED (1)	SOLUTION 1 (2)	SOLUTION 2 (3)	DIFFERENCE (2)-(1)	DIFFERENCE %	DIFFERENCE (3)-(2)	DIFFERENCE %
INTAKE							
Slope	87.39 (185,170)	87.48 (185,355)	85.61 (181,406)	0.09 (185)	0.10	- 1.86 (- 3,949)	- 2.13
Skip Shaft	11.31 (23,970)	11.05 (23,408)	10.81 (22,910)	-0.27 (-562)	-2.34	- 0.24 (- 498)	- 2.13
No. 2 Shaft	155.81 (330,136)	156.43 (331,460)	146.51 (310,432)	0.62 (1324)	0.40	- 9.92 (- 21,028)	- 6.34
No. 3 Shaft	95.87 (203,134)	95.40 (202,152)	95.05 (201,393)	-0.46 (-982)	-0.48	- 0.36 (- 759)	- 0.38
No. 4 Shaft	121.97 (258,440)	122.41 (259,377)	114.96 (243,580)	0.44 (937)	0.36	- 7.46 (- 15,797)	- 6.09
TOTAL	472.35 (1,000,850)	472.77 (1,000,752)	452.94 (959,721)	0.43 (902)	0.09	-19.84 (- 42,031)	- 4.20
RETURN							
No. 1 Shaft	61.63 (130,580)	61.62 (130,560)	61.99 (131,358)	-0.01 (20)	-0.02	0.38 (798)	0.61
No. 2 Shaft	167.35 (354,600)	165.77 (351,246)	145.07 (307,390)	-1.58 (-3354)	-0.95	-20.70 (- 43,856)	-12.49
No. 3 Shaft	99.98 (211,847)	104.91 (222,283)	105.18 (222,858)	4.93 (10,436)	4.93	0.27 (575)	0.26
No. 4 Shaft	140.68 (298,087)	140.48 (297,663)	140.70 (298,116)	-0.20 (-424)	-0.14	0.21 (453)	0.15
TOTAL	469.64 (995,114)	472.79 (1,001,752)	452.94 (959,722)	3.13 (6638)	0.67	-19.84 (- 42,030)	- 4.20
WORKING FACE							
7 South	30.98 (65,647)	31.04 (65,765)	28.83 (61,094)	0.06 (118)	0.18	- 2.20 (- 4,671)	-7.10
8 Right	22.47 (47,619)	22.55 (47,791)	21.36 (45,267)	0.08 (172)	0.36	- 1.19 (- 2,524)	-5.28
9 Right	21.67 (45,924)	21.79 (46,176)	20.12 (42,627)	0.12 (252)	0.55	- 1.67 (- 3,549)	-7.69
10 Right	20.06 (42,507)	20.54 (43,521)	18.41 (39,000)	0.48 (1014)	2.39	- 2.13 (- 4,521)	-10.39
154 Road	22.80 (48,304)	22.78 (48,277)	22.50 (47,667)	-0.01 (- 27)	-0.06	- 0.29 (- 610)	-1.26
18 Road	14.25 (30,194)	14.25 (30,197)	14.25 (30,195)	0.00 (3)	0.01	0.00 (- 2)	-0.01
8 Road	14.72 (31,180)	14.69 (31,127)	14.70 (31,143)	-0.03 (- 53)	-0.17	0.01 (16)	0.05
TOTAL	146.95 (311,375)	147.65 (312,854)	140.17 (296,993)	0.70 (1479)	0.47	- 7.49 (- 15,861)	-5.07

Table 2. Fan Pressure and Air Power

	Fan Pressure, Pa (in.)			Air Power, kW (hp)						
	Measured	Sol. 1	Sol. 2	Measured	Sol. 1	Sol. 2	DIFFERENCE		DIFFERENCE	
	(1)	(2)	(3)	(4)	(5)	(6)	(5)-(4)	%	(6)-(5)	%
No. 1 Fan	796 (3.200)	*	*	49.1 (65.84)	49.1 (65.83)	49.4 (66.24)	0.0 (-0.01)	-0.02	0.3 (0.41)	0.62
No. 2 Fan	1244 (5.000)	*	1011 (4.063)	208.3 (279.38)	206.4 (276.74)	146.7 (196.78)	-2.0 (-2.64)	-0.94	-59.6 (-79.96)	-28.89
No. 3 Fan	560 (2.250)	*	*	56.0 (75.11)	58.8 (78.81)	58.9 (79.01)	2.8 (3.70)	4.93	0.1 (0.20)	0.25
No. 4 Fan	1729 (6.950)	*	1636 (6.575)	243.4 (326.45)	243.1 (325.99)	230.3 (308.88)	-0.3 (-0.46)	-0.14	-12.8 (-17.11)	- 5.25
TOTAL				556.9 (746.78)	557.3 (747.37)	485.4 (650.91)	0.4 (0.59)	0.08	-71.9 (-96.46)	-12.91

NOTES: 1. See "RETURN" in Table 1 for air quantities.

2. "*" indicates no change from the last (left) column.

settings of the fans, thereby gaining a significant savings in power consumption and eliminating an overheating of bearings problem at number 4 fan during summer months. Solution 2 represents the results obtained by making the changes described. Table 1, column (3), shows the predicted quantity changes at critical locations, and Table 2 predicts the resulting power savings. It was concluded from these results that a significant power savings of 12.91% could be achieved by making the proposed changes and still maintain adequate face ventilation. At a later date, the predictions were verified. In solution 2, new fan curves were generated for fans number 2 and 4 for input, while fans number 1 and 3 were maintained at previous settings.

Coincident to the development of the model, projections for mining the 8 Road area were being drawn up. Two major sets of projections were proposed, one requiring the sinking of a new air shaft. The ventilation model was used in the decision making process. The decision was to be based on power costs, the capital costs for a new shaft, and, most importantly, the quantity of air reaching each working section at the critical point in development of the area. Eventually, it was decided to forego the shaft, and this decision has been found correct.

CONCLUSION

The Nemacolin ventilation model was invoked twice and management decisions were made based upon its authenticity and predicted results. It was an effective model, but models become obsolete after years of development and retreat alter the configuration of the mine. Before it is used again, a new pressure-quantity survey will be needed to update the model. Results as predicted by the original model should be compared to results obtained by the later survey. This comparison should go a long way in model verification.

REFERENCES

Hartman, H.L., and Trafton, B.O., 1963, "Digital Computer May Find New Use in Determining Mine Ventilation Networks," Mining Engineering, Vol. 15, Sep., pp. 39-42.

Holland, C.T., and Skewes, W.J., 1962, "Investigating Construction Materials and Methods for Stoppings in Coal Mine Ventilation Systems," Trans. SME-AIME, Vol. 223, pp. 21-30.

Kawenski, E.M., and Mitchell, D.W., 1963, "An Evaluation of Stopping Construction," Mining Congress Journal, Vol. 49, Sep., pp. 57-61.

Kharkar, R., Stefanko, R., and Ramani, R.V., 1973, "Analysis of Leakage and Friction Factors in Coal Mine Ventilation Systems," Special Research Report Number 99, Coal Research Board, Pennsylvania State University.

Luxbacher, G.W., Ramani, R.V., and Stefanko, R., 1977, "Application of PSU/MVS to the Analysis of Modification to Existing Ventilation Systems," Final draft report to USBM on Project H0133040.

McElroy, G.W., and Kingery, D.S., 1957, "Making Ventilation-Pressure Surveys With Altimeters," Bureau of Mines Information Circular 7809.

Trafton, B., and Hartman, H.L., 1964, "The Use of the Digital Computer for Mine Ventilation Problems," Trans. SME-AIME, Vol. 229, pp. 313-318.

Wang, Y.J., and Hartman, H.L., 1967, "Computer Solution of Three-Dimensional Mine Ventilation Networks with Multiple Fans and Natural Ventilation," Int. J. Rock Mech. Min. Sc., Vol. 4, pp. 129-154.

Wang, Y.J., and Saperstein, L.W., 1970, "Computer-Aided Solution of Complex Ventilation Networks," Trans. SME-AIME, Vol. 247, pp. 238-250, and discussion Vol. 250, March 1971, pp. 45-46.

Chapter 15

MEASURING AND MODELING NATURAL DRAFT IN UNDERGROUND MINE VENTILATION SYSTEMS

William E. Bruce

Chief, Ventilation Division
Denver Safety and Health Technology Center
Mine Safety and Health Administration

ABSTRACT

This paper is intended to provide an improved understanding of natural draft (natural ventilating pressure) and the effect of natural draft on underground mine ventilation circuits. Attention is focused on measurement of natural draft pressures during ventilation surveys in mechanically ventilated mines and on different methods for modeling natural draft using the measured data.

A number of examples are presented which utilize data collected during absolute air pressure surveys made by personnel of the Ventilation Division of the Mine Safety and Health Administration's Safety and Health Technology Center, Denver, Colorado. Examples are used to illustrate how the data from absolute air pressure surveys may be interpreted to determine whether natural draft pressure is present and what its magnitude is. In addition, two different methods are described for mathematically estimating the magnitude of the natural draft pressure. Comparisons are made of results obtained using the different methods.

Different procedures are described for modeling natural draft using the MSHA-modified Penn State computer program. Use of the routines is described and illustrated relative to drafts that aid or oppose the mechanical ventilation system. Tabulations of results facilitate comparison of outcomes obtained using the different modeling procedures.

INTRODUCTION

The primary mission of the Mine Safety and Health Administration (MSHA) is to ensure safe and healthful working conditions in the nation's mines. The Technical Support organization supports that mission by providing mine operators with recommendations based on the most up-to-date technology available. The most up-to-date technology applicable to optimum mine ventilation includes digital computer modeling of mine ventilation circuits. Proper computer modeling of ventilation circuits must, at times, account for the presence of natural draft in the ventilation system. This paper is written to serve as a guide for persons involved in measuring, modeling, designing, or analyzing underground mine ventilation circuits. The procedures presented are for application to natural drafts (natural ventilating pressures) that are caused by differences in temperatures of air columns within underground mines; natural drafts also result from the wind blowing into mine openings, but wind-induced natural drafts are not addressed in this paper.

BACKGROUND

Natural draft pressure and the calculation of natural draft have been described by Kingery[1] and by Hartman[2]. Natural draft is caused primarily by temperature differences in air columns that extend over the same vertical distance (figure 1).

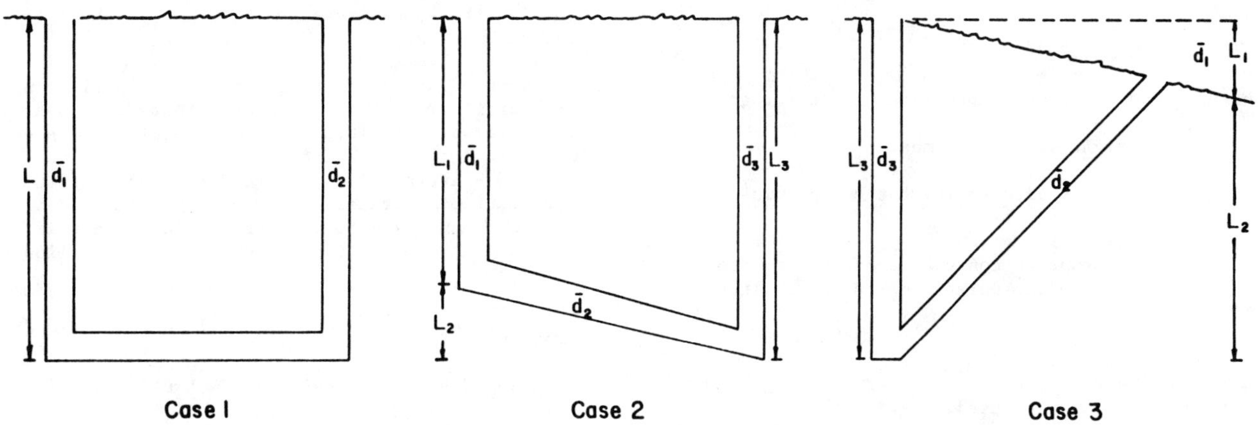

Figure 1. Cross sections of three mines with natural draft.

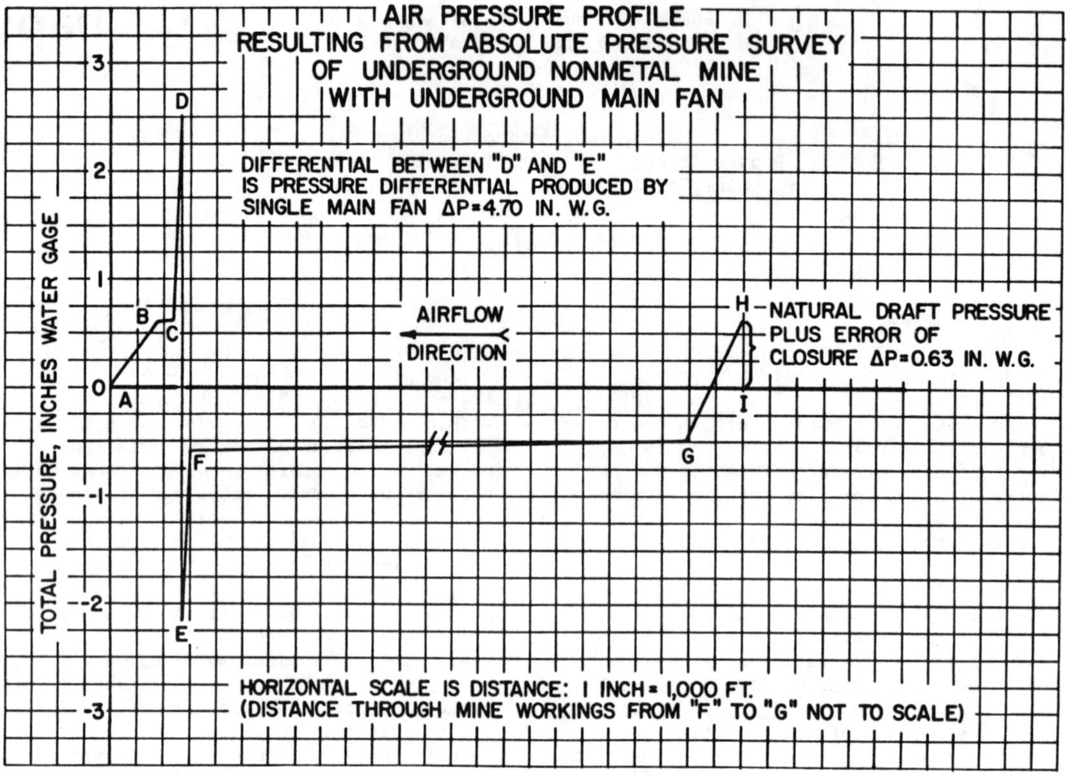

Figure 2. Results of pressure survey when survey and calculations are begun at top of return air shaft and closed at top of intake air shaft.

Although there are a number of sophisticated methods for calculating natural draft pressures, a relatively simple method using air densities has been found satisfactory for the cases described in this paper. In the simple U-shaped mine, as illustrated by case 1 in figure 1, the collar elevations and shaft bottom elevations are equal for both shafts. For such a mine, the difference in average densities between the two air columns can be used to approximate the natural draft pressures as follows:

$$H_n = \frac{L}{5.2} (\bar{d}_1 - \bar{d}_2) \quad , \qquad (1)$$

where H_n = natural draft pressure, in. water gage,

L = height of air columns, ft,

$\bar{d}$ = average air density for one column of air, pcf,

and 5.2 = constant to convert pounds per square foot to in. water gage, psf/in. water gage.

In mines illustrated by cases 2 and 3 of figure 1, a similar formula may be used.

$$H_n = \frac{1}{5.2} \{L_3 \bar{d}_3 - (L_2 \bar{d}_2 + L_1 \bar{d}_1)\} \quad . \qquad (2)$$

Air density may be calculated using the formula

$$d = \frac{1.327 \ (B - 0.378f)}{460 + T} \quad , \qquad (3)$$

where d = air density, pcf,

B = barometric pressure, in. Hg,

T = dry bulb air temperature, ^{o}F,

and f = dewpoint vapor pressure from tables, in. Hg.

It should be mentioned that wet and dry bulb air temperatures and the barometric pressure are required to find the vapor pressure, f, using psychrometric tables.

There is a rule of thumb method for estimating natural draft pressures. It states that at or near sea level air density (0.075 pcf), natural draft pressure is approximately equal to 0.030 in. water gage for each 10°F average temperature difference per 100-ft increment of vertical elevation. Application of the rule of thumb will be illustrated later in this paper. The rule may be written mathematically as follows:

$$H'_n = 0.030 \left(\frac{L}{100}\right) \left(\frac{\Delta T}{10}\right) \quad , \qquad (4)$$

where H'_n = estimated natural draft pressure, in. water gage,

L = height of air column, ft,

and ΔT = difference between average dry bulb temperatures of air columns, ^{o}F.

Ventilation pressure measurements constitute an important part of a complete mine ventilation survey.

The pressure survey data are used to diagnose and correct a number of different problems that may be present in a mine ventilation system. Total pressure data may be used, for example, to pinpoint areas of excessive pressure loss in the mine ventilation circuit. In addition, the pressure data is often used to develop ventilation system models which are analyzed using digital computer techniques.

Three U.S. Government publications provide in-depth discussion and analysis of the subject of underground mine ventilation pressure surveys. The McElroy-Kingery report[3] describes the techniques of making a pressure survey as well as describing the reduction of data step-by-step. The report by Harris et al.[4] describes a program for quicker, more accurate data reduction of pressure survey data using a programmable desk-top calculator. The Anderson-Nugent[5] publication describes development of additional programs for pocket programmable calculators for speedy, accurate reduction of pressure survey data.

As previously mentioned, data collected during pressure surveys is often used in modeling of ventilation systems using digital computer techniques. The basic ventilation system modeling techniques presently used by MSHA have been described in detail in MESA Informational Report 1066 by Anderson and Dvorkin[6] and will not be reiterated herein. The existence of a significant natural draft during a pressure survey will require that some additional special techniques be used in modeling such ventilation systems.

NATURAL DRAFT PRESSURE IN MINE VENTILATION PRESSURE SURVEYS

The presence of natural draft air pressures in mechanically ventilated underground mines is not at all uncommon. Such natural draft pressures, however, often confuse and complicate the final reduction and interpretation of total pressure data collected during a ventilation pressure survey. An example of data collected during a ventilation pressure survey is shown in figure 2.

Survey Methods

A few precautions should be observed when one decides to determine natural draft pressures during a pressure survey in a mechanically ventilated mine. The survey method used can be the base-and-rover method, the leapfrog method, or a combination of both methods. The base altimeter may be located outside the mine in the atmosphere or an underground base altimeter may be used once a total pressure value has been established for the proposed underground base location. The survey should begin and end in the atmosphere outside the mine. This can be accomplished by starting the survey in the atmosphere outside the mine intake opening, proceeding downstream along the intake system and into the return system, and ending in the atmosphere outside the mine return opening. Alternately the foregoing procedure can be reversed, proceeding from return to intake. The easiest way to calculate the survey is in the direction that it was run, and the apparent closure error if any, will be noted at the end of the calculations. In the absence of survey errors, the apparent closure error will consist of natural draft pressure.

As with any pressure survey, the elapsed time for completion should be kept to a minimum. Best results will be obtained when in-mine and atmospheric temperatures remain relatively constant during the course of the survey. Main fan performance measurements should be made either concurrently with the pressure survey or when airway and atmospheric temperatures are the same as during the pressure survey.

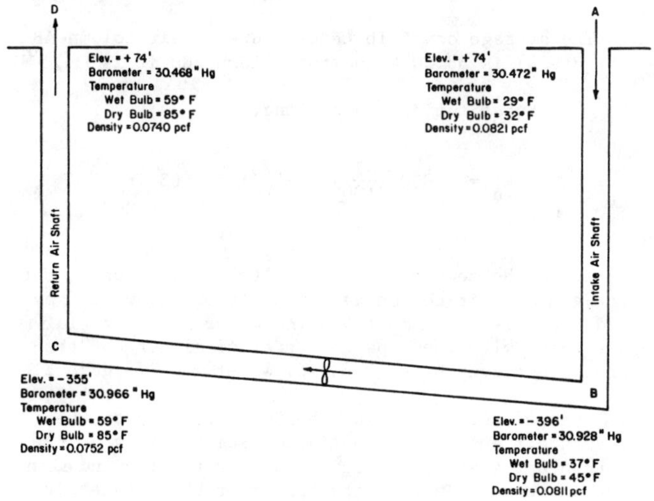

Figure 3. Conditions causing natural draft pressure in nonmetal mine.

Nonmetal Mine Example

Figure 2 is the total pressure profile that was developed using data from a pressure survey in a nonmetal mine. The error of closure (from "H" to "I") is 0.63 in. water gage. Figure 3 shows the temperature relationships in the mine, and calculations show that the apparent error of closure is almost entirely due to natural draft pressure.

Air densities have been calculated using equation 3, and the results are shown in figure 3. Equation 2 was used to calculate the natural draft pressures as follows:

$$H_n = \frac{1}{5.2}\ \{L_3\,\bar{d}_3 - (L_2\,\bar{d}_2 + L_1\,\bar{d}_1)\},$$

where $\bar{d}_3 = \frac{1}{2}\ (.0821 + .0811) = 0.08160$ pcf,

$$\bar{d}_2 = \frac{1}{2}\ (.0811 + .0752) = 0.07815 \text{ pcf},$$

$$\bar{d}_1 = \frac{1}{2}\ (.0752 + .0740) = 0.07460 \text{ pcf},$$

$$L_3 = 470 \text{ ft},$$

$$L_2 = 41 \text{ ft},$$

and $L_1 = 429$ ft.

Substituting and solving

$$H_n = \frac{1}{5.2}\ \{470(.08160)-41(.07815)-429(.07460)\}$$

$$= 0.60 \text{ in. water gage.}$$

Using the rule of thumb (equation 4),

$$H'_n = 0.030\ (\frac{L}{100})\ (\frac{\Delta T}{10}).$$

The average dry bulb temperature of air column AB (figure 3) is 38.50°F and for column BCD it is 83.25°F.

Substituting and solving,

$$H'_n = 0.030\ (\frac{470}{100})\ (\frac{44.75}{10}) = 0.63 \text{ in. water gage.}$$

It can be seen that the results obtained using both the average air column densities (0.60 in. water gage) and the rule of thumb (0.63 in. water gage) are in agreement with the apparent error of closure of the pressure survey, i.e. 0.63 in. water gage (figure 2).

The pressure survey was begun at the top of the return air shaft, was continued from the foot of the return air shaft and progressed past the fan and along the intake to close at the collar of the intake air shaft. The return air temperatures exceeded the intake air temperatures, and, as a result, the natural draft was assisting the fan. Those individuals who are familiar with total pressure profiles may notice on figure 2 that the positive pressure (closure error) noted at point "H" or "I" is much the same condition that one would expect if a blowing fan, operating at plus 0.63 in. water gage, had been located at the top of the intake air shaft.

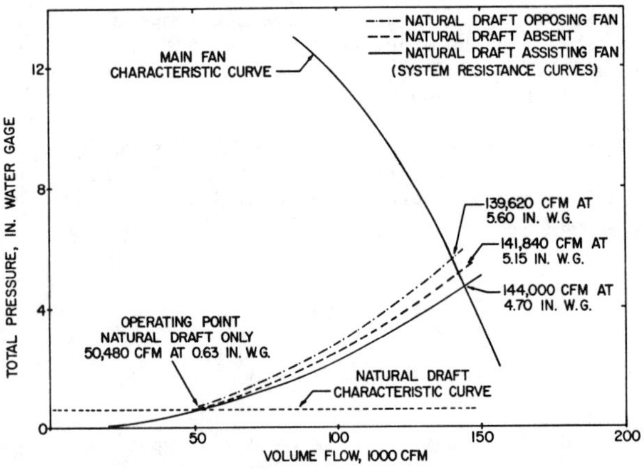

Figure 4. Fan and natural draft characteristic curves with apparent mine resistance curves and operating points.

NATURAL DRAFT PRESSURE IN VENTILATION SYSTEM MODELING

As mentioned earlier in this paper, the basic techniques for digital computer modeling have been described in detail in MESA (now MSHA) Informational Report 1066 by Anderson and Dvorkin.

Two basic but slightly different problems in modeling mine ventilation systems with natural draft present are envisioned by the author. The first type of modeling problem would typically be encountered by an individual who has made an air pressure-air quantity ventilation survey complete with pitot tube measurements made to determine main fan performance. This individual is aware that a natural draft is influencing the ventilation system and wishes to model the system with and without natural draft present. In this case, the model would be developed by entering the measured total pressures (identified as absolute pressures on the MSHA modified Penn State computer format) and air quantities, as well as the fan data and measured natural draft data.

The second type of modeling problem would typically be encountered by an individual who is applying digital computer modeling to the design of a mine ventilation system. This individual also may wish to determine the effects of natural drafts on the ventilation system model. In this second type problem, the model would be developed by entering estimated airway resistance values, as well as projected fan and natural draft data.

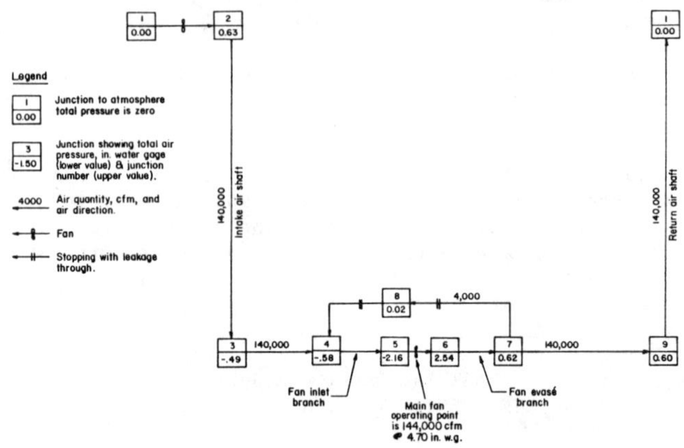

Figure 5. Mine ventilation system model with natural draft present and assisting mechanical ventilation, constructed using data collected during pressure-quantity ventilation survey.

Both problem types are very similar. The illustrations to follow begin with an example (table 1) which is representative of the case where each airway branch is defined by two absolute pressures and a quantity. However, data files set up for input in the form of absolute pressures and quantities may be transformed to the format of a data file using only resistances. The program used to make the transformation is named ROUT. The ROUT program reads the data files, and if those files are set up with input other than resistance (e.g. absolute pressures and quantities), then ROUT changes the files so that they will have resistance values in place of absolute pressure values. The main use of ROUT is to simplify the editing of the data file. The data file with airway resistance values is easier to edit than one with absolute pressures. Editing is required when digital computer projections are made to show results of proposed ventilation system changes.

Example 1

The first example is presented to show how to model a ventilation system where a natural draft is assisting an underground main fan. Figures 4 and 5 present data such as would be collected during a mine ventilation pressure-quantity survey. Figure 4 shows the measured main fan performance of 144,000 cfm at 4.70 in. water gage total pressure. Figure 5 shows the measured main fan operating point, the measured total air pressures, and the balanced air quantities for the entire mine ventilation circuit. Junction 2 on figure 5 represents a point in the atmosphere near the top of the intake air shaft. The pressure survey was closed at junction 2. The total pressure value of 0.63 in. water gage noted at junction 2 is the natural draft pressure. The natural draft pressure is represented by the symbol for a fan located between junctions 1 and 2 (figure 5). All junctions designated "1" are at atmospheric pressure. The ventilation branch that connects junctions 4 and 5 represents the inlet conditions for that particular fan installation. In reality, the fan inlet conditions were poor as a result of a missing nose piece. The branch that connects junctions 6 and 7 represents the discharge conditions for the fan installation. Discharge conditions were less than ideal because only one evase' section was used, whereas two are recommended. The branches connecting junctions 7 and 8 and junctions 8 and 4 represent the leakage path that existed through the fan bulkhead. The bulkhead was relatively well-constructed.

The ventilation system data entered into the computer (plus some computed data) are shown in table 1. Airway data actually entered include the junction numbers listed in the columns headed "J1" and "J2," total air pressures corresponding to each junction listed in the columns headed "ABS1" and "ABS2" and air quantities (in units of 100,000 cfm) in the column headed "QUANT." The first two branches in table 1 are the fan branches (junction 5 and 6 and junctions 1 and 2) and no total pressures are assigned to fan branches which are assumed to have zero resistance. Fan data actually entered are shown following the designation "FAN 1" and "FAN 2." Coordinates for the mechanical fan characteristic curve (figure 4) are entered following "FAN 1" with the fan air quantities (in units of 100,000 cfm) entered in the column headed "X" and the fan total pressure (in in. water gage) entered in the column headed "FX." Coordinates for the natural draft fan curve (figure 4) are entered in a similar manner following "FAN 2."

Probably the most significant element of the procedure followed in the example is the method for modeling the natural draft. It can be seen in figure 4 that the natural draft characteristic is that of a constant-pressure air-moving device.

Example 2

This example (table 2 and figure 6) is presented to show how to model the example mine with the natural draft removed so that the mine is ventilated only with the mechanical fan. The input data (table 2) has been transformed using the ROUT program described previously. The input fan air quantity (branch 1) need be only a very rough estimate (more iterations are required for a poor estimate), and the other input air quantities may have any value including zero. A text-editing procedure has been used to remove the natural draft fan from the circuit. The resistance of branch 2 (connecting junctions 1 and 2), which was formerly the natural draft branch, is unchanged, remaining at zero resistance. Branch 2 has been left in the circuit only to simplify editing and it does not represent an in-mine airway. The resistances of all the other branches remain the same as in example 1. The results of the computer run are shown in figure 6. The new operating point for the fan is also illustrated on figure 4 where it occurs at the intersection of the fan characteristic curve and the resistance curve for the system with natural draft absent.

Table 1. ORIGINAL DATA DIGITAL COMPUTER INPUT-OUTPUT FOR MECHANICALLY VENTILATED MINE MODELED WITH NATURAL DRAFT PRESENT AND ASSISTING FAN.

```
*************************************************************************

    NB       NJ     MAXJ       NM       NF   MAXIT   NFIXB          E
    10        9        9        2        2      50       0      .0001

 IFANCO    IRES   ITPRES    INPRT     IBAL
      3       1        1        1        1

BRANCH  J1  J2            R      ABS1      ABS2     QUANT         H

     1   5   6     0.000000     0.000     0.000   1.44000     0.000
     2   1   2     0.000000     0.000     0.000   1.40000     0.000
     3   2   3      .571429      .630     -.490   1.40000     1.120
     4   3   4      .045918     -.490     -.580   1.40000      .090
     5   4   5      .761960     -.580    -2.160   1.44000     1.580
     6   6   7      .925926     2.540      .620   1.44000     1.920
     7   7   8   375.000000      .620      .020    .04000      .600
     8   8   4   375.000000      .020     -.580    .04000      .600
     9   7   9      .010204      .620      .600   1.40000      .020
    10   9   1      .306122      .600     0.000   1.40000      .600

FAN        J1      J2       KEY     POINT          X         FX

 1          5       6         0
                                        1      .8500    13.0000
                                        2     1.1000    10.4925
                                        3     1.3000     7.4177
                                        4     1.4400     4.7000
                                        5     1.5600     2.0000

 COEFFICIENTS

     1.042096E+01   1.312908E+01  -1.187636E+01
         QMAX          1.560   FQMAX      2.000
         QMIN           .850   FQMIN     13.000

FAN        J1      J2       KEY     POINT          X         FX

 2          1       2         0
                                        1      .1000      .6300
                                        2     1.4000      .6300
                                        3     2.0000      .6300

 COEFFICIENTS

     6.300000E-01  -2.273737E-13   0.
         QMAX          2.000   FQMAX       .630
         QMIN           .100   FQMIN       .630

                      * * * OUTPUT * * *

      1 ITERATIONS             SUMD =  .2426351E-07


 BRANCH  J1  J2            R         Q         H

      1   5   6     0.000000    1.4400     0.000
      2   1   2     0.000000    1.4000     0.000
      3   2   3      .571429    1.4000     1.120
      4   3   4      .045918    1.4000      .090
      5   4   5      .761960    1.4400     1.580
      6   6   7      .925926    1.4400     1.920
      7   7   8   375.000000     .0400      .600
      8   8   4   375.000000     .0400      .600
      9   7   9      .010204    1.4000      .020
     10   9   1      .306122    1.4000      .600


 ALL TOTAL PRESSURES HAVE BEEN CALCULATED.
 (THE PROCESS REQUIRED   2 ITERATIONS.)
 THE NON-ZERO VALUES ARE PRINTED BELOW.
 (JUNCTION NO. ON 1ST LINE, TOT. PRES. BELOW IT ON 2ND).

      2         3         4         5         6         7
    .630     -.490     -.580    -2.160     2.540      .620

      8         9
    .020      .600


 FAN  J1  J2         Q        FQ

   1   5   6    1.4400     4.700
   2   1   2    1.4000      .630
```

Example 3

This example is presented (table 3 and figure 7) to show how to model the example mine with the mechanical fan turned off so that the mine is ventilated only with natural draft. In this example, the natural draft is represented by fan 1; editing procedures were used to delete the mechanical fan data. As mentioned previously, the only data required in the column headed "QUANT" is a rough estimate of the air quantity moved under natural draft conditions. In this case, an estimate of 1.44 (144,000 cfm) was used. The other

Table 2. DIGITAL COMPUTER INPUT-OUTPUT FOR MECHANICALLY VENTILATED MINE MODELED WITH NATURAL DRAFT REMOVED.

NB	NJ	MAXJ	NM	NF	MAXIT	NFIXB	E
10	9	9	2	1	50	0	.0001

IFANCO	IRES	ITPRES	INPRT	IBAL
3	0	1	1	1

BRANCH	J1	J2	R	QUANT
1	5	6	0.000000	1.44
2	1	2	0.000000	1.40
3	2	3	.571429	1.40
4	3	4	.045918	1.40
5	4	5	.761960	1.44
6	6	7	.925926	1.44
7	7	8	375.000000	.04
8	8	4	375.000000	.04
9	7	9	.010204	1.40
10	9	1	.306122	1.40

FAN	J1	J2	KEY	POINT	X	FX
1	5	6	0			
				1	.8500	13.0000
				2	1.1000	10.4925
				3	1.3000	7.4177
				4	1.4400	4.7000
				5	1.5600	2.0000

COEFFICIENTS

1.042096E+01	1.312908E+01		-1.187636E+01
QMAX	1.560	FQMAX	2.000
QMIN	.850	FQMIN	13.000

* * * OUTPUT * * *

9 ITERATIONS   SUMD = .1547338E-05

BRANCH	J1	J2	R	Q	H
1	5	6	0.000000	1.4184	0.000
2	1	2	0.000000	1.3701	0.000
3	2	3	.571429	1.3701	1.073
4	3	4	.045918	1.3701	.086
5	4	5	.761960	1.4184	1.533
6	6	7	.925926	1.4184	1.863
7	7	8	375.000000	.0483	.876
8	8	4	375.000000	.0483	.876
9	7	9	.010204	1.3701	.019
10	9	1	.306122	1.3701	.575

ALL TOTAL PRESSURES HAVE BEEN CALCULATED.
(THE PROCESS REQUIRED 5 ITERATIONS.)
THE NON-ZERO VALUES ARE PRINTED BELOW.
(JUNCTION NO. ON 1ST LINE, TOT. PRES. BELOW IT ON 2ND).

3	4	5	6	7	8
-1.073	-1.159	-2.692	2.457	.594	-.283

9
.575

FAN	J1	J2	Q	FQ
1	5	6	1.4184	5.149

numbers in the column may be replaced with zeros or may have any convenient value. The remaining values in the "QUANT" column are those that were present during the example 1 computations. The results are similar to those previously described. The resulting air quantity and total pressure values are shown in figure 7. The natural draft performance is illustrated graphically in figure 4 where the operating point of 50,480 cfm occurs at the intersection of the natural draft characteristic curve and the resistance curve for the system.

Example 4

The purpose of this example (table 4 and **figure** 5) is to illustrate a different technique for handling the same case discussed in example 1. The conditions are the same as those in example 1, i.e. the mine is mechanically ventilated by a single fan, and the natural draft is aiding the fan. The modeling method uses the existing natural draft routine in the MSHA-modified Penn State Program.

To use the natural draft routine, one must design the amount of natural draft that will be present in each branch of the ventilation circuit. Natural draft data used as input for this example are taken from the total pressure differences between junctions in figure 7.

The format for the data input is illustrated in table 4, where the natural draft input data appear in the column headed "NVP" following the data for general branches. When the natural draft data are entered at the computer terminal, the junctions (nodes) must be entered in the same order as was used for the general branches, i.e., J1 and J2 shall not be reversed. If the direction of pressure loss is different for natural draft than it is for the mechanical ventilation system, then the sign must be reversed. It may be noted (figures 5 and 7) that in branches connecting junctions (nodes) 7 to 8 and 8 to 4 that the natural draft pressure loss is in a different direction than in the mechanically ventilated system, and for that reason the natural draft pressure differential values entered in branches 7 and 8 in table 4 are preceded by minus signs. The results obtained using this alternative method are identical to the results described in example 1.

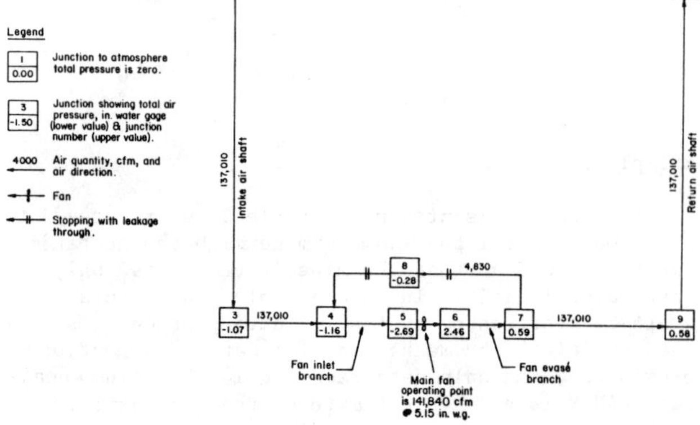

Figure 6. Mine ventilation system model with natural draft removed, mechanical ventilation only.

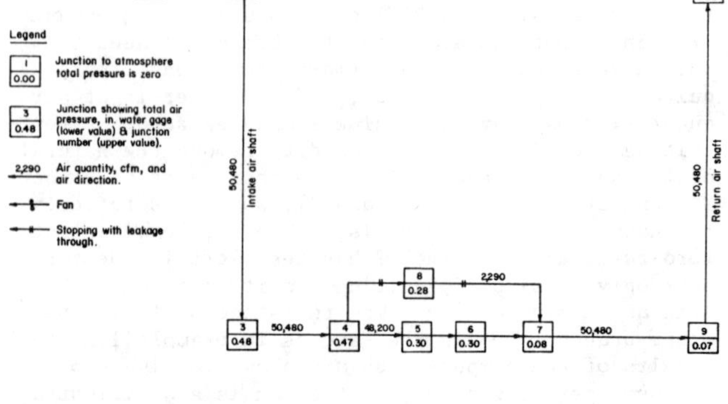

Figure 7. Mine ventilation system model with ventilation provided only by natural draft.

```
*****************************************************************

Table 3. DIGITAL COMPUTER INPUT-OUTPUT FOR MINE MODEL
VENTILATED ONLY BY NATURAL DRAFT.

*****************************************************************

   NB      NJ    MAXJ      NM      NF   MAXIT  NFIXB        E
   10       9       9       2       1      50      0    .0001

IFANCO    IRES  ITPRES   INPRT    IBAL
     3       0       1       1       1

BRANCH  J1  J2           R            QUANT

     1   1   2    0.000000             1.44
     2   5   6    0.000000             1.40
     3   2   3     .571429             1.40
     4   3   4     .045918             1.40
     5   4   5     .761960             1.44
     6   6   7     .925926             1.44
     7   7   8  375.000000              .04
     8   8   4  375.000000              .04
     9   7   9     .010204             1.40
    10   9   1     .306122             1.40


FAN        J1   J2      KEY    POINT          X        FX

 1          1    2        0
                                   1      .1000     .6300
                                   2     1.4000     .6300
                                   3     2.0000     .6300

 COEFFICIENTS

    6.300000E-01   -2.273737E-13   0.
         QMAX       2.000   FQMAX       .630
         QMIN        .100   FQMIN       .630


                 * * * OUTPUT * * *

    9 ITERATIONS          SUMD =  .5373324E-04


BRANCH  J1  J2           R          Q        H

     1   1   2    0.000000      .5048    0.000
     2   5   6    0.000000      .4820    0.000
     3   2   3     .571429      .5048     .146
     4   3   4     .045918      .5048     .012
     5   4   5     .761960      .4820     .177
     6   6   7     .925926      .4820     .215
     7   7   8  375.000000     -.0229    -.196
     8   8   4  375.000000     -.0229    -.196
     9   7   9     .010204      .5048     .003
    10   9   1     .306122      .5048     .078

ALL TOTAL PRESSURES HAVE BEEN CALCULATED.
(THE PROCESS REQUIRED   2 ITERATIONS.)
THE NON-ZERO VALUES ARE PRINTED BELOW.
(JUNCTION NO. ON 1ST LINE, TOT. PRES. BELOW IT ON 2ND).

     2          3         4         5          6         7
   .630       .484      .473      .296       .296      .081


     8          9
   .277       .078


FAN  J1  J2          Q         FQ

  1   1   2      .5048       .630
```

```
*****************************************************************

Table 4. DIGITAL COMPUTER INPUT-OUTPUT FOR MECHANICALLY
VENTILATED MINE MODELED WITH NATURAL DRAFT PRESENT AND
ASSISTING THE FAN.  USING EXISTING NATURAL DRAFT ROUTINE
IN THE MODIFIED PENN STATE PROGRAM.

*****************************************************************

   NB      NJ    MAXJ      NM      NF   MAXIT  NFIXB        E
   10       9       9       2       1      50      0    .0001

IFANCO    IRES  ITPRES   INPRT    IBAL
     3       0       1       1       1

BRANCH  J1  J2           R            QUANT

     1   5   6    0.000000             1.44
     2   1   2    0.000000             0.00
     3   2   3     .571429             0.00
     4   3   4     .045918             0.00
     5   4   5     .761960             0.00
     6   6   7     .925926             0.00
     7   7   8  375.000000             0.00
     8   8   4  375.000000             0.00
     9   7   9     .010204             0.00
    10   9   1     .306122             0.00

   BRANCH          NVP

        3     .1460000
        4     .0110000
        5     .1770000
        6     .2150000
        7    -.1960000
        8    -.1960000
        9     .0030000
       10     .0780000

FAN        J1   J2      KEY    POINT          X        FX

 1          5    6        0
                                   1      .8500   13.0000
                                   2     1.1000   10.4925
                                   3     1.3000    7.4177
                                   4     1.4400    4.7000
                                   5     1.5600    2.0000

COEFFICIENTS

    1.042096E+01   1.312908E+01  -1.187636E+01
         QMAX       1.560   FQMAX     2.000
         QMIN        .850   FQMIN    13.000


                 * * * OUTPUT * * *


    8 ITERATIONS          SUMD =  .3636845E-04


BRANCH  J1  J2           R          Q        H

     1   5   6    0.000000     1.4400    0.000
     2   1   2    0.000000     1.4000    0.000
     3   2   3     .571429     1.4000    1.120
     4   3   4     .045918     1.4000     .090
     5   4   5     .761960     1.4400    1.580
     6   6   7     .925926     1.4400    1.920
     7   7   8  375.000000      .0400     .600
     8   8   4  375.000000      .0400     .600
     9   7   9     .010204     1.4000     .020
    10   9   1     .306122     1.4000     .600


ALL TOTAL PRESSURES HAVE BEEN CALCULATED.
(THE PROCESS REQUIRED   5 ITERATIONS.)
THE NON-ZERO VALUES ARE PRINTED BELOW.
(JUNCTION NO. ON 1ST LINE, TOT. PRES. BELOW IT ON 2ND).

     2          3         4         5          6         7
   .630      -.490     -.580    -2.160      2.540      .620

     8          9
   .020       .600


FAN  J1  J2          Q         FQ

  1   5   6     1.4400      4.700
```

Example 5

This final example (table 5 and figure 8) is presented to show how to model a natural draft that is opposing the mine fan. The existing natural draft routine in the MSHA-modified Penn State Program is the only technique that was found to be satisfactory for the situation where the draft opposes the fan.

As in the previous example, it is necessary to design the natural draft differential that will be present in each branch of the ventilation circuit prior to entering the data. The values used for natural draft differentials were exactly the same as those used in the previous example, except that the signs were reversed because flow direction was tending to reverse.

Table 5. DIGITAL COMPUTER INPUT-OUTPUT FOR MECHANICALLY VENTILATED MINE MODELED WITH NATURAL DRAFT PRESENT BUT OPPOSING THE FAN. USING EXISTING NATURAL DRAFT ROUTINE IN THE MODIFIED PENN STATE PROGRAM.

```
    NB       NJ    MAXJ       NM       NF   MAXIT   NFIXB         E
    10        9       9        2        1      50       0     .0001

IFANCO    IRES  ITPRES    INPRT     IBAL
     3       0       1        1        1

BRANCH  J1  J2             R            QUANT

     1   5   6      0.000000             1.44
     2   1   2      0.000000             0.00
     3   2   3       .571429             0.00
     4   3   4       .045918             0.00
     5   4   5       .761960             0.00
     6   6   7       .925926             0.00
     7   7   8    375.000000             0.00
     8   8   4    375.000000             0.00
     9   7   9       .010204             0.00
    10   9   1       .306122             0.00

  BRANCH          NVP

       3    -.1460000
       4    -.0110000
       5    -.1770000
       6    -.2150000
       7     .1960000
       8     .1960000
       9    -.0030000
      10    -.0780000

FAN        J1    J2       KEY    POINT          X         FX

 1          5     6         0
                                     1      .8500    13.0000
                                     2     1.1000    10.4925
                                     3     1.3000     7.4177
                                     4     1.4400     4.7000
                                     5     1.5600     2.0000

 COEFFICIENTS

     1.042096E+01   1.312908E+01  -1.187636E+01
          QMAX         1.560   FQMAX      2.000
          QMIN          .850   FQMIN     13.000


                   * * * OUTPUT * * *

     9 ITERATIONS           SUMD =  .5334764E-05


BRANCH  J1  J2             R        Q         H

     1   5   6      0.000000   1.3963     0.000
     2   1   2      0.000000   1.3408     0.000
     3   2   3       .571429   1.3408     1.027
     4   3   4       .045918   1.3408      .083
     5   4   5       .761960   1.3963     1.485
     6   6   7       .925926   1.3963     1.805
     7   7   8    375.000000    .0555     1.154
     8   8   4    375.000000    .0555     1.154
     9   7   9       .010204   1.3408      .018
    10   9   1       .306122   1.3408      .550


ALL TOTAL PRESSURES HAVE BEEN CALCULATED.
(THE PROCESS REQUIRED   5 ITERATIONS.)
THE NON-ZERO VALUES ARE PRINTED BELOW.
(JUNCTION NO. ON 1ST LINE, TOT. PRES. BELOW IT ON 2ND).

     2         3         4         5         6         7
  -.630    -1.657    -1.740    -3.225     2.374      .569

     8         9
  -.586      .550

FAN  J1  J2        Q         FQ

  1   5   6   1.3963      5.599
```

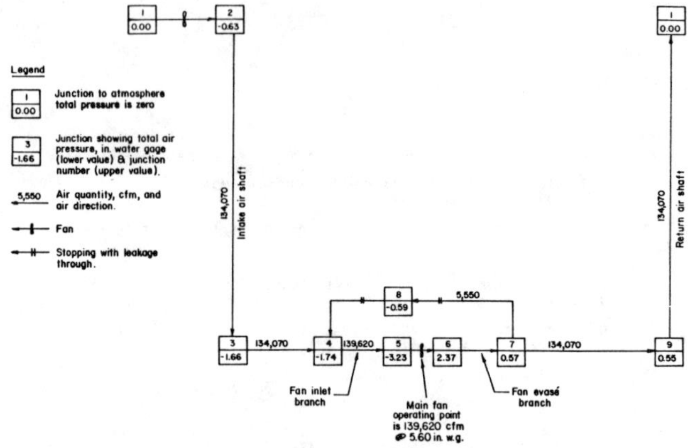

Figure 8. Mine ventilation system model with natural draft present and opposing mechanical ventilation; modeled using modified Penn State routine for natural draft.

The computer output presented in figure 8 shows the mine model with natural draft opposing the fan. The main fan operating point is shown in figure 4 where it occurs at the intersection of the fan characteristic curve and the system resistance curve for the system when natural draft opposes the fan.

CONCLUSION

The subject of natural draft relative to underground mine ventilation circuits has been and continues to be a troublesome and controversial subject. This paper has presented different methods of calculating and measuring natural draft pressures using actual field data in the accompanying examples. It has been shown that there is good agreement of results for the various methods discussed and that the simple rule of thumb method gives a quick and accurate estimate of the magnitude of natural draft pressure.

Methods for digital computer modeling of natural draft in the MSHA-modified Penn State Program have been discussed and illustrated. Methods applicable to both measured and designed ventilation circuits have been presented. It was shown how the differing approaches produce identical results. However, only one method was presented for handling an opposing natural draft; development of an alternate simpler method which could be easily applied to measured ventilation circuits remains a high-priority need.

REFERENCES

1. Kingery, D.S. Introduction to Mine Ventilating Principles and Practices. BuMines Bull. 589, 1960, 54 pp.

2. Hartman, Howard L. Mine Ventilation and Air Conditioning. The Ronald Press Co., New York, 1961, 398 pp.

3. McElroy, G.E., and D.S. Kingery. Making Ventilation - Pressure Surveys With Altimeters. BuMines Inf. Circ. 7809, 1957, 20 pp.

4. Harris, E.J., R.W. Dalzell, R.J. Kline, and E.J. Miller. A Method for Calculating Mine Ventilation Pressure Losses Using Computers and Desk-Top Calculators. BuMines Inf. Circ. 8594, 1973, 22 pp.

5. Anderson, Thomas C., and Joseph W. Nugent. Mine Ventilation Pressure Differentials With a Programmable Hand Calculator. MESA Inf. Rept. 1050, 1977, 29 pp.

6. Anderson, Thomas C., and David Dvorkin. Mine Ventilation Digital Simulation and Analysis Capabilities at MESA's Denver Technical Support Center. MESA Inf. Rept. 1066, 1978, 108 pp.

Chapter 16

COMPUTER-ASSISTED VENTILATION PLANNING AT A COAL MINE

Duk-Won Park

Associate Professor of Mineral Engineering
The University of Alabama
University, Alabama

ABSTRACT

The PSU/MVS program, one of the computer programs for ventilation network analysis most recently developed at The Pennsylvania State University, was applied to the analysis of modifications to an existing ventilation system for a coal mine located in southern West Virginia. The program is well designed and has diversified capabilities which allow the users to make decisions on the optimum size and number of fans, the location and number of airshafts, and the optimum operational procedures.

The mine had 10 working faces, from which the average production rate was about 5900 tons of clean coal per day. Two vane-axial mine fans were used in parallel at a shaft, and both of them operated simultaneously. One additional shaft was in the final construction stage and it would be used in four months to alleviate the load for the existing fans and distribute intake air more evenly.

The overall ventilation surveys, including the altimeter pressure survey and the air quantity survey, were carried out in the mine entries, shafts, slopes, and gob areas. Computations were made using the above-mentioned computer program on the present condition and after one additional fan was placed in operation at the new shaft. Also, recommendations were made on the air route and fan blade settings for the most economical and efficient ventilation system.

Further studies were carried out to decide the proper period when the third air shaft would be required according to the long-term mining plan. Additional branches were added for the next 4 years of the production plan, and the ventilation conditions were tested at the end of each year. Numerous combinations of blade settings and modifications of air routes were tested. It was recommended that the third air shaft be added at the end of the third year so as not to experience fan stall.

INTRODUCTION

In recent years, the development of larger machines for the higher coal production rates has contributed to the complexity of ventilation planning, and more careful ventilation planning is required to meet the stringent regulations being imposed. "Rules of Thumb" or hand calculations can be applied to a certain degree, but the complexity of the system requires better tools, such as computer-aided analysis.

Digital computer technology has been utilized for mine ventilation network analysis since the 1950's. Although numerous computer programs[1,2,3] have been developed, computer-assisted ventilation planning is not yet a common practice.

One of the computer programs most recently developed at The Pennsylvania State University (PSU/MVS program) was applied to the analysis and modifications to an existing ventilation system of a coal mine.

The computer program was utilized to serve the objectives of this study which are as follows:

1. To provide detailed information on present ventilation characteristics in the mine;
2. To modify the mine ventilation system in order to make it most efficient and economical; and
3. To provide adequate future mine ventilation plans.

CURRENT VENTILATION SYSTEM

The mine, located in the southern part of West Virginia, was 8 years old. The average overburden thickness was about 700 feet and the average coal seam thickness was 60 inches. The excessive horizontal stresses[4] in this area caused frequent roof falls which in turn blocked the airways partially or sometimes almost completely. According to the previous studies[5], the main fracture system in this area is oriented north-south, therefore, the entries oriented north-south were experienced with more roof ralls than those oriented east-west.

The average production rate of this mine was about 5900 tons of clean coal per day. From the statistical data, the methane emission rate from the coal seam was found to be 531 ft^3 per ton of clean coal, which means that the mine liberated over 3 million ft^3 of methane gas per day.

As shown in Figure 1, the mine had 10 continuous mining sections, two of which were developing entries of the longwall panels. Mine management decided to mine the coal thicker than 30 inches. Because the coal located in the western part of the area is especially good quality and was of regular seam thickness, the longwall mining method was planned. The rest of the area was being planned to continuously use the room-and-pillar mining method.

A main of 16 parallel entries ran north-south. In the middle of these entries, A shaft, B shaft and a slope were located. Both A and B shaft were 18 feet in diameter. A shaft was used as a supply and intake air shaft and B shaft was a return air shaft where two vane-axial fans were connected in parallel and operated simultaneously.

The mains of 9 entries branched out to the east and west from the north-south mains. One entry was neutral for a belt entry, and the rest of them were

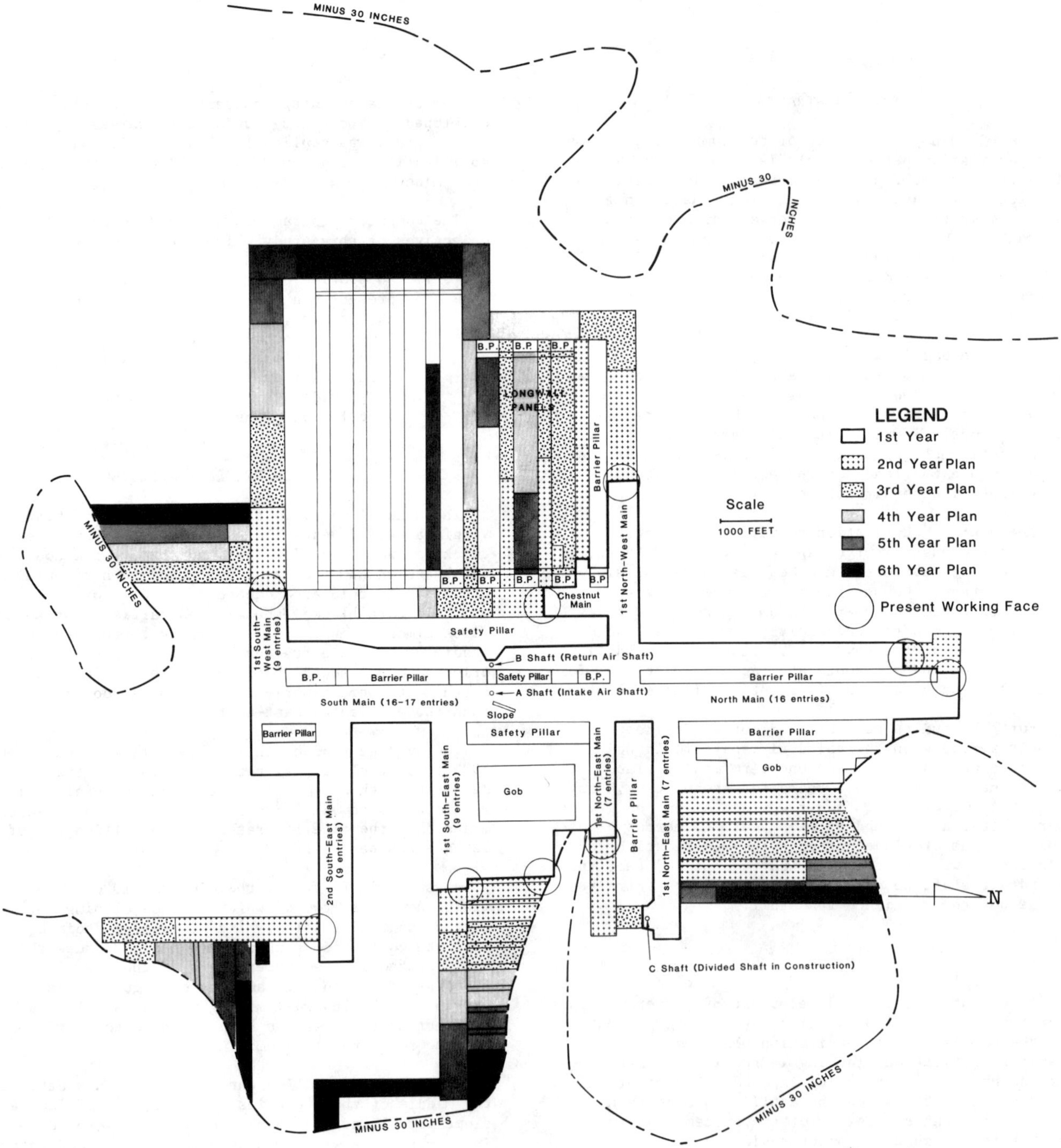

Figure 1.
FIVE YEAR MINING PLAN

divided evenly for intake and return entries. The amount of air carried in the track entries was small, but it was considered as intake air. A typical pattern of the 9 entry system is shown in Figure 2.

Most of the pillars were spaced at 100-foot intervals both between entries and between crosscuts. The entry size was mostly 18 feet wide. According to the mining plan, pillars in the panel sections were extracted subsequently. In the first year, there were two gob areas as shown in Figure 1.

At the end of the first northeast main, C shaft was in construction. C shaft had an elliptical shape with a dimension of 20 x 12 feet. The shaft was to be divided into two compartments - one for intake air and the other for return air. The shaft was used as an intake air shaft during the period of construction.

Two vane-axial mine fans (Joy Axivane mine fans, Model M-108-58-9' diameter) were used in parallel at B shaft, and both of them were operated simultaneously. Two identical fans were purchased for C shaft. After the construction of C shaft was completed, one fan was to be used at each shaft, while the other was to be reserved as a back-up fan.

During the first year, 10 working faces were in operation. For face ventilation, the regular brattice method was used.

MINE VENTILATION SURVEY

The ventilation survey was conducted in a total of 8 working shifts for 5 days. It included the altimeter pressure survey and the air quantity survey for the purpose of calculating the mine resistance and friction factors (K) in every section of the entries.

An altimeter was placed at the bottom of B shaft and regularly read, and a roving altimeter was read at every junction split and face. Along with the altimeter survey, air quantity was measured using an anemometer, and a psychrometer was read and the entry sizes were measured. The pressure drop across every regulator was measured using manometer.

It was found that the track entries carried a small amount of intake air but still were isolated from the intake entries. Therefore, they were considered as parallel entries to the main intake entries. The belt entries were surveyed but the air quantities were negligible, and also isolated from the major ventilation system. So belt entries and the slope where a small size fan was connected to ventilate belt entries were neglected for the network analysis to reduce complexity.

NETWORK ANALYSIS

The PSU/MVS program has diversified capabilities and advantages[6], such as:

1. Extensive report generation on air quantity in each branch, methane concentration, total horsepower required, etc.
2. Multiple fan network analysis and automatic fan selection.
3. Network revision routines for the future mine planning.
4. Standardized input data, with options for the use of several resistance formulas, parallel or series branches, parallel or series fans, changes in air density or velocity, evasés, etc. Using limited quantity data format, required air quantities for each face area can be input.
5. Dummy branches can be used for the simulation of gas emission, and leakage and fixed quantity branches can be used as needed.
6. Application for ventilation planning of new mines as well as operating mines.

These capabilities assist the user in making decisions on optimum fan size, number of fans, locations and numbers of air shafts, and optimum procedure of operations can be made.

Network analysis has been carried out in 4 different stages as follows:

Stage 1: Ventilation condition as of August of the first year when the initial ventilation survey was carried out.
Stage 2: Prediction of the ventilation condition as of December of the first year when a fan at C shaft is in operation.
Stage 3: Prediction of the ventilation condition at the end of the second year, and recommendation for the best ventilation condition.
Stage 4: Prediction of the time when a new air shaft is required.

Stage 1: Ventilation Condition as of August of the First Year

The objective of this stage is to simulate the present ventilation conditions. All the input parameters were calculated from the data obtained from the ventilation survey.

A computer program was used to calculate head loss at every branch which connects two nodal points. The program was capable of correcting the raw altimeter data for the base reading (fluctuation of atmospheric pressure), temperature effect and elevation change. The calculated head loss subsequently was converted into resistance of the branch by dividing by the square of air quantity:

$$R = \frac{H}{Q^2}$$

The simplified network of the initial condition is shown in Figure 3. The questionable spots were resurveyed and corrections were made.

There were presently 10 working faces, which make a total of 26 working face-shifts a day. The average production of every panel was 226.8 tons of clean

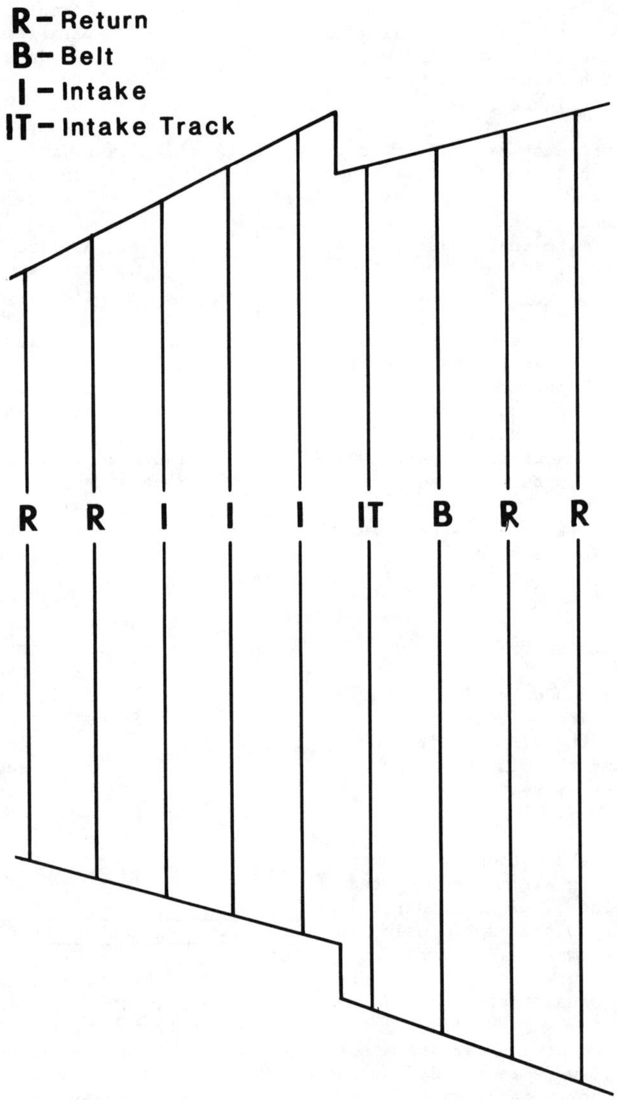

Figure 2

Typical Pattern of Nine Entry System

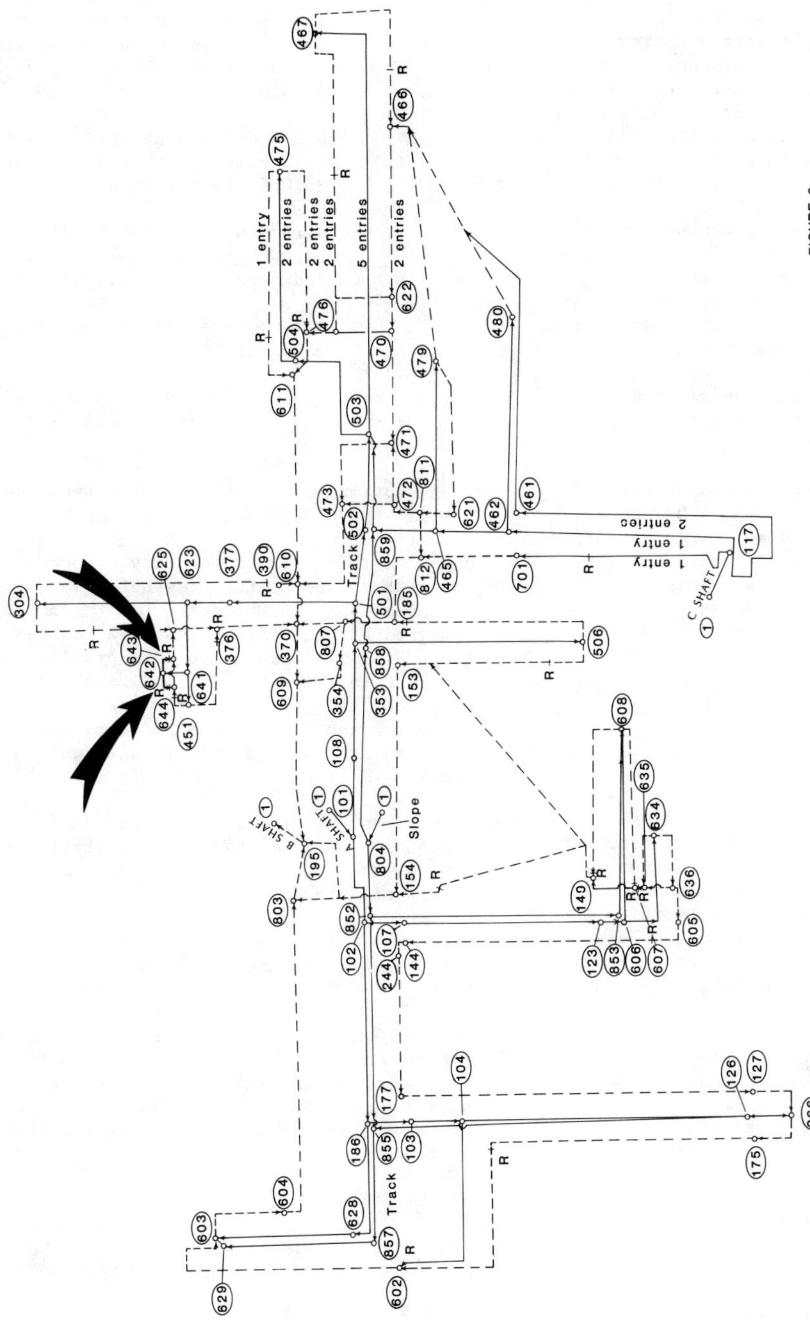

FIGURE 3
VENTILATION NETWORK
AS OF AUGUST OF 1st YEAR

→ INTAKE
- - → RETURN
o NODE
(1) ATMOSPHERE
R REGULATOR

coal per shift, which makes the total production 589.7 tons per day per face.

From a state's statistical data, the methane emission rate from the coal seam in this mine was found to be 531 ft^3 per ton of clean coal. From this data, the average methane emission rate was calculated to be 217 cfm per face. For the arbitrary maximum allowable methane content (MAC) of 0.5%, the required air quantity becomes 43,400 cfm per face. It if is assumed that 60% of fresh air arrives at the working faces, the air supply by the fans to a working face should be 72,000 cfm. This means that 720,000 cfm of air is enough to ventilate all 10 working faces.

The fan data was taken from the Joy Fan Data Book. Two Joy Axivane Mine Fans (Model M-108-58-9' diameter) were used in parallel connection at B shaft. The identical fans were also to be in operation at C shaft sometime in December of the first year. For each blade setting, a family of curves was used according to the different fan speed in rpm. In Stage 1, the fan data for the dual fan was used. The air quantity is multiplied by a factor of two for each fan head because they were connected in parallel.

The input data inlucded fan H-Q relationships, entry resistance, entry dimensions, etc. The dummy branches were used for gas emission and normal branches for all other branches. The result is shown in Table 1. There were no branches with excessive gas concentration except two branches connecting the nodal points 642-644 and 642-643 near the longwall panels, which have methane gas concentrations of 1.394%.

Stage 2: Prediction of Ventilation Condition as of December of the First Year When the Fans are in Operation at C Shaft

For this stage, analyses were conducted for the condition that the dual fans are connected in parallel at C shaft. The C shaft was elliptical shaped and it was to be divided into two compartments by putting a divider in the middle. One compartment will be used as the intake air shaft, the other as the return air shaft. The simplified network for Stage 2 is shown in Figure 4.

The items to be changed for Stage 2 from the condition in Stage 1 are listed as follows:

1. Two fans are connected in parallel (dual fan) at C shaft, but one fan is in operation.

2. All the 2nd north-east mains are used as intake entries in Stage 1, but two entries located in the north side and one entry in the south side are return in this stage.

3. Working faces have advanced two or three blocks since Stage 1.

4. During the ventilation survey, it was found that several overcasts transmit very little air. The branches connecting nodal points 633-604, 631-632 and 144-802 were either blocked by roof falls or stopings. Those overcasts should be open for a better ventilation condition with the fans installed at C shaft. In the computer input, these branches are considered open and clean. Therefore, these overcasts should be inspected and cleaned to function as soon as possible.

5. Limited branches were used for the face branches assigning mostly 90,000 cfm per face.

6. Air quantities are limited to 12,000 cfm in the branch 475-611 and branch 475-476.

7. A face is added on the south-west mains.

8. An entry at the Chestnut mains is added (see dotted line in Figure 4).

The optimum condition was made when the blade angle was 23° for the fans at B shaft and 13° for a fan at C shaft as summarized in Table 2. The air quantities supplied to the limit quantity branches are sufficient except the face at the Chestnut Main (Node 642), which has only 36,000 cfm. Limit quantity is assigned air quantity. Node 642 has gas concentrations of 0.97%. To solve this problem, air quantity supplied to the other working face nearby (Node 451) should be reduced and induce more air to Node 642. Another remedy is to make only one working face at the Chestnut mains until it is interconnected to B shaft.

Gas concentration of 1.46% is indicated for the branches 475-611 and 475-476. Actually, these entries are inactive, so there is little chance of dangerous levels of methane concentration.

Stage 3: Prediction of Ventilation Condition at the End of the Second Year

Two different series of study were carried out for this stage. The first series is based on the original planning laid out by mine management. The second series is based on the planning proposed by the author for better efficiency for the ventilation system.

Original Plan

The simplified network system of the activities in the second year, as shown in Figure 5, includes:

1. Developing longwall mining panels at Chestnut mains. Two or three panel entries will be completed at the end of the second year.

2. During the second year, the face on the north mains (Nodal point 467) will advance to connect point 475 shown in Figure 4.

Table 1

SUMMARY OF RESULTS FOR VENTILATION CONDITIONS
AS OF AUGUST OF THE FIRST YEAR

Mine Ventilation Network Analysis

Fan Blade Setting: B shaft - 2 fans - 25°
Fan speed - 900 rpm

Air HP Analysis

Static AHP	1227.808	Fan AHP	1229.264
Regulators	0.000	NVP	0.000
		Added AHP	0.000
Total	1227.808	Total	1229.264

Branch Class	AHP		Quantities	
Intake 1	482.160	39.27	Q Into the Mine	857.969
Intake 2	0.000	0.00	Q Out of the Mine	861.469
Intake 3	0.000	0.00	Q Air Injected	0.000
Return 1	643.481	52.41	Q Gas Injected	3.500
Return 2	0.000	0.00	Faces	524.497
Return 3	0.000	0.00	Q Leak 1	0.000
Leak 1	0.000	0.00	Q Leak 2	0.000
Leak 2	0.000	0.00	Q Leak 3	0.000
Leak 3	0.000	0.00	Total Leakage	0.000
Face	42.855	3.49		
Gob	59.316	4.83		
Total	1227.808			

Table 2

SUMMARY OF OPTIMUM CONDITION AS OF DECEMBER OF THE FIRST YEAR
WHEN A FAN AT C SHAFT IS IN OPERATION

Fan Blade Setting: B shaft - 2 fans - 23°
C shaft - 1 fan - 13°
Fan speed - 900 rpm

Air HP Analysis

Static AHP	1232.505	Fan AHP	1553.115
Regulators	315.865	NVP	0.000
		Added AHP	0.000
Total	1548.370	Total	1553.115

Branch Class	AHP		Quantities	
Intake 1	654.721	53.12	Q Into the Mine	915.337
Intake 2	0.000	0.00	Q Out of the Mine	919.187
Intake 3	0.000	0.00	Q Air Injected	0.000
Return 1	435.036	35.30	Q Gas Injected	3.850
Return 2	0.000	0.00	Faces	749.352
Return 3	0.000	0.00	Q Leak 1	0.000
Leak 1	0.000	0.00	Q Leak 2	0.000
Leak 2	0.000	0.00	Q Leak 3	0.000
Leak 3	0.000	0.00	Total Leakage	0.000
Face	77.340	6.28		
Gob	63.238	5.13		
Total	1232.505			

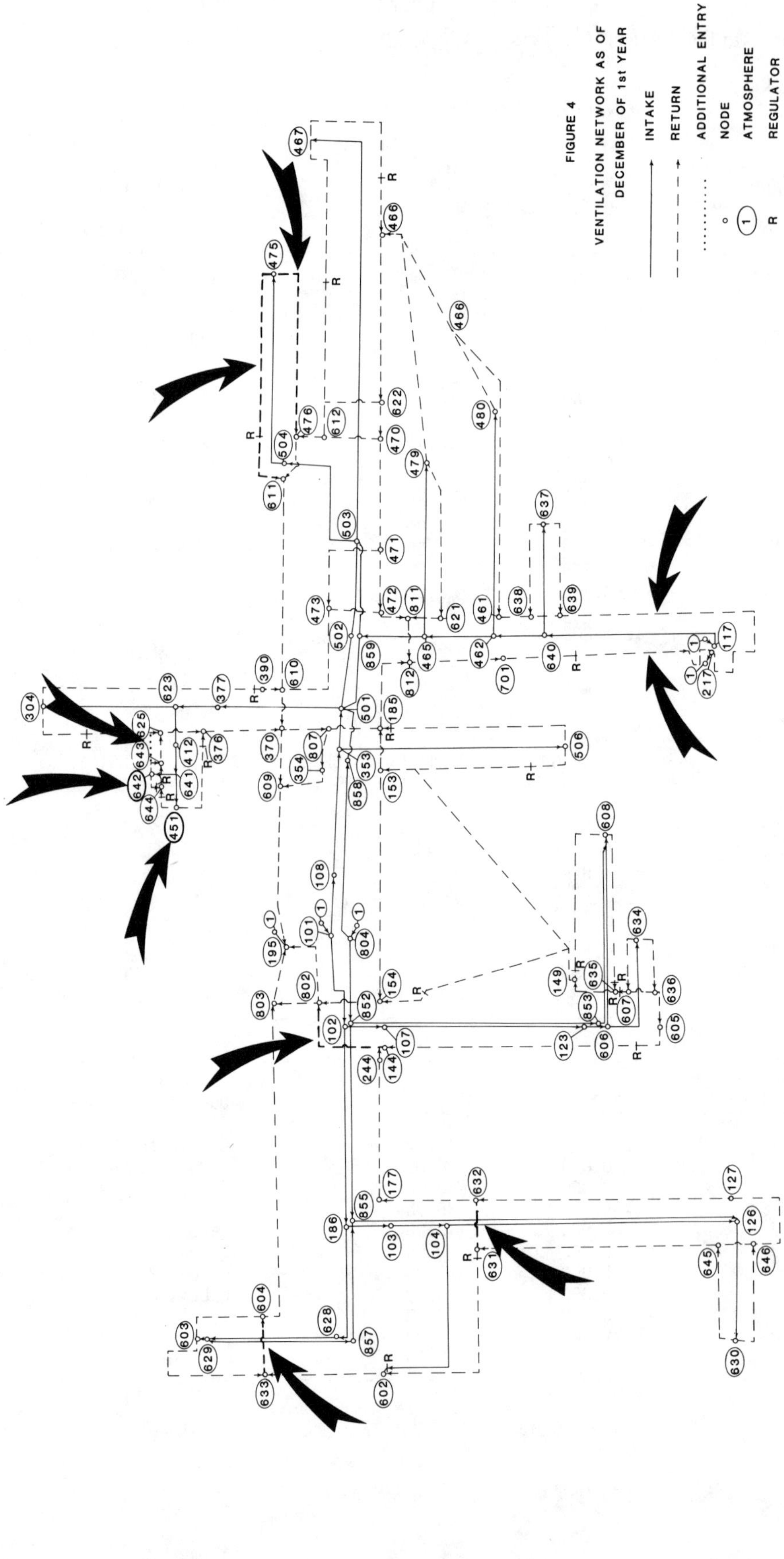

FIGURE 4
VENTILATION NETWORK AS OF DECEMBER OF 1st YEAR

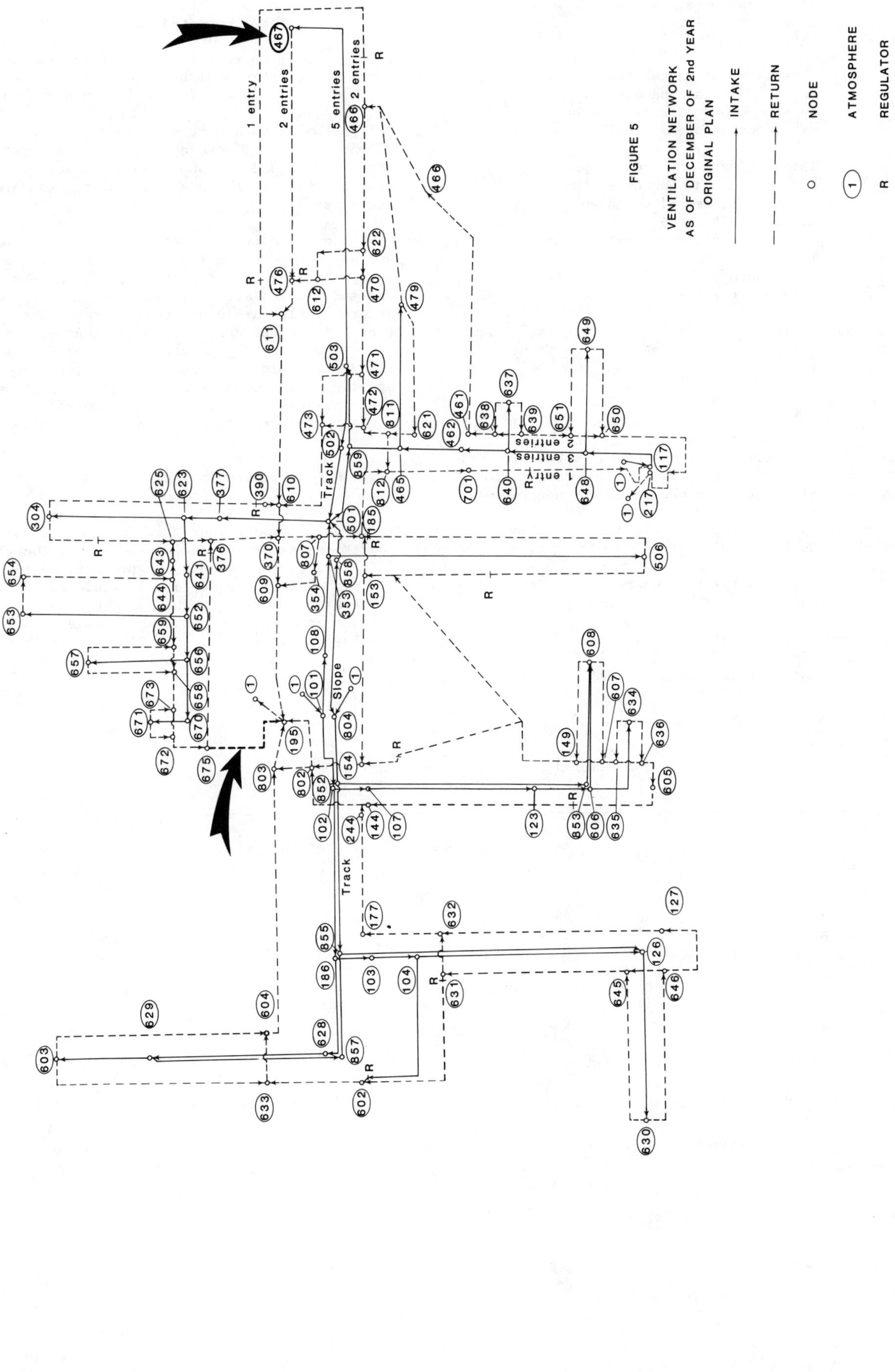

FIGURE 5
VENTILATION NETWORK
AS OF DECEMBER OF 2nd YEAR
ORIGINAL PLAN

3. Two or three panels along the second north-east mains will be mined and pillars will be extracted.

4. At south-east mains, two or three additional panels will be opened and pillars will be extracted from two panels.

5. The entries will be developed and extended at second south-east mains, second south-west mains, first north-east mains and first north-west mains.

6. The branch 675-195 is open for return air entry.

The blade angle for the fans at B shaft is 23° and that a C shaft is 13°. The summary of the results is shown in Table 3.

Proposed Plan

To improve the ventilation system, consideration was given to reducing the total length of the air flow path and the number of intake and return entries to C shaft, above all other possible means of improvements.

The conditions changed from the original plan are as follows (see Figure 6):

1. Additional return entries are connected. They are 506-117, 605-506, and 127-605, as indicated with dotted lines in Figure 6. Five entries are used for each branch.

2. The return entries 812-701-117 become intake entries, and they are connected to 501.

3. The return entry 506-185 (shown in Figure 5) becomes intake entry and is added to the intake branch 353-506.

The two cases are studied depending on the number of the additional return entries. The first case is that 506-117 and 650-506 are connected. The second case is that all three additional returns are connected.

The outputs for the best combination of blade angles are listed in Table 3, which compares the performance between the original planning and the revised plan. Table 3 explains that as the additional return entries are connected, more air quantity is supplied and less horsepower is required, and the fans have operating points in the safe range of the fan characteristic curves. However, the cost of opening those return entries might be high because the coal seam thickness if low. Unless it is not extremely difficult, the connection of 605-506-217 is recommended.

Stage 4: Prediction of the Time When a New Air Shaft is Required

Computer simulations were carried out for the conditions as of the end of the third, fourth, and fifth year, respectively, to predict the time when a new air shaft is required. The calculated air quantities supplied to the limit quantity branches (face branches) were compared for each year, and a conclusion was made for the time a new air shaft is required.

Table 3

COMPARISON OF THE OPTIMUM CONDITION FOR DIFFERENT PLANNINGS FOR THE SECOND YEAR

	Original Plan	Revised Plan 605-506-217 connected (Output #4)	 127-605-506-217 connected (Output #5)
Blade Setting Angles (and fan head in an inch of water)			
Fans at B Shaft	23° (10.87)	18° (10.62)	18° (10.70)
Fans at C Shaft	8° (9.59)	18° (10.26)	18° (9.95)
Total Intake Air Quantity (cfm)	854,589	859,320	861,837
Total HP	1436.029	1425.200	1420.870
Remarks	Operating point close to margin of fan curve.	Recommended	

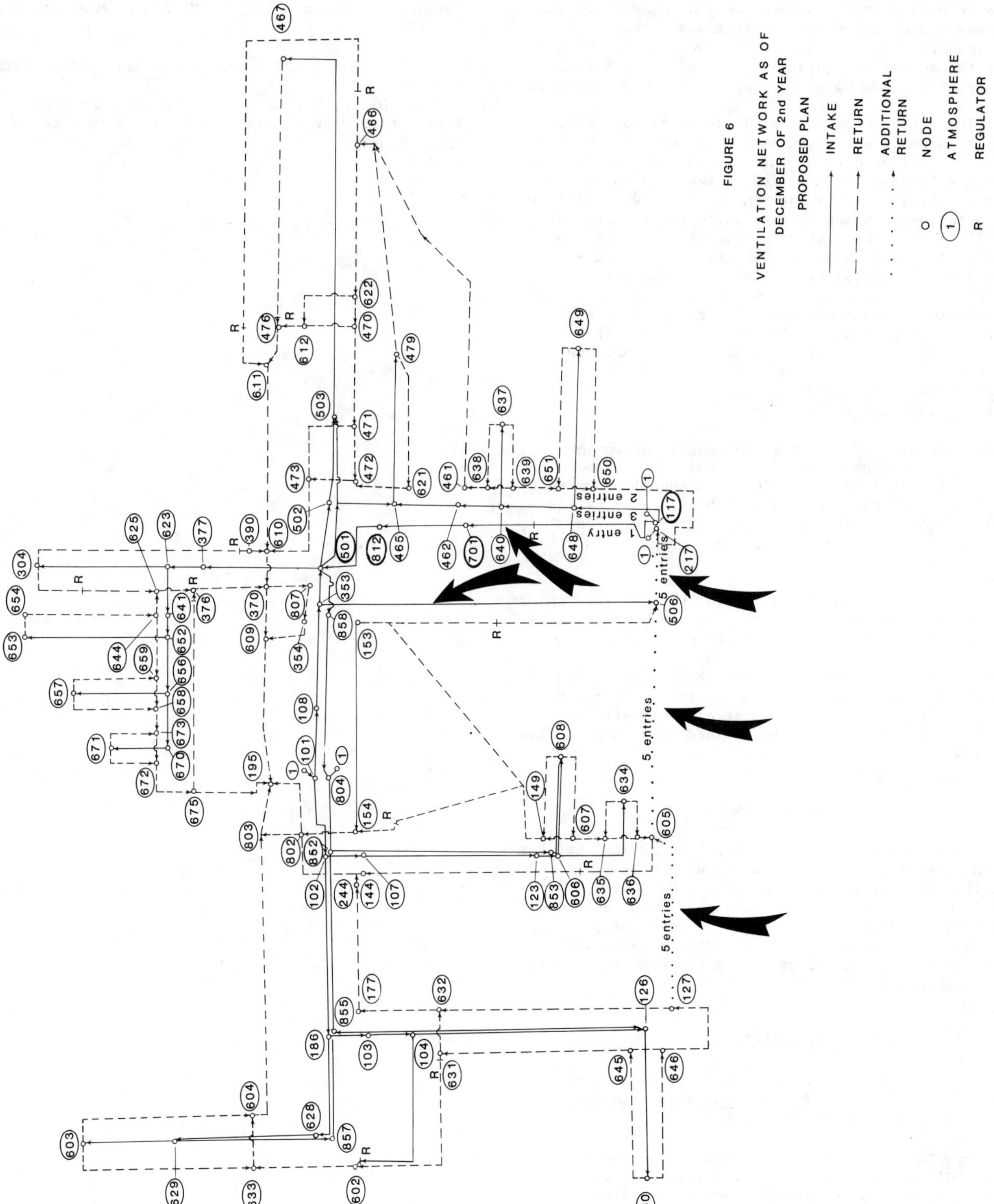

FIGURE 6
VENTILATION NETWORK AS OF DECEMBER OF 2nd YEAR PROPOSED PLAN

Ventilation Condition at the End of the Third Year

The additional nodal points for those three years are shown in Figure 7. Seventy thousand cfm of air was assigned at each active working face, and different air quantities were assigned at each gob area depending on the sizes. It should be noted that because the assigned air quantity was without consideration of leakage, it will be safe to consider that the actual air quantity supplied to each face may be 50-60% of those the computer outputs indicate.

Various combinations of blade settings for the 2 fans at B shaft and 1 fan at C shaft were checked to find the best condition, which means the condition when the largest air quantity is supplied to the mine with the fans in operating range. The best condition is as follows: blade setting angle for the fans at B shaft is 13° and that at C shaft is 18°. The total intake air quantity is 723,000 cfm. Mostly, the quantities of air supply to the limit quantity branches are more than 50,000 cfm, which is considered satisfactory. Only one face at nodal point 727 (Chestnut main area) has below 20,000 cfm. This face should be frequently checked for methane content.

Ventilation Condition at the End of the Fourth and Fifth Years

The simplified map of the 4th year is shown in Figure 8. The best condition of this year (blade angle - 13° at B shaft and 18° at C shaft) shows that there are five faces supplied with air quantity less than 50,000 cfm. They are nodal points 711, 720, 727, 729, and 735. The total intake air quantity is 728,000 cfm.

The analysis of the condition for the 5th year shows similar results as the 4th year. There are six faces supplied with air less than 50,000 cfm.

It is concluded that a new air shaft is required at the end of the third year or early in the fourth year. The locations of the shaft, size of the shaft, and the specifications should be decided with some further studies.

CONCLUSIONS

This project demonstrated the practical applications of the computer programs to solve a complex ventilation network of a coal mine. Various conditions were tested for finding the most efficient and economic means of existing mine ventilation system. Also, the most adequate future ventilation plan can be achieved using the computer analysis with high accuracy and relatively low cost.

REFERENCES

1. McPherson, M.J., "Mine Ventilation Network Analysis by Digital Computer," The Mining Engineer, Volume 126, No. 73, October 1966.

2. Wang, Y.J., and Saperstein, L.W., "Computer-Aided Solution of Complex Ventilation Networks," Transactions-AIME, Volume 247, September 1970.

3. Didyk, M., "Mine Ventilation and Atmospheric Environment Simulation," unpublished Mineral Engineering Management M. Eng. Report, The Pennsylvania State University, 1974.

4. Aggson, J.R., and Curran, J., "Coal Mine Ground Control Problems Associated with a High Horizontal Stress Field," AIME Annual Meeting, March 2, 1978.

5. Holland, C.J., personal communcation, 1979.

6. Didyk, M., et al, "Advancement of Mine Ventilation Network Analysis from Art to Science, Final Report, Contract H0133040, U.S. Bureau of Mines.

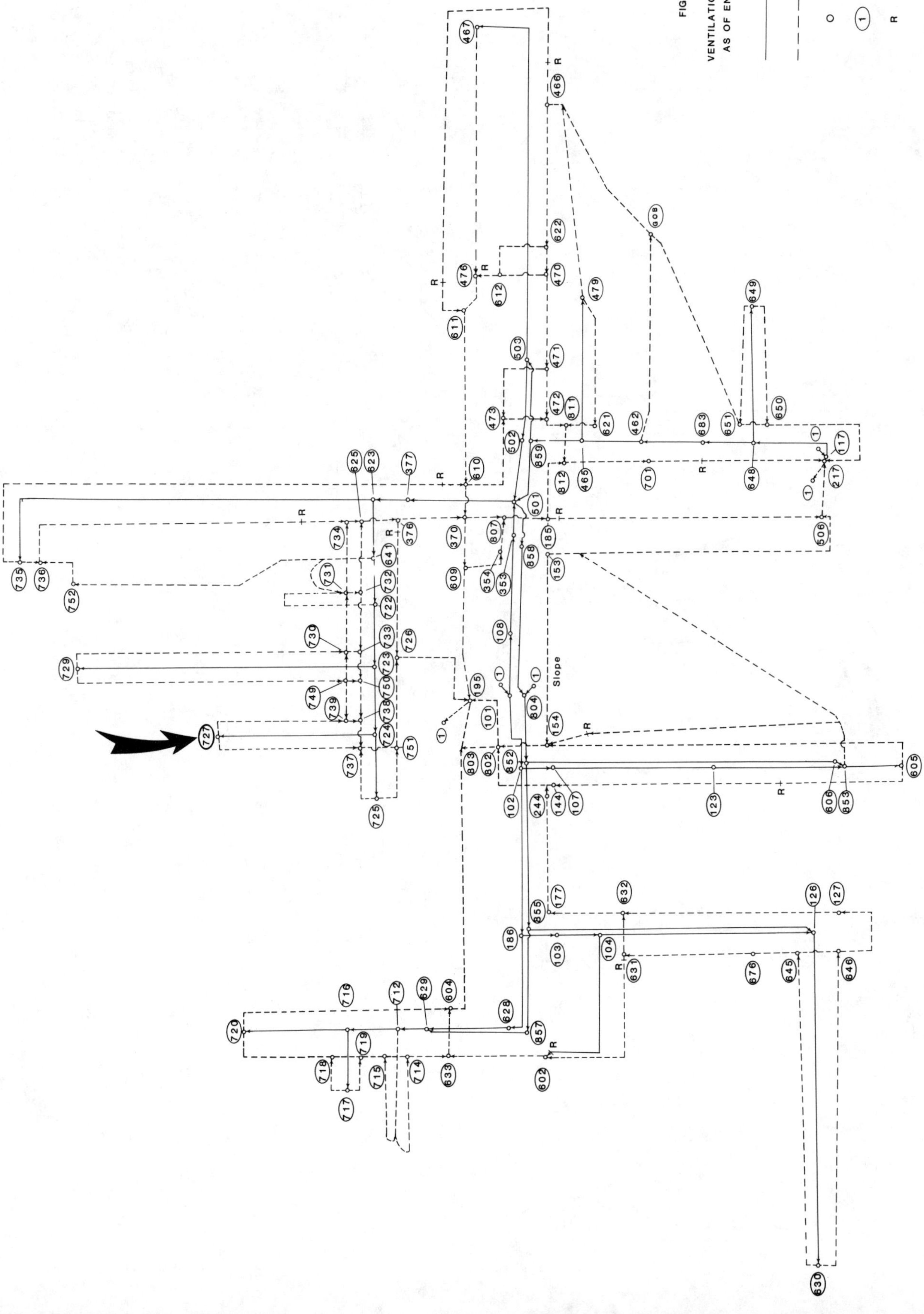

FIGURE 7
VENTILATION NETWORK AS OF END OF 3rd YEAR

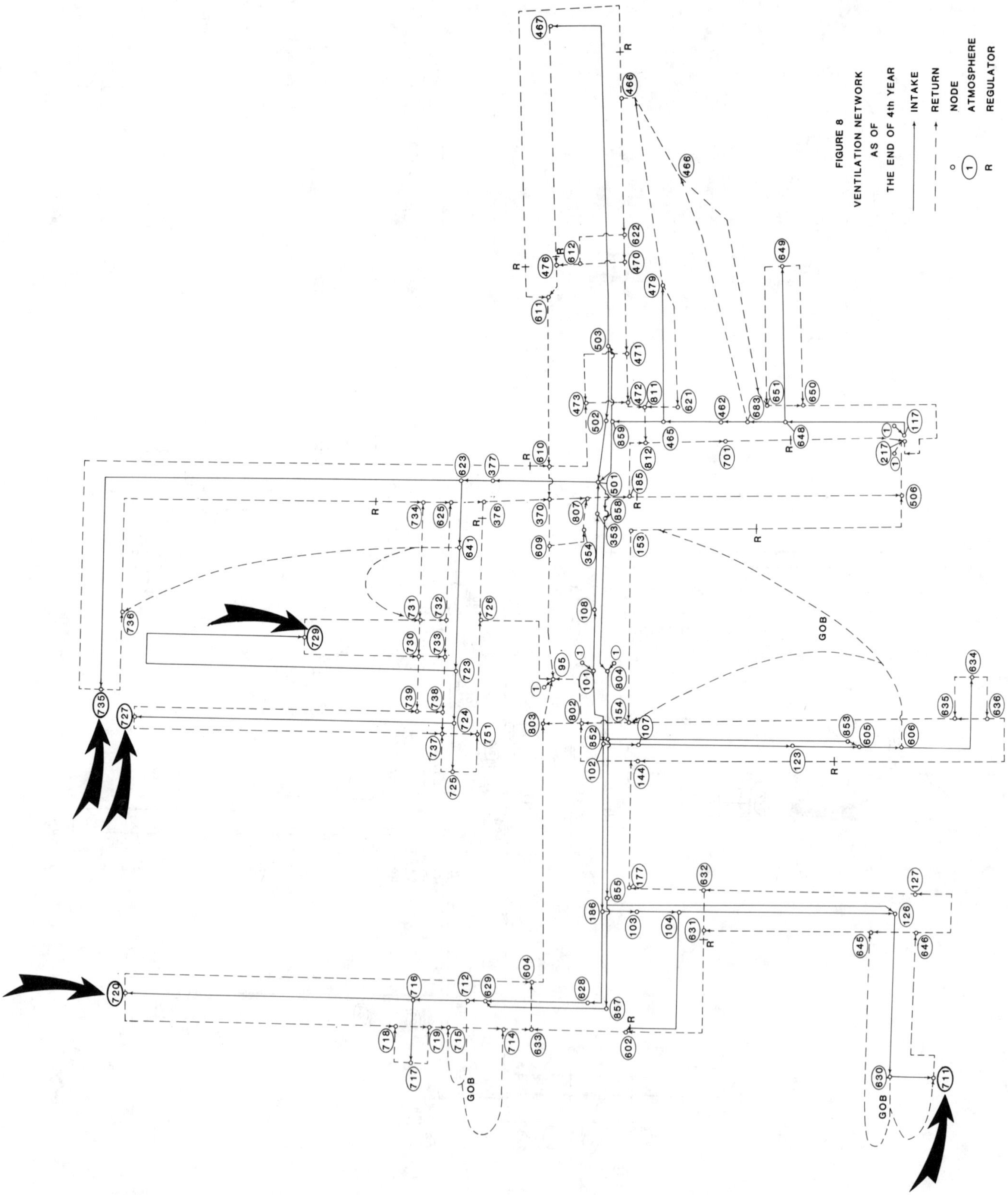

FIGURE 8
VENTILATION NETWORK
AS OF
THE END OF 4th YEAR

Chapter 17

COMPUTER APPLICATIONS FOR HENDERSON MINE VENTILATION PLANNING

Tribhawan N. Srivastava

Engineer
Climax Molybdenum Company, Amax Inc.
Empire, Colorado

ABSTRACT

The Henderson Mine started production in 1976 utilizing a unique, push-pull, mine ventilation plan. As the mine expands in size, the mine ventilation requirements will increase and a more comprehensive system for maintaining ventilation will be needed. A computer simulation study was completed and has resulted in a revised, systematic, mine ventilation plan which is being implemented as mine development proceeds.

Based on the computer simulation study, the new ventilation plan takes into account mining of the orebody simultaneously from both the 8100 and 7700 production levels. Present experience and empirical data indicate that volumes of air needed to adequately ventilate both production levels cannot be obtained utilizing the existing mine layout. The total volume of air needed for the mine excluding haulage tunnel will increase from 708 m^3/s (1 500 000 cfm) to 1652 m^3/s (3 500 000 cfm), thus a new No. 5 exhaust shaft and connecting drifts are being developed to expand the mine ventilation system to service both production levels.

The original ventilation plan for the 7500 haulage level was to bring air from the No. 2 and No. 3 intake shafts and exhaust it through the 94X raise to the 8035 exhaust level and then through the No. 1 exhaust shaft to surface. Control of air volumes in loading drifts was initially planned to be achieved by air doors and air curtains. Air doors on the 7500 level require large cutouts, expensive installations and constant maintenance.

The computer simulation study indicates that airflow in the 7500-level production drifts can be controlled exclusively by air curtains which are less expensive to install and more effective in regulating airflows than air doors. Air from the 7500 level can be exhausted through a new raise to the new 7635 exhaust level and then through the new No. 5 exhaust shaft.

INTRODUCTION

Every mine has its own unique problems and the Henderson mine is no exception. The 1976 mine ventilation plan had to be upgraded and planned air volumes increased to be able to ventilate the new 7700 production level in the early 1990's. In reviewing studies of the ventilation plan, it appeared that a new ventilation exhaust shaft and additional intake and exhaust access drifting would be required.

Several computer simulation studies were completed. The purpose of the first study was to determine the diameter of the proposed exhaust shaft, the number and size of additional return airways, and the overall pressure loss in the proposed ventilation system.

The purpose of the second computer study was to develop a procedure to control the haulage ventilation system utilizing air curtains.

EXISTING HENDERSON MINE VENTILATION SYSTEM

The Henderson mine ventilation system is divided into two major integrated systems, namely, the mine ventilation system and the railroad haulage-tunnel ventilation system. The No. 4 exhaust shaft is located midway in the 15.13 km (9.4 miles) haulage tunnel. The two exhaust fans at No. 4 shaft pull fresh intake air from the tunnel portal and also from the mine and exhaust it through the vertical shaft to surface. At the mine, fresh air is pushed down No. 3 intake shaft by two surface intake fans and pulled up No. 1 exhaust shaft by three identical surface fans. The No. 2 service shaft supplies a limited amount of fresh air to both the 8100 production and 7500 railroad haulage levels. The airflow at No. 2 shaft is produced by the differential in air volume between the fanned intakes and exhausts. Specifications for the exhaust and intake fans and the different shafts are shown in Table 1.

Within the mine, the concept of fresh-air distributor and exhaust collector levels is used for each production level. The 8050 fresh-air level receives air directly from the No. 3 shaft via fresh-airway drifts and distributes it to production as well as development levels through vertical raises. The 8035 exhaust collector level along with completed portions of the 7635 exhaust level receive exhausted air from the various levels and return it directly to No. 1 shaft by three exhaust airway drifts on 8035 level and two exhaust drifts on 7635 level (see Fig. 1 and 2).

In the production operation on 8100 level, 3.82 m^3 (5 cu yd), 126.77 kW (170 hp) LHD units gather ore from drawpoints and dump it into 2.44 m (8 ft) diam ore passes at an average rate of 250 buckets per shift per loader. Dust generated during this operation is captured at the ore passes by exhausting them to the 8035 level. The ore passes each carry 9.44 m^3/s (20 000 cfm) of air under negative pressure, which is produced by 1.83 m (6 ft) diam, 74.57 kW (100 hp) fans located in the 8035 exhaust ventilation laterals. Makeup air is supplied to the 8100 level from the 8050 level through a series of 1.52 m (5 ft) diam regulated raises each carrying a volume of 9.44 m^3/s (20 000 cfm). Positive pressure is produced by 1.83 m (6 ft) diam, 44.742 kW (60 hp) fans located in the 8050 laterals. The system results in a partial exchange of air every 97.54 m (320 ft) along each production drift. The 8050 and 8035 ventilation laterals are driven perpendicular to and below the 8100-level production drifts. Ventilation intake raises and ore passes are located every 97.54 m (320 ft) along each of the 8100-level production drifts.

TABLE 1

SHAFT DESCRIPTIONS | FAN DESCRIPTIONS

Number	Diameter	Depth	Number of	Type	Diameter	Power	Location	Air Volume	Pressure
1	7 m (23 ft)	799 m (2556 ft)	3	Vane axial	3.05 m (10 ft)	932 kW (1250 hp)	Surface	283 m^3/s (600 000 cfm each)	2488 Pa (10 in. wg) at altitude
2	8.5 m (28 ft)	946 m (3105 ft)	None	-	-	-	-	-	-
3	7 m (23 ft)	698 m (2291 ft)	2	Vane axial	3.05 m (10 ft)	932 kW (1250 hp)	Surface	378 m^3/s (800 000 cfm each)	747 Pa (3.0 in. wg) at altitude
4	3.35 m (11 ft)	483 m (1585 ft)	2	Vane axial	2.13 m (7 ft)	373 kW (500 hp)	Under-ground	94 m^3/s (200 000 cfm each)	1120 Pa (4.5 in. wg) at altitude

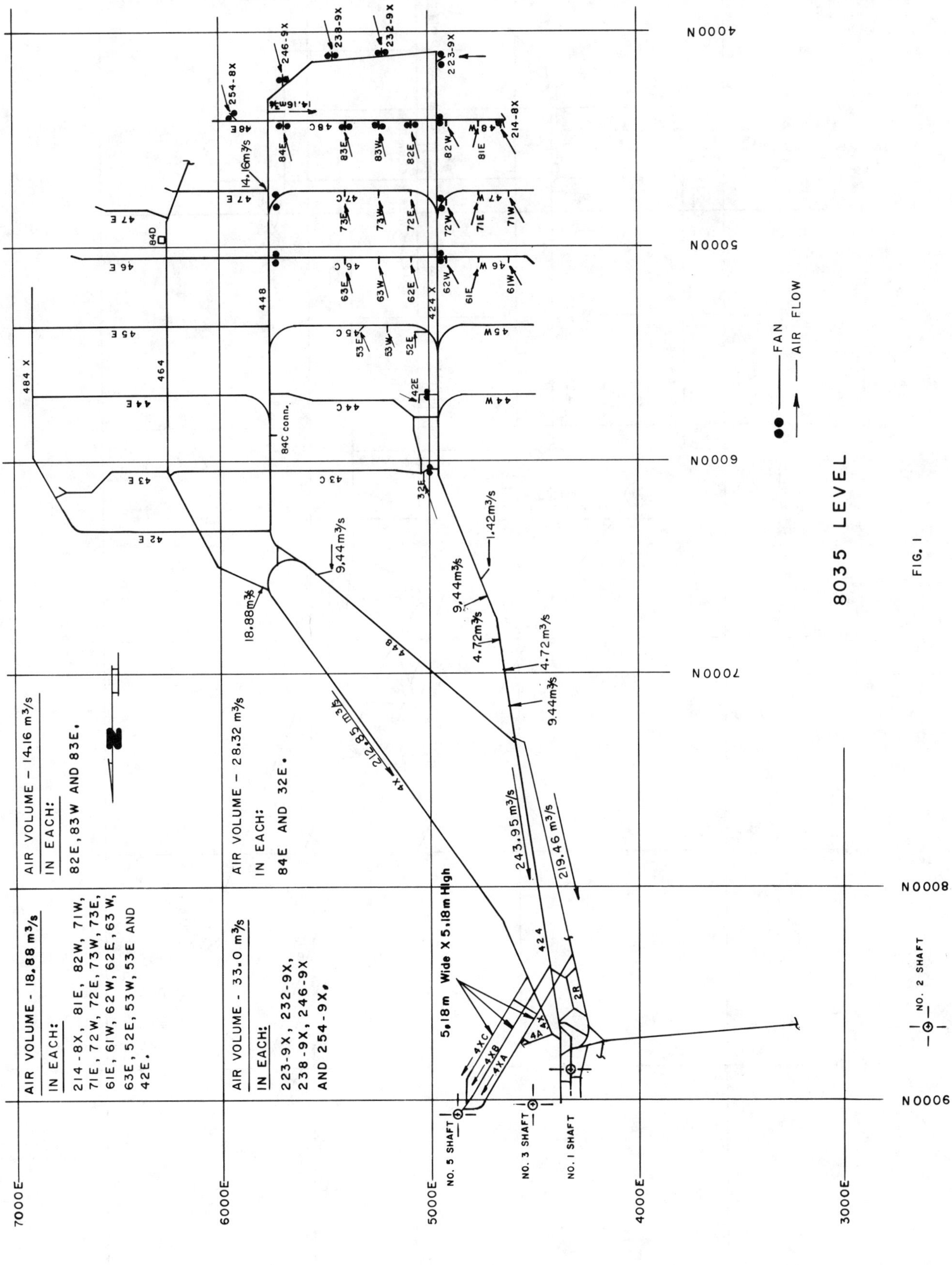
8035 LEVEL
FAN
AIR FLOW
AIR VOLUME - 18.88 m³/s
IN EACH:
214-8X, 81E, 82W, 71W, 71E, 72W, 72E, 73W, 73E, 61E, 61W, 62W, 62E, 63W, 63E, 52E, 53W, 53E AND 42E.
AIR VOLUME - 14.16 m³/s
IN EACH:
82E, 83W AND 83E.
AIR VOLUME - 33.0 m³/s
IN EACH:
223-9X, 232-9X, 238-9X, 246-9X AND 254-9X.
AIR VOLUME - 28.32 m³/s
IN EACH:
84E AND 32E.
5.18m Wide X 5.18m High
NO. 5 SHAFT
NO. 3 SHAFT
NO. 1 SHAFT
NO. 2 SHAFT
243.95 m³/s
219.46 m³/s
212.85 m³/s
7000E
6000E
5000E
4000E
3000E
4000N
5000N
6000N
7000N
8000N
9000N

FIG. 1

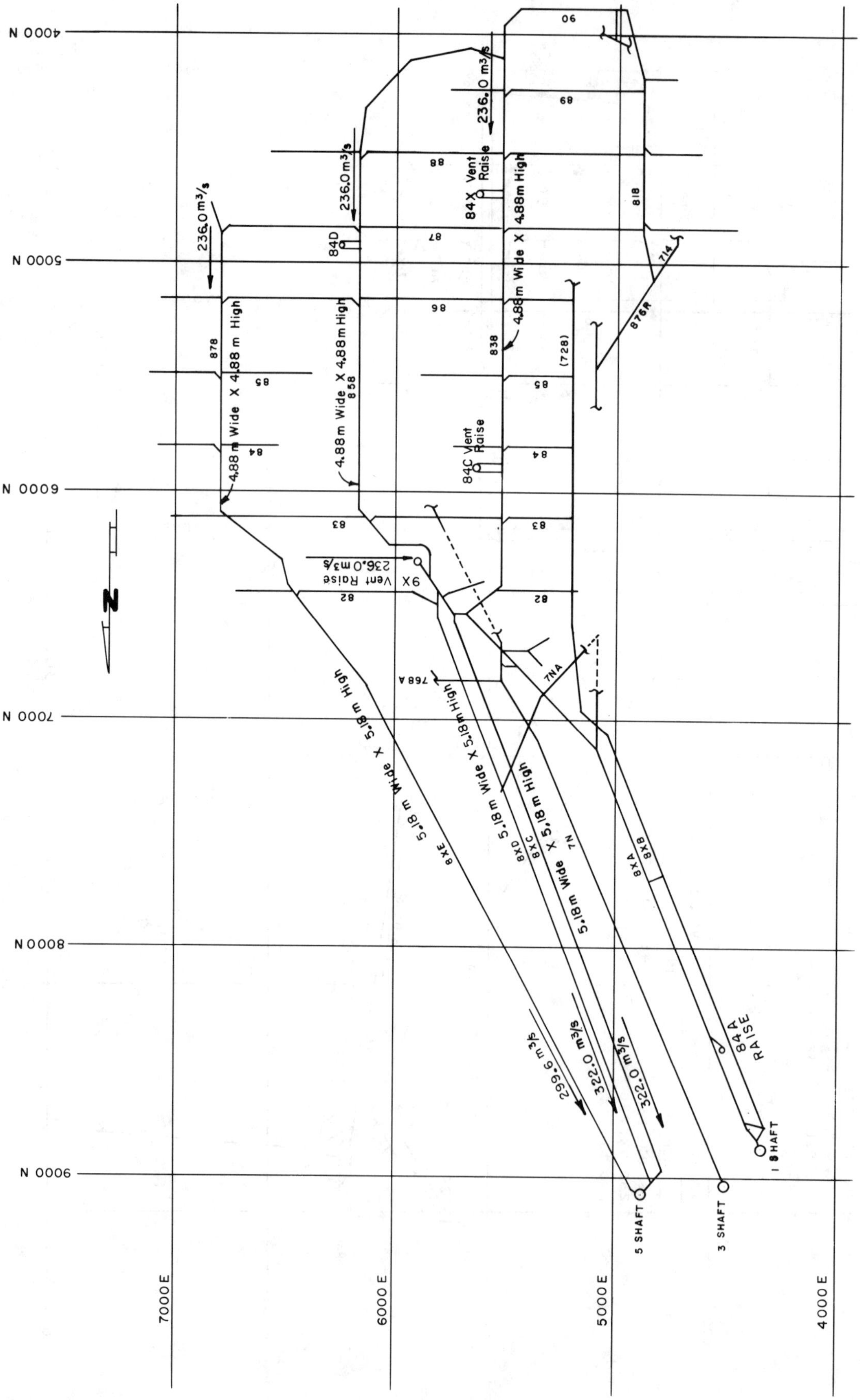

7635 LEVEL

FIG. 2

During the winter, the fresh-air temperature is raised to make the mine environment comfortable and prevent ice buildup in the No. 2 and 3 shafts. At No. 3 shaft, this is done by utilizing the waste heat from two sources. A heat exchange system traps the heat from the mine discharge water at 26 - 32°C (80 - 90°F) and a supplemental waste-oil-fired boiler. To raise the temperature of fresh intake air through the No. 2 shaft, outgoing cooling water from the mine air compressors is further heated by an external gas-fired boiler and the heated water is passed through a heat exchanger.

COMPUTER SIMULATION

Instead of developing a new mine ventilation program, an effort was made to use existing computer programs. One such program called the 'Pennsylvania State Mine Ventilation Simulation' developed by the Pennsylvania State University was found to be suitable. This program evaluates the pressure loss and airflow in each branch of the network and, also, the overall pressure loss of the ventilation system.

The ventilation simulation package consists of two FORTRAN programs, operating on a Decsystem-20 computer. The first program allows the user to enter network data interactively at a video terminal and builds an input file in NAMELIST format for the second program which performs the analysis of the ventilation system and prints the results of the simulation. At the user's request, the input program will display the data in small segments and allow limited changes to the file.

The first (input) program allows three methods for the input of resistance values:

1. To enter the resistance of each branch.

2. To compute resistances from two absolute pressure or temperatures and quantity for each branch.

3. To compute resistances from a friction factor by giving height, width, and length of each branch.

A number of methods are also available for computing fan curves:

1. Newton's finite divided-differences method with interpolation.

2. The original 'Pennsylvania State' technique.

3. The original 'Pennsylvania State' technique with input scaled by 100.

4. Least squares fit.

The second (outout) program which performs the analysis uses the iterative process known as the Hardy Cross method to determine the flow rates. The program follows a sequence of steps as follows:

1. Reads the input file created by the first program and lists the data.

2. Arranges the branches in order of resistance (high to low) and determines which branches to use as starting points for generating the meshes.

3. Generates the meshes required for the analysis.

4. Prints a table of natural ventilation pressures if any have been specified.

5. Prints the mesh table, simultaneously computing the natural ventilation pressures.

6. Reads fan characteristics and computes fan curves.

7. Performs Hardy-Cross iterations.

8. Computes pressure loss.

9. Generates a listing of each branch with corresponding pressure, quantity, and resistance. Total pressures and fan pressures are also printed.

A number of assumptions are made by this simulation package:

1. The input must describe a network which is closed, i.e. there must be N independent loops (meshes) where:

   N = (# of branches) - (# of junctions) + 1.

2. Every fixed-quantity branch and every fan branch must appear in a loop which contains, in addition to that fixed-quantity or fan branch, only general branches, i.e. no other fixed-quantity or fan branches.

3. The total number of fixed-quantity and fan branches cannot exceed N defined above.

4. All branches into a single junction cannot be fixed-quantity branches.

At the Henderson Mine, the resistance factor of each branch was calculated from the common pressure-loss formula for mine ventilation and entered into the input file with the fixed quantities of airflow, if any.

The common pressure-loss formula for mine ventilation is:

$$H = \frac{KLOQ^2}{5.2A^3}$$ at standard density of 0.075 lb per cu ft

where H is the pressure loss in inches water,
K is airway friction factor expressed in whole number,
L is the length of airway in feet,
O is the perimeter of the airway in feet,
Q is the quantity of airflow, cu ft per minute expressed in units of 100 000,
A is the cross-sectional area of the airway in sq ft

substituting R for $\frac{KLO}{5.2A^3}$ - the resistance factor of the airway.

The formula becomes $H = RQ^2$.

FIRST COMPUTER SIMULATION STUDY

A reevaluation of the 1976 ventilation plan in light of both our operating experience and the 1980 and 1981 long-range mining plans resulted in formulating the following parameters:

1. The original concept of exhaust and intake distributor levels should be retained in the plan.

2. The original concept of a changeout of air in the production drifts should be retained, but the air volume per exchange should be increased.

3. The production would be scheduled from the 8100 and 7700 production levels simultaneously. Although mine tonnage would not necessarily increase, the extent of the area to be ventilated would essentially double and, therefore, the volume of air required would also nearly double.

Based on these parameters, the future mine ventilation requirement was determined to be 1652 m^3/s (3 500 000 cfm) distributed as follows: 3100 production and sublevels - 708 m^3/s (1 500 000 cfm); 7700 production and sublevels - 779 m^3/s (1 650 000 cfm); and 7500 rail haulage level - 165 m^3/s (350 000 cfm). By simulating the original ventilation system using the Pennsylvania State ventilation program, it was possible to verify the need for a new ventilation shaft as well as determine the size requirements of the shaft and the airway drifting connecting it to the mine. Furthermore, in scheduling the No. 5 shaft ventilation program, it was determined what steps could be taken early in the program to increase mine ventilation with our existing shafts and fans. These steps included deepening No. 1 shaft, completing portions of the 7635 exhaust collector level and using it to parallel the 8035-level exhaust system. The predicted results of these steps compared with actual measured mine volume increases were as follows.

DEEPENING NO. 1 SHAFT (8XA airway drift completed):

Predicted - 37 - 42 m^3/s (80 - 90 000 cfm)

Actual - 47 m^3/s (100 000 cfm)

8XB AIRWAY DRIFT COMPLETED:

Predicted - 33 - 37 m^3/s (70 - 80 000 cfm)

Actual - 42 m^3/s (90 000 cfm)

USING THE 7635 LEVEL TO PARALLEL 8035 LEVEL:

Predicted - 37 - 42 m^3/s (80 - 90 000 cfm)

Actual - work in progress

Based on the anticipated future ventilation requirements, the computer simulation resulted in a future ventilation system as follows.

1. The Nos. 1, 2 and 3 shafts will all become intake shafts and the new No. 5 shaft will serve as the single exhaust shaft.

2. The haulage tunnel exhaust will remain unaltered.

3. The No. 3 shaft will eventually be connected to the 7700 level via 7N drift.

4. The 8100 production-level intake will not change.

5. The air from the 8035 exhaust level will go through three 5.18 m (17 ft) wide by 5.18 m (17 ft) high drifts (4XA, 4XB and 4XC) to the No. 5 exhaust shaft.

6. The air from the 7635 exhaust level will be returned through three 4.88 m (16 ft) wide by 4.88 m (16 ft) high lateral drifts (838, 858 and 878) connected to three return exhaust drifts (8XC, 8XD and 8XE) 5.18 m (17 ft) wide by 5.18 m (17 ft) high and then to No. 5 shaft.

7. The net exhaust-air return from all levels will be pulled up the No. 5 shaft.

8. The No. 5 shaft, which is currently being developed, is 9.75 m (32 ft) in diameter, smooth lined with concrete.

9. The predicted mine static head was 30 X 10^2Pa (12 inches water) at standard conditions. Four 3.66 m (12 ft) dia., 1305 kW (1750 hp) vane axial fans with continuously adjustable blades were chosen for the project.

SECOND COMPUTER SIMULATION STUDY

All ore is transported from the 7500 ore collection level to the mill through a 15.13 km (9.4 mile) haulage tunnel using an electrified rail haulage system. Locomotives are powered from an overhead trolley which is maintained at 1400 V (dc) in the tunnel and 600 V (dc) within the 7500 level. The typical 22-car train with three 45 359 kg (50 short-ton) locomotives is continuously loaded from the side-loading chutes in the 7500-level loading drifts (LD's). The chutes are connected to 183 m (600 ft) long, 2.44 m (8 ft) diam ore passes coming from the 8100 production level above.

Operating experience to date has shown that an airflow of 19 m^3/s (40 000 cfm) per loading drift will control dust during the chute-pulling/car-loading portion of the haulage cycle as well as provide adequate cooling for the locomotives. The Berry-type air curtains have been used in the loading drifts on a limited basis and have been successful in regulating airflows. The use of air curtains is effective at Henderson because problems associated with air doors such as trolley interference, damage caused by moving trains, and sizeable excavations are eliminated. The use of air doors in crosscuts between loading drifts is, however, viable.

The purpose of the computer simulation was to determine:

1. If air curtains would provide an operable regulation system when 7500 level was completely developed.

2. The number of air curtains and air doors required.

3. The location of the air curtains and air doors.

Several simulations were run to answer these questions. The results are shown on Figure 3. First, the air distribution with no regulators in the loading drifts was established. Then, required airflow values were assigned to the loading drifts west of the 9X crosscut and air doors arbitrarily located in the crosscuts. The model was modified by changing air-door locations until simulated flows in the loading

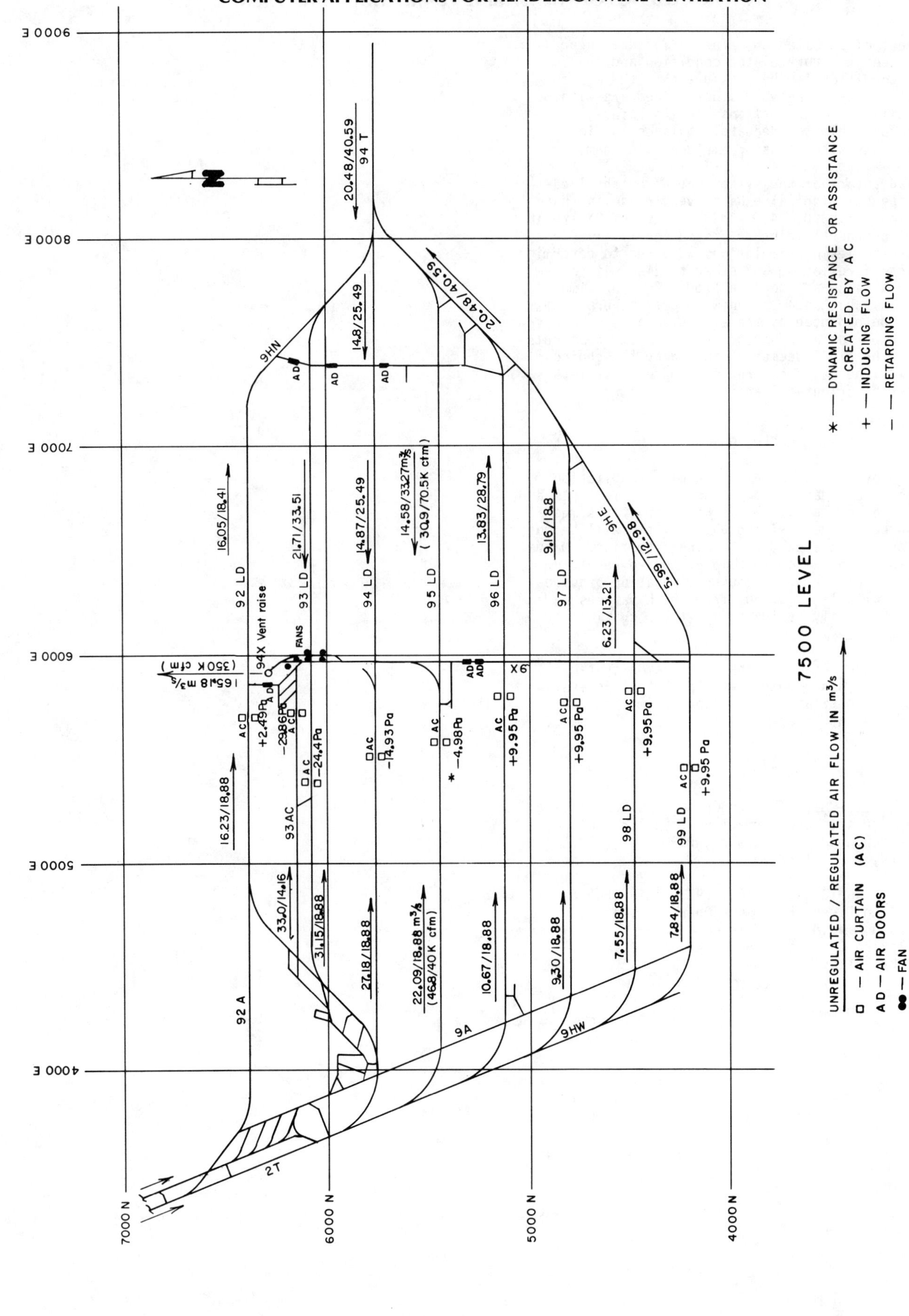
7500 LEVEL
UNREGULATED / REGULATED AIR FLOW IN m³/s
□ — AIR CURTAIN (AC)
AD — AIR DOORS
●● — FAN
* — DYNAMIC RESISTANCE OR ASSISTANCE CREATED BY AC
+ — INDUCING FLOW
− — RETARDING FLOW
9000 E
8000 E
7000 E
6000 E
5000 E
4000 E
7000 N
6000 N
5000 N
4000 N
94 T
20.48/40.59
9HN
9HE
9A
9HW
2T
92 A
93 AC
92 LD
93 LD
94 LD
95 LD
96 LD
97 LD
98 LD
99 LD
94X Vent raise
FANS
165.18 m³/s
(350 K cfm)
16.05/18.41
21.71/33.51
14.87/25.49
14.58/33.27 m³/s
(30.9/70.5K cfm)
13.83/28.79
9.16/18.8
6.23/13.21
5.99/12.98
14.8/25.49
16.23/18.88
33.0/14.16
31.15/18.88
27.18/18.88
22.09/18.88 m³/s
(46.8/40K cfm)
10.67/18.88
9.30/18.88
7.55/18.88
7.84/18.88
+2.49Pa
−29.86Pa
−24.4Pa
−14.93 Pa
−4.98Pa
+9.95 Pa

FIG. 3

drifts east of 9X balanced. The difference in pressures between the unregulated condition and the balanced condition is the pressure that the air curtain must supply. The values determined are within the capability of our curtains and, therefore, it should be possible to adequately balance loading-drift airflows without using air doors in them.

A disadvantage of the system caused by the location of the existing 94X exhaust vent raise is that the portions of 93LD, 94LD and 95LD west of 9X are in series with other loading drifts on the level. An additional series of simulations were run to determine the effect of replacing 94X exhaust raise with a new 93X exhaust raise as shown in Figure 4. The simulation shows that under the new condition, pressures that must be produced by the air curtains are reduced but the number of air doors required in the crosscuts doubles. The final decision concerning the choice of systems will be based on the anticipated dust levels in the series-oriented loading drifts.

CONCLUSIONS

The improvement of an existing mine ventilation system to accommodate a second production level is a multidecision process. Different alternatives must be evaluated before determining which system will be most cost effective in a particular mine situation. Manual calculations are extremely complex toward resolving multinetwork problems. Computer simulation provides quick and reliable solutions from which various parameters can be easily evaluated for their criticality and effectiveness.

Computer simulation of the ventilation system at Henderson Mine has provided a helpful tool in evaluating future ventilation plans and optimizing the performance of our present system.

REFERENCES

Steinhoff, J., 1981, "The Henderson Mine Ventilation System," Design and Operation of Caving and Sublevel Stoping Mines", Stewart, D.R., ed., AIME, New York, pp. 821-824.

Sadik, W.A., and Jensen, E.B., 1979, "Air Quality Control in Diesel Equipment Operations at the Henderson Mine," Second International Mine Ventilation Congress, Mousset-Jones, P., ed., AIME, New York, pp. 195-200.

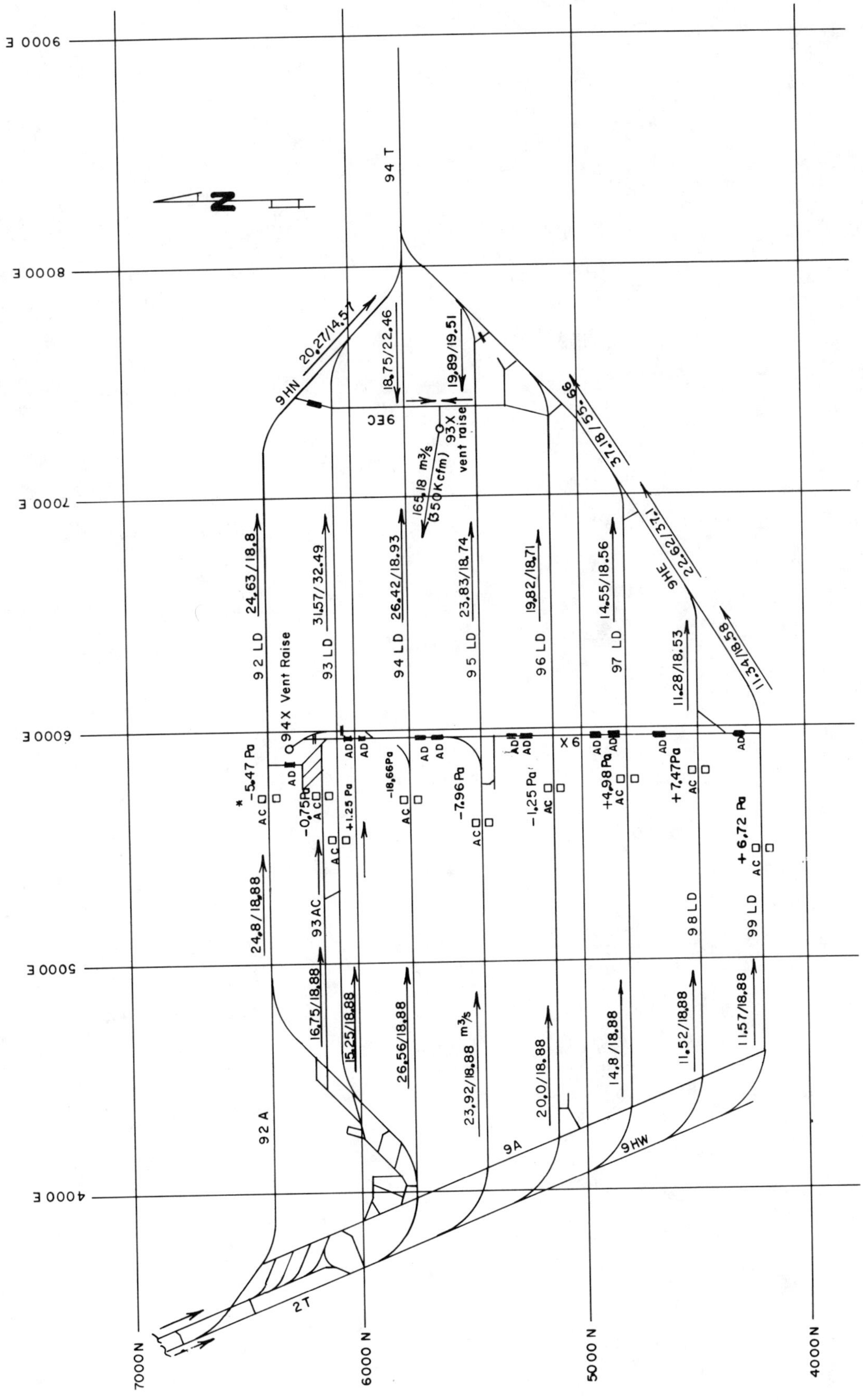

7500 LEVEL

UNREGULATED / REGULATED AIRFLOW IN m³/s

□ — AIR CURTAIN (AC)

AD — AIR DOORS

* — DYNAMIC RESISTANCE OR ASSISTANCE CREATED BY AC.

\+ — INDUCING FLOW

− — RETARDING FLOW

FIG. 4

COST-EFFECTIVENESS OF INCREASING AIRFLOW IN UNDERGROUND COAL MINES

Chapter 18

Sandip K. Mukherjee
Anthony W. Laurito
Madan M. Singh

Engineers International, Inc.
Downers Grove, IL

and

Jon C. Volkwein

U. S. Bureau of Mines
Pittsburgh, PA

ABSTRACT

In the past, little attention has been paid to the costs of ventilation, since it seldom represented a significant portion of the total mining cost. However, in recent years adverse mining conditions and increases in power charges have caused ventilation costs to rise. The federal regulations have set minimum standards for volumetric airflow rates in underground coal mines. As a mine develops, the air volume and pressure requirements increase. There are several methods by which the quantity of airflow to the mine and working faces can be increased. Some of the methods include (1) increasing main fan output or replacing the main fan by a new higher capacity fan, (2) reducing leakage across stoppings and overcasts, (3) adding a new ventilation shaft or borehole, (4) reducing resistance to airflow by reducing the resistance factor of the individual airways, increasing the number of airways (especially the parallel airways), or rearranging the air courses by using the available entries, and (5) using auxiliary ventilation systems.

This paper addresses the cost-benefit analysis of several of these alternatives as applied to a large operating coal mine located in southern Illinois. Pressure and quantity surveys were conducted underground, and measurements were taken at the main exhaust fans on the surface. The existing ventilation network was modeled by using the Penn State ventilation simulator. The input data included the configuration of branches, resistance of each branch (including a component for shock losses), and changes in pressure and volume. The results of the simulation were compared with the survey data to check the merits and accuracy of the model network. The network was designed to take into account shafts, major entries, overcasts, regulators, gob areas, working faces, and major leakage paths.

The model network served as the basis for the cost analyses of networks designed for increased airflow. The applicable alternatives for increasing airflow were simulated. Performance data obtained from these simulation runs were utilized to conduct the cost-effectiveness analysis. The costs for each of the alternatives included capital, operating and maintenance costs. These costs were expressed as annual costs in terms of dollars per cfm of circulating air, taking into account the economic life associated with each of the alternatives. The benefits were expressed as increased airflow quantity in the critical sections. The result of these analyses show that for the short-range plans evaluated, the most cost-effective way of increasing airflow at this mine would be to reduce leakage through stoppings and overcasts.

INTRODUCTION

Ventilation is considered the most essential ancillary operation in coal mining and a key factor in mine projections. It can have a major impact on mining costs, and often on the physical capacity to continue production. Modern technology has contributed to the design and efficiency of ventilation systems, but effectiveness and economy of ventilation still depend on good planning and maintenance by ventilation engineers and mine personnel.

Broadly speaking, ventilation planning involves estimating the quantity of air that must be circulated through the mines and portions thereof to "dilute, render harmless, and to carry away, flammable, explosive, noxious, and harmful gases, and dust, and smoke and explosive fumes" (CFR 30, Part 75.301). Federal regulations have further specified minimum standards for volumetric airflow rates in underground coal mines. However, because of changing geologic and environmental conditions and operating characteristics (leakage through stoppings and overcasts, condition of roadways, and mine extensions), adequate air may not be reaching the section. In such situations, the ventilation engineer must consider and evaluate all the available alternatives to increase airflow to the faces and improve ventilation system performance.

Several methods can be used to increase airflow in underground mine workings, including (1) increasing main fan output, (2) replacing the main fan by a new, higher capacity fan, (3) reducing leakage through stoppings and overcasts, (4) adding a new ventilation shaft or borehole, (5) reducing the resistance to airflow by reducing the resistance factor of the individual airways, increasing the number of airways (especially the parallel airways), or rearranging the air courses by using the available entries.

This paper discusses the results of an economic analysis of increasing airflow in an underground coal mine located in southern Illinois. A detailed ventilation survey was conducted underground, and measurements were taken at the main fans to establish fan operating parameters. These data were utilized to model the existing ventilation network by using the modified Penn State ventilation simulator (Wang, 1981). The applicable alternatives were simulated by making the necessary modifications to the network. The results (increased airflows) were then equated with the associated costs to assess the relative merits of each of the alternatives.

MINE DESCRIPTION

The investigation took place at a large coal mine in southern Illinois working the Herrin No. 6 seam. The seam is relatively flat, averaging 96 in. thick (2.44 m), and is under approximately 660 ft (201 m) of cover. Access to the seam is through four shafts, two in the northern area of the property and two in the southern area.

The northern shafts are oval, with the production shaft having an area of 182 sq ft (16.9 sq m) and a perimeter of 63 ft (19.2 m) with two skips. The man, material, and ventilation shaft is a split-shaft with a return air cross-sectional area of 146 sq ft (13.56 sq m) and a perimeter of 50 ft (15.24 m). The intake area is 292 sq ft (27.13 sq m) with a perimeter of 82 ft (25 m). The cage in the man, material, and ventilation shaft occupies 137 sq ft (12.72 sq m) with a perimeter of 48 ft (14.6 m).

The southern shafts are circular and consist of a man, material, and ventilation shaft and a return air shaft. The ventilation shaft has an intake area of 142 sq ft (13.19 sq m) with a 48-ft (14.6 m) perimeter and a cage that has a cross-sectional area of 92 sq ft (9.53 sq m) with a 39-ft (11.89 m) perimeter. The return air shaft has an area of 143 sq ft (13.28 m) and a perimeter of 42 ft (12.8 m).

The mine is ventilated by a pair of fans located at the northern and southern portion of the mining property. The north fan was operating at -5.90 in. (1468 Pa) total water gauge pressure producing approximately 298,000 cfm (141 m^3/s) at 880 rpm. The south fan produced about -7.60 in. (1891 Pa) water gauge total pressure and 308,000 cfm (145 m^3/s) at 880 rpm. Curves for these fans are shown in Figures 1 and 2, respectively.

The mine layout (Figure 3) consists of the mains oriented north-south and submains driven east-west. The mains consist of five intake airways with two sets of three-entry returns on either side of the main intakes, the intakes and the returns being separated by barrier pillars. Because the mine was opened before the Federal Coal Mine Health and Safety Act of 1969 went into effect, the belt line is not isolated. The submains also have five intakes and six return airways except where the trolley and belt entries must be isolated, in which case it necessitates the driving of seven intake entries. Panels are developed to the north and south of the submains approximately every 500 ft (452 m). Panel dimensions are approximately 500 ft wide by 2100 ft (640 m) long.

The north-south mains extend for more than 11,000 ft (3353 m). The northeast submains had advanced over 13,000 ft (3962 m) before succumbing to poor ventilation. At the time of the survey, pillars were being extracted, but currently no mining is taking place (although a substantial amount of exploitable coal remains). Because a major portion of the northeastern part of the property is overlain by a lake, a new shaft or borehole to provide additional air is not feasible. The western submains and the southeastern submains have advanced just over 7,000 ft (2134 m) and have adequate ventilation at the present time.

Continuous miners, boring machines, and one longwall shearer were used to sustain production. At the time of the survey, there were 11 active sections. (The longwall section was being moved to a new panel.) Of the active sections, seven were continuous miner sections and four were borer sections. Face haulage for the sections was done by shuttle cars. Section belts fed the submain belt, which in turn fed the main belt in the north-south mains. All coal was hoisted to the surface through the northern production shaft.

The ventilation method for the mains and submains is the basic two-split fishtail system. Ventilation in the panels is accomplished through a three-entry system, with the middle entry maintained neutral for belt haulage.

UNDERGROUND VENTILATION SURVEY AND FAN TESTING

Before the ventilation survey was started, the mine ventilation map was studied carefully, and stations were provisionally selected at all critical points in the distribution system, such as splits, junctions, and where physical characteristics of the airways changed. The survey was then executed by four three-man crews and one other person located at the altimeter base station to record periodic altimeter readings. Each crew was equipped with a Paulin model M-1* altimeter capable of reading to the nearest foot, one low-speed and one high-speed Davis anemometer, one Dwyer magnehelic gauge capable of indicating up to 1 inch of water and one capable of indicating 4 inches of water, one Taylor sling psychrometer for wet and dry bulb readings, and a tape measure for determining the dimensions of the entries. The altimeters were set to a known elevation on the surface before being taken underground; the base station was established near the shaft bottom. The methods of conducting pressure surveys and the computational procedure have been discussed thoroughly in several publications (McElroy and Kingery, 1957; Harris et al., 1973) and will not be repeated here. The underground pressure and quantity survey was completed in four "graveyard" shifts; an additional shift was necessary for the fan measurements.

The fan measurements were made with a pitot tube and manometer in accordance with Air Moving

*Reference to specific equipment is made to facilitate understanding and does not imply endorsement by the U. S. Bureau of Mines or Engineers International, Inc.

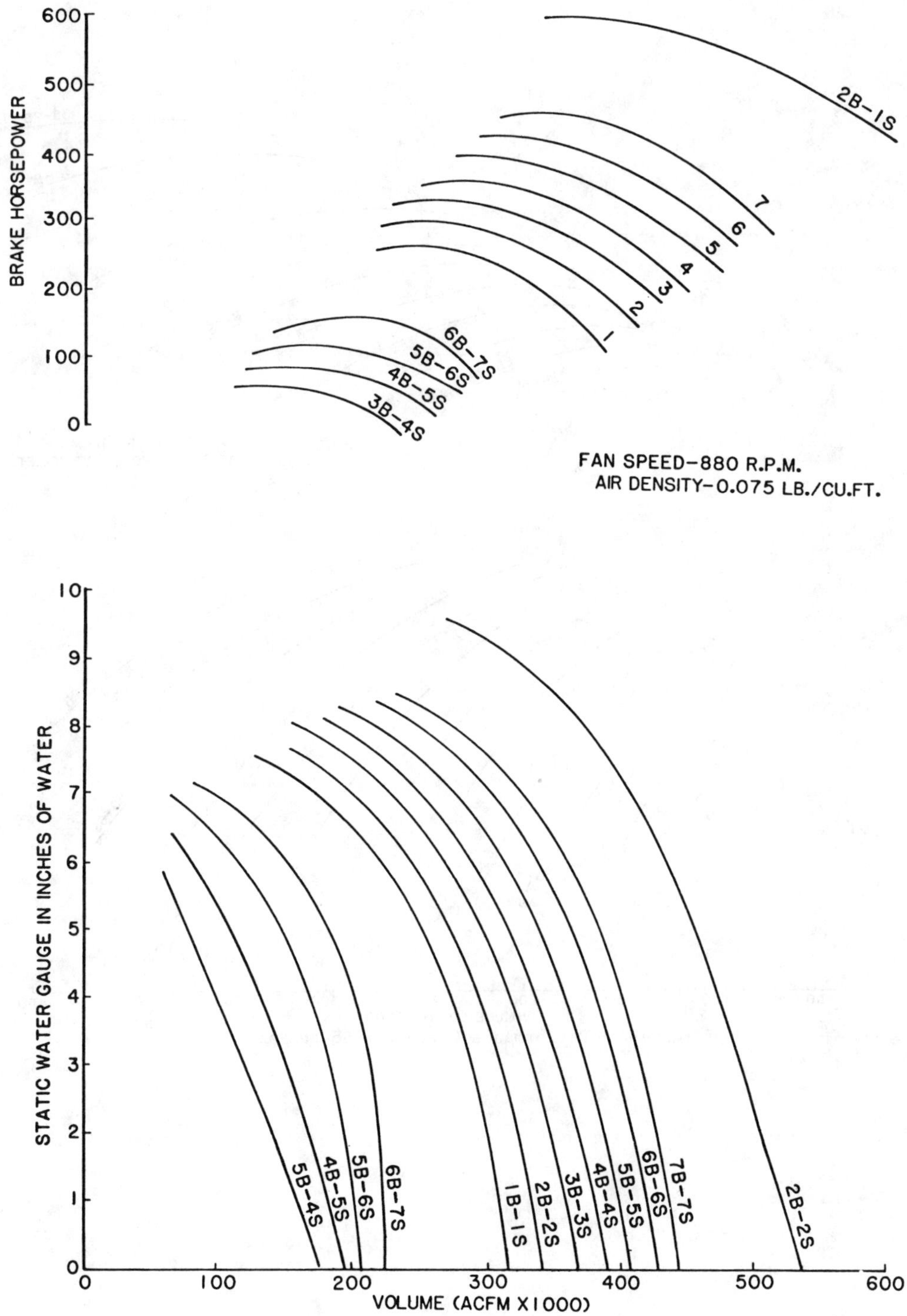

Figure 1 - Performance Curve for North Fan

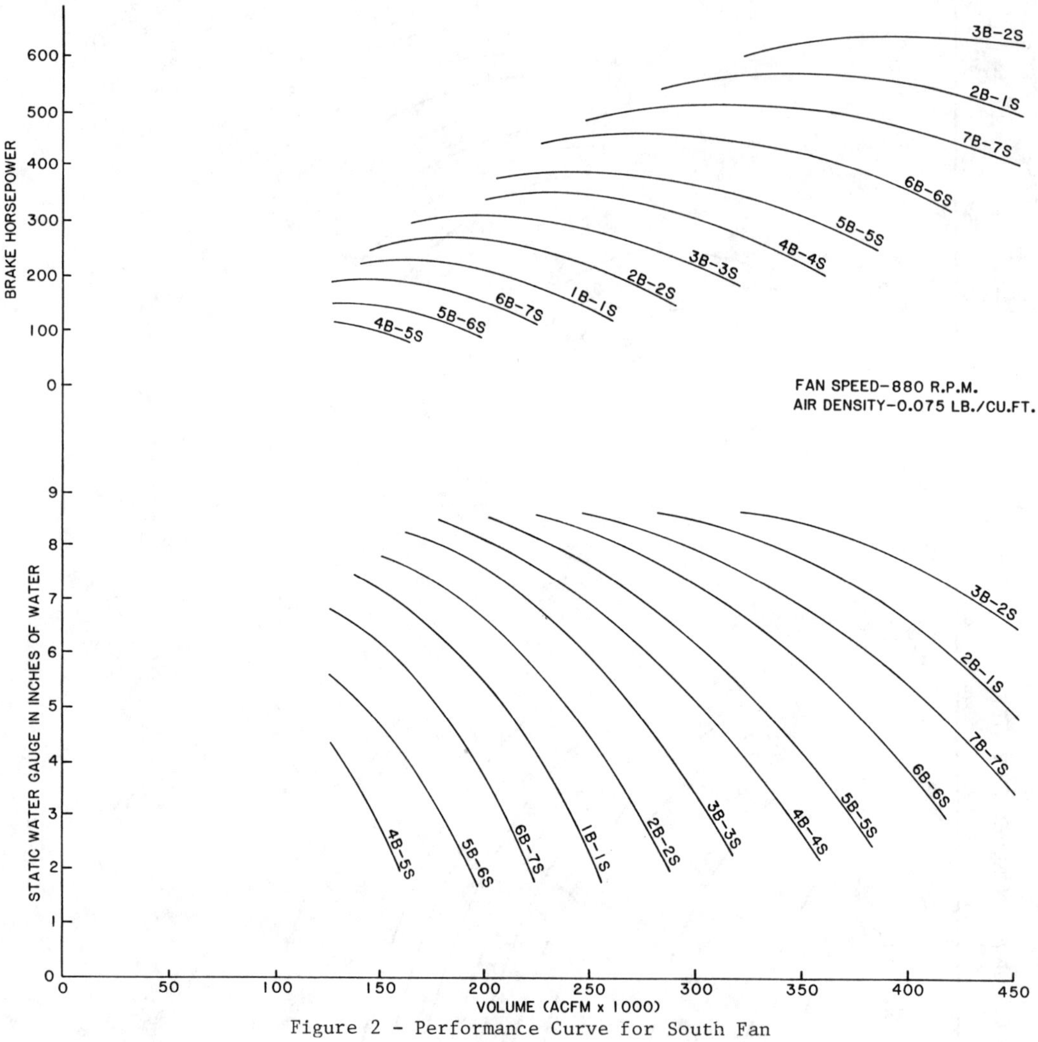

Figure 2 - Performance Curve for South Fan

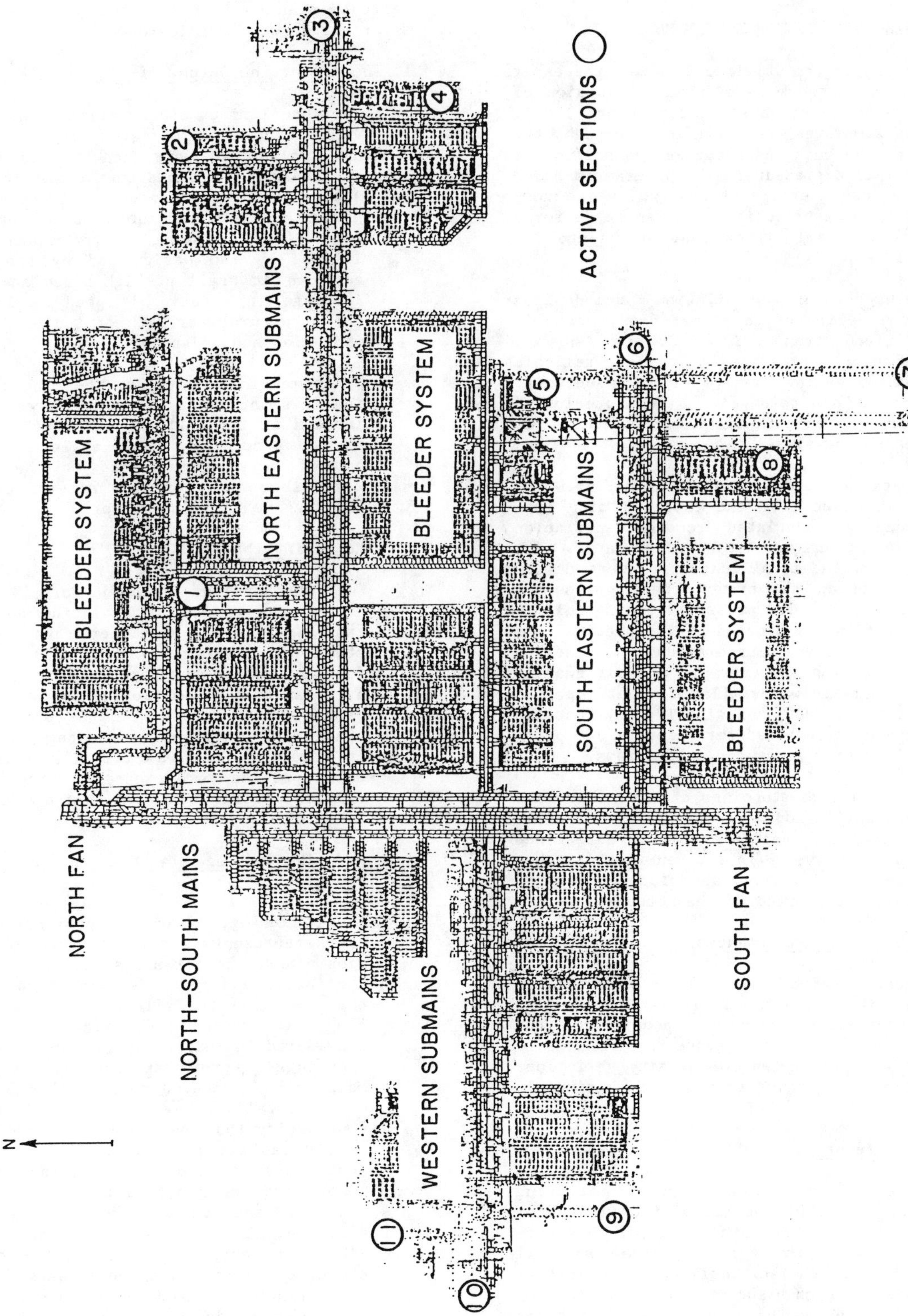

Figure 3 - Layout of Mine

and Conditioning Association (AMCA) standards. The average velocity head measurement was used to compute the air velocity through the fan duct. This average air velocity and the effective cross-sectional area at the fan was used to estimate the quantity of air exhausted by the fan.

EXISTING VENTILATION SYSTEM MODELING

The computer program utilized to model the entire mine ventilation network is a modified version of the Penn State ventilation simulator (Wang, 1981). The program simulates the airflow through the branches of the network under the action of fans or natural ventilation pressure. The program is based on the Hardy Cross iteration technique, which computes the airflow distributions and fan heads for a given network with all branch resistances and fan curves given.

Since the success of a ventilation model depends mainly on the quality of the input data, which is basically derived from the ventilation survey, great care should be exercised to collect reliable data prior to modeling any ventilation network. Reliable surveys help to develop a good predictive model to study the effects of proposed changes or modifications to the system.

The schematic of the mine ventilation network (Figure 4) was drawn from the existing mine map. Air volume readings calculated from the ventilation survey data were transferred to the schematic. For parallel airways, the total quantities were determined and placed on the branch that represented them. Each absolute total pressure reading relative to the base station was plotted at the precise location at which it was taken underground. The method adopted for balancing the pressure and air quantities in the schematic was similar to that reported recently (Harrison, 1980). Basically, any ventilation network can be balanced for quantity and pressure using Kirchoff's first and second laws.

Air leakage through stoppings and overcasts are common in any ventilation network. To a reasonable extent, these leakage quantities can be determined from the quantity survey. In the ventilation schematic, leakages are idealized as point sources and were reasonably distributed in the network.

Computation of Airway Resistances

The balanced data in the schematic were used to compute all of the branch resistances using Atkinson's equation. Since the density of the mine air was different from standard, the resistance values were corrected for density variation. These corrected resistance values were ultimately used as computer input data.

Preparation of Input Data

The input data format and sequence for executing the modified version of the ventilation simulator is shown in Table 1. The main input for this program is the fan characteristic curves and all branch data in the schematic defined by an initial and final node and each branch resistance. In about 90 percent of the cases, the modeled network resulted in air quantities that were within 5 percent of the measured values; however, in the remainder of the cases, the modeled network deviated more than 10 percent.

ALTERNATIVES TO INCREASE AIRFLOW

The balanced network was modified and simulated for each plan to increase airflow. The following alternatives were considered.

1. Increasing the Output of the Main Fan System

   Most coal mines use axial flow fans. For a constant system resistance, the pressure and volume of air can be augmented by increasing the blade angle and/or rotation speed. However, it requires increased horsepower, which should be supplied by the fan motor. The increase of the blade angle is limited by fan design and by the possible aerodynamic stall at high blade angles. The increase in rotational speed is limited by the structural strength of the blades and by vibration and noise levels.

   As mentioned earlier, the mine considered in this analysis is ventilated by fans located in the northern and southern portion of the property. Blade angle settings for both fans were increased to the limit to study the effect of this alternative on the existing ventilation network.

2. Replacing the Main Fan

   If the topology of the network is kept constant, the effect of replacing an existing fan with one of a higher capacity can be studied. In this analysis, only the north fan was under consideration for replacement by a fan that would provide approximately the same airflow but at a higher pressure. The characteristic curve of the new fan is shown in Figure 5. It was felt that this change could provide an increase in airflow to the critical section located in the northeastern submains.

3. Reducing Leakage by Coating Stoppings and Overcasts

   Air leakage through stoppings and overcasts represents a serious waste of ventilation power. It forms a short circuit between the intake and return entries outby the face thus reducing the effectiveness of the airflow circulating in the mine. For the mine considered in this analysis, leakages through stoppings and overcasts in the mains and submains alone amounted to more than 30 percent of the total airflow circulating in the mine. Considering that about 600,000 cfm (283 m^3/s) of air was being exhausted by the north and south fans, it was felt that the savings in power cost could offset the cost of coating stoppings and overcasts in the high-pressure differential areas of the mine. From the mine plan, estimates were made of the total number of stoppings and overcasts (including equalizing overcasts) in the mains and submains that needed to be coated. Table 2 shows the number of stoppings and overcasts in the

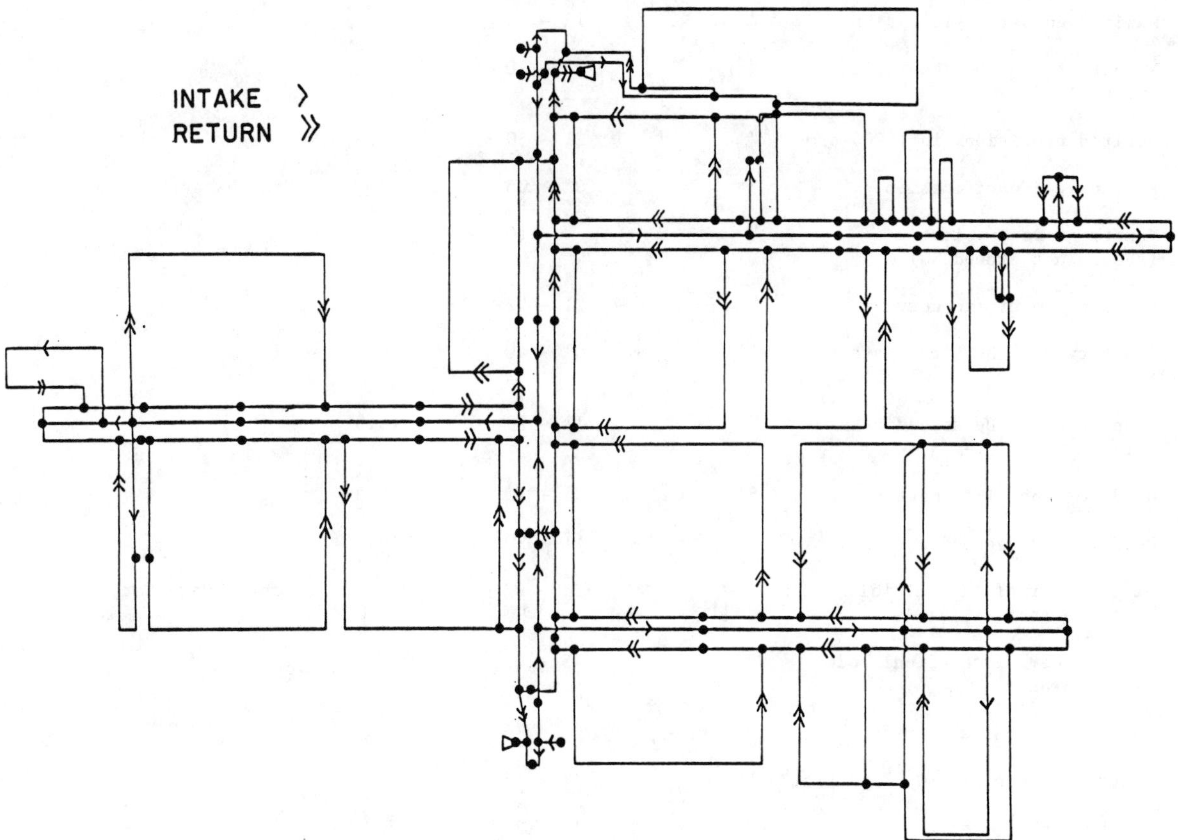

Figure 4 - Schematic of Mine Ventilation Network

TABLE 1 - Data Input Format and Sequence

VARIABLES	CARD NO.	COLUMN	FORMAT	REMARKS
Problem Identification	1	01 - 80	10 A8	Prob. ID Card
Number of branches	2	03 - 05	I3	Network parameter card
Junction number for a junction at atmospheric datum	"	08 - 10	I3	"
Maximum number of iterations	"	13 - 15	I3	"
Number of fan characteristic curves	"	18 - 20	I2	"
Desired precision, in 1,000 cfm	"	21 - 30	F10.3	"
Fan curve identification	3*	02 - 05	A4	Fan characteristic
Number of Data points (should not exceed 80)	"	10	I1	"
Description of fan curve	"	21 - 52	4A8	"
Quantity for the first data point	4*	01 - 10	F10.3	"
Quantity for the second data point	"	11 - 20	F10.3	"
Head for the first data point	5*	01 - 10	F10.3	"
Head for the second data point	"	11 - 20	F10.3	"
Node number of the initial node of the branch	6**	01 - 03	I3	Branch Data Card
Node number of the final node of the branch	"	05 - 07	I3	"
Number of parallel airways	"	09 - 10	I2	"
Friction factor in 10^{-10}lb/m^2/ft^4	"	11 - 15	F5.2	"
Height, in feet	"	16 - 20	F5.2	"
Width, in feet	"	21 - 25	F5.2	"
Length, in feet	"	26 - 30	F5.0	"
Perimeter, in feet	"	31 - 35	F5.2	"
Area, in square feet	"	36 - 40	F5.2	"
Resistances in 10^{-10}in.-m^2/ft^6	"	41 - 45	F5.2	"
Fixed quantity in 1000 cfm	"	46 - 50	F5.2	"
Fan identification (referring to Col 02-25 and 3) to be left unpunched if no fan is installed	"	52 - 55	A4	"
Natural Ventilation Pressure	"	56 - 60	F5.3	"

* Three cards required (Card No. 3, 4, and 5) for each characteristic curve, not needed if no fan is present.

** One card for each branch

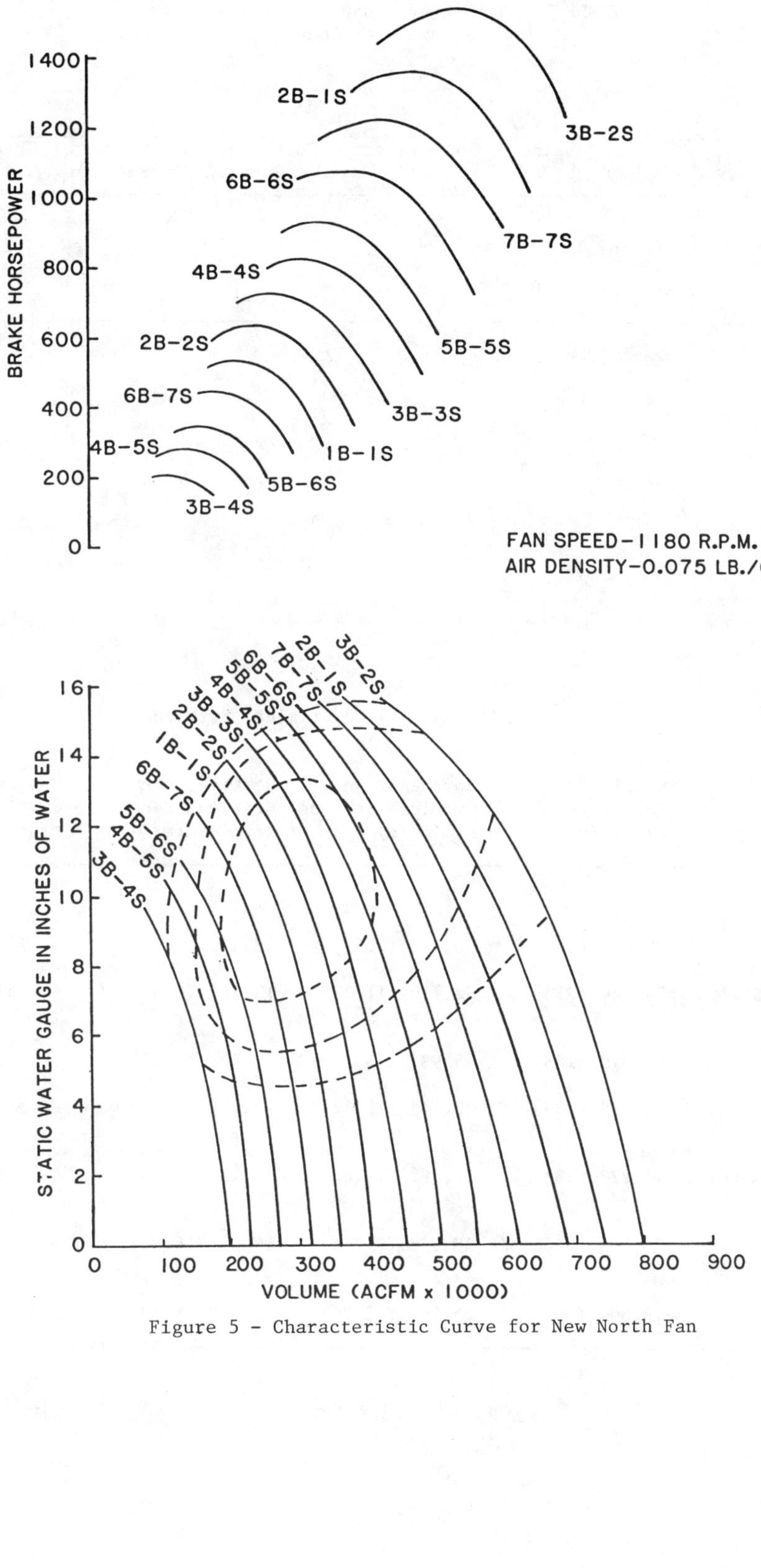

Figure 5 - Characteristic Curve for New North Fan

TABLE 2 - Number of Stopping and Overcasts Considered for Coating

	Mains	Northeastern Submain	Southeastern Submain	Western Submain	Total
Number of Stoppings	118	60	110	64	353
Number of Overcasts	30	8	10	10	58

TABLE 3 - Air Quantity to Critical Sections

	Total Air Quantity, cfm(m^3/s)	Northeastern Critical Section cfm(m^3/s)	Southeastern Critical Section cfm(m^3/s)	Western Critical Section cfm(m^3/s)
Base Case	600,031 (283)	20,418 (9.64)	12,510 (5.90)	36,702 (17.32)
Increased Blade Settings	674,140 (315)	23,790 (11.23)	13,543 (6.39)	40,657 (19.19)
Reducing Leakage	557,619 (263)	45,937 (23.38)	17,593 (8.30)	44,940 (21.21)
Cleaning Up Returns	613,079 (289)	22,177 (10.47)	13,708 (6.47)	40,512 (19.12)
Replacing North Fan	715,223 (338)	27,292 (12.88)	13,234 (6.25)	40,736 (19.23)
Increasing Parallel Entries	623,863 (294)	20,628 (9.74)	13,332 (6.29)	39,018 (18.41)
Exhaust Shaft	678,000 (320)	25,964 (12.25)	13,412 (6.33)	-

mains and principal submains considered in the economic analysis.

From the modeling of the existing ventilation network, the pressure difference between the intake and returns was determined. The sealant (Block Bond) was assumed applied 1/8 in. (0.0032 m) in thickness with stoppings construction of solid blocks, wet laid. Leakage rate for the newly coated stoppings and overcasts was considered to be 32 cfm/100 ft^2 per inch of pressure differential (0.015 m^3/s per 9.929 m^2/Pa). This rate was chosen based on information received from an existing Bureau contract studying the effectiveness of various sealants used in underground coal mines (Kissell, 1981). New and lower leakage values were computed based on the known leakage rate of the freshly coated stoppings, the existing pressure difference, and the size of the stoppings and overcasts. These new leakage splits were treated as fixed quantity branches in the network.

4. Reducing Airway Resistance

The resistance to airflow in a ventilation network can be reduced by (1) decreasing the resistance factor of the individual airways, (2) increasing the number of airways (especially the parallel airways), or (3) rearranging the air courses by using the available entries. For the economic analysis, cleaning up high quantity main and submain returns and increasing the number of parallel return entries from three to five in the mains were considered. It was estimated that cleaning up the returns would result in about a 26-percent decrease in airway resistance.

5. Addition of a New Ventilation Shaft

When the system resistance on which a fan is operating is high, the adjustment or replacement of the fan will result in excessive pressure, but with only a small increase in the air quantity. To effectively increase the airflow, the system resistance must be reduced, which can be accomplished by decreasing the resistance factor for individual airways or by sinking a new ventilation shaft.

If a new ventilation shaft is installed with a suitable fan, in addition to the airflow through the new fan, the reduction of the system resistance will result in lower pressures and higher volumes for the old fans. The total effect will be a significant increase in the air quantity flowing into the mine. However, it should be emphasized that sinking a new shaft and installing a fan involve a major capital investment for the mine.

The shaft location is considered to be one of the most important factors determining the effectiveness of a new ventilation shaft. Another important factor is the depth of the shaft, which is determined by the depth from the ground surface to the coal seam. The resistance factor for the shaft varies directly with the shaft depth. Cost of sinking the shaft also increases with depth.

The effect of adding a new shaft with an exhaust fan was studied by locating the shaft at the west end of the property as planned by the mining company. A 14 ft (4.27 m) diameter, 200,000 cfm (94.39 m^3/s), concrete lined, conventionally sunk shaft was chosen to study the effect of this alternative.

All the above alternatives were simulated by making the necessary modifications to the input data. Air quantities up to the last open crosscut of three critical sections were obtained from each of the alternative plans simulated. These critical sections were located in the northeastern submains, southeastern submains, and the western submains. Table 3 summarizes the airflow quantities obtained for each critical section and the total quantity of air circulating in the mine.

ECONOMIC ANALYSIS OF ALTERNATIVES

The economic analysis for comparing the various alternatives is based on the computation of equivalent uniform annual cost for each of the alternatives. In computing this cost, only the capital and operating costs of the two fans and the annual cost associated with each of the alternatives were considered. The cost of the shafts already in existence and ventilation control devices were not included in this analysis. All costs are in 1981 dollars.

Annual Capital Cost of Fans

The purchase price of each fan was about \$201,500 installed. This includes housing, motor and instrumentation. To compute the annual capital cost, the following assumptions were made:

Interest rate on capital:	16%
Sinking fund interest rate:	12%
Economic life:	20 years
Taxes and insurance:	3%

The annual cost was determined by using the following relationship (Parks, 1949; Hartman, 1961):

$$A = V_p \frac{r}{(1+r^n)-1} + r' + r_t$$

where
V_p = present value of investment, dollars
r = sinking fund interest rate, percent
r' = interest rate on capital, percent
r_t = taxes and insurance rate, percent

By applying the above formula, the annual cost of each fan is about \$41,082.

Annual Operating Cost

The operating cost of the fans is mainly the power cost. Unit power costs obtained from the utility company serving the mine were translated into yearly cost per brake horsepower (kilowatts) per year. This cost was about \$303/bhp per year (\$325/kw/per year) and includes an average annual power, energy, and fuel adjustment charge.

Cost for Coating Stoppings and Overcasts To Reduce Leakage

The material cost for sealing stoppings and overcasts has been based on \$11.40 per 100 ft^2 (9.3 m^2) of application. The coating area for each stopping and overcast was estimated to be about 100 ft^2 (9.3 m^2) and 816 ft^2 (75.8 m^2), respectively, based on their average dimensions. Material cost for coating only one side of the stopping and the intake surface of the overcast was considered. Labor cost for coating was based on 2 hours per 100 ft^2 (9.3 m^2) of application. The total hours for coating the required number of stoppings were determined, and this was then converted to the number of worker-shifts by considering an effective shift time of 250 minutes. Labor rate was assessed at \$145.00 per worker-shift and includes salary and benefits. The total number of worker-shifts for coating stoppings and overcasts was estimated to be 169 and 228 worker-shifts, respectively. Annual costs were computed on the basis of a 3 year economic life for the coating.

Cost for Cleaning Up Returns

A modest program of cleaning up the high quantity return airways was chosen to investigate the effectiveness of this alternative. Heavy roof falls in the return airways are not uncommon in this mine. But whenever these occur, the mine management makes an effort to clean up the fall and bolt the area. However, due to prolonged deterioration of these entries, there was still a substantial amount of muck obstructing the airways, which if eliminated offered potential improvements to the ventilation system of the mine. Consequently, it was decided to include only cleanup costs in this analysis and not consider rebolting costs.

The ownership cost of a battery-operated scoop, based on a 5-year straight-line depreciation schedule, operating one shift per day is about \$112.50. Maintenance cost is about \$15.00 per shift over the life of the scoop. Considering only one operator at the rate of \$145.00 per shift, the total cost of cleanup per shift is \$272.50. Based on the conditions prevailing in the mine, it was estimated that for the three-entry return system, up to three crosscuts could be covered per shift. At this rate, it would take about 57 shifts to clean up the mains and 41 shifts for each of the submains.

Replacing North Fan

The major change involved in this replacement was the acquisition of a new rotor and higher capacity motor. This meant an additional cost of about \$75,000, which includes installation cost. The annual cost of the new fan was computed in the same manner as outlined earlier, taking into account this additional cost. The annual capital cost for the new fan is about \$56,475.

Exhaust Shaft With Fan

The total cost associated with sinking the shaft at the west side of the property was estimated at \$1,250,000. The annual cost was determined in a manner outlined recently (Wang, 1979); specifically, the annual investment recovery charge considering an interest rate of 12 percent and a life of 20 years is \$167,348. Insurance and taxes account for \$19,500, assuming an insurance rate of 3 percent and taxes at 48 percent.

Estimation of Section Ventilation Costs

For each of the critical sections, the total ventilation costs per year were estimated. This was accomplished in the following manner:

- The yearly capital cost of the fans was distributed evenly amongst the total number of active sections to determine the capital cost per section.
- The total operating cost of the fans was first distributed among the three submains in proportion to the quantity of air entering each submains. This cost was further divided by the number of active sections in the submains to get the operating cost per section.
- In the case of reducing leakage and cleaning up returns, the costs associated with the mains were spread equally among the three submains since any improvements in the mains would also benefit the submains. This cost, together with the cost associated with the submain, was then distributed evenly amongst the active sections in the submains to obtain the cost for each section.

Costs computed in this manner for the critical sections together with the air quantities are shown in Tables 4, 5, and 6. The effect of the exhaust shaft on the critical section in the west submains is not shown because the air quantity entering this section was regulated. The section cost for the increased parallel entry system takes into account only the capital and operating costs of the mine fans. It is assumed in this case that the cost of this drivage will equal the revenue generated for the production obtained.

DISCUSSION OF RESULTS

Table 7 shows the ratio of the increased costs to the increased air quantity on a section basis for all the alternatives. Reducing leakage along mains and submains has caused substantial increases in airflow to all the critical sections despite the fact that the total quantity of air circulating in the mine has decreased. Clearly, this alternative has a decisive cost advantage over all the other alternatives concerned. Cleaning up the return airways has increased airflows up to 10.4 percent and appears to be the next most attractive alternative for all the sections. However, rebolting and helper costs were not included in the evaluation of this alternative.

For the critical section in the northeastern submains, replacing the north fan is the next best alternative, followed by increasing the blade angle, installing exhaust shaft option, and increasing the

TABLE 4 - Air Quantity and Section Ventilation Costs for Critical Section in Northeastern Submains

	Air Quantity, cfm (m^3/s)	Increase in Air Quantity, cfm (m^3/s)	Cost/year $	Increase in cost/year $
Base Case	20,418 (9.64)	-	28,336	-
Increased Blade Settings	23,790 (11.23)	3,372 (159)	40,230	11,894
Reducing Leakage	43,633 (23.38)	23,215 (10.96)	33,090	4,754
Cleaning Up Returns	22,177 (10.47)	1,759 (0.83)	29,830	1,494
Replacing North Fan	27,292 (12.88)	6,874 (3.24)	45,060	16,724
Increased Parallel Entries	20,628 (9.74)	210 (0.10)	33,418	5,082
Exhaust Shaft	25,964 (12.25)	5,276 (2.49)	75,878	47,542

TABLE 5 - Air Quantity and Section Ventilation for Critical Section in Southeastern Submains

	Air Quantity, cfm (m^3/s)	Increase in Air Quantity, cfm (m^3/s)	Cost/Year, $	Increase in Cost/Year, $
Base Case	12,510 (5.90)	_	22,625	_
Increased Blade Settings	13,543 (6.39)	1,033 (0.49)	29,542	6,917
Reducing Leakage	17,593 (8.30)	5,083 (2.40)	24,804	2,179
Cleaning Up Returns	13,708 (6.47)	1,198 (0.57)	24,413	1,788
Replacing North Fan	13,234 (6.25)	724 (0.34)	29,402	6,777
Increased Parallel Entries	13,332 (6.29)	822 (0.39)	27,373	4,748
Exhaust Shaft	13,412 (6.33)	902 (0.43)	57,401	34,776

TABLE 6 - Air Quantity and Section Ventilation Costs for Critical Section in Western Submains

	Air Quantity, cfm (m^3/s)	Increase in Air Quantity, cfm (m^3/s)	Cost/Year, $	Increase in Cost/Year, $
Base Case	36,702 (17.32)	-	33,812	-
Increased Blade Settings	40,657 (19.19)	3,955 (1.87)	45,773	12,961
Reducing Leakage	44,940 (21.21)	8,238 (3.89)	39,957	7,145
Cleaning Up Returns	40,512 (19.12)	3,810 (1.80)	43,030	10,218
Replacing North Fan	40,736 (19.23)	4,034 (1.90)	60,513	27,701
Increasing Parallel Entries	39,018 (18.41)	2,316 (1.09)	52,514	19,702
Exhaust Shaft	-	-	-	-

TABLE 7 - Ratio of Increased Costs to Increased Airflow (Dollars Per Increased cfm)

Alternatives	Northeastern Critical Section	Southeastern Critical Section	Western Critical Section
Increased Blade Settings	3.52	6.7	3.28
Reducing Leakage	0.20	0.43	0.87
Cleaning Up Returns	0.85	1.49	2.68
Replacing North Fan	2.43	9.36	6.87
Increasing Parallel Entries	24.2	5.78	8.51
Exhaust Shaft	9.0	38.55	-

number of parallel entries. Increasing the number of parallel return entries has provided the least increase in airflow to this section.

For the critical section located in the southeastern submain, besides reducing leakage and cleaning up returns, all the other alternatives have provided a marginal increase in airflow. In fact, the percentage increase is substantially less than 10 percent for all these cases. However, in decreasing order of importance, the ranking would be as follows: increasing the number of parallel entries, followed by increasing the blade angle, and replacing the north fan. Sinking an exhaust shaft is the least beneficial for this section.

For the critical section located in the western submains, increasing blade settings is the third best alternative. From the data presented in Tables 6 and 7, it appears that after increasing the blade angle, replacing the north fan would provide better results than would increasing the number of parallel return entries.

CONCLUSIONS

Based on the analysis presented herein and on the assumptions made in the evaluation of the alternatives, the following conclusions can be drawn for the mine under investigation:

- When considering short-range ventilation improvements, reducing leakage through stoppings and overcasts appears to be the most cost-effective way of increasing airflow to the critical sections. Increases in airflow up to 114 percent have been achieved at a cost of $0.20 for each cfm increase in air quantity.
- Cleaning up major return airways is the next best alternative if rebolting costs are not included. However, the increase in airflow is substantially less than that obtained by reducing leakage.
- Increasing the blade settings increases airflow to all sections, but the increased power costs for running the main fans resulted in a higher dollar cost per increase in air quantity when compared with reducing leakage and cleaning up returns. A significant quantity of the increased airflow from the fan is not utilized at the face area because of leakage in the mains and submains. For the critical section located in the western submains, increased blade angle appears to be the next attractive alternative, but this is not true for the other two critical sections analyzed.
- Replacing the existing north fan with a higher capacity fan provided more airflow to the critical sections located in northeastern and western submains when compared with the alternative of cleaning up the returns. For the critical section in the northeastern submains, the fan replacement resulted in a lower cost per increased quantity of airflow when compared with increased blade angle; however, the reverse is true for the critical section located in the western submains.
- Increasing the number of parallel return entries provides little improvement in the quantity of airflow to the critical sections. Even though no additional costs were assumed to be associated with the extra drivage of the entries, this alternative does not appear to be attractive when compared with the other short-range alternatives. For the critical sections located in the northeastern and southeastern submains, the additional shaft at the western part of the mine has not proved as cost effective as the other alternatives. However, significant benefits can accrue to the sections located in the western submains. The shaft location, therefore, is an important criterion in determining its effectiveness.

Although the results of this analysis cannot be generalized for other mines, they could be used as a guide in the evaluation of various proposals to improve ventilation system performance.

ACKNOWLEDGMENTS

The authors wish to express their sincere appreciation to the U. S. Bureau of Mines, Pittsburgh Research Center, for providing the funds to carry out this work. The authors also wish to express their gratitude to Mr. John C. LaScola, USBM, Pittsburgh Research Center, for his valuable help, counsel, and encouragement during a major part of this study. Finally, the cooperation from mining company and the mine personnel involved during the ventilation survey is deeply appreciated. The authors also wish to express their sincere appreciation to Professor Y. J. Wang of West Virginia University and to Mr. M. O. Rahim, Ventilation Engineer, Engineers International, Inc., for their critical input to model the mine ventilation network.

The views and conclusions expressed in this paper are those of the authors and should not be interpreted as necessarily representing the official policies or recommendations of the Interior Department's Bureau of Mines or the U. S. Government or Engineers International, Inc.

REFERENCES

1. Harris, E. J., R. W. Dalzell, R. J. Kline, and E. J. Miller, 1973. "A Method for Calculating Mine Ventilation Pressure Losses Using Computers and Desk Top Calculators" BuMines Information Circular 8594, 1973.

2. Harrison, S. P., 1980. "Ventilation Surveys and Their Use in Mine Planning." Proceedings of the Second International Mine Ventilation Congress, Reno Nevada, November 1979. (Editor: Pierre Moussett-Jones).

3. Hartman, H. L., 1961. Mine Ventilation and Air Conditioning. The Ronald Press Co.-John Wiley & Sons, Inc., New York.

4. Kissell, F. N., 1981. Private Communication.

5. McElroy, G. E. and D. S. Kingery, 1957. "Making Ventilation Pressure Surveys With Altimeters." BuMines Information Circular 7809.

6. Parks, R. D., 1949. Examination and Valuation of Mineral Property. Addison-Wesley Publishing Co., Reading, Massachusetts.

7. Wang, Y. J., 1981. Personal Communication

8. Wang, Y. J., J. M. Mutmansky and G. H. Walrod, 1979. "Optimal Sizing of Conventionally-Sunk Ventilation Shafts Based Upon Capital and Operating Criteria." Mining Engineering, v. 31 n. 1, January 1979.

APPLICATION OF CPM PROCEDURES IN MINE VENTILATION

Chapter 19

Y. J. Wang

Associate Professor of Mining Engineering
West Virginia University
Morgantown, West Virginia

Jan M. Mutmansky

Associate Professor of Mining Engineering
The Pennsylvania State University
University Park, Pennsylvania

INTRODUCTION

Mine ventilation analysis is an engineering discipline that can be considered a branch of the body of science known as network analysis. Likewise, the group of engineering procedures known as the critical-path method (CPM) is considered a branch of network analysis. It is therefore not surprising that mine ventilation network analysis and CPM have many similarities. These similarities are useful in analyzing several types of mine ventilation problems and will be utilized in this paper.

The analogy between the free split in a ventilation circuit and the critical path in a scheduling network has been recognized by Owili-Eger (1973). While this was recognized as a property of a general ventilation network, many important applications to controlled-splitting problems also exist. The mathematical procedures necessary to apply CPM and network methods have previously been presented (Wang, 1981; Wang, 1982). This paper will illustrate the implementation of these methods by application to mine ventilation networks.

RELATIONSHIP TO MINE VENTILATION PLANNING

The mathematical procedures that we will use are applicable to controlled-splitting mine ventilation problems, i.e., only to problems in which the ventilation air quantities in some or all of the mine airways are fixed, resulting in known air quantities and directions for all mine airways. Under these conditions, the analogy between CPM and mine ventilation analysis can be defined as shown in Table 1.

Table 1.

Analogous Entities in CPM and Mine Ventilation Networks Utilizing Controlled Splitting.

Ventilation Network	CPM Network
(1) airway	(1) activity or job
(2) junction	(2) event
(3) airway head loss	(3) activity duration
(4) junction head loss (measured with respect to entrance)	(4) event time (measured from start of the project)
(5) free split	(5) critical path
(6) regulator head loss	(6) free float

One of the primary objectives of CPM is to find the minimum time to complete a project with many interrelated activities or jobs. As a result, the same calculation procedure can also be implemented in determining the minimum fan head in a ventilation network with many interrelated airways. Thus, if a ventilation network employs a single fan, the fan head will be analogous to the minimum completion time in a CPM solution.

For those not familiar with the terminology in CPM, a quick study of any basic text will provide the necessary background. A good elementary reference with a slant toward management is the volume by Wiest and Levy (1977). The analogy between CPM and mine ventilation networks can be readily understood by referring to such a reference and noting the similarity between a CPM planning network and a ventilation network. This paper will attempt to explain these methods in terms a practicing engineer will understand without the use of a mathematical format.

APPLICATION OF CPM TO NETWORKS WITH A SINGLE ENTRANCE AND A SINGLE EXIT

To illustrate the applications of CPM and network theory to ventilation problems, we will first work with elementary ventilation networks. For example, assume a ventilation layout like the one shown in Figure 1, which is typical for a small, noncoal, multi-level, vein mining operation. Assume also that the air quantities moving through some or all of the airways are known as a result of regulatory or practical constraints in supplying air to the production centers. In many cases, this will allow us to determine the air quantities for every airway. This is the controlled-quantity situation that we will examine.

To initiate analysis of the network, we must number the junctions of the layout. Since the directions of the air in all the airways are known, it is possible to number the junctions in an orderly manner so that the air always flows from a junction that has a lower number than the junction to which the air flows. This is accomplished for the ventilation network given above by logically moving from entrance to exit of the network. One such numbering result for the network in Figure 1 is shown in Figure 2 where the junction numbers are given beside the circles representing the junctions. It is easy to verify that, for a given junction, every predecessor junction or the junctions coming before it in the flow of air, have lower junction numbers. For example, the predecessor junctions corresponding to junction (7) are junctions (4) and (6) and each has a lower number. This must be true for every junction in the network.

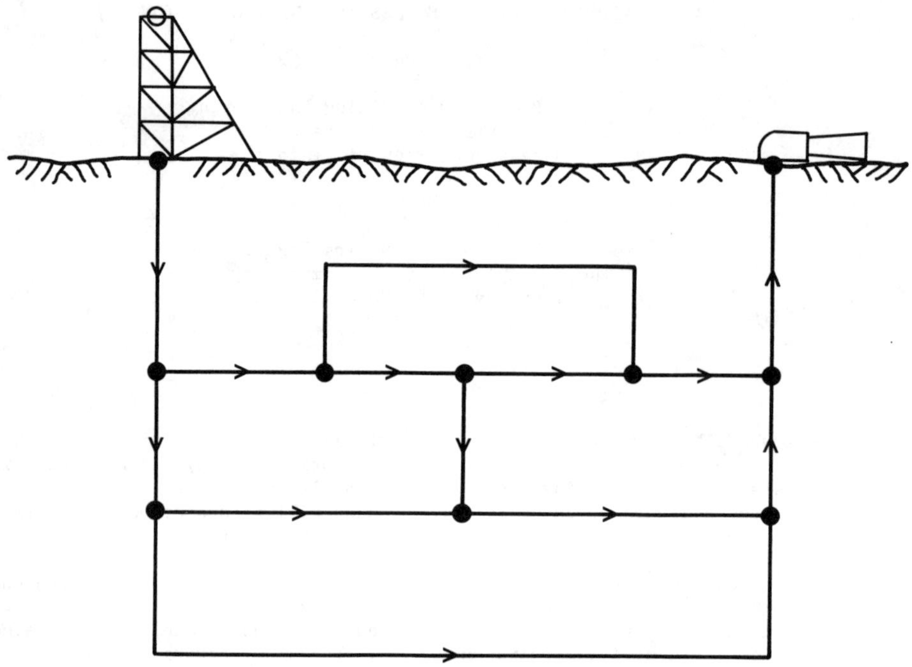

Figure 1. Schematic of a Ventilation Network for a Small Multi-Level Vein Mining Operation.

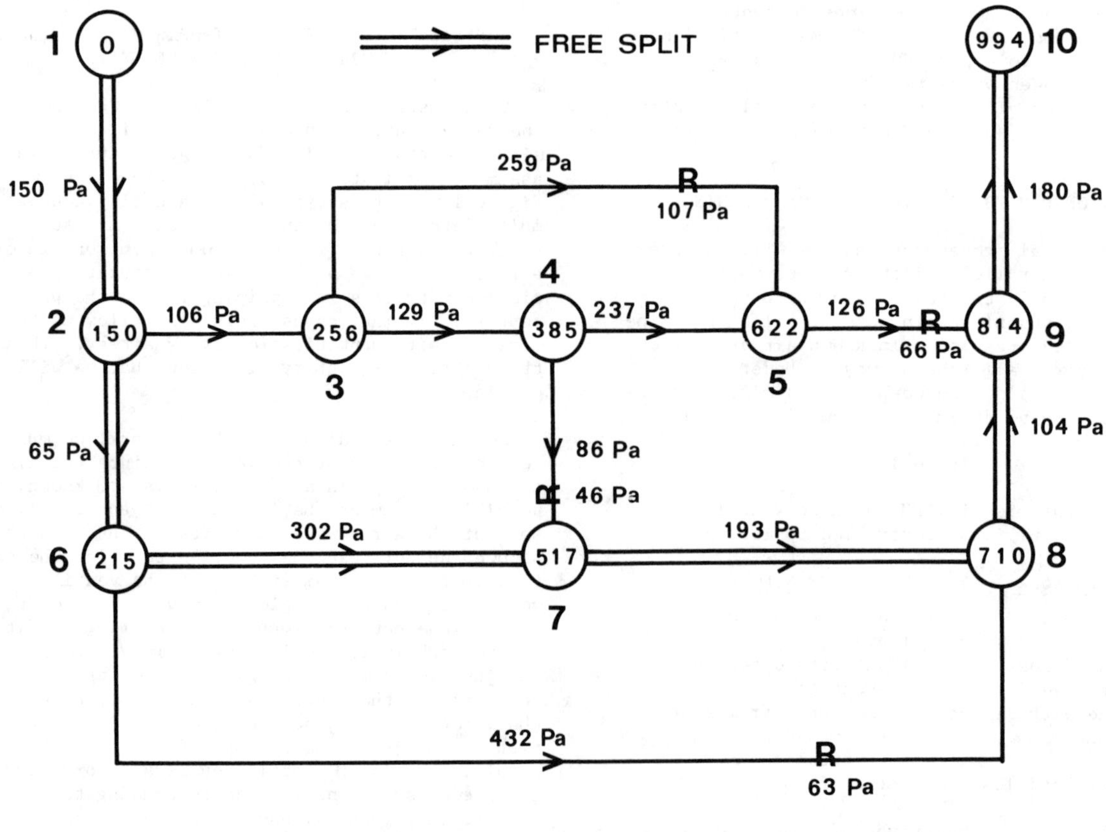

Figure 2. Revised Schematic Showing Junction Numbers, Friction Heads, Cumulative Junction Head Losses from the Entrance, and Locations of the Regulators for the Initial Forward-Pass Solution.

The next step in the procedure is to determine the airway head losses. If the resistance factor of airway $\underline{i}$, denoted R_i, and the quantity of air through airway $\underline{i}$, denoted Q_i, are both known, then the head loss H_i can be determined for each airway. These are done for the example network in Table 2. The heads, calculated here in pascals (Pa), are then placed on the network diagram as shown in Figure 2. The heads shown are airway head losses and are positive; it follows that fan heads, when fans are used, are negative.

Table 2.

Calculation of Airway Head Losses for the Example Network of Figure 1.

Airway	R_i (kg/m^7)	Q_i (m^3/s)	H_i (Pa)
(1,2)	0.015	100	150
(2,3)	0.035	55	106
(2,6)	0.032	45	65
(3,4)	0.143	30	129
(3,5)	0.414	25	259
(4,5)	0.731	18	237
(4,7)	0.597	12	86
(5,9)	0.068	43	126
(6,7)	0.755	20	302
(6,8)	0.691	25	432
(7,8)	0.188	32	193
(8,9)	0.032	57	104
(9,10)	0.018	100	180

The next step in the process is to assign the cumulative head losses within the system to each junction or node in the order specified by the junction numbers. The ventilation engineer can easily perform this job on the schematic of the system by moving from entrance to exit while cumulating heads. The proper cumulative head to be assigned to any junction is the highest cumulative head achieved when moving from entrance to the junction by any path consistent with the directions of airflow in Figure 2. The cumulative head losses at the nodes are provided in Pa within the circles representing the junctions. Note that the cumulative head for junction 7 is 517 Pa. The head cumulated along path (1,2), (2,6), (6,7) is 517 Pa and the head cumulated for path (1,2), (2,3), (3,4), (4,7) is 431 Pa. The junction is assigned a head of 517 Pa and the consistency in the network is kept intact by placing a regulator in airway (4,7). This makes the cumulative head losses independent of the path from the entrance to the given junction. Alternately, one can locate a booster fan in airway (2,6) or (6,7) to make the cumulative head loss at junction 7 consistent, but this alternative will not be considered until later. For now assume that only regulators will be used to balance the head loss in any given network loop.

Using the procedure established, all the junction values can be calculated and the necessary regulators inserted into the network. For the example network provided, the locations and values of head loss for the regulators established following the above procedures are shown in Figure 2. When the necessary regulators, have been inserted, the free split or critical path can be identified as any direct path from entrance to exit that does not involve the regulators. There will normally be only one such direct path but multiple critical paths are possible. Only one free split or critical path exists for the example problem and that is shown in Figure 2.

The above information can also be readily obtained from the typical output of a CPM computer program. The free split or critical path is identified as those airways where the total float is zero. The location of regulators corresponds to airways with a positive value of free float. The airways in the free split plus the additional airways containing no regulators form the longest-path spanning tree. From this longest-path spanning tree, alternative solutions to be discussed later can be generated.

The CPM method is not necessary for solution of the simple example network provided above. However, as the complexity of the network increases, the CPM procedures for finding the critical path become more useful. CPM codes for use on digital computers can handle the most complex of networks and the element of human error is minimized.

The logical procedure of moving from entrance to exit in assigning the junction head values in Figure 2 is called the forward-pass technique of CPM. The backward-pass technique, another common CPM procedure, can likewise be used. For the network shown in Figure 1, the backward-pass solution is provided in Figure 3. The junction head values are determined for this solution by moving backward from the known head value of 994 Pa at the exit of the network. At each junction, the lowest calculated head resulting from moving from the exit backward to the junction counter to the flow of air is assigned to the junction. Regulators are then added where necessary to balance the loops.

The backward-pass solution has the property that the junction values along the critical path are identical to that of the forward-pass solution. However, the other junction values and the locations of the regulators in the solution vary. The reason for obtaining this second solution is the additional information that may be extracted from the second solution. This will be discussed in more detail later. A further difference between the forward-pass and the backward-pass solution is that the forward-pass procedure tends to locate the regulators closer to the exit while the backward-pass procedure tends to locate the regulators closer to the entrance.

APPLICATION OF CUTSET OPERATIONS TO VENTILATION NETWORKS

After the original network solution by the forward-pass or the backward-pass technique has been obtained, it is perhaps logical to again think of the option of using booster fans instead of regulators in the network. To evaluate all possible alternatives, a procedure is established using the concepts of a cutset. A cutset is defined as a minimal set of branches (airways) whose removal from the network results in exactly two subnetworks.

To investigate alternative solutions, call the subnetwork of junctions containing the entrance N_1 and the subnetwork of junctions containing the exit N_2. Then the cutsets applicable to investigating the addition of booster fans are chosen so that they cut at least two airways and so that all the airways in the cutset are positive (i.e., flow from N_1 to N_2). In this case, we assume that the return dummy branch is not included. As a result of the above rules, it should be noticed that any cutset chosen

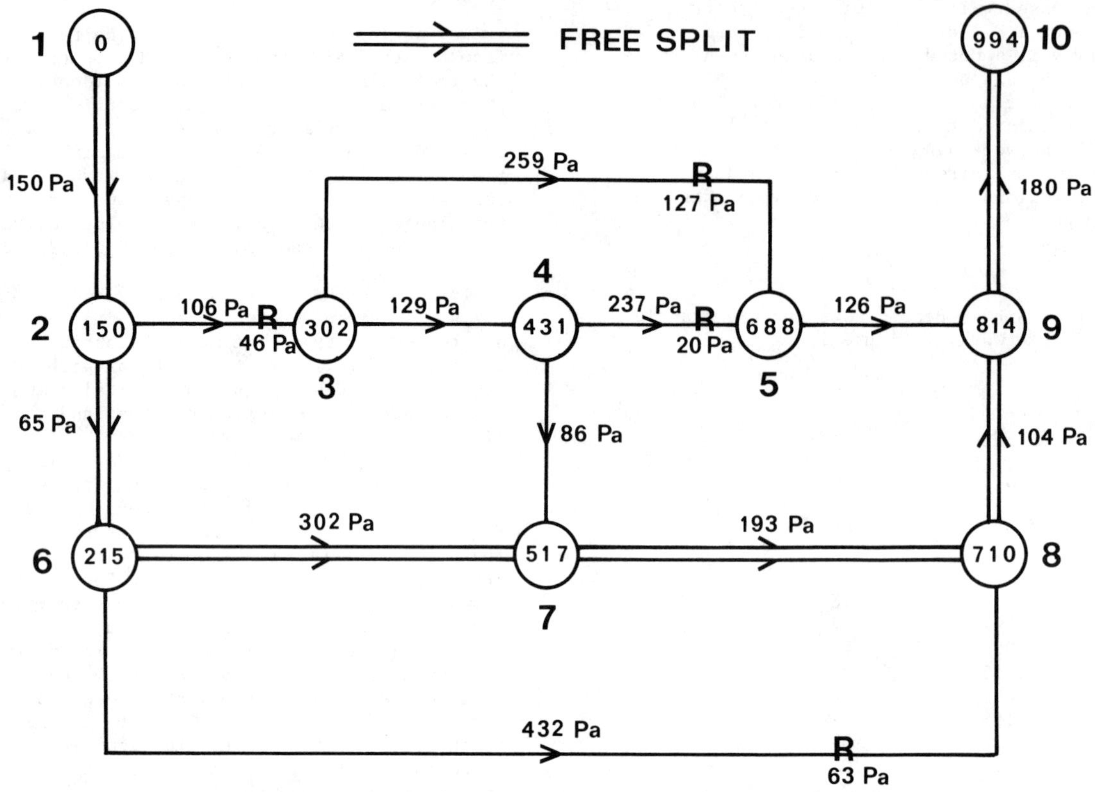

Figure 3. Revised Schematic Showing the Initial Backward-Pass Solution.

automatically contains an airway from the free split. The cutsets applicable to the solutions found in Figures 2 and 3 are shown in Figure 4 as dotted lines running through the network.

The evaluation of the individual cutsets are performed by making an exchange of booster fans for regulators in accordance with Kirchhoff's voltage (head) law. As an example, study cutset K shown in Figure 4. Cutset K cuts three airways with characteristics as follows in the solution of Figure 2:

Airway	Head Loss Without Regulator (Pa)	Artificial Head (Pa)
(5,9)	126	66
(7,8)	193	0
(6,8)	432	63

Assuming that we wish to place a single booster fan in the system, the logical position for the fan is in airway (7,8). Since the smallest regulator head loss affected by the cutset is 63 Pa, a booster fan located in airway (7,8) must supply -63 Pa of head while the regulators in airways (5,9) and (6,8) are reduced by 63 Pa. The resulting solution gives:

Airway	Head Loss Without Regulator (Pa)	Artificial Head (Pa)
(5,9)	126	3
(7,8)	193	-63
(6,8)	432	0

The solution with a single booster fan is shown in Figure 5. Note that the negative value of artificial head indicates a fan while a positive value indicates a regulator. To complete the solution, it is necessary to recompute the junction values for junctions 8, 9, and 10.

As most ventilation engineers know, the total power consumed by the fans in the system should be reduced as a result of adding a booster fan in the manner indicated above. The power saved by such an action can be calculated as follows:

$$W_c = (63 \text{ Pa})\ (43 \text{ m}^3/\text{s} + 25 \text{ m}^3/\text{s}) = 4284 \text{ W}$$

where W_c is the power conserved compared to a solution without booster fans. Similarly, the same cutset can be used to evaluate the addition of two booster fans. In this case, the resulting solution gives the cutset airways the following values:

Airway	Head Loss Without Regulator (Pa)	Artificial Head (Pa)
(5,9)	126	0
(7,8)	193	-66
(6,8)	432	-3

The two fans in the network will result in the main fan having a reduced head of -928 Pa and a net power savings of:

$$W_c = (63 \text{ Pa})(25 \text{ m}^3/\text{s}) + (66 \text{ Pa})(43 \text{ m}^3/\text{s}) = 4413 \text{ W}$$

Thus, additional power savings are accrued.

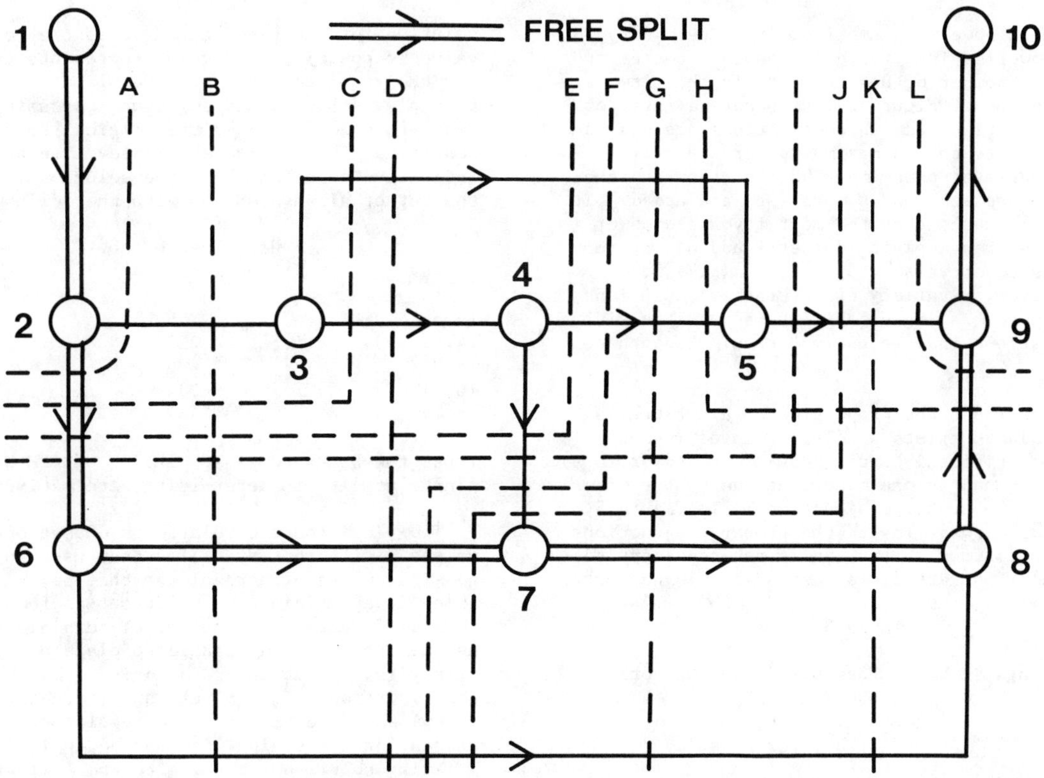

Figure 4. Cutsets Applicable to the Solutions of Figures 2 and 3.

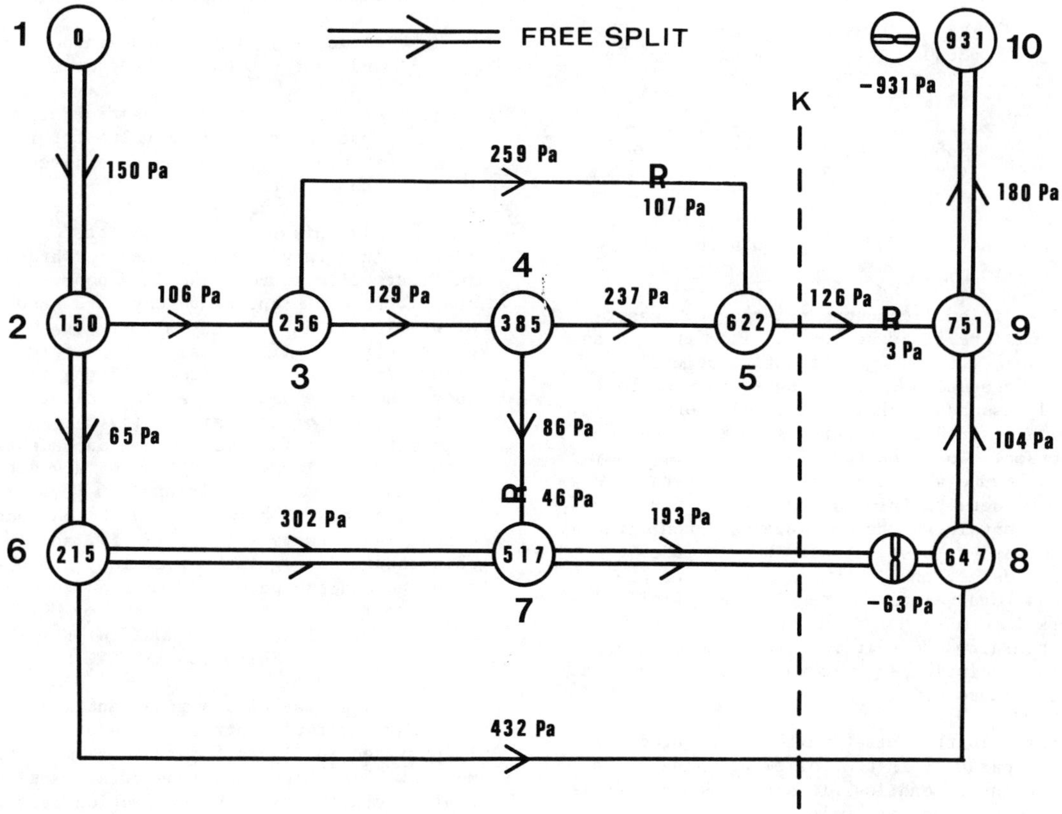

Figure 5. Solution Obtained by Adding One Booster Fan Using Cutset K in the Solution of Figure 2.

To optimize power consumption, it is necessary to determine power savings for each possible cutset and each possible number of booster fans in the network. For the solution of Figure 2, the power savings for all possible cutset and fan combinations are provided in Table 3. Note that a maximum power savings, 7640 W, is achieved using four booster fans and that more savings are accrued as more fans are added. It should also be noted that the most favorable power savings arise when applying cutset F and adding four booster fans to airways (4,5), (4,7), (6,7), and (6,8). However, a variety of alternative solutions with equal power-saving capability is possible with cutset operations and the analyst should keep these in mind.

One of the important properties of a network is that there always exists a solution involving no more than M artificial heads (regulators and/or fans). Here, M is defined as one more than the number of airways less the number of junctions. In Figure 5, the number of airways is 13, the number of junctions is 10, and M is four. Thus, it is not necessary to use more than four artificial heads in this network.

Table 3.

Power Savings in Watts Possible in the Network of Figure 2.

Cutset	Number of Booster Fans Added			
	1	2	3	4
A	2530	-	-	-
B	3680	4105	-	-
C	2530	5705	-	-
D	3680	4530	7280*	-
E	2530	3390	6065	-
F	3680	4836	4965	7640*
G	4284*	4413	7088	-
H	2838	5513*	-	-
I	2530	3390	-	-
J	3680	4836	4965	-
K	4284*	4413	-	-
L	2838	-	-	-

*Maximum power savings for the given number of booster fans.

Several additional concepts relating to cutset operations will be discussed here. The first is an added procedure of finding alternative solutions using cutset operations. This procedure uses both the forward-pass and backward-pass solutions to generate the desired alternative solution based on a given cutset. To illustrate this concept, study cutset F in Figure 4. To assemble the alternative solution, the network information for the portion of the network containing the entrance is taken directly from the forward-pass solution. This includes the airway head losses, the junction values, and the location and head values for regulators. Similarly, the network information for the portion of the network containing the exit is taken from the backward-pass solution. These values are assembled as shown in Figure 6.

The airways in the cutset must be evaluated for regulators separately in the following manner. The value of regulator head loss in any cutset airway is calculated as the junction value at the end of the airway minus the junction value at the beginning of the airway minus the airway head loss. If the resulting value is zero, no regulator exists in this solution for the given airway. If the resulting value is positive, the value represents the head loss for the regulator that must be placed in that airway to achieve the desired solution. Assuming that only regulators were used in the original solutions, the resulting airway value should never be negative, indicative of a fan. For the solution in Figure 6, the cutset airways end up with the following values:

Airway	Head Loss Without Regulator (Pa)	Artificial Head (Pa)
(3,5)	259	173
(4,5)	237	66
(4,7)	86	46
(6,7)	302	0
(6,8)	432	63

Thus, the backward-pass solution of CPM has a definite utility in determining alternative solutions.

The final concept related to cutset operations involves a different type of cutset used in evaluating possible moves of a regulator that is in an objectionable airway. This type of cutset is formed according to a different set of rules. To explain these rules, assume that you had planned to place a regulator in airway (s, t) but later found it to be an objectionable location. To evaluate an alternative airway (u, v) as a location for the regulator, the cutset chosen for the evaluation must be applied to any existing solution according to the following rules:

(1) the directions of airway (u, v) and airway (s, t) must be opposite in direction with respect to the cutset,

(2) airway (u, v) must contain no regulator and must not be on the free split, and

(3) the regulator head loss for airway (s, t) must be the minimum value for regulators in branches with the same direction in the cutset.

To apply this method, assume that the regulator located in airway (5, 9) of the forward-pass solution in Figure 2 is found to be in a poor position with respect to equipment travelways. Assume also that airway (4, 5) is a likely alternative location for the regulator. If a cutset of the form shown in Figure 7 is used, it follows all the rules listed above and hence can be used to evaluate the removal of the regulator from airway (5, 9). As a result of this cutset, it is possible to conclude that the regulator in airway (5, 9) can be moved to airway (4, 5) provided that an additional 66 Pa of regulation is also placed in airway (3, 5) to balance the loop containing airways (3, 4), (4, 5), and (3, 5). Under this change of regulator location, the head of the main fan and the power consumption does not change.

APPLICATION TO NETWORKS WITH MULTIPLE ENTRANCES AND EXITS

In many situations, a mine ventilation network will have multiple entrances and/or exits. This is illustrated in Figure 8 for a coal mine ventilation network. In this type of network, particularly one that is characterized by production centers that are distant from each other, multiple entrances and exits are common. To analyze such a network by CPM methods, it is generally helpful to restructure the network.

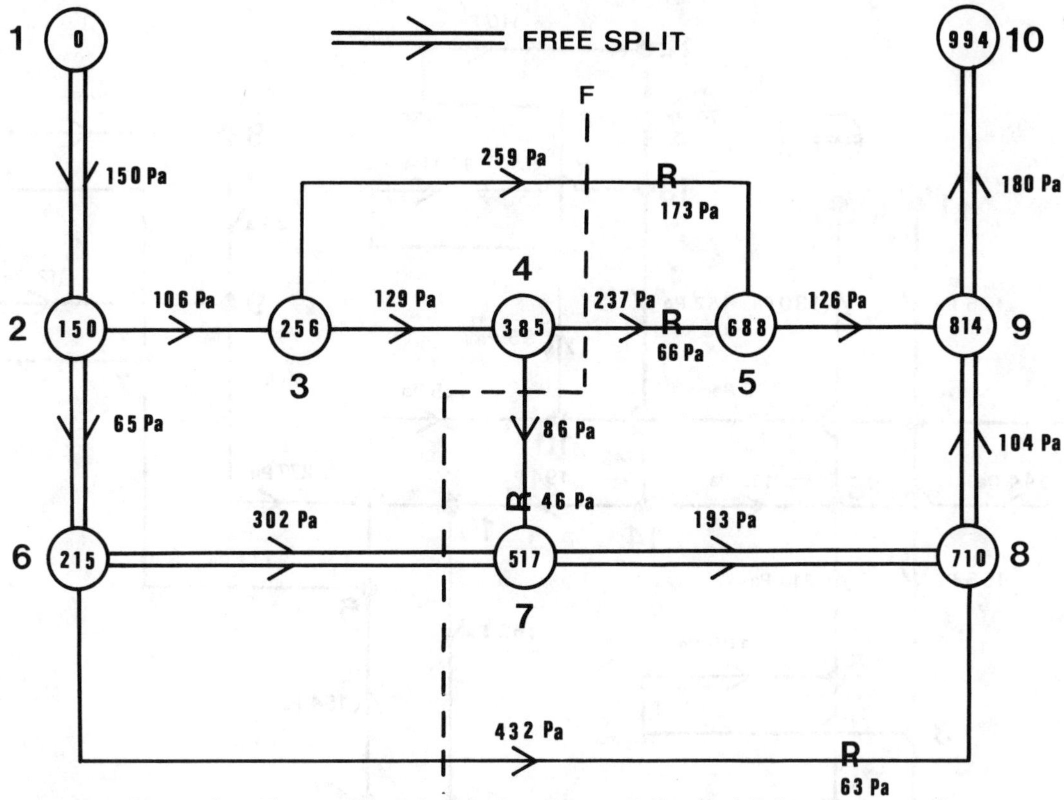

Figure 6. Alternative Solution for Cutset F Obtained Using Elements of the Forward-Pass and Backward-Pass Solutions.

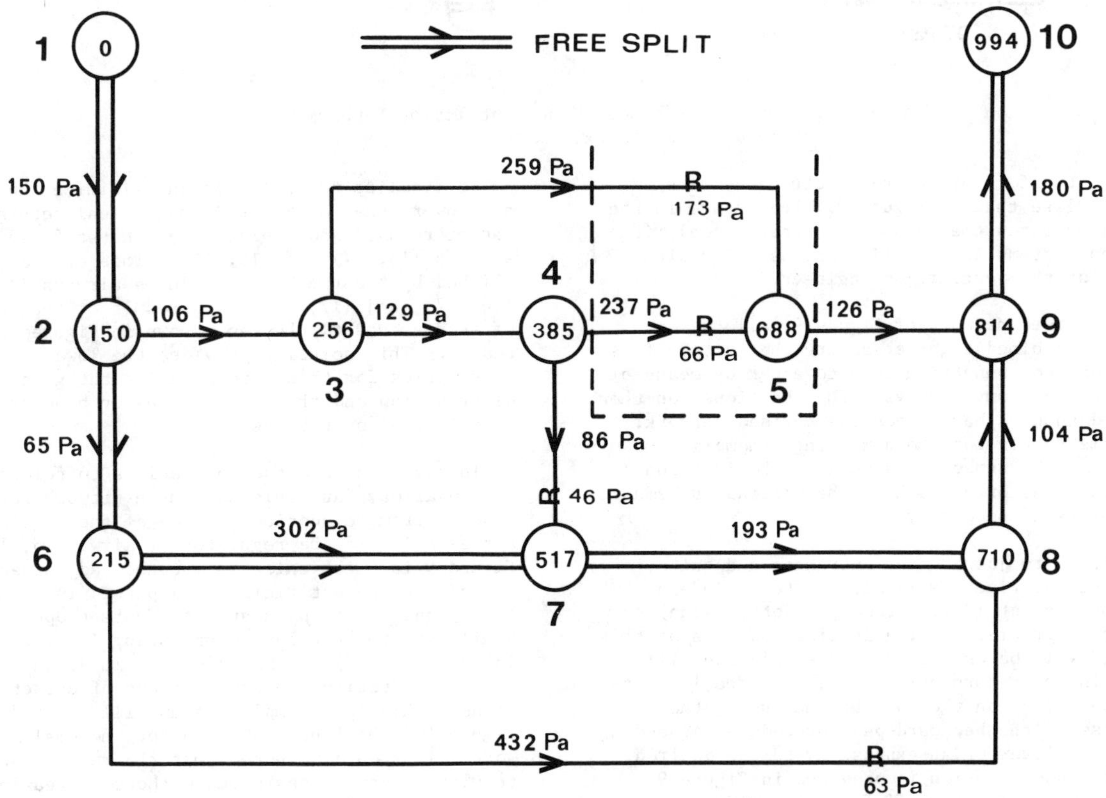

Figure 7. Alternative Solution to the Forward-Pass Solution of Figure 2 Indicating the Removal of the Regulator from Airway (5, 9).

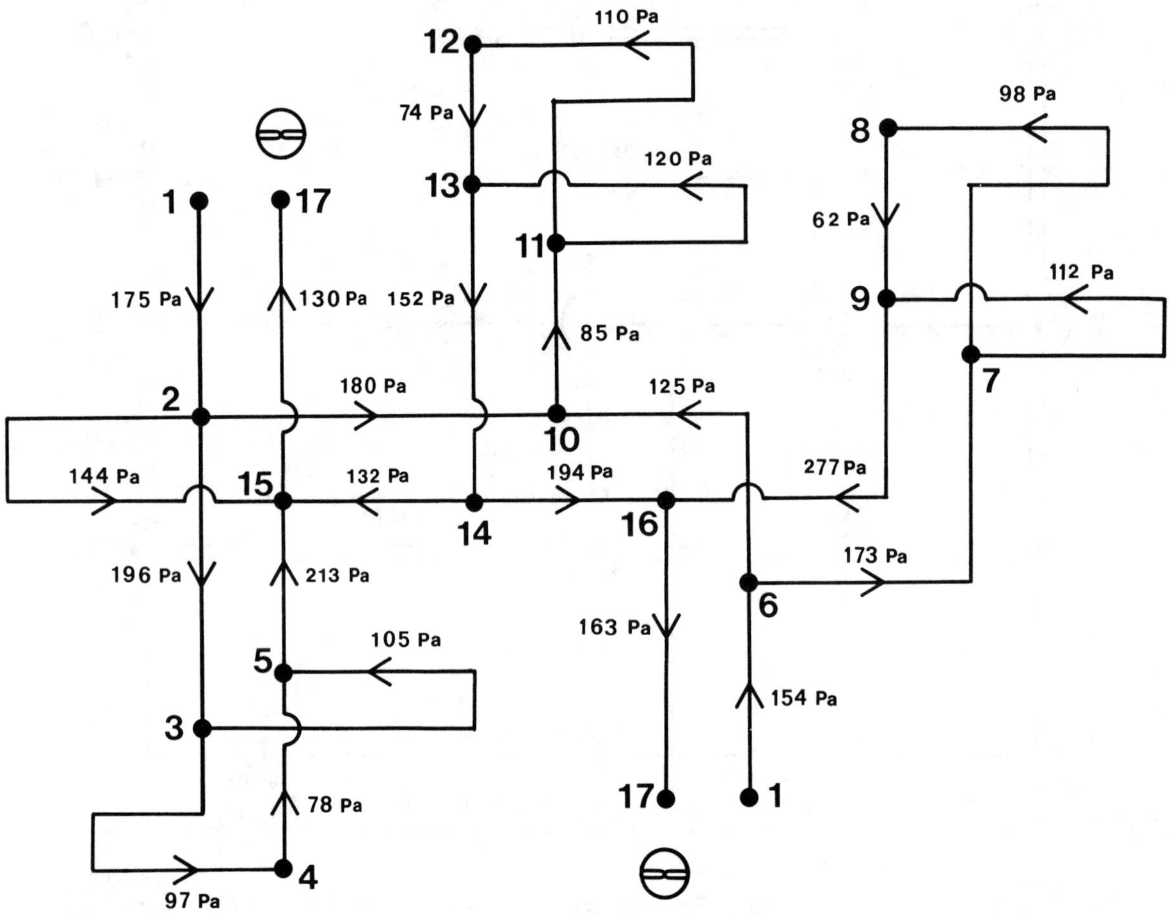

Figure 8. Schematic of a Coal-Mine Ventilation Network with Two Entrances and Two Exits.

In general, the form of the restructured network should look like that in Figure 9. The change in the network illustrates the parallel nature of coal mine ventilation systems and provides an easier working schematic for the ventilation engineer.

In the revised network of Figure 9, all the entrances are combined together as are the exits. This is equivalent to connecting them together by means of zero-resistance dummy airways. The junctions can then be assigned numbers based upon the revised network. One possible result for the numbering scheme is shown in Figure 9. The numbers assigned to the junctions can then be transferred back to the original schematic of the network if necessary.

In addition to joining the entrances and exits, the schematic of Figure 9 is altered to include a dummy return branch (designated by a dotted line) to complete the circuit. In the initial analysis of this network, it will be assumed that a single fan will be placed in the return airway (17, 1). The longest-path CPM solution can then be obtained using the forward-pass or the backward-pass procedure, depending on whether a blower or an exhaust system is desired. The forward-pass solution is provided in Figure 9 where a fan pressure of -1133 Pa is shown in the return airway.

To finalize the solution and minimize the total air power, the fan in the hypothetical return branch can be removed and replaced by exhaust fans in return airways (15, 17) and (16, 17). This can be accomplished by a cutset that includes airways (15, 17), (16, 17), and (17, 1). As a result of the analysis of such a cutset, the fans shown in Figure 10 will result. This solution provides the lowest air consumption for this mine assuming the given air distribution and the restrictions on booster fans in coal operations.

In addition to cutset operations to relocate the return airway fan, this coal mine network may also benefit from relocation of some of the regulators. For example, if the regulator of airway (5, 15) in Figure 9 is inconveniently located, adjustments can be made in the artificial head placed in airways (4, 5) and (3, 5) by means of a cutset operation. Similarly, the regulators appearing in airways, (9, 16) and (15, 17) can be relocated in airways closer to the production centers by means of cutset operations. This is accomplished and illustrated in Figure 10, the final solution for the coal mine network. It should be noted that the relocation of the regulators can be performed without increasing the power consumption in the network.

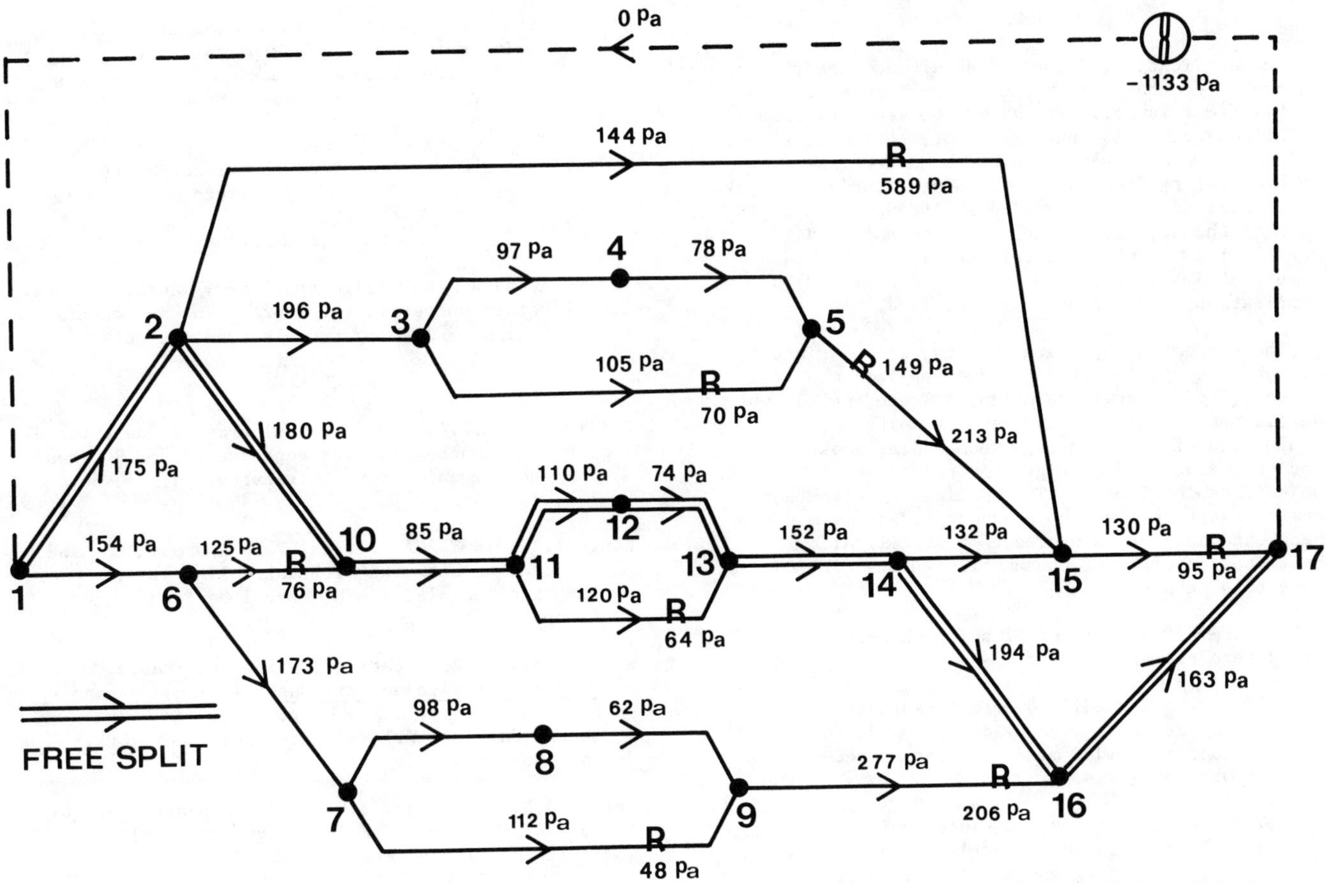

Figure 9. Revised Schematic of the Coal-Mine Ventilation Network Showing the Initial Forward-Pass Solution Assuming a Single Fan in the Dummy Return Airway.

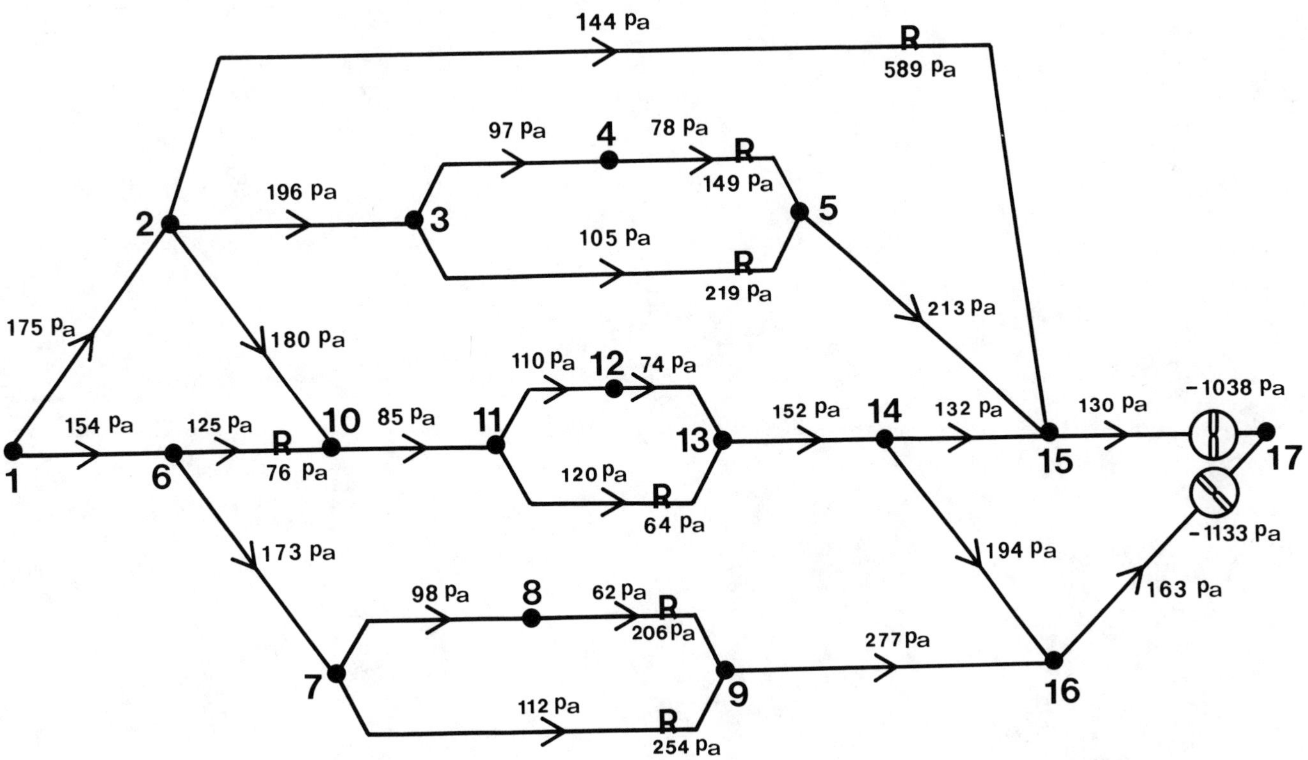

Figure 10. Final Solution of the Coal-Mine Ventilation Network with Fans in the Exhaust Airways and Relocated Regulators.

SUMMARY

The application of tools from CPM to mine ventilation problems has been presented here in a non-mathematical manner. The analogy between CPM methods and mine ventilation network theory has been used to illustrate the forward-pass and backward-pass procedures for finding solutions to ventilation network problems involving controlled splitting. The usefulness of the analogy between the free split in mine ventilation and the critical path in CPM has been illustrated. Overall, the application of CPM in mine ventilation planning has been outlined.

The concept of applying cutsets, which arises from network theory, is explained and applied to a variety of ventilation network problems. These include the development of alternative network solutions, the evaluation of power savings when adding booster fans, and the evaluation of regulator relocation from objectionable airways. The application of cutset operations to evaluation of booster fans in mines has been extended into a procedure that will allow optimization of power consumption when adding booster fans to a network.

The primary advantages of using the methods outlined here are as follows:

(1) CPM is a readily-available software package that is convenient to use for computer-aided mine ventilation network analysis.

(2) network analysis procedures using cutset operations provide the basis for optimizing power consumption in mine ventilation networks,

(3) the analysis of regulator relocation using cutset operations permits the ventilation planner a maximum of freedom without increasing power costs, and

(4) the number of devices to introduce artificial head in the network can be minimized.

ACKNOWLEDGEMENT

The authors acknowledge the Energy Research Center, West Virginia University, for providing support for the initial work leading to this publication.

REFERENCES

Owili-Eger, A.S.C., 1973, "Simulation of Quality and Quantity Control in Mine Ventilation," M.S. Thesis, The Pennsylvania State University, pp. 55, 56, 123, 124.

Wang, Y.J., 1981, "A Critical Path Approach to Mine Ventilation Networks with Controlled Flow," SME Preprint 81-23, AIME Annual Meeting, Chicago, 27 pp.

Wang, Y.J., 1982, "Ventilation Network Theory," Ch. 17, Mine Ventilation and Air Conditioning, Hartman, H.L., Mutmansky, J.M., Wang, Y.J., eds., Second Edition, Wiley-Interscience, New York, pp. 483-516.

Weist, J.D., and Levy, F.K., 1977, A Management Guide to PERT/CPM, Second Edition, Prentice-Hall, Englewood Cliffs, New Jersey.

4
Mine Fires

Chapter 20

MINE SHAFT FIRE AND SMOKE PROTECTION SYSTEMS-- AN UPDATE ON HARDWARE DEVELOPMENT AND IN-MINE TESTING

Guy A. Johnson

Mining Engineer
Twin Cities Research Center
Minneapolis, Minnesota

ABSTRACT

In 1976, The Bureau of Mines developed a prototype system to sense and extinguish fires in shafts and shaft stations in underground metal and nonmetal mines. Subsequent work modified this technology to include fueling areas, spontaneous combustion zones and coal mines. This paper updates IC-8783 "In-mine Fire Tests of Mine Shaft Fire and Smoke Protection Systems," which was published in 1978 and summarized the design and in-mine, actual fire testing of the first prototype mine shaft fire and smoke protection system. This paper also updates related work from IC-8775 "Spontaneous Oxidation and Combustion of Sulfide Ores in Underground Mines," (also published in 1978) and IC-8808 "In-mine Evaluation of Underground Fire and Smoke Detectors," (published in early 1979).

INTRODUCTION

As metal and nonmetal mines become deeper, more complicated, and more mechanized, the fire danger increases because of increasing use of highly combustible materials (fuel, solvents, wiring, etc.) and restrictions on miner egress. This fact was unfortunately illustrated in May of 1972 by the Sunshine mine fire at Wallace, Idaho, in which 91 miners died.

To help solve this growing safety problem, and to help protect the production capabilities of our nation's mines, the Bureau of Mines expanded its mine fire research and development work to develop a mine shaft fire and smoke protection technology for metal and nonmetal mines. The methodology utilized for this program followed that which had been previously used in the Bureau's successful effort to develop automatic fire protection systems for mobile mining equipment; i.e., the ruggedizing of existing technology to solve a specific mining safety problem. The new objective was to evaluate noncoal mine fire and smoke hazard problems, then develop and demonstrate a reliable mine shaft fire and smoke control prototype system. This hardware was to be flexible in design so it would be adaptable for use in the majority of metal and nonmetal mine shafts and adjoining shaft stations.

SYSTEM DESIGN

Because electrical shorts, welding, and torch cutting are the chief causes of underground noncoal mine fires, early detection of the fire (so it can be easier to extinguish) is an important design requirement. Also, because more miners die from smoke than from burns, smoke control is another important requirement for an underground fire protection system.

The system's design utilized thermal, carbon monoxide, and ionized-particle smoke detectors. Remotely controlled smoke doors and sprinklers were used to protect both the shaft and shaft station areas of a mine. The system's surface control unit receives underground fire warning signals via multiplexed electronics sent through two separate twisted-pair routings. The system's underground control units can be activated from the surface control unit to warn the miners in the shaft station area of fire danger. One underground control unit would be needed in each shaft level protected by the system. Surface personnel also would activate a stench system to warn miners of danger. The sprinklers and smoke control doors could be opened or closed by either local controls, or remote actuation from the surface control unit. Automatic sprinkler actuation currently was not a feature of the system, but such a design alternative could easily be added if so desired.

IN-MINE FIRE TEST

During the winter of 1975, a shaft/shaft station mockup (see fig. 1) was built at the contractor's facilities in Santa Clara, Calif. Lab testing of system components was highlighted on March 20, 1975, when the system's hardware was demonstrated to Bureau, MSHA, and industry representatives by an actual fire test in the mockup.

After these preliminary evaluations, the first generation mine shaft fire and smoke protection prototype hardware was installed at the 3000 level of the Silver Summit shaft near Wallace, Idaho (see fig. 2). The system was successfully demonstrated to Bureau, MSHA, and mine personnel by actual in-mine fire testing in April of 1976 (fig. 3 through 7). The test fire was lit in a 4- by 2-foot steel box and was subsequently sensed and extinguished (by remote control) after warning devices had actuated. For this test, the carbon monoxide sensors were set to detect 200 parts per million and the smoke sensors 2 pct per foot obstruction.

Follow-up work was conducted to modify and reinstall the mine shaft fire and smoke protection prototype system at a second location for long-term testing. This took place in the summary of 1976 at Union Carbide's Pine Creek mine in Bishop, Calif. (fig. 8). This second generation hardware was also subjected to an actual in-mine fire testing (fig. 9 through 11). The long-term in-mine test demonstrated that the hardware was rugged enough for reliable operation in an underground environment.

Concurrent work by the Bureau to in-mine test the reliability of underground fire sensors was initiated (and is ongoing) at Hecla's Lakeshore mine in Casa Grande, Ariz. Figure 12 shows a sensor package currently undergoing long-term evaluation at Lakeshore. A charged-particle smoke sensor produced in South Africa is proving itself to be most reliable. The CO monitors and optical flame detectors still need to be made more rugged in order to be applicable underground.

ADVANTAGES AND COST OF THE SHAFT SYSTEM

The mine shaft fire and smoke protection technology gives a mine quick response capabilities in underground fire situations. Also, having remotely controlled sprinklers and smoke control doors in the shaft and shaft station areas is more effective in extinguishing shaft fires than the presently used shaft-collar water rings. The cost of a shaft fire system is estimated to be from $50,000 to $100,000, depending upon the sophistication desired. This is a substantial investment, but compared with the millions of dollars a shaft costs, the increased protection provided the miner, this technology is becoming an important contribution to improved underground fire safety.

SUMMARY

The development of improved shaft fire protection hardware (fig. 13) involved the delineation of the mine shaft fire and smoke hazard problem, and the design, lab testing, and in-mine testing of a system that better protects miners (and mines) than the shaft-collar water rings now commonly used. The shaft fire system utilizes thermal, carbon monoxide, and smoke detectors, remotely controlled smoke doors at the shaft stations, and remotely controlled sprinklers. It was successfully demonstrated by fire tests in the Silver Summit shaft near Wallace, Idaho, and in Union Carbide's Pine Creek mine in Bishop, Calif. Long-term, in-mine testing showed the system to be a reliable improvement in mining health and safety. More information about this technology's development and in-mine testing is available from the Bureau's new 9-minute Technology Transfer film "Mine Shaft Fire and Smoke Protection System." This film is available from: Motion Pictures, Bureau of Mines, 4800 Forbes Ave., Pittsburgh, Pa., 15213.

Three companies now commercially supply hardware based on the Bureau's work. They are: FMC Corporation, Santa Clara, Calif. (the system's original developer), Allison Controls, Co., Fairfield, N.J. (who have designed and installed a system for the Diamond Crystal salt mine near New Iberia, La., Aquatrol Corp., St. Paul, Minn. (who are installing a fire system in FMC's Green River trona mine, Green River, Wyo.) and the Ansul Co., Marinette, Wis. (who has also developed an underground fueling area fire protection system (fig. 14): two prototypes of which have been in-mine tested at the Pine Creek mine and AMAX's Boss lead-zinc mine, Boss, Mo. Building upon earlier work, FMC is also cooperating with Magma Copper Co.'s Superior mine at Globe, Ariz., to in-mine test spontaneous combustion fire warning hardware, (fig. 15) and a shaft fire system for coal mines (fig. 16).

It should be noted that the National Fire Protection Association (writer of the Life Safety Code and the National Electric Code) is now developing industry consensus stands for mine fire protection with its new Mining Facilities technical committee. Also, the Mine Safety and Health Administration's monthly magazine Mine Safety and Health recently wrote an article on Deep Mine Fires (pages 10 to 17 of the March-April 1981 issue) summarizing much of the Bureau's work.

More information on the above subjects can be obtained from IC-8783 "In-mine Fire Tests of Mine Shaft Fire and Smoke Protection Systems," by Johnson and Forshey (1978), IC-8775 "Spontaneous Oxidation and Combustion of Sulfide Ores in Underground Mines," by Ninteman (1978), and IC-8808 "In-mine Evaluation of Underground Fire and Smoke Detectors," by Griffin (1979). Also, a recent news release (dated October 27, 1980), by the Department of the Interior outlines how the Spontaneous Combustion Fire Warning System prototype successfully sensed a "heating event" at the Magma mine in the fall of 1980.

Anyone interested in analyzing the fire protection hardware components is invited to the Twin Cities Research Center in Minneapolis, Minn. We have working demonstration units (fig. 17) of most of the hardware we have evaluated in the mines.

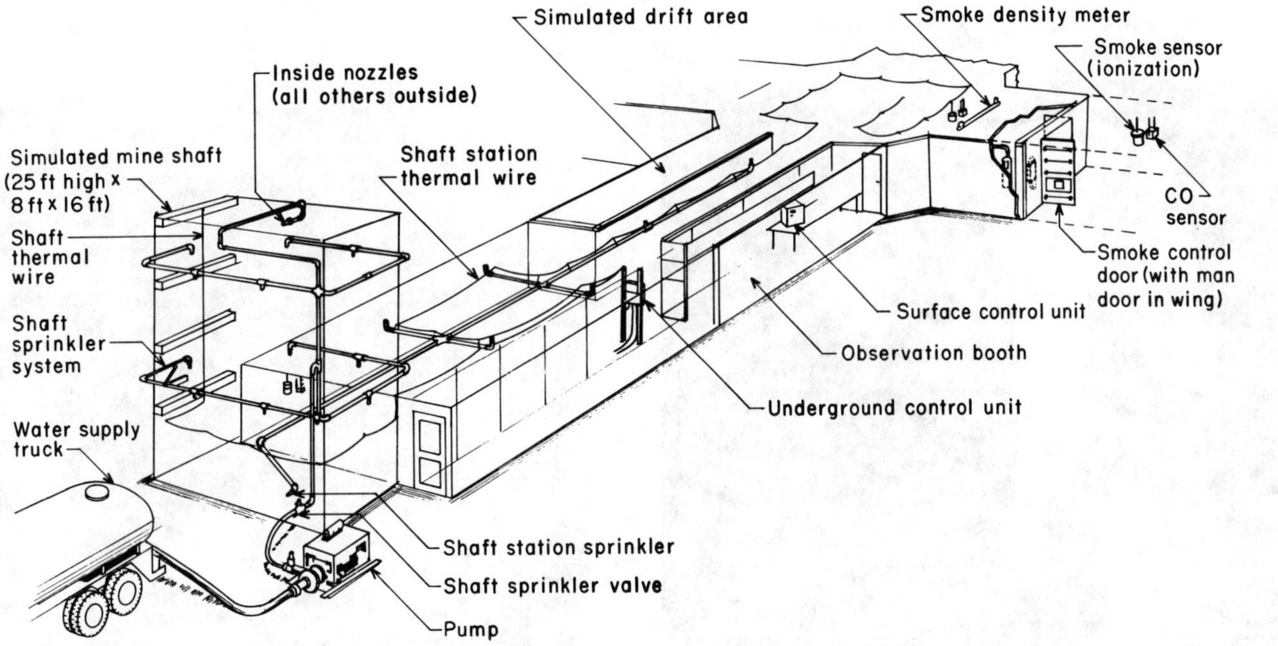

Fig. 1.- Shaft/shaft station mock-up.

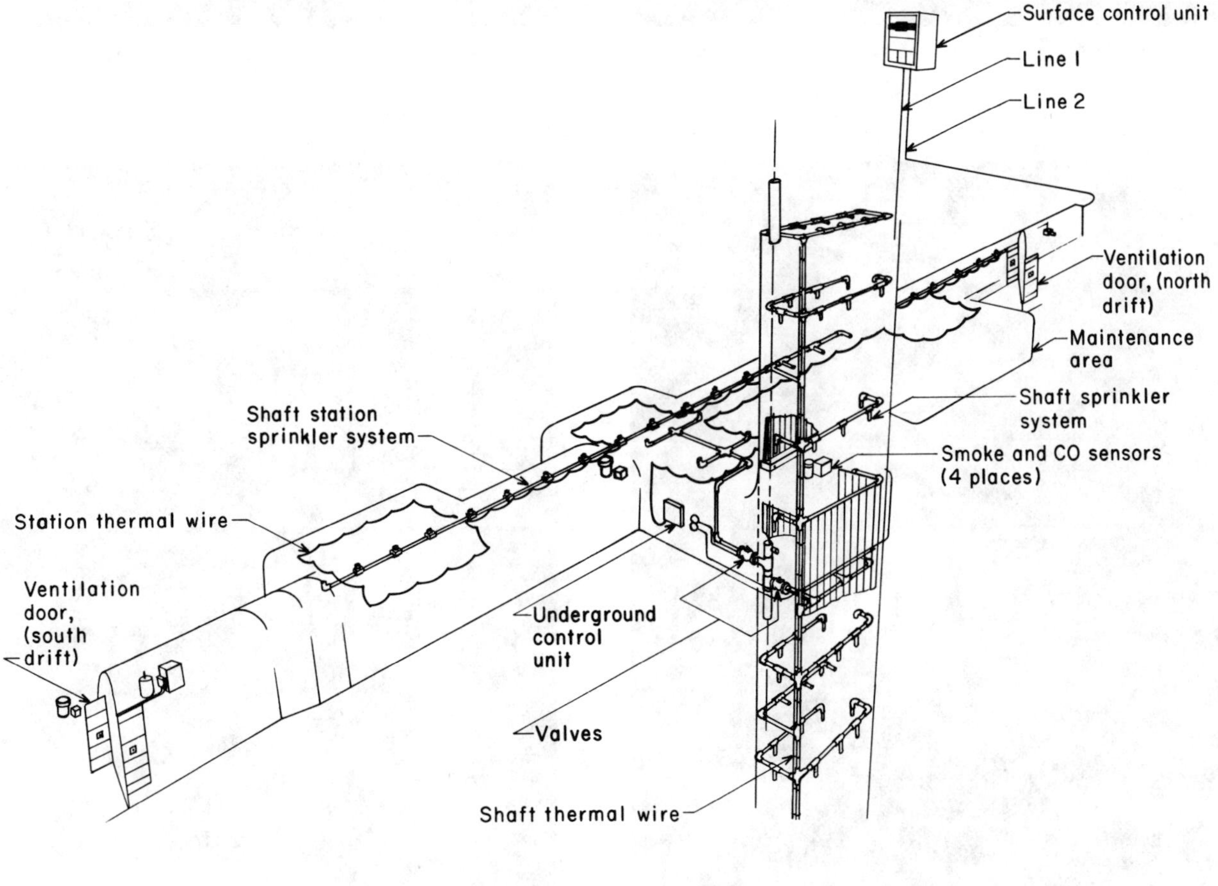

Fig. 2.- First Generation Prototype system at the Silver Summit shaft, Coeur d'Alene--1975.

Fig. 3- Lighting the in-mine test fire.

Fig. 4- Smoke from the in-mine test fire.

Fig. 5- In-mine test fire.

Fig. 6- In-mine test fire being estinguished by the system.

Fig. 7-Test fire completely extinguished.

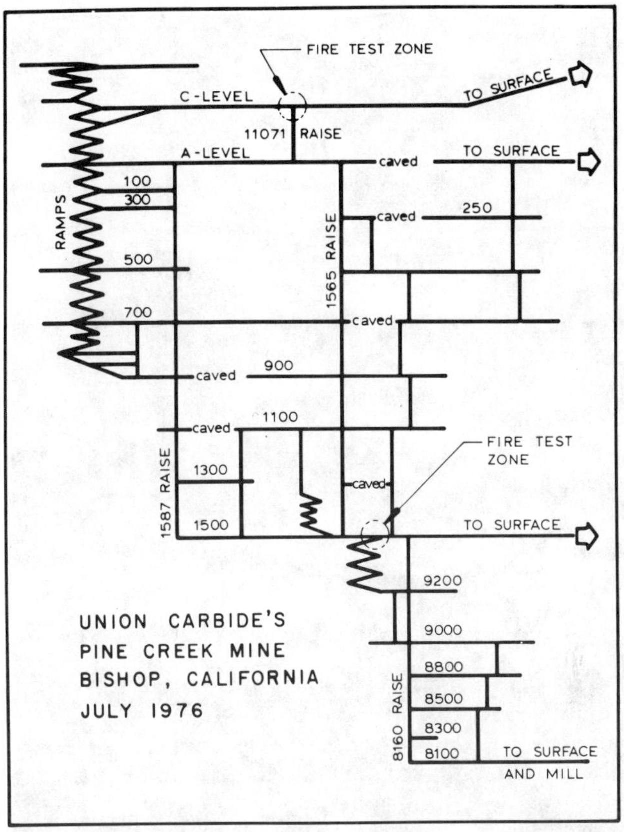

Fig. 8-Second-generation system at the Pine Creek mine, Bishop, California. 1976.

Fig. 9-Starting the in-mine test fire.

Fig. 10-In-mine test fire.

Fig. 11- In-mine test fire extinguished by the second-generation system.

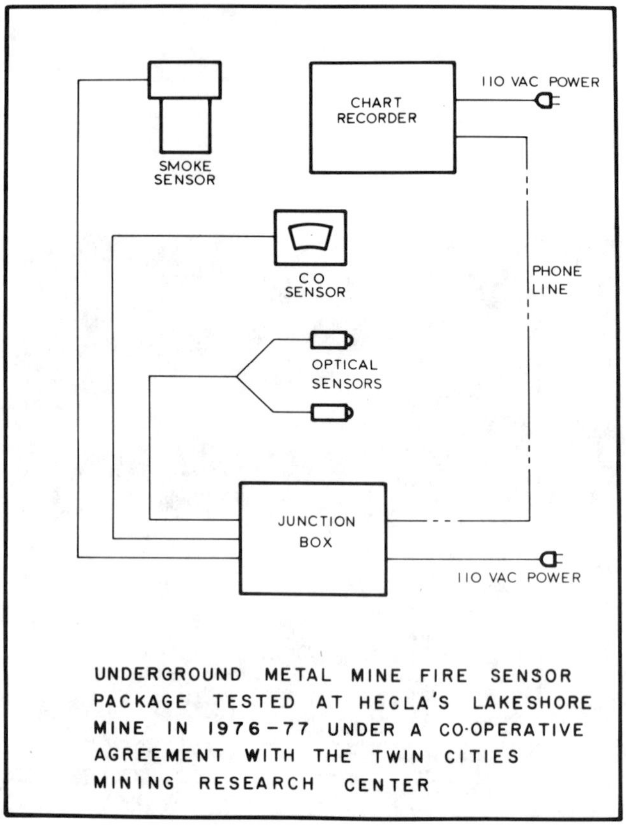

Fig. 12- Lakeshore mine sensor package.

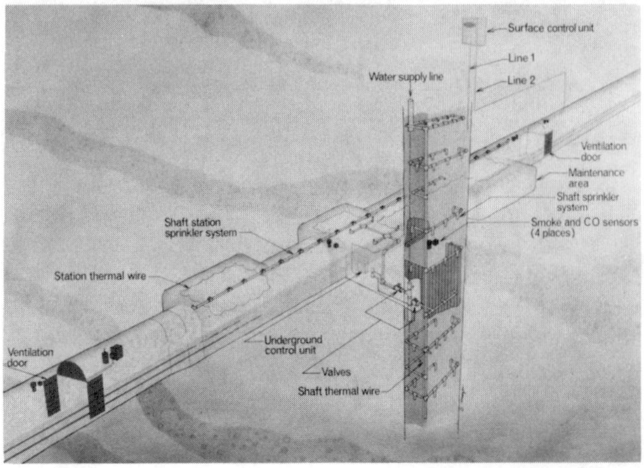

Fig. 13- Mine shaft fire and smoke protection system.

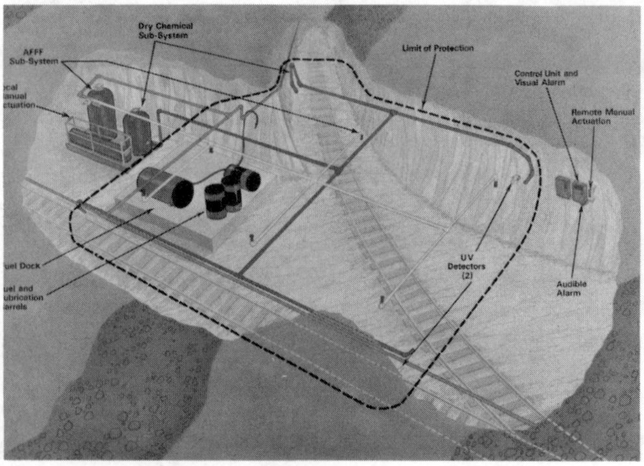

Fig. 14- Underground area fire protection system.

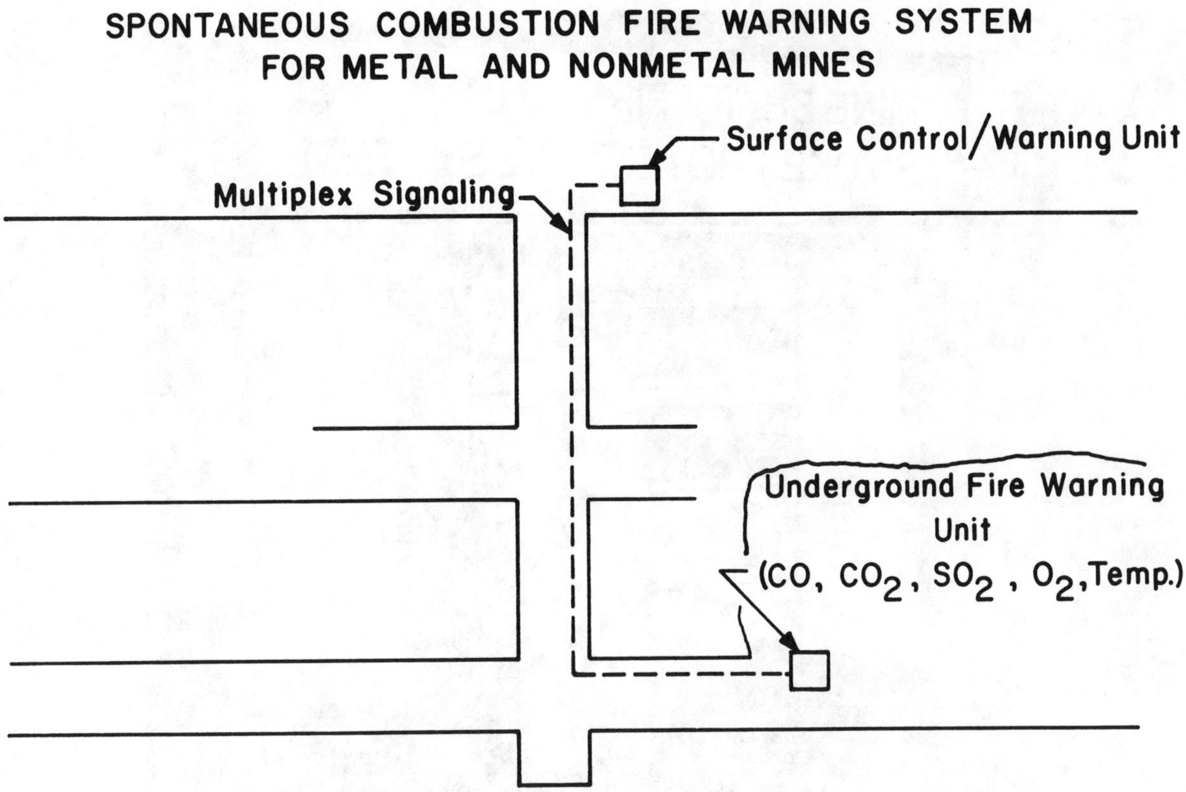

Fig. 15- Spontaneous combustion fire warning system metal/nonmetal mines.

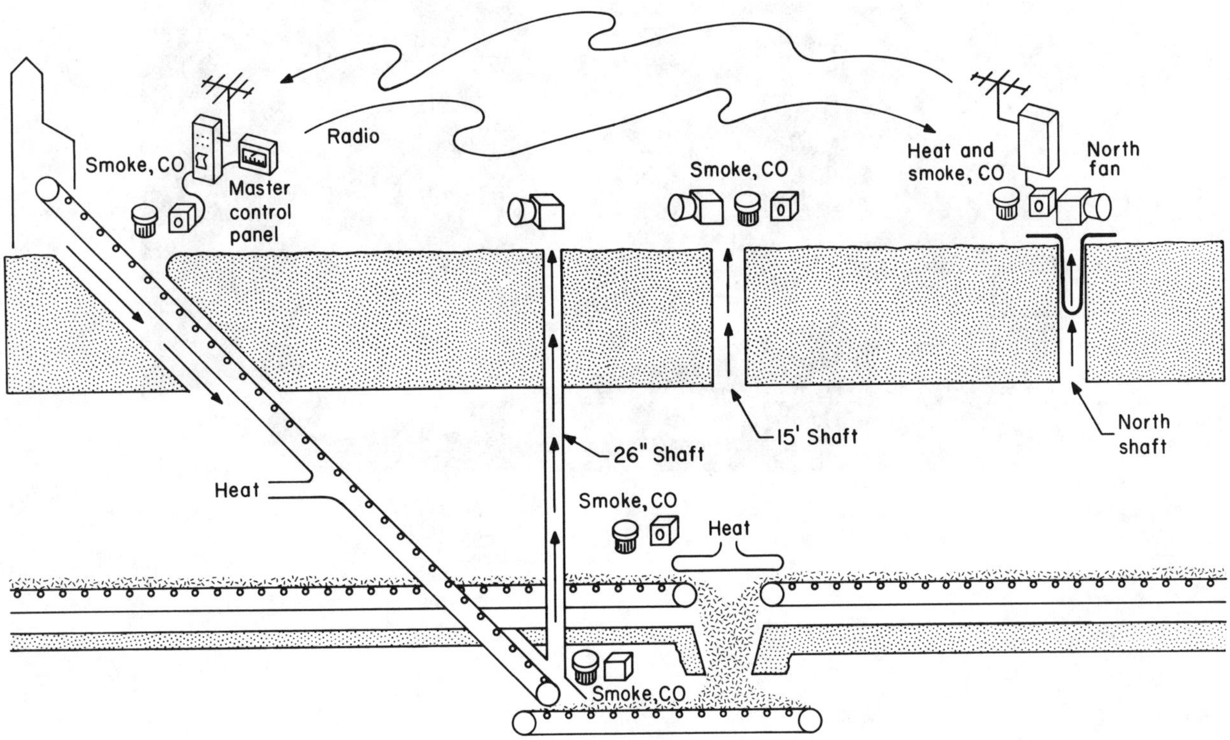

Fig. 16- In-mine tested prototype shaft fire protection system for coal mines, at Peabody #10 mine, Paunee, Ill.

Fig. 17- Mine shaft fire and smoke protection system demonstration unit at TCRC.

Chapter 21

INSTRUMENTATION FOR MINE SAFETY: FIRE AND SMOKE PROBLEMS AND SOLUTIONS

Ralph B. Stevens

Project Engineer
ESD Corporation
Energy Group
Santa Clara, California

INTRODUCTION

Underground fires continue to be one of the most serious hazards to life and property in the mining industry. Although underground mines are analogous to high-rise buildings where persons are isolated from immediate escape or rescue, application of technology to locate and control fire hazards while still in their controllable state is slow to be implemented in underground mines.

Even in large surface structures such as hotels, often only fire protection systems which meet minimal laws are implemented due to the high cost of adding extensive extinguishing systems, isolation barriers, alternate ventilation, escape routes and alarm systems. Incomplete and ineffective protection occasionally is evidenced where costs would not seem to be a factor, such as the $211 million MGM Grand Hotel fire November 21, 1980[1].

Paramount in increasing fire safety and decreasing the threat of serious fire is early warning followed by proper decision analysis to perform the correct action. However, very complex fire situations can be produced in structures such as high-rise buildings and underground mines simply because of the distances between the numerous fire-potential locations and fire safe areas. Other complexities arise when normal activities occur that emit products of combustion signaling a fire condition to a sensitive fire/smoke sensor. For example, the operation of diesel equipment or the performance of regular blasting can produce combustion products that reach the sensitive alarm points of many sensors[2].

Smoke detectors for surface installations provide fire warning when occupants are at a distant location or when sleeping, thus greatly reducing injuries and property damage. However, when installed in the harsh environments of underground mines, fire and smoke detection equipment soon becomes inoperative, unreliable, or requires excessive maintenance. The U.S. Bureau of Mines has performed many studies and tests to improve fire and smoke protection for underground mine workers[3].

This paper describes several USBM safety programs which included in-mine testing with mine fire and smoke sensors, telemetry and instrumentation to develop recommendations for improving mine fire safety. It is hoped that the technology developed during these programs can be added to other programs to provide the mining industry with the necessary fire safety facts. By recognizing fire potentials and being provided with cost-effective, proven components that will perform reliably under the poor environmental conditions of mining, mine operators can provide protection for their working life and property equal to that which they provide for themselves and their families at home.

The basis of this report is two USBM programs for fire protection in metal and nonmetal mines[4,5] and one coal program[6]. The data was collected beginning in May 1974 and continuing through the present with underground tests of a South African fire system installed at Magma Mine in Superior, Arizona, and a computer-assisted, experimental system at Peabody Coal Mine in Pawnee, Illinois. The conduct of each program was as follows:

- Define the problem and its magnitude in the industry
- Develop concepts to solve or diminish the problem
- Review available hardware or systems approaches to fit the concepts
- Install and demonstrate the performance of a prototype system through fire tests in an operating mine.

MINE FIRE FACTS

Whether in coal or metal and nonmetal mines, the potential severity of fire hazard is directly related to location. As shown in Figure 1, fire in intake air at zones A, B, C or D can cause contamined air to route throughout the mine quickly if not detected, isolated or rerouted. Causes and location of former metal and nonmetal fires are represented in Table 1; the cause and location of fatalities and injuries is shown in Table 2.

Coal-related fires and their impact on deaths and injuries are graphed in Figure 2; their locations are described in Table 3[7]. Significantly the table shows that the hazard to personnel was three times greater for fires occurring in shaft or slope areas, and the percentage of deaths and injuries was four times that of other areas.

Number of Persons Affected

A 129-mine sample indicated that from 8 to 479 employees per shift work in underground metal and nonmetal mines, and that deeper mines have larger populations, as shown in Figure 3. Coal mining relates similar employment, and a 16-state sample of 670 mines employing at least 25 persons shows the distribution in Figure 4. Drift mines accounted for 58 percent of the sample but employ only 45 percent of the underground workers.

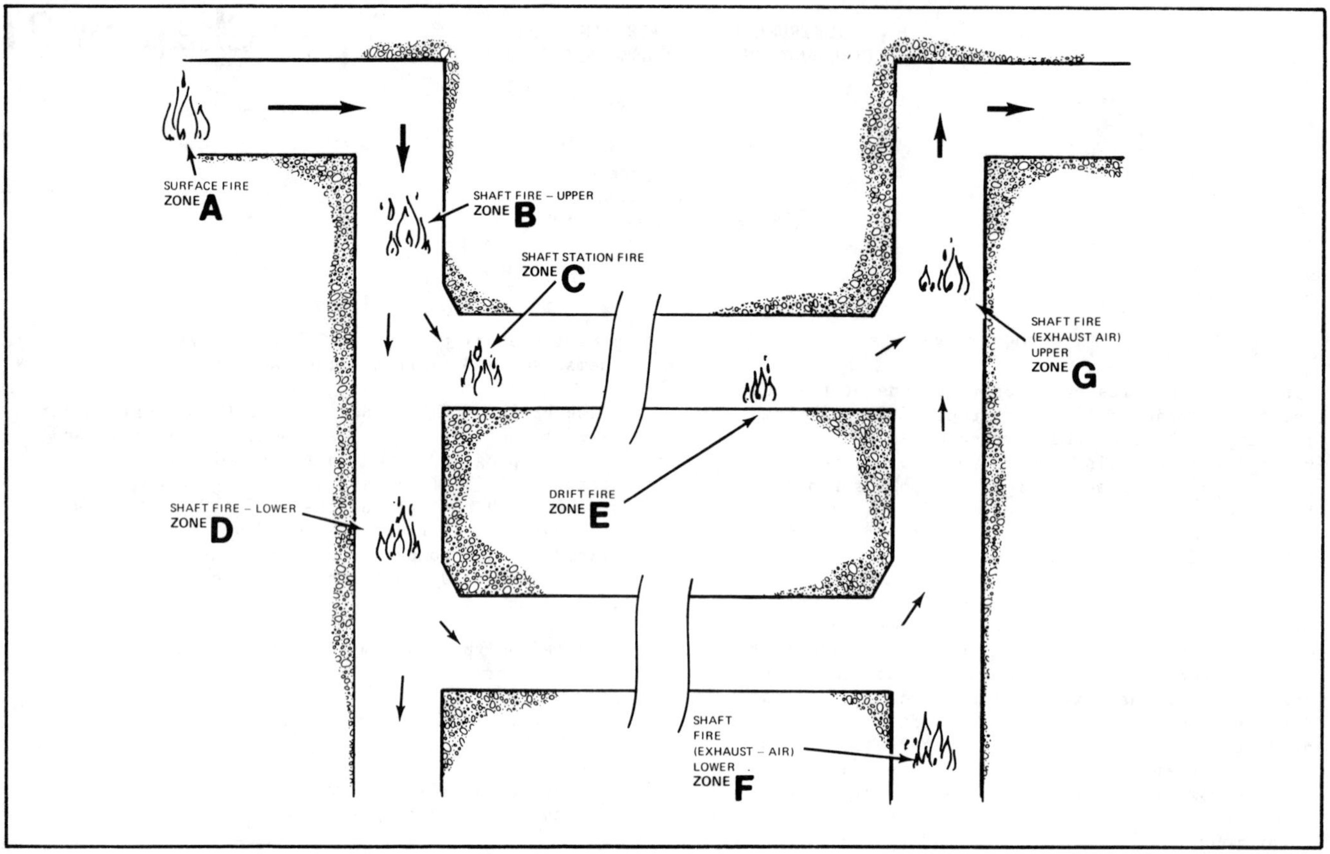

Figure 1 HAZARD ZONES TYPICAL OF UNDERGROUND MINES

Table 1 CAUSE AND LOCATION OF FIRES AND EXPLOSIONS IN UNDERGROUND METAL AND NONMETAL MINES (1945-1974)

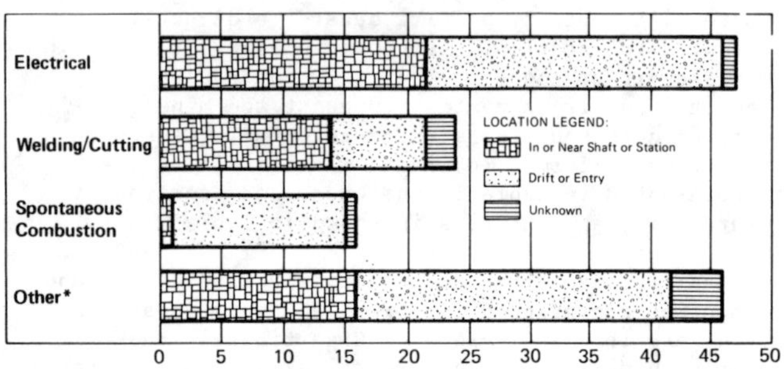

*includes friction, combustible waste ignition, smoking, combustible gas accumulation, blasting, and unknown.

Table 2 CAUSE AND LOCATION OF FATALITIES AND INJURIES FROM FIRES AND EXPLOSIONS IN UNDERGROUND METAL AND NONMETAL MINES (1945-1974)

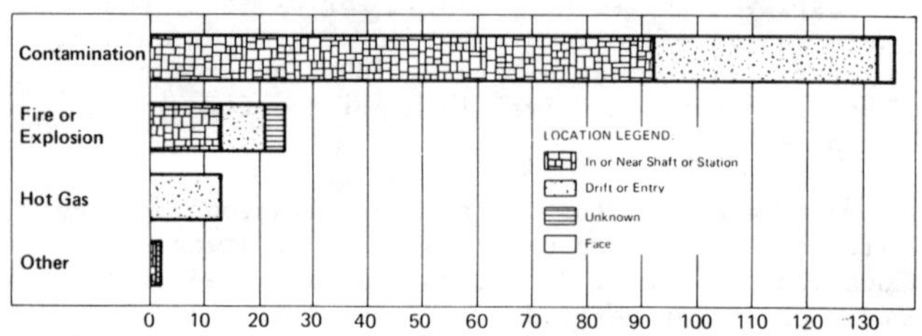

Table 3 DISTRIBUTION OF FIRES, INJURIES, AND DEATHS BY LOCATION

Underground location	1 Number of fires	2 Number of injuries	3 Number of deaths	4 Number of deaths and injuries	5 Number of fires resulting in death or injury	6 Percent of all fires	7 Percent of all deaths and injuries	8 Percent of all fires resulting in death or injury
Non-shaft/slope	877	154	73	227	96	86	79	87
Location not specified	93	24	0	24	11	9	8	10
Shaft/slope	44	37	0	37	3	5	13	3
Total	1,014	215	73	288	110	100	100	100

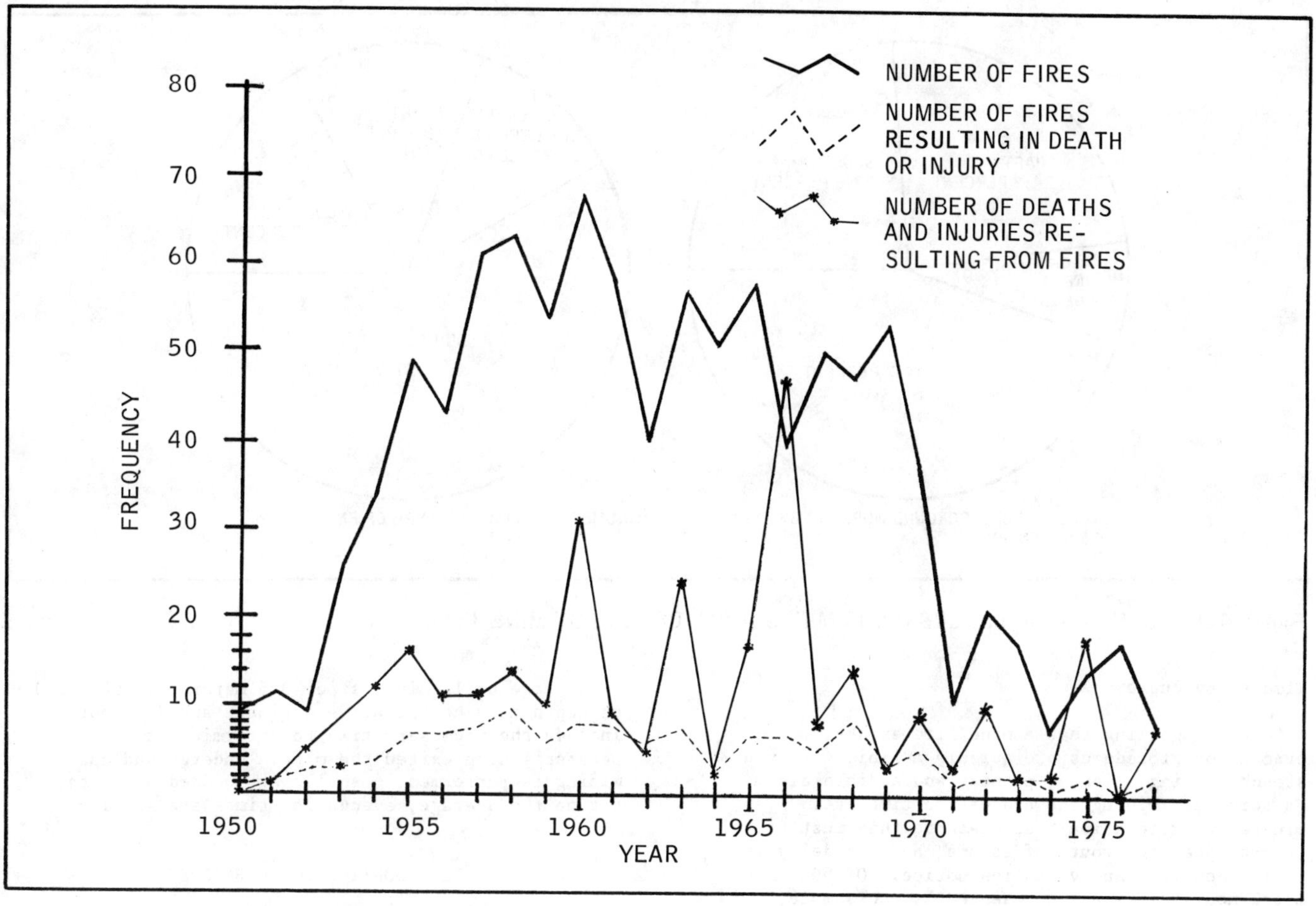

Figure 2 DISTRIBUTION OF COAL-RELATED FIRES AND DEATHS/INJURIES

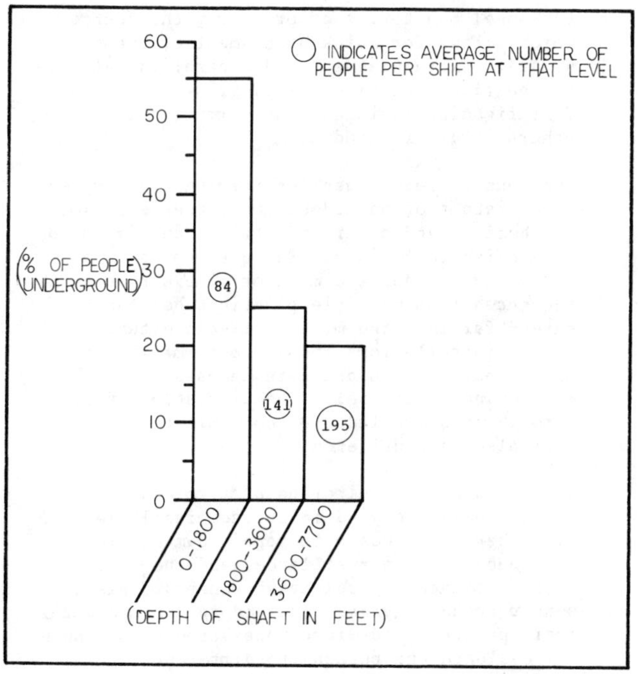

Figure 3 NUMBERS OF PEOPLE UNDERGROUND AS A FUNCTION OF WORKING LEVEL DEPTH

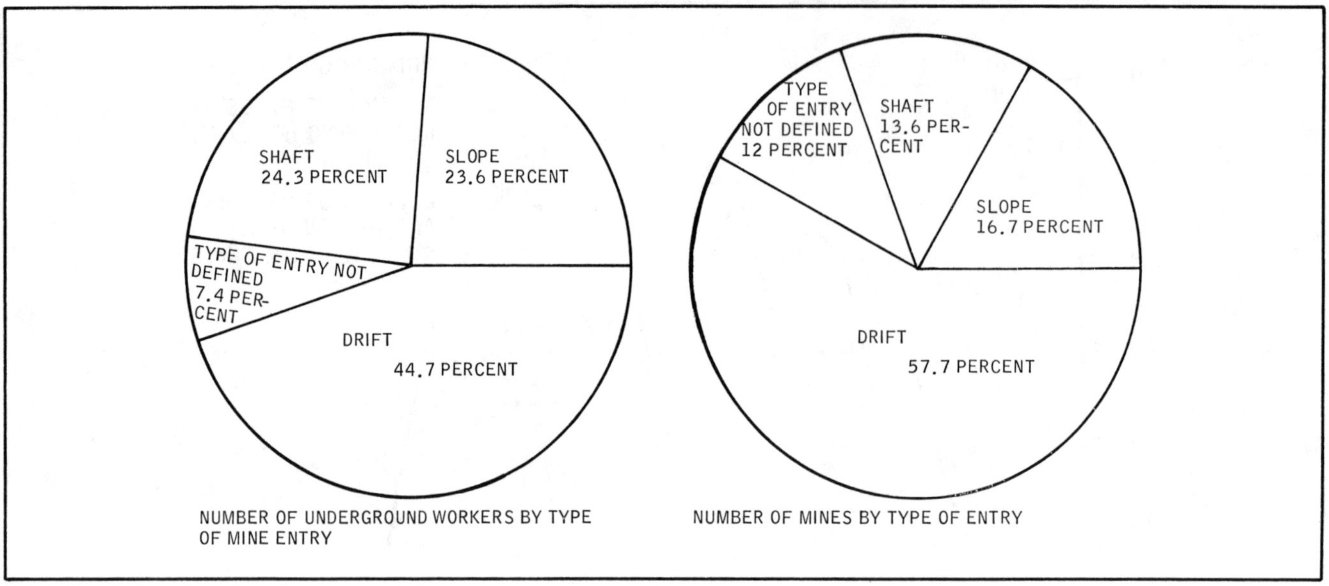

Figure 4 PERCENTAGE OF UNDERGROUND MINES AND WORKERS BY MINE TYPE

Time to Evacuate

Once recognizing that a mine fire exists, evacuation procedures are quickly administered via stench warning, lights, phones, and audible alarms. Factors determining the evacuation time are: miners' work location; the distance they must travel to evacuate; the route of escape; and the delay in their receiving an evacuation notice. Of 50 metal/nonmetal mines represented by one sample, average evacuation time after receiving evacuation notice was related to depth as shown in Figure 5.

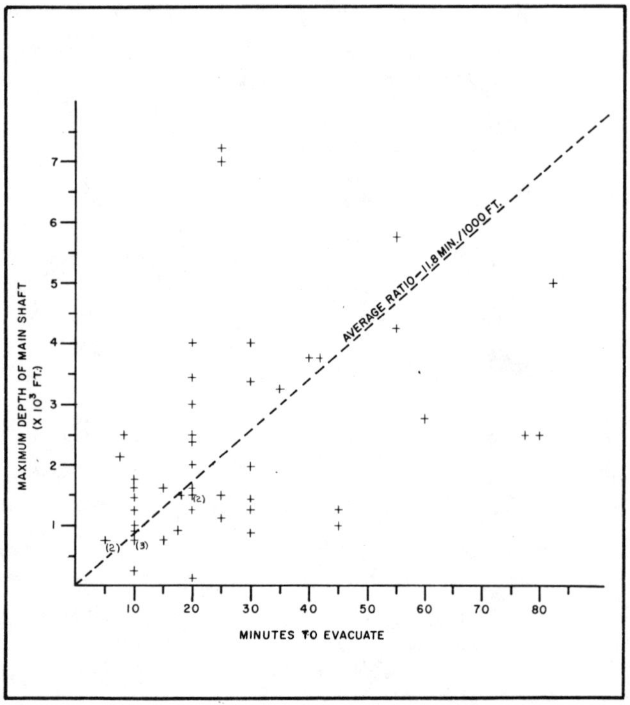

Figure 5 EVACUATION TIME VS. MAXIMUM DEPTH

An average of 12 minutes per 305 meters (1,000 feet) of depth is a conservative estimate and does not include the necessary time to determine if all personnel have exited the mine. Underground coal mining is performed in shallower, bedded deposits, but on the average, evacuation times are similar.

FIRE MONITORING TECHNIQUES

Three general types of monitoring techniques are used in underground mines, and each has its strengths and weaknesses. These techniques are:

- Personnel monitoring at or nearby the source which allows for quick assessment of the hazard. Extinguishment and control is relative to the fire size, intensity, growth, available fire fighting equipment and communication to others if aid is needed.
- Tube bundle techniques for drawing air samples from distant or hazardous locations is popular in chemical and other industries, and is common in English coal mines. Sensors remain in a clean environment and monitor mine air as it is sequenced from multiple plastic tubes that extend far into the mine. Effective tube bundle installations require a commitment by mine management to dedicate necessary maintenance personnel. The alteration of gas through moisture-laden or contaminated tubes must also be considered[8,9].
- Instrumentation monitoring of fire and smoke is being widely accepted in surface facilities due to higher labor costs to repair damage, more continuous fire surveillance, and the unbiased facts presented by selective sensors. Also, sensor technology has improved so that computer logic programs and microprocessor control can now evaluate and respond to different conditions with pre-determined and more accurate results[10,11]. This technology is being transferred into underground mining

gradually and with caution, primarily due to the harsh mine environments, limited availability of maintenance-free equipment, initial cost of a total system, and the level of technical support needed to operate the equipment.

Fire recognition is commonly and most accurately assessed when personnel are at the point of ignition. Such situations often are resolved by personnel in the immediate area, without the need to sound a general alarm and cause mine evacuation. Many factors can interfere with fire recognition however. Incorrect assessment of the problem, or the assumption that it is a production- or maintenance-oriented condition, or that someone else has taken corrective action may prevent the performance of proper safety actions. Since mines are extensive networks of multiple rooms, drifts, stopes and shafts, personnel cannot be everywhere. In addition, there are regular periods of time when no workers are underground, such as between shifts, during blasting, or on holidays.

Mine operators recognize that certain mine areas are bigger fire hazaards than others and employ various techniques to monitor these areas. Standard home-type thermometers are occasionally located in "hot-spots", and regular gas checks are made of sealed areas, abandoned workings, and at faces and bleeder entries of coal mines. One metal mine continuously monitored intake and exhaust air into a working stope due to levels of 2,000 ppm CO and 170,000 ppm CO_2 being emitted from the sulfide ore fill in the area. Proper ventilation and these continuous gas checks allowed production operations to continue. Charting a rise in hazardous gas concentrations or in temperatures have caused other mines to permanently close and seal mine sections to inhibit the combustion process.

Mine personnel are also aware of the classic fire triangle that controls a fire. Although fuel and air are intimately mixed almost everywhere, a fire does not start unless sufficient energy is present to ignite a full-oxidizer system. A fire spreads because the heat feeds into new fuel and raises it to the ignition point.

In sealed areas, oxygen buildup or CO, CO_2 and methane increases can bring serious consequences. In open areas, exhaust shafts and workings, CO is the primary gas of interest in the U.S. and many other countries. However, in many South African mines CO_2 gas and particle sensing from ionization sensors are considered the significant factors to monitor due partly to extensive wood cribbing and high air flows[12].

MINE PRODUCTION-MONITORING METHODS

Ten of 14 mines sampled during early 1980 had operational data-recording systems between remote sensors and a surface control center. These systems are production oriented and provide quick maintenance or corrective action for failed conveyor belts, hoppers, pumps, fans and hoists. The addition of fire safety devices to these systems would be the next step toward a complete mine fire safety system. Highlights of findings from several mines are discussed in the following paragraphs.

FMC Trona Mine, Green River, Wyoming[13]

Figure 6 shows a system developed by Aquatrol Corporation to receive production information. Each of sixty-eight belt lines has 12 functions monitored. This system uses a frequency division, multiplex telemetry system and microprocessor control that interrogates each station in sequence. A data-fail signal is given if three, 3-second interrogations of any sensor get no response. All data is transmitted on a pair of wires, however, the system is divided into three zones and multiple conductors are available to switch pairs and minimize overall system failure. A large expansion of this system is planned which will include fire and smoke sensors.

Inland Steel Coal Mine, Sessor, Illinois[14]

A computer-based information system and advanced voice communication system is operating successfully at this large mid-western coal mine. The system receives full management commitment and has contributed to improved production since 1977. Four areas are addressed within the system: production status and recording; general underground data; spare parts or maintenance; and safety.

Underground supervisors and maintenance foremen have computer terminals and CRT screens to receive and input information. Status of all working sections, quantity of coal mined, production delays and reasons for delay are recorded by foremen calling the central coordinator who inputs the data into the system. When added, the Inland safety system will support production and management to provide information about mine conditions, potential violation areas, and corrective actions. The existing monitoring system provides a distribution and data collection network which can easily be expanded for fire and smoke safety components.

Republic Steel Mines

Three eastern mines of Republic Steel were visited to review different forms of remote data collection from underground to surface. As with many other coal mines, the belts and fans receive high priority. Trouble spots had to be quickly located to allow maintenance personnel to take corrective action before any serious consequences occurred. The Republic mines were using automatic, voice-recorded messages coupled to the mine phone/paging system when an alarm condition existed. The belt and fan voice systems were made by Jabco, and used a standard 8-track tape. Inputs from water-flow of the belt sprinkler system or the centrifugal belt-slip switch would trigger an alarm. The system output stops the belt transmitting the signal and a continuous voice annoncement from the tape recorder identifies the problem belt over the main phone/paging system. Republic, as well as other mines, expressed maintenance problems with dust on recorder tapes, and moisture affecting operation of thermal switches.

Jim Walter Resources

During mid-1980, the Number 4 mine of Jim Walter Resources was using a Giangarlo gas monitoring system to measure carbon monoxide in belt-air used in face ventilation. MSHA granted approval to evaluate ways of diluting methane buildup in this deep Alabama mine. Seven Ecolyzer 4000 CO analyzers with electrochemical cell sensors are coupled to an RFL W405 UF telemetry system. Time, location and CO

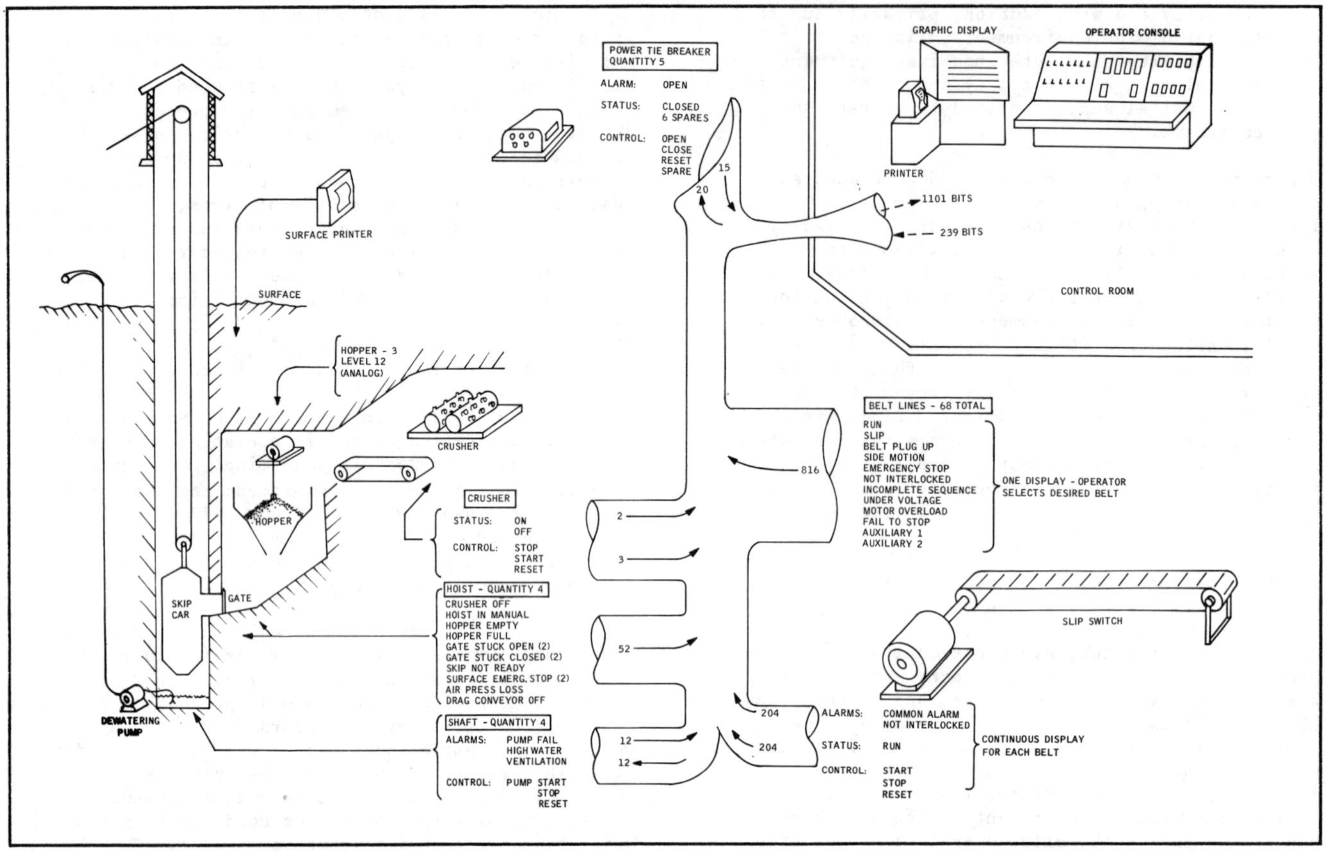

Figure 6 BLOCK DIAGRAM OF AQUATROL CORPORATION'S DATA COLLECTION SYSTEM AT FMC TRONA MINE

concentration in volts is recorded on a data logger every 20 minutes from each of the seven sensors. The system provides warning at approximately 15 PPM and sounds an alarm at approximately 20 PPM CO concentration.

Plans were in process for installing a J-Tec Associates, Inc. microprocessor-controlled system to monitor CH_4 and total flow out of a CH_4 harvesting system.

EARLY MINE SHAFT FIRE AND SMOKE PROTECTION SYSTEMS IN METAL AND NONMETAL MINES

The goals of a 1974 through 1976 USBM effort[15] to control fire and the spread of contaminated air in metal and nonmetal mines were:

- To provide cost-effective, reliable sensors
- To evaluate and test an extinguishing system for shafts and shaft stations
- To study, select and test ventilation barriers to isolate the area
- To develop a remote-control system that could receive and transmit signals to and from different zones of an expanding underground metal mine.

Information acquired from 129 mines regarding their 201 shafts determined a variety of mine profiles and an existing need for an improved fire and smoke protection system for most of them. Of highest priority were mines 609.6 meters (2,000 feet) or deeper with wood-lined shafts. Sixty-seven percent of the 201 shafts were within this category. Maintenance shops, refuse dumps, electrical transformer and power distribution stations, supply depots, weld and repair shops, fuel storage and battery charging stations are often located in or near shaft stations, and were found to be the origin of several mine fires.

The single station shaft fire test at Silver Summit Mine and the multiple location fire tests at Pine Creek Mine which are discussed in the following pages were part of this effort.

Shaft Station Fire Test at Silver Summit Mine

A remote-controlled fire protection system as shown in Figure 7 was developed. This diagram represents the four-compartment, wood-lined shaft at the 914.4-meter (3000-feet) level of the Silver Summit Mine in Idaho. Mine water was falling constantly inside the shaft and saturating the interwood structure of sets and lagging. The outside of the lagging, which varied between zero and four feet to the rock wall, was dry and dusty. Typical ambient conditions in the shaft stations were 32°C (90°F) and 95 percent relative humidity during the one-month installation and preparation for the fire test.

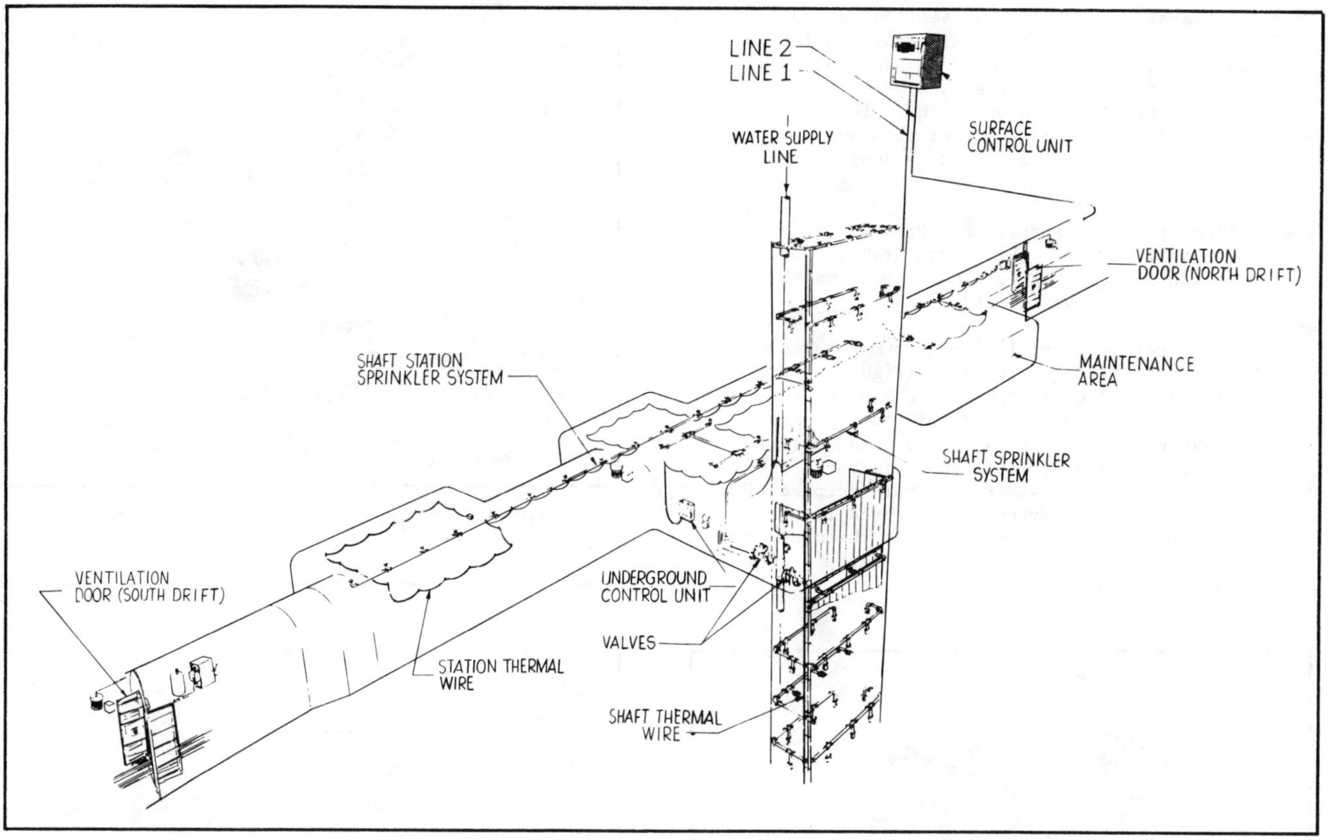

Figure 7 SILVER SUMMIT MINE FIRE PROTECTION SYSTEM

Seven hundred equipment manufacturers were sampled during this effort; 219 responded; 73 had products that appeared useful; and only 35 had products actually used in underground mines. The eventual selection of detection equipment, isolation barriers (doors), extinguishing systems and remote alarm/control systems resulted in the following items being purchased, assembled and fire tested at the Silver Summit mine:

- Detectors which were categorized as photoelectric, ionization, Wilson cloud chamber, carbon monxide gas (MOS, electro-chemical cell, catalytic oxydation and IR), carbon dioxide gas (solid state, electro-chemical cell, IR), thermal (fixed, rate-of-rise, point and line), and radiant energy (infrared, ultraviolet or visible light ratios)

- Three types of sensors (see Figure 8), each having a particular function to allow adequate decisions by personnel at a remote control monitoring station. The three sensor packages were used in groups near the shaft, in the shaft station, and on the inby side of isolation barriers extending into the mine.

- A Becon MK II ionization detector from Anglo American Electronics Laboratories in South Africa, shown in Figure 8. This detector is designed to operate in air velocities up to 365.8 meters (1200 feet) per minute, in super-saturated atmospheric conditions and in corrosive environments, and has a wide temperature range. It obtains its power from low voltage telemetry lines, is easily calibrated by unskilled labor, and requires minimum maintenance.

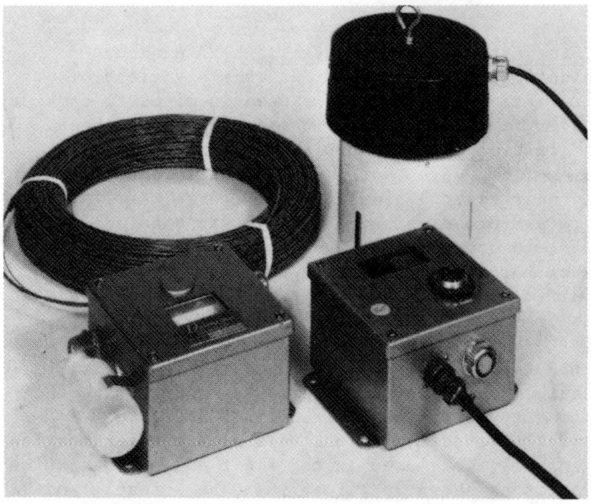

Figure 8 SELECTED SENSORS: LINE-TYPE HEAT SENSOR (UPPER LEFT); BECON MK II IONIZATION DETECTOR (UPPER RIGHT); DYNAMATION GAS SENSOR (LOWER LEFT); AND CARBON MONOXIDE SENSOR (LOWER RIGHT).

- A solid-state (metal oxide semiconductor) carbon monoxide gas sensor from Dynamation Inc. and Inmet, shown in Figure 8. These units are similar in design and are used primarily for surface installations other than mining. They require more technical calibration and maintenance and are affected by humidity and temperature.

- A line-type heat sensor from Protecto-Wire shown in Figure 8. This inexpensive, backup heat sensor commonly is found around machinery and conveyor drive motors.

- A standard mine ventilation door manufactured by American Mine Door Company (Figure 9) was selected after evaluation of roll-up type, sliding, single hinge, and inflatable doors because of its proven in-mine use and ability to withstand and operate under pressures from different directions. The primary purpose of the ventilation barrier was to check the flow of contaminated air from entering the mine from a shaft or to isolate the shaft if the fire was within the mine.

Figure 9 SELECTED MINE VENTILATION DOOR

Fire Extinguishing Systems. NFPA Handbook guidelines were reviewed, as were a variety of fire-extinguishing systems used in mines to protect shafts, shaft stations and other critical areas. The initial fire-extinguishing agent was foam. Three foams were tested with 11 different sprinklers and nozzles on a simulated shaft area; similar tests were performed using water. It was concluded that foam did not provide significant advantages over water for protection of timbered areas, and the additional equipment required for foam would add to the cost and reduce overall reliability.

NFPA Handbook 13 was followed as closely as possible and other work reviewed to determine the necessary density and duration of the water sprinkler system. Three types of sprinkler nozzles eventually were used in the shaft and shaft station systems as shown in Figure 10. The Grinnell nozzle used in the shaft station and drifts provided an upward spray to reach roof support timber and lagging. The Spraying Systems nozzle and the Viking horizontal sidewall sprinkler were used inside and around the outside of the shaft (see Figure 11). Water densities of 1.63 gpm/0.09 square meter (one square foot) inside the shaft [12 square meters (130 square feet)], and 360 gpm over the exterior shaft surfaces [246 square meters (2,650 square feet)]

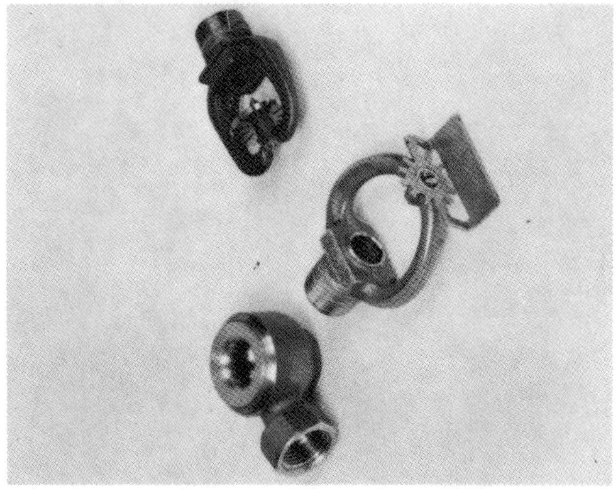

Figure 10 SELECTED SPRINKLER COMPONENTS

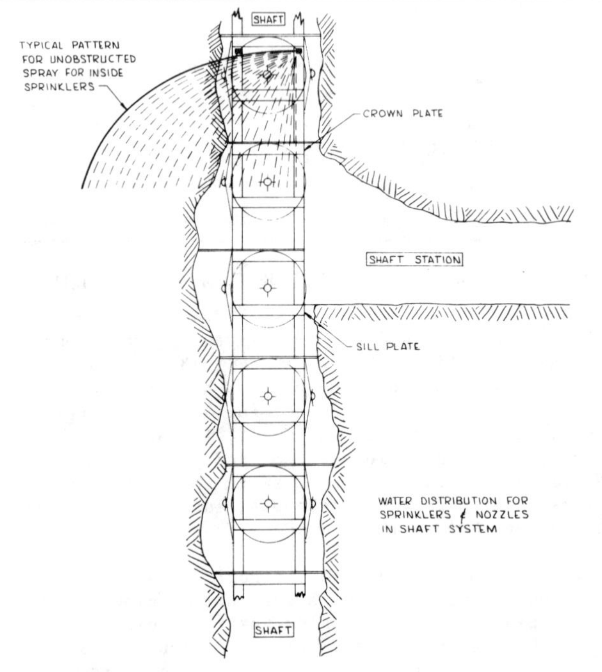

Figure 11 EXTERIOR OF MINE SHAFT MOCK-UP SHOWING SPRINKLER RINGS AND SPRINKLER COVERAGE OF IN-MINE TEST.

provided an average density of 0.14 gpm/0.09 square meter. The shaft station and drift areas were classified as light hazards by NFPA Handbook 13, and densities of 0.066 gpm/0.09 square meter were designed. Guidelines of water flow for 30 to 120 minutes duration could not be met due to limited water supply at the test location, however, wood-pile fire tests indicated reasonable control after 4 to 6 minutes.

Remote Monitoring and Control. Control of detection, isolation barriers (doors) and extinguishing systems is exercised best from a position remote from the potential trouble zone, preferably on the surface where rescue and safety personnel coordinate critical decisions during emergency situations. Fully automatic, semiautomatic, manual and a combination of these types of controls were evaluated with different types of multiple-function signaling systems. The candidates reviewed were: hard wire, or point-to-point using individual conductors for each sensor; and frequency-division and time-division multiplex systems. Hard wire systems using a pair of wires for each transducer are used widely for simplicity, but they are limited by the number of conductors and longer distances rapidly increase costs.

The availability of mine-proven hardware in an existing package was limited since the design goal was for deep mines (609.6 meters minimum), or multilevel mines (up to 32 levels) with multiple signals being transmitted to and from each level. Existing systems marketed for coal-mine conveyor monitoring and similar systems did not have the capacity or flexibility desired. The time-division, multiplex (TDM) systems of American Multiplex Systems, Larse Corporation and Universal Sentry Systems were evaluated. The AMS hardware had been used in underground, coal-mine conveyor systems, and Larse hardware had been tested by USBM as part of a mine monitoring system supplied by RFL Inc.

The selected system would be for up to 32 underground, remote stations using the Larse time-division, multiplex telemetry system. Each remote station would transmit 16 digital signals and receive 8 digital signals as shown in Figure 12. Concern was expressed that if the single pair of wires leading to the surface was broken, operation of the entire system would be inhibited. Therefore a second pair was routed to ensure reliability.

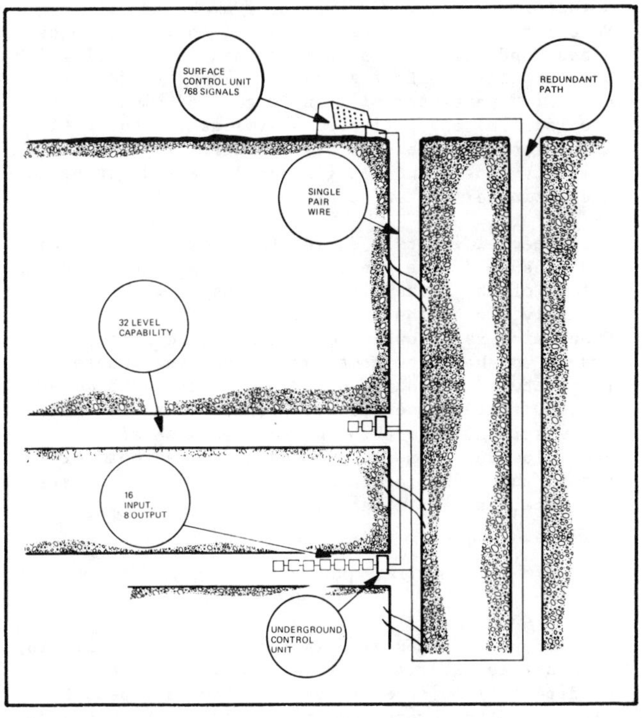

Figure 12 SIGNAL COMMUNICATION SYSTEM, TIME-DIVISION MULTIPLEX

Field Test Demonstrations. An actual fire test of the complete system was limited to two hours because of the adjacent mining operations of the Sunshine Silver Mine. The sensors, doors, extinguishers and controls were tested individually during installa tion. Temperature, humidity and inadequate water supply contributed to problems encountered during the test and the time required to install and test the underground installation was almost four times that needed for a surface facility. The fire test scenario was as follows:

- Two fire cribs made with 2 by 2 and 2 by 4 lumber weighing approximately 13 kilograms (28 pounds) were ignited and placed in a metal fire box having a lid to choke free burning and create smoke
- Fire safety crews from Hecla Mining Company and MSHA, and test personnel wore Dreger breathing apparatus in the fire zone
- An additional mine rescue crew and other test personnel were located at the surface control station 914.4 meters (3000 feet) above the test zone
- Intake air was to carry the products of combustion in the shaft station into the mine, past the ventilation barriers (doors).

It was anticipated that fire and smoke sensors in the shaft station and in the drift would signal an alarm. Acknowledgement of at least two sensors in the shaft station, or one sensor in the station and one in the drift, were to cause surface personnel to isolate the station and turn on the extinguishing system. A time-line orientation of fire condition and sensor responses is shown in Figure 13.

Smoke obscuration was determined by a special meter built in accordance with Underwriters Laboratory Standards 167 and 168. Additional data was recorded on an 18-channel oscillograph, an Ecolyzer Model 2100 and continuous voice and signal response on cassette tape recorders. As can be

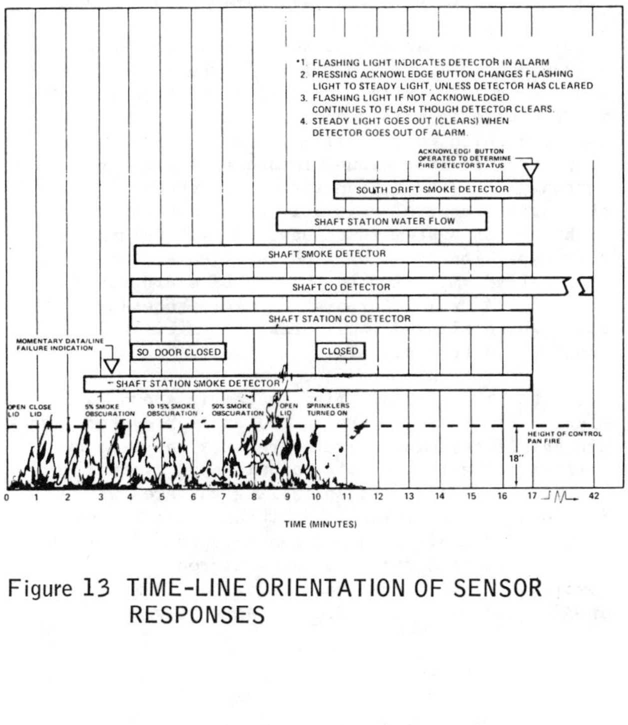

Figure 13 TIME-LINE ORIENTATION OF SENSOR RESPONSES

noted in Figure 12, the first smoke alarm was from the shaft station at two and one-half minutes after start of the fire, with the station CO detector alarming at four minutes. The south door was closed after the second alarm was received and remained closed for three minutes. It was cycled again to test the opening and closure functions from the surface controls. After almost nine minutes of crib burning, the shaft station sprinklers were turned on for six minutes at which time the burning cribs were completely extinguished.

Operation of ventilation doors, the sprinkler system, and remote audible alarms were provided by surface personnel in response to signals identified on the graphic display as shown in Figure 14.

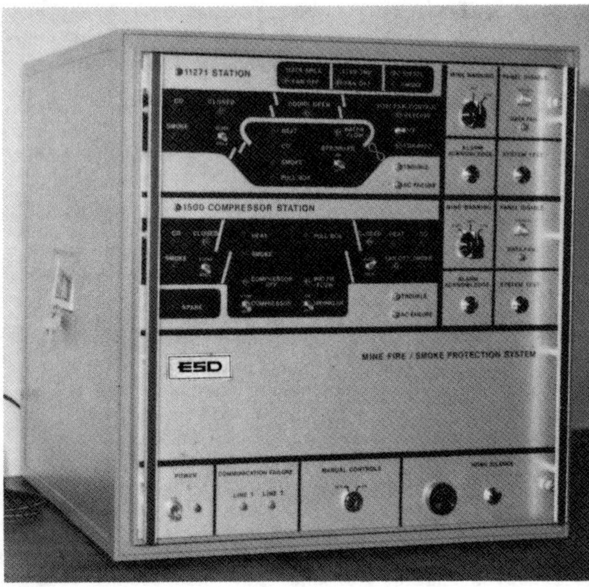

Figure 14 SURFACE CONTROL UNIT

Multiple Location Fire Tests at Pine Creek Mine

Following the short but successful fire tests at Silver Summit Mine in April 1975, the system was modified and installed in the Pine Creek Mine. The major change in this second prototype system was the controls which were modified so that more than one mine level or underground remote unit could be tested. Interconnection wiring and changes within the surface control unit quickly became burdensome with only 38 system functions. It was determined that expansion to a full 768 function system would be greatly simplified by the use of microprocessor control. A Prolog Corporation microprocessor was added which allowed considerable flexibility of input and output signals.

The total Pine Creek mine system is shown in Figures 15 and 16. As with the earlier system, the process of securing certain areas of the mine by first sensing the trouble, opening and closing ventilation doors, and sprinkling the area was followed. In addition, the microprocessor control allowed for monitoring and remote operation of two remotely located fans and an underground air compressor. Technical results of the 14 month in-mine test demonstration are as follows:

- Nearly 50 percent of the 752 signals received was due to a broken wire of one of the multiplex address cards
- Lightning prompted several design changes which gave additional protection. Benefits of the flexible design were demonstrated when test personnel witnessed a power outage and immediately restarted two fans located over 20,000 feet away. Several other fans not on the monitoring system were shut down until mine personnel could travel the long distances to the fans to restart them.
- The Becon ionization smoke detector in the diesel maintenance bay regularly alarmed due to vehicle activity. The Becon detector at the compressor station was readjusted due to similar alarms.
- Inadequate water supply continued to be a problem when installing water sprinkler systems
- Mine personnel elected to add larger standby batteries for the underground remote units to allow up to two-day operation when the primary AC mine power was shut off
- The automatic, redundant line switching of the alternate pair of wires between the two underground units was seldom used, but is still considered necessary.

Fire Test Demonstration. The in-mine fire tests conducted in July 1976 were similar to those at Silver Summit. MSHA and mine safety rescue crews were present underground and at the surface, and the fire crib and fire test scenerio were identical. A wood crib fire contained in a metal pan was started in two mine areas and allowed to burn until detectors sensed the presence of smoke or gas. Ventilation doors opened or closed to isolate mine areas, and sprinkler systems turned on to extinguish the crib fires. During the fire, sensors detected up to 100 parts per million CO gas and smoke (ionization) and activated alarms at the surface. After receiving the two alarms, surface personnel remotely closed the ventilation doors and activated the sprinkler system to extinguish the fire.

Of special interest during the slow fire growth and closure of the ventilation door was the layering affect of the smoke within the closed area. Possibly due to the inverted mine operations (Pine Creek Mine is an upside down mine where miners travel up shafts to go to work), and the inherent temperature/ pressure conditions (over 3,353 meters (11,000 feet) above sea level), the smoke from the fires stratified within the secured area after the door closure. Sensors mounted near the roof were not in the smoke layer and therefore failed to alarm as quickly as anticipated.

SPONTANEOUS COMBUSTION IN UNDERGROUND MINES

Spontaneous combustion is the outbreak of fire in combustible material that occurs without application of direct flame or spark and usually is caused by slow oxidation processes under conditions restricting the dissipation of heat. Historically, several disastrous mine fires have been attributed to the spontaneous combustion of wood and ores in remote,

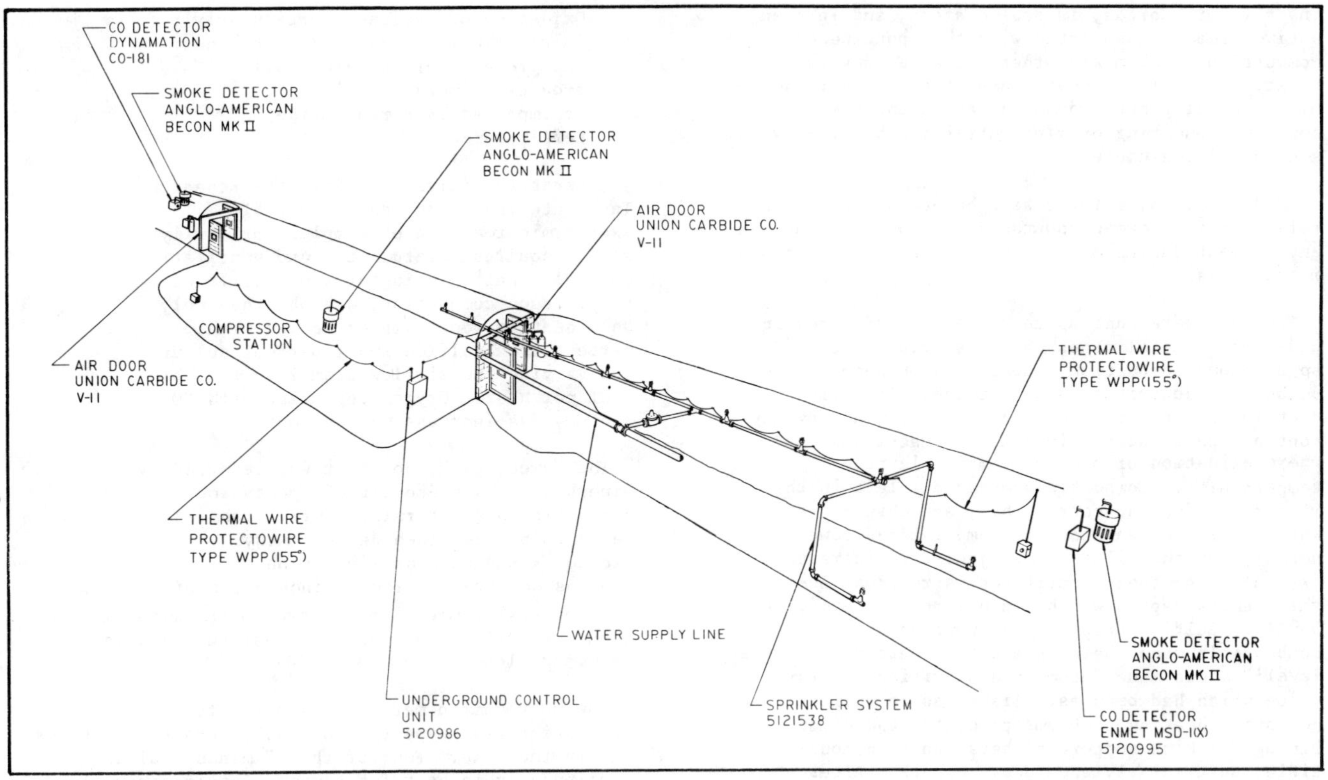

Figure 15 PINE CREEK MINE INSTALLATION (View 1)

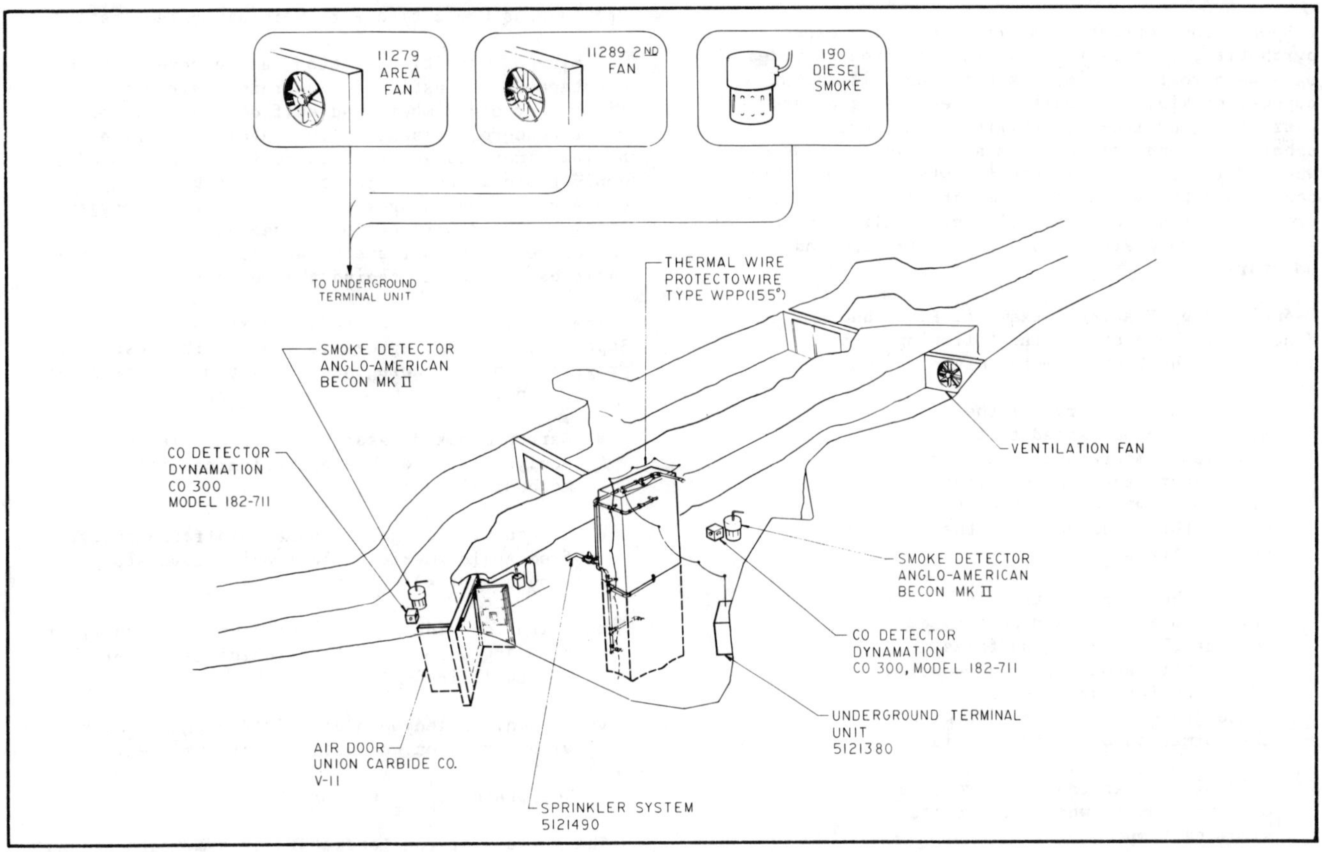

Figure 16 PINE CREEK MINE INSTALLATION (View 2)

inactive mine areas, in sealed areas, and in open, active areas. Associated with the spontaneous combustion problem are other instances involving ignition of sulfide dust, combustion of An-fo heated by surrounding hot ground surfaces, and the continuous heating of mines utilizing backfill with a high sulfide content.

In early 1978, efforts were begun to define the relationship between spontaneous combustion and the environment during mining conditions in underground metal mines[16].

Findings were that up to 20 percent of fires in underground metal mines can be attributed to spontaneous combustion. Several spontaneous combustion contributors were found. Significant contributors are the changes in airflow and oxygen content, particularly in sealed, inactive areas where oxidation of ore and timber elevates temperatures. Combustible materials left in the mine respond as do cotton, hay, and chip piles in which ignition can occur at temperatures lower than normal when the piles are subjected to increased oxygen[17]. Another contributor is residual An-fo which emits vapors which can become exothermic at 50°C[18]. Sulfide ores can be spontaneously combustible when there is a 3 to 7 percent moisture level[19]. One mine documented a critical sulfur value which had been established as an indication of potential heating problems prior to production mining[20]. Mine support timbers can be combustible, particularly when decay occurs. Alternating wet and dry mine conditions contribute to wood decay and weakening of timbers, and increases wood surface causing increased oxidation.

Spontaneous oxidation of iron sulfides, especially pyrrhotite, is a serious problem, as iron sulfides oxidize almost ten times faster than sulfides of copper, nickle, or cobalt. In decreasing order, pyrrhotite, pyrite-arsenopyrite, chalcopyrite, sphalerite, and galena are most active in areas where moisture is available[21]. Other factors that contribute to spontaneous combustion in metal mines are the sulfide quantity, elemental sulfur, particle size, available air, insulation, mass size and moisture.

Articles by Flan, Luzaszewski, Ryss, and Vinogradova have stated the following regarding spontaneous heating in metal mines:

> It is not necessary for the material to be heated to its ignition temperature, but to a temperature that is sufficiently high for heat of oxidation to exceed that dissipated to the surroundings.
>
> If this heat of reaction is delivered rapidly and conserved, so that the reacting surfaces become hot, spontaneous combustion of the sulfur can occur at temperatures as low as 50°C with the formation of sulfur dioxide.
>
> Acid mine water caused hydrolysis of mine timber, which lowers its kindling temperature.
>
> Oxidation of pyrites, particularly in the finely pulverized state and the presence of certain small amounts of moisture, is accompanied by a rapid rise in temperature.

Laboratory analysis by Gillette Research Institute indicated that repeatable exothermic reaction of samples (primarily wood) requires careful heat and ventilation control. Self heating of wood, followed by spontaneous combustion, was obtained only when heating wood samples seeded with 5 percent iron sulfide and 1 percent sulfur and varying the airflow from 2 to 10 cubic feet per hour. Oxygen depletion with CO and CO_2 gas increase was evident.

The "sweet gas" odor that was reported by miners to be a precursor of spontaneous combustion was determined during laboratory tests to be acetaldehyde and methyl alcohol. Spontaneous combustion of wood samples occurred following incubation of 160 hours at temperatures above normal mine ambience (160°C to 210°C), but below established levels for wood alone (280°C to 500°C).

The most common method of sensing for spontaneous combustion was inspection of background temperatures within the mine. Four of the 17 mines visited continuously inspect for temperature increases in selected areas and a potential test area at one mine was eventually sealed.

Spontaneous Combustion Fire Warning System Tests

A mine on standby operation was selected for the spontaneous combustion fire warning system test. Fifty-year-old timbers and sulfide ores were of major concern, presenting low-level combustion behind older sealed areas where the mine planned to conduct block-caving operations. Of greatest concern to mine management was keeping the oxygen level below 13 percent in the sealed area. Therefore the sensor system was designed to measure the pressurized air behind the bulkhead.

When this mine was closed, the Magma Mine in Superior, Arizona, agreed to conduct the test program. The following sensor system was installed in the mine:

- Carbon monoxide gas: Energetic Science Ecolyzer 4125 with electrochemical cell and mechanical pump
- Carbon dioxide gas: Spanair infrared analyzer from Anglo American Electronics Laboratory (South Africa)
- Sulfur dioxide gas: Spanair infrared analyzer from Anglo American Electronics Laboratory (South Africa)
- Oxygen: Teledyne Model 320AX oxygen analyzer with diffusion collection microfuel cell
- Temperature: Hy Cal BA-500 RTD.

The Anglo American Laboratory transmitters and receivers were used to multiplex the analog information from underground to a surface recorder. The telemetry method has a capacity of 64 channels per wire pair, was reported to provide high noise immunity and can be transmitted over ordinary mine power cable. Each channel is assigned a carrier frequency within the band 315 to 10,000 Hz which is switched by a modulator at a proportionate rate of 5 to 15 Hz, corresponding to a signal input of zero to -1.0 milliamp. The modulated carrier is coupled to a balanced line that is typically 40 to 80 kilometers, filtered, and demodulated. The original analog signal is displayed on a six-channel Elnik Instruments chart recorder Model NSC6-1 with a chart speed of 1.25 inches per minute and approximately 2 second channel cycle time.

Field Test Demonstrations. Chart paper records from 131 days testing were analyzed prior to the final demonstration on July 26, 1979. Highlights of the five-month demonstration were as follows:

- A total of 10 days recording was lost because of disengagement of the chart paper on the Elnik recorder.
- CO and CO_2 traces consistently indicated blast times during shift changes. The CO_2 trace deflected only 9 percent full scale, as compared to 58 percent for the CO analyzer, however, full-scale measurements had to be interpreted accurately for concentration levels.
- Underground power failures also were routinely recorded by the chart traces going to zero or full scale depending on the sensor.
- Appearance of a fire condition occurred on July 5 between 8:20 A.M. and 1:30 P.M. with a notable decrease in oxygen, and increases in CO_2, CO, and temperature. The cause of the fire indicators was the restarting of the mine after the July Fourth holiday.

Several operational problems occurred. The early failure of the microfuel cell in the Teledyne oxygen analyzer probably was caused by the 34°C to 36°C temperature and high humidity of the area. Dust or erratic power conditions may have caused the pump from the carbon monoxide analyzer to fail prematurely. Laboratory analysis previously had confirmed that pump failure was caused by low voltage and pump crank design.

Typical chart traces and the anticipated spontaneous combustion chart trace are depicted in Figure 17. As noted in the timing marks to the left of the trace, mine blasts occurred regularly at midnight, noon and approximately 4 P.M. on the carbon monoxide curve that extends significantly through its 0 to 50 ppm range. Temperature and oxygen traces are quite steady with the full scale of temperature -17°C to 66°C (0 to 150°F), and oxygen scale of zero to 25 percent. Difficulty in chart paper interpretation was experienced due to the direction of the needle movement. Three sensors were designed to read zero percent on the left side of the chart, but the South African CO_2 and SO_2 sensors use zero as saturated and 100 percent to indicate low concentrations. As noted in the chart, the CO_2 regularly monitored the mine blast, but with smaller excursions toward the left of the paper. Chart B of Figure 18 shows what can be anticipated from an actual, slow-growth spontaneous

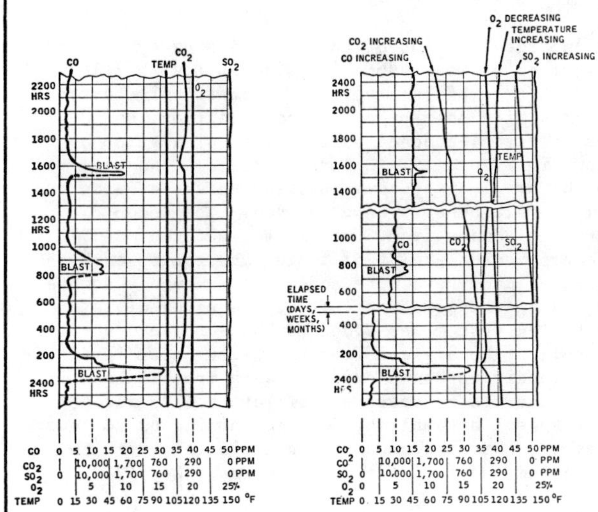

A. Normal Background

B. Anticipated Spontaneous Combustion

Figure 17 CHART PAPER TRACES AT 3000-81 WINZE AREA, MAGMA MINE

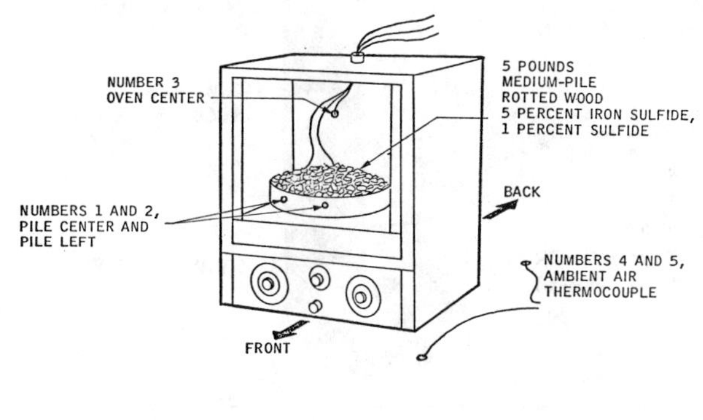

Figure 18 TEST OVEN THERMOCOUPLE LOCATION

combustion occurrence with the carbon monoxide and temperature increasing to the right of the chart, and the CO_2, oxygen and SO_2 moving to the left.

An in-mine spontaneous combustion test demonstration was performed in July 1979 by incubating 5 pounds of moist and rotted mine wood with 5 percent iron sulfide and 1 percent sulfur in a standard laboratory oven (see Figure 17). The test was terminated 16 hours after start of incubation by opening the oven door and raising the lid over the wood pile. The entire pile was heavily charred and flamed easily when the door opened allowing oxygen into the oven. Thermocouple recordings exhibited spontaneous combustion within the center of the pile at 244°C (472°F) while the oven air temperature was 168°C (334°F).

Second-Generation Spontaneous Combustion Fire Warning System[22]

An enlarged version of the fire warning system was installed at Magma Superior Mine on November 16, 1981. It combines a variety of previously tested sensors with the Anglo American Electronics Laboratory 64-channel surface control station as shown in Figure 19. Several modifications were made to the original system. Oxygen, sulfur dioxide and temperature sensors were removed. The Becon particle sensor from South Africa and the MSA 571 electrochemical-cell, diffusion absorption carbon monoxide analyzers were added. A frequency counter to aid in the alignment of the frequency-division multiplex system was added to the surface control console. The signals from the Becon particle sensor and Spanair carbon dioxide sensors from South Africa were inverted to operate from zero to 100 percent. Four sensor assemblies, each containing up to four sensors, are scheduled for installation, with two installed and fire tested on February 26, 1982.

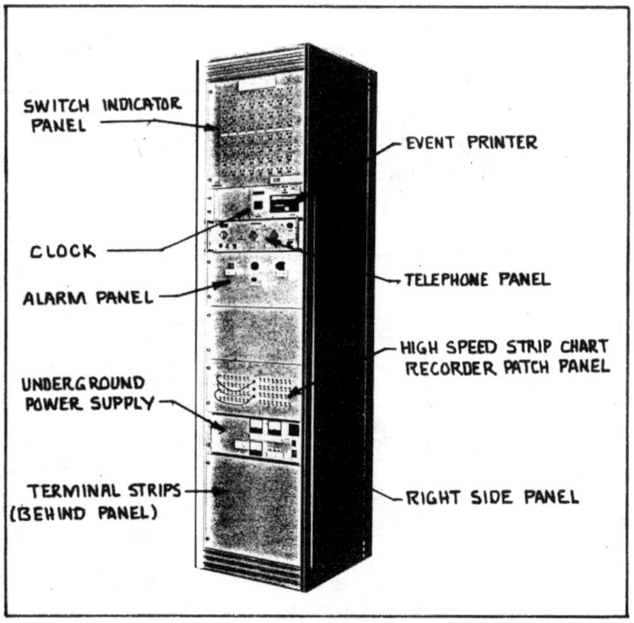

Figure 19 FRONT VIEW OF ANGLO AMERICAN ELECTRONICS SURFACE CONTROL CONSOLE

The system alarm point had been pre-established at 0.28 volts by the South African designer; this conforms well with the zero to 100 ppm full-scale reading of both the MSA and Ecolyzer CO analyzers to be 28 ppm. The stable output of the Becon and Spanair analyzers was adjusted to give an ambient, non-alarm condition of 0.25 to 0.27 volts. Therefore, the lesser deflections from these sensors during smaller fire tests will place them in alarm also.

Standard equipment with the surface console from South Africa is an event printer (see Figure 20 for sample output) that records when each channel goes in or out of an alarm condition. The plus symbol (+) indicates the sensor has gone over 0.28 volts and into alarm. The second two digits are the channel number; the last four digits are the time of day.

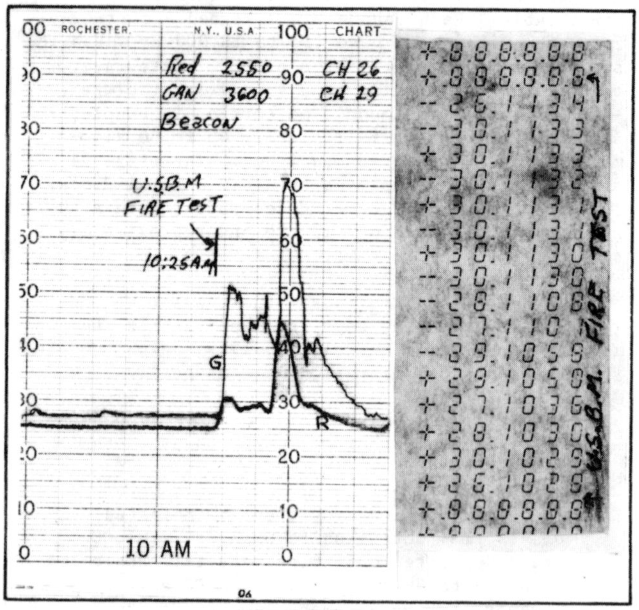

Figure 20 SAMPLE OUTPUTS FROM EVENT PRINTER

Since adding more sensors in the future will increase time-consuming analysis of chart recorder traces, a modification of this procedure is being considered. The AAEL surface control console has provisions to patch any three of the 64 channels into a recorder during fire condition. Since this allows the more stable and reliable sensors to be monitored only as necessary, only those sensors with erratic behavior (such as the MSA exhibits) will be connected to continuous chart recorders.

COAL MINE FIRE AND SMOKE PROTECTION SYSTEM

Information collected during early 1981 on fire histories of underground coal mines was similar to that for metal and nonmetal mines. Of the 221,000 coal mine workers in the United States, 66 percent (146,000) are employed in underground mines. These workers were exposed to a total of 1,014 underground coal mine fires between 1950 and 1977[24]. Forty-four of the underground fires occurred in mine shafts and slopes, accounting for 14 percent of the deaths and injuries related to fires. Eighty-nine percent of the reported 1,014 underground coal mine fires caused no injuries, however, costs in terms of equipment and production loss can be assumed to be significant. Additional data indicates that direct loss from deaths and injuries, excluding equipment and capital investment loss, total $3.4 million annually[25].

Microprocessor-Based Fire Warning System at Peabody Coal Mine

Based on state-of-the-art technology and information gathered from visiting mines, and with a background of similar developments at the Silver Summit and Magma Superior Mines, a microprocessor-based system with 24-hour memory retention of all analog signals was developed and installed at the Peabody Number 10 mine during August 1980 (see Figure 21). System capabilities include:

- CO and particle (smoke) sensing at the slope collar to monitor intake air, at the slope-bottom or undergound hopper-bottom area, at the underground hopper-top area, at a nearby exhaust shaft, and at a distant ventilation shaft (5 miles)

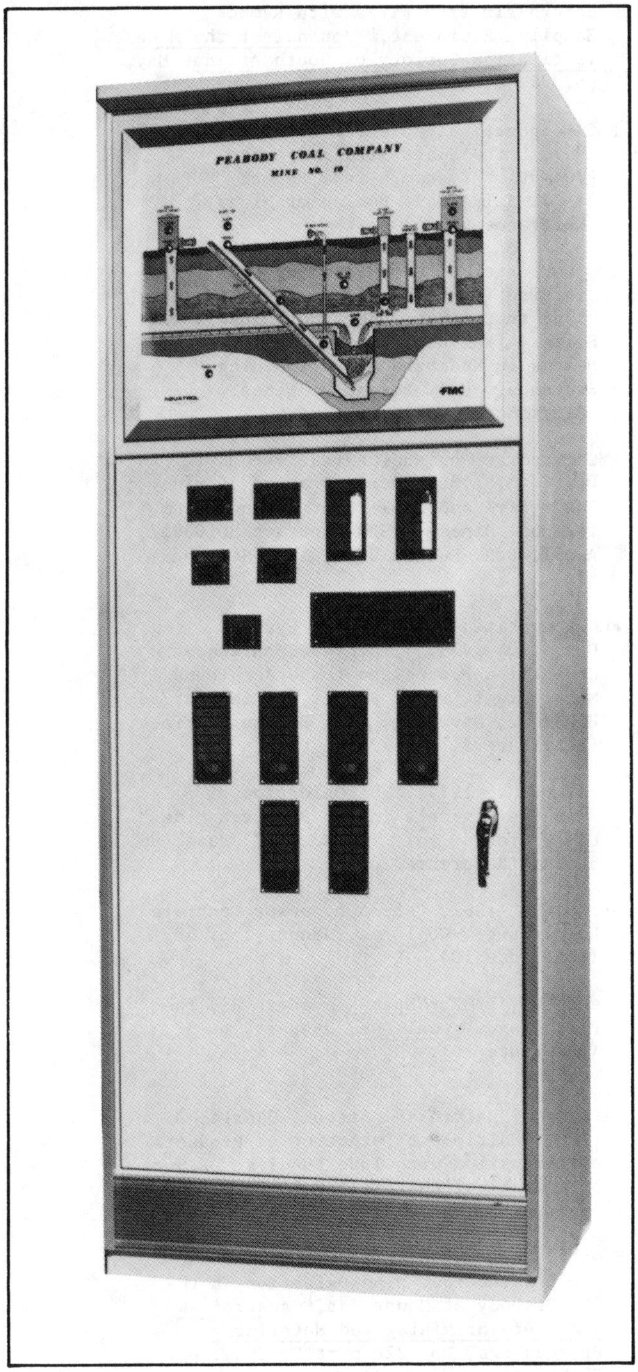

Figure 21 CONTROL UNIT FOR PEABODY COAL MINE SYSTEM

- Frequency-division, multiplex telemetry between all sensors and a surface control station

- Radio telemetry of multiplexed analog signals from a 5-mile distant exhaust shaft

- Surface control console readout and adjustment for monitoring and control by the hoist operator

- A printer to record all signals including CO, smoke, heat, power failure, data failure, fire-alarm pull handle and fan operating condition. The printer also records time and date of all events including changes in sensor alarm levels, and periodic manual tests.

The surface control console, remote station and telemetry equipment are Aquatrol Corporation products. Aquatrol has performed similar installations for monitoring conveyor belts at the FMC Trona Mine in Green River, Wyoming. Their standard hardware was modified to include the fire and smoke sensors, the graphic display and the 24-hour memory of analog signals. Figure 22 is an example of the printer output when the system operator selects and activates the channel and graph-function switch. The 24-hour charted history of any sensor eliminates the need for continuous chart recorders and the resultant time-consuming chart-paper analysis. The memory array of the prototype system was only 4K RAM, thus limiting the number of sensors, maximum poll rates and chart resolution. Expansion of the system will include larger memory.

System Testing. On four occasions during the 12-month in-mine tests, the accurate performance of the system was demonstrated. The hoist operators' responses exemplified the ease of operation of the total system. Specific test results are as follows:

- A night watchman's truck parked near the slope air intake caused the CO analyzer at the slope collar to alarm. The hoist operator observed the persistant alarm and location on the surface control graphic display, and investigated the situation. When the truck was moved, the alarm stopped. The response by the hoist operator reflected the general awareness of Peabody mine personnel to respond to conditions that might adversely affect underground air quality.

- The Becon smoke detector located in the underground hopper gave continuous alarms. The hoist operator called the shift superintendent to investigate the alarm. He found an overheated and smoking 150-horsepower, belt-drive motor located 15 crosscuts from the Becon smoke detector. There were no injuries to personnel, and the conveyor was repaired and in operation for start of the next shift.

- The hopper-top Becon smoke detector alarmed. It was determined that a small pile of coal found in the hopper was self-heating. The coal was removed, thus prohibiting a potentially serious heating incident.

- Mine evacuation occurred due to a smoldering, heating condition in a worked-out area of the mine. The CO gas and smoke (particle) sensors monitoring this area of the mine had not received MSHA approval and were drawing air through a tube in the exhaust shaft. The tubing had been blocked due to condensation in the line, making the sensors unresponsive.

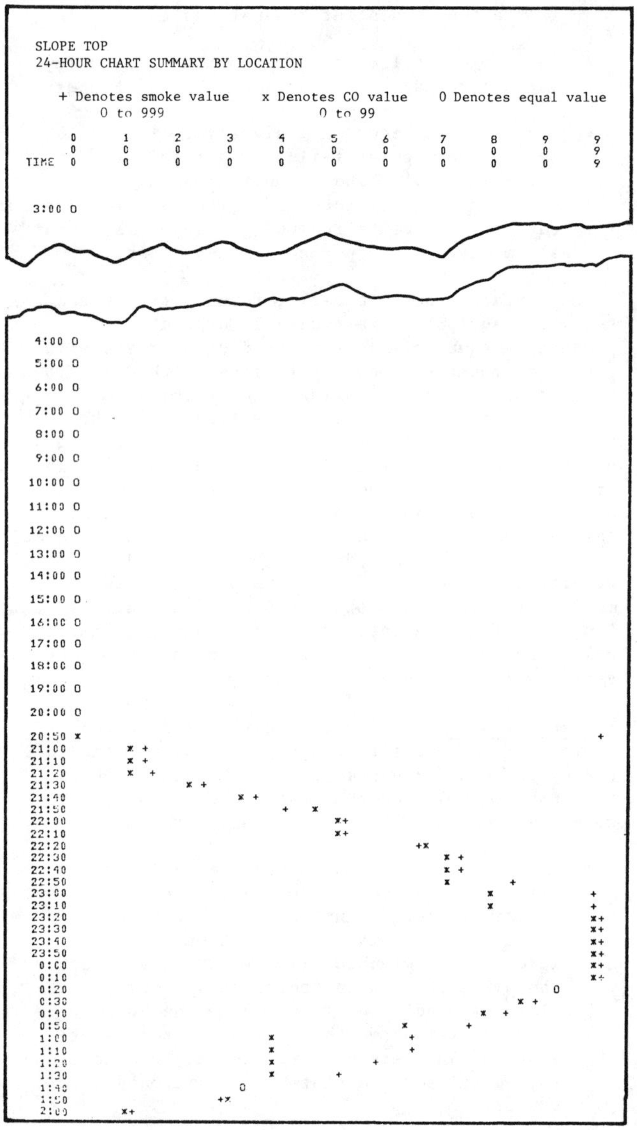

Figure 22 SAMPLE OUTPUT FROM PEABODY MINE SYSTEM PRINTER

SUMMARY

Vast improvements in fire and smoke sensors, in remote-monitoring, and data-collection systems have been made over the past decade. Although most of the efforts have been directed toward the surface facilities of petrochemical industries, computer centers, offices, and resident structures, the use of this technology is progressing slowly into underground mines. Mine-proven components exhibiting high reliability and low maintenance in the extremely harsh environments of underground mines eventually will become commonplace in protecting underground mine workers from the hazards of fire and smoke. Many mines already have become aware of the benefits of remote monitoring systems in lowering overall production costs and have installed such systems. Reliable fire and smoke sensors can be added easily to these systems.

REFERENCES

An Annotated Bibliography of Coal Mine Fire Reports, Allen Corporation of America, Alexandria, Virginia, 1979, (Reference 24).

Eschenburg, H. M. W., 1978, "Monitoring Return Air from Fires with Remote Sampling Equipment," Journal of the Mine Ventilation Society of South Africa, May, pp. 94-96, (Reference 8).

FMC Corporation, Advanced Products Division, 1975, "Mine Shaft Fire and Smoke Protection System, Final Report," USBM 24-77, July, U.S. Bureau of Mines, (Reference 4, 15).

FMC Corporation, Engineered Systems Division, 1976, "Accident Cost Indicator Model to Estimate Costs to Industry and Society from Work-Related Injuries and Deaths in Underground Coal Mining," September, U.S. Bureau of Mines, (Reference 7, 25).

FMC Corporation, Engineered Systems Division, 1980, "Demonstration of Mine Shaft Fire and Smoke Protection System for Coal Mines," USBM Contract H0100017, April, U.S. Bureau of Mines, (Reference 6).

FMC Corporation, Engineered Systems Division, 1979, "Improved Spontaneous Combustion Protection for Underground Metal Mines, Final Report," USBM Contract H0282002, November, U.S. Bureau of Mines, (Reference 5, 16, 22).

Good, B. H., 1977, "The Oxidation of Sulphide Minerals in the Sullivan Mine," CIM Bulletin, Vol. 70, No. 782, June, pp. 83-88, (Reference 20).

Green, P., 1982, "Microprocessor Controls HM Circuit," Coal Age, January, p. 58, (Reference 10).

Green, P., 1982, "Space Invaders Hit the Prep Plant," Coal Age, January, p. 52, (Reference 10).

Hertzberg, Martin and Litton, Charles D., 1976, "Multipoint Detection of Products of Combustion with Tube Bundles," Report of Investigations 8171, U.S. Bureau of Mines, (Reference 9).

Hewett, R. L., 1968, "Deep Leaching and Accelerated Post-Mine Oxidation in the 500 Orebody at Mount Isa," Australian Institute of Mining and Metallurgy Proceedings, No. 226 part 2, June, pp. 73-88, (Reference 18).

Jackson, Dan, 1979, "FMC Mine Adopts Monitoring System," Coal Age, January, (Reference 13).

Klassen, V. I. and Mokrousov, V. A., 1963, An Introduction to the Theory of Flotation, Butterworths Publishers, London, pp. 12-13, (Reference 21).

Mitchell, D. W. and Burns, F. A., 1979, "Interpreting the State of a Mine Fire," Information Report IR 1103, Mine Safety and Health Administration, (Reference 2).

Nielsen, George F., 1980, 1980 Coal Mine Directory, McGraw-Hill, Inc., New York, (Reference 23).

San Jose Mercury News, February 19, 1982, (Reference 1).

Schmidt, Fred L., 1969, "Observations on Spontaneous Heating Toward Combustion of Commercial Chip Piles," Tappi, Vol. 52, No. 9, September, pp. 1700-1701, (Reference 17).

Stiefel, Al, 1980, "Underground Communications and Information Systems," Mining Congress Journal, July, pp. 31-41, (Reference 14).

Sudbury, M. P., and Petkovich, F., 1961, "Exothermic Hardening of Cu-Ni Sulfide Agglomerates," Transactions of AIME, Vol. 220, pp. 477, 484, (Reference 19).

"Technical Highlights: Health and Safety Research, 1970-1980," March 1981, U.S. Bureau of Mines, (Reference 3).

Van der Walt, N. T., et. al., 1980, "An All-Analogue Fire Detection System for South African Gold Mines," Journal of the Mine Ventilation Society of South Africa, Vol. 33, No. 1, January, pp. 2-13, (Reference 12).

Chapter 22

VENTILATION CONTROL MEASURES TAKEN AND INTERPRETATION OF CHROMATOGRAPH ANALYSES DURING RECENT ARIZONA MINE FIRES

Ralph K. Foster

Mining Engineer
Ventilation Division
Denver Safety and Health Technology Center
Mine Safety and Health Administration
Denver, Colorado 80225

ABSTRACT

During five underground mine fires which occurred in Arizona during 1980 and 1981, sealing the fire area became necessary. To minimize leakage into the fire area, ventilation control measures were instituted to balance pressure across the bulkheads. Problems encountered in sealing the fire areas and measures taken to improve these conditions are discussed. The routing of the ventilation and ventilation control to ensure the workers' safety were also important considerations in reopening the mines. The use of continuous CO monitors is discussed. One important consideration was the condition of the fire, and chromatographic analysis was used to establish this. Graphs comparing the Trickett ratio, carbon monoxide, carbon dioxide, and oxygen concentrations are given. The Trickett ratio was useful but needed to be augmented by the comparison of CO, CO_2, and O_2. Problems with getting good samples from sealed areas and recommended procedures that were developed during these fires are discussed.

INTRODUCTION

Four mine fires which occurred in Arizona between May 1980 and April 1981 required sealing fire areas and curtailing production. One additional fire, which occurred during the 1980 copper strike, did not require ceasing production but did require extensive sealing and ventilation control. Three of the fires were at Magma Copper Company's Superior Division Magma Mine, Superior, Arizona, and one at the same company's San Manuel Mine, San Manuel, Arizona. The fifth fire was at Noranda Exploration Incorporated's Lake Shore Mine, Casa Grande, Arizona.

During all the fires, Mine Safety and Health Administrations personnel in both Metal and Nonmetal and the Denver Safety and Health Technology Center worked closely with mine personnel in a cooperative effort to insure the safety of people taking part in the sealing and reopening of the mines. Ventilation control was important in this effort and in bringing the fire under control. This paper describes some of the problems encountered and methods used to solve these problems.

The gas chromatograph for analysis of samples has been used extensively at the mine site for coal mine fires, but, with the exception of the Sunshine Mine fire, the incidents discussed in this paper are the first metal mine fires where on-site chromatograph analysis has been used as an aid to making decisions regarding the effectiveness of the sealing and the safety of production people returning to work.

During the course of the five fires, improvements were made in the procedures used in gas sampling through bulkheads so that during the last fire at the Lake Shore Mine, the bulkhead sampling results were uniform. Portable and continuous carbon monoxide were used with success during three of the fires and were found to offer many advantages in safety and economy over using stain tubes.

SAMPLING FROM BEHIND BULKHEADS

Getting uniformly good samples from behind bulkheads has presented problems over the years. The standard method has been breaking off the tip of a vacuum bottle in a plastic or rubber sampling line. A great deal of patience and practice is necessary to insure that the broken tip of the bottle does not block the gas sample from filling the bottle.

MSHA's Denver Safety and Health Technology Center uses a stainless steel disc-type sampler called a "cricket." These have offered a number of advantages. They are easy to carry and ship. The biggest problem with the cricket has been people failing to screw the cap down tightly enough to seat the lead ball against the cricket top.

During the fires at the Magma and San Manuel Mines, different sampling techniques were tried. As with the vacuum bottle, the problem with the cricket is that when the sample is taken, the inrush of air is extremely rapid. With the vacuum bottle, it is essential that the tube fit snugly around the neck, or the sample can be diluted by outside air leaking along the neck. With the cricket, the problem is to keep the outside air from leaking along the threads. The design which proved most successful is shown in Figure 1.

Fred Ryan, Denver Safety and Health Technology Center Chemist, developed the basic O-ring and tube design for use with the chromatograph and this was adapted to the sampling T. The taking of duplicate samples is recommended as a check, in the event a sample is lost due to a loose cap or malfunction of the chromatograph.

During the fires, the installation of adequate sampling lines during the building of bulkheads and plugs was a problem. In most cases the crews were working against time, under oxygen, in an extremely difficult environment, and installing a sampling line was often forgotten under those conditions. A 0.25-inch tubing extending through the bulkhead 25-feet (7.6m) or more into the fire area is recommended. It is difficult to evacuate a 2- or 3-inch line (5.1 to 7.6cm) from which to get good samples, particularly if the bulkhead is incasting.

BULKHEAD SAMPLING
USING A CRICKET GAS COLLECTOR

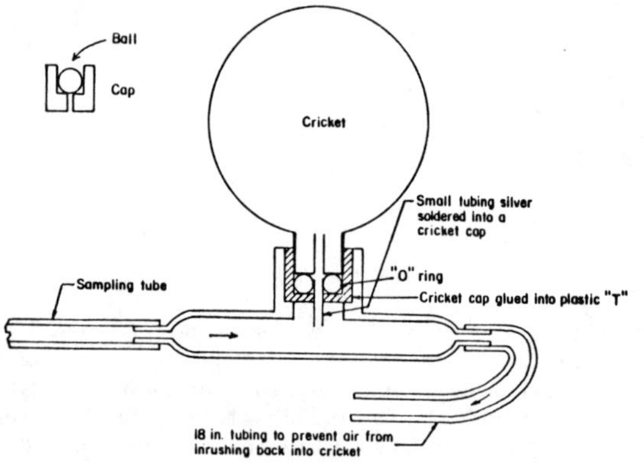

1. Remove cap with ball from cricket.
2. Screw cricket into the T so it is snug against the O ring.
3. Evacuate the cricket a minimum of 3 times and then take sample. (make sure both sides are convex)
4. Remove cricket from the T and screw on cap with ball tightly.

Figure 1

The Trickett ratio, which was used in all the fires, is a relationship of CO_2, CO, and O_2 and has been used for many years for evaluating fires and self-heating in coal mines, and to determine whether or not a fire exists in sealed areas containing CO. A report on the subject by Donald W. Mitchell and Frank Burns has been published by the Department of Labor as MSHA IR 1103, "Interpreting the State of a Mine Fire."

The formula is:

$$\frac{\%CO_2 + 0.75\%\ CO - 0.25\%\ H_2}{0.265\ N_2\% - O_2\%}$$

For use in calculation in this mine, 20.93%, which is the O_2 in standard air, was used in place of the 0.265 N_2.

MAGMA MINE FIRES

Ventilation

The ventilation of the Magma Mine is divided into two systems (Fig. 2). Ventilation to the working section of the mine is coursed down the No. 9 shaft, which is also the production shaft. Fans on each level distribute the air to the stopes, and the return air is directed to the west to ventilation raises and the No. 6 shaft. On the 2550 level, some of the air is split to the No. 4 shaft. The older section of the mine is maintained for water collection and pumping from the No. 3 shaft, 3600 level, and as a second escapeway from the mine. The No. 2, 3, and 5 shafts are on intake air, and the No. 8 shaft exhausts essentially all of the air from this section of the mine.

May 1980 Fire

Smoke was detected May 2, 1980, on the 3320 sub-level of the Magma Mine (Fig. 3). The first indication was a gray haze along the roof, which was thought to be the result of blasting. The area was watched closely until May 7, 1980, when with additional smoke present and an increase of carbon monoxide from 50 to 300 ppm, the decision was made to evacuate the mine.

Following the evacuation of the mine, an attempt was made to get to the fire area to see if the fire could be controlled by a direct attack. Because the smoke was too dense for access to the fire, the decision was made to seal the area. One of the first actions taken, because of the smoke and gases being directed up the 3320 timber slide by the pressure from the 3400 level fan, was to pour a concrete plug in the slide. Sand lines extending through the plug were capped and small tubes were installed in caps for sampling. The pressure across the plug measured 8 in. wg (1990 Pa), with the fire side being the high pressure side. Fortunately, this plug was extremely tight and did not leak onto the 3300 level. This remained the best sampling point throughout the fire.

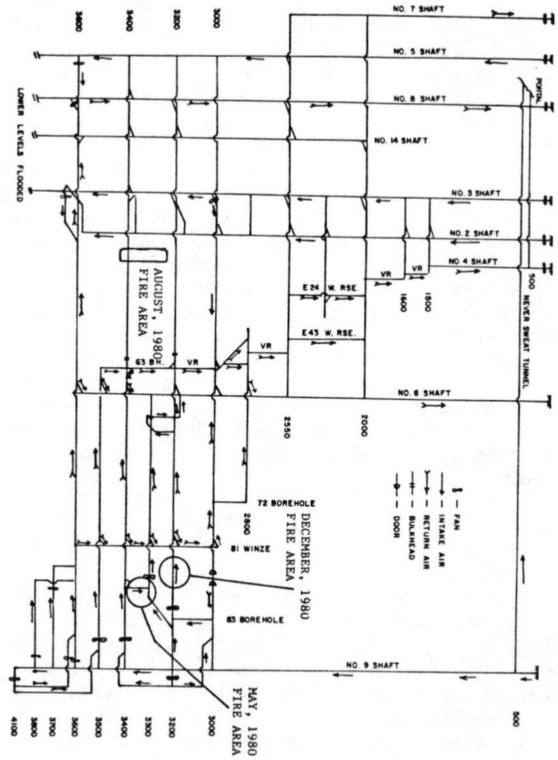

MAGMA MINE, SUPERIOR DIVISION, MAGMA COPPER CO.
SUPERIOR, AZ.
VENTILATION SCHEMATIC ON MAY 7, 1980

Figure 2

Bulkheads were built on the north edge of the 83 footwall drift and both ends of 81 footwall drift on the 3300 level. Additional bulkheads were built on the 3400 level and on the 3200 level. A total of eleven bulkheads were built and sand filling behind the bulkheads progressed on a priority basis.

Bulkheads were of substantial wood construction and were shotcreted. Although a 1- to 2-inch (2.5 to 5.1 cm) thickness of shotcrete was applied to the bulkheads and ribs, some leakage still occurred through the pores of the shotcrete. Also, some leakage occurred through cracks in the rock around the bulkheads. Plaster and polyurethane foam were used for sealing the shotcrete leaks and cracks in the walls and back.

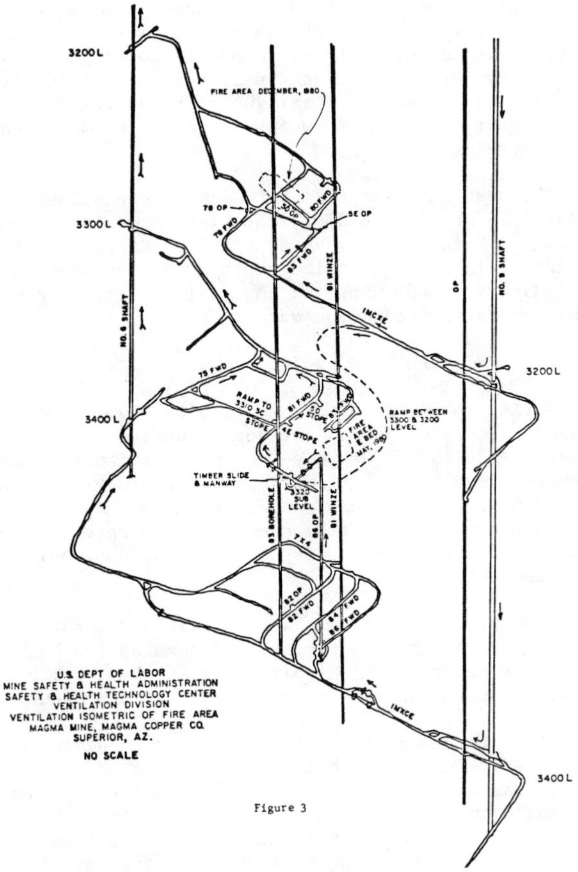

Figure 3

The old stopes extended along the dip up to the 2800 level, with connections on the 3200 and 3000 levels. Sealing the 3200 level with bulkheads and sand fill was fairly effective, but the timbered raises and sublevels on the 3000 level proved difficult to seal effectively. Openings were sealed as well as could be done, with the primary emphasis on sand filling the area above and around the fire.

Very little was done in balancing pressures across the bulkheads during this particular fire. Once bulkheads were installed, pressure readings were taken to watch for any sudden change of pressure from the fire. After the area was sealed, monitoring for CO in the drift and behind the seal was performed, using indicator tubes.

Cricket samples for the chromatograph were taken at the 3320 timber slide plug and other bulkheads until they were sand filled. Gas samples taken on the 2800 level were useful in determining an estimated total amount of burning, and the samples taken in the 8E and 6D raises on the 3000 level also served as an excellent index to the fire condition.

A ventilation survey was made to guarantee the direction of air, fire containment was evaluated, and scenarios were worked out concerning the possible danger to personnel if the fire were to burn through a bulkhead.

Considerations of paramount concern before workers were allowed to return underground were:

1. What was the condition of the fire? Was the fire dying down or building up? Until some background was established that the fire intensity was diminishing, people were not allowed underground except for fire fighting purposes.

2. Had the fire been contained so that it could not expand rapidly and trap people?

3. Had the ventilation been controlled so the pressure against bulkheads were toward the fire and the intake airstream would not be contaminated by leakage from the fire?

4. Had monitoring been set up so a change could be noted and mine management alerted to the possibility of a developing problem?

Once it was decided that the fire was under control and was contained by bulkheads and plugs so that an expansion of the fire could not trap people, the mine was put back into production in stages. The stopes on the 3400 level and below were the first to be put back into production, and after monitoring for two days, two stopes on the 3300 level were returned to operation. A third stope on the 3300 level was held back as a precaution for another two days. This last stope was in the same E bed where the fire was thought to be located. A fire watch was continued on these levels for a period of time.

Chromatograph Interpretations

Once the sealing was decided on, a request was made for a chromatograph and chemist from MSHA's Denver Safety and Health Technology Center. The chromatograph was dispatched and arrived on May 10.

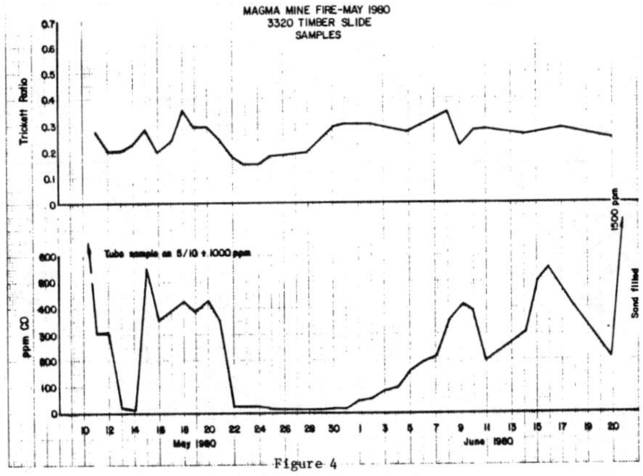

Figure 4

In the meantime, indicator tube measurements from the sampling pipe in the plugged 3320 timber slide decreased from 700 ppm CO on May 8 to 360 ppm CO on May 11, and all levels were free of CO except for the 8E area near 81 winze on the 3000 level, which read 5 to 20 ppm.

The chromatograph was set up at the mine, and cricket samples analyzed rapidly. Samples from the timber slide sampling pipe indicated a decrease of CO to between 0 and 37 ppm and a CO_2 value of 0.3%. The Trickett ratio calculation was 0.25, which indicated no fire (Fig. 4). On the night of May 14, however, the CO, as monitored from the timber slide by indicator tubes, suddenly started to increase. Mine management and production supervisors were alerted, and a close watch was kept of the fire area. The chromatograph samples were continued, and although they varie considerably, they gave indications of the condition of the mine fire.

Following this fire and the August 29 fire, the conclusion was reached that sand filling affected the CO reading. This is particularly noticeable (Fig. 4) toward the end of sand filling the plug area. The importance of finding a location which gives an indication of total burning cannot be overemphasized.

The chromatograph analyses of samples taken on the 8E raise, 3000 level, as shown in the figure 5 graph, were used to evaluate the fluctuations on the 3320 timber slide sample. The graph shown for the 3320 timber slide (Fig. 4) shows a decrease to practically no CO on May 13 and 14, and a sudden rise in CO on May 15. There was a second rapid drop of CO on May 22, leveling off until it started increasing June 1. The graph (Fig. 5) for the 8E raise reflected some rise, but not to the extent that the 3320 timber slide graph showed. The fluctuations caused concern, and close monitoring was maintained.

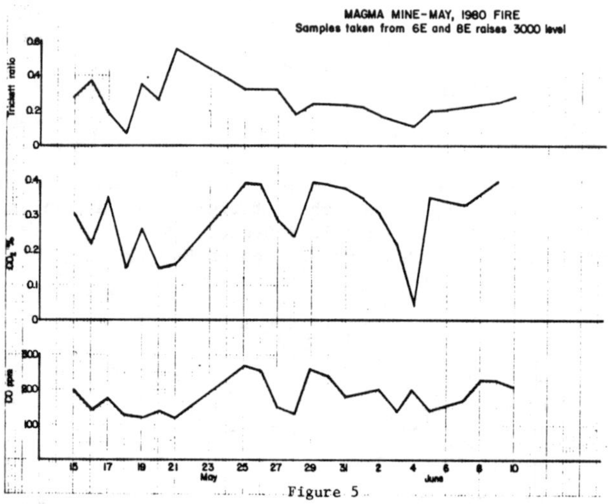

Figure 5

Calculations based on the CO concentrations and airflow on the 2800 level indicated only a small amount of wood burning. These calculations used a figure of approximately one cubic foot of CO produced by every pound of wood burned. The rate-of-burning figure is important in evaluating the condition of the fire.

The Trickett ratio never approached the 0.6 minimum number for burning wood. For evaluating the burning condition, the trend of the Trickett ratio graph was watched rather than the actual number.

Extra monitoring being done at No. 6 shaft because of the May 2 fire indicated an elevation of CO found in a drift near No. 6 shaft on the 3000 level. Investigation indicated that a door to an old section of the mine had been opened to augment air furnished by auxiliary ventilation and ventilation tubing to a section being rehabilitated in that area. The old section was immediately sealed, and the CO in the drift disappeared. The monitoring and early sealing probably prevented a fire from developing.

Although a recirculation pattern affecting the entire mine, such as occurred during the 1972 Sunshine fire, would not have developed, the burning through of the airlock doors on the 3320 sublevel could have caused a very rapid expansion of the fire on the sublevel, presenting problems to people working in stopes downstream of the fire on the 3300 level. This situation was changed by installing a fan on the 3200 level near No. 9 shaft, which replaced the 3200 and 3300 level fans.

This was the second time the chromatograph had been sent to a metal mine fire, the first being the Sunshine fire. The rapid analyses of the fire gases were invaluable in making knowledgeable decisions concerning the safety of the mine and the advisability of returning people to production work.

August 1980 Fire

A rise in CO_2 at No. 3 shaft, 3600 level on August 29, 1980, during the copper strike, was investigated (Fig. 6). This rise was not particularly unusual, since the lower portion of the old mine below the 3600 level had been allowed to flood, and CO_2 was often flushed out of this area when the water level rose; however, a check for CO showed 30 ppm, which was unusually high.

Investigation showed that the bulkheads on the 3200 level had rotted through and some of the bulkheads needed repair. Sampling was started, using the Magma Copper Company's chromatograph at San Manuel.

Other bulkheads found to be in poor condition were repaired. One of the crosscut bulkheads on the 3000 level was torn out because there was not enough space to build another bulkhead on the drift side. The drift was left open for a shift, which resulted in the CO reading at the 3200 level monitoring point increasing to 1300 ppm.

Because the airflow was allowed into the area that was already either on fire or unstable, the fire increased in intensity. The conclusion was reached that where old areas have been sealed for some time, a temporary brattice or bulkhead should be installed to minimize the airflow to the area while the permanent bulkhead is being built.

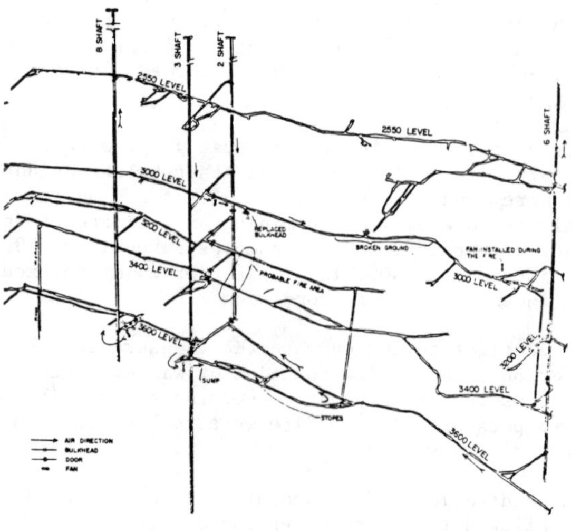

MAGMA MINE
LEVELS BETWEEN NO. 3 AND NO. 6 SHAFTS
AUGUST, 1980 FIRE
Figure 6

Another difficulty in controlling this fire was that the stope area on the 3600 level was not accessible for sealing because of the sump located in the drift off No. 3 shaft and the water draining through the level from the No. 6 shaft. Further, because the drift handled the main drainage from the working section of the mine, some airflow had to be maintained across the level to prevent spalling, which occurs in the mine when the cooling from ventilation is stopped. There was no good way to seal off the bottom of the fire and the best that could be done was to control the air by sealing off the top of the fire and sand filling wherever possible.

The need for balancing bulkhead pressures to control leakage was also indicated. Some work (Fig. 7) was done by the Magma ventilation engineer in determining fan combinations that would best balance the pressures across the bulkheads and reduce the leakage through the bulkheads on the 3200 and 3400 levels.

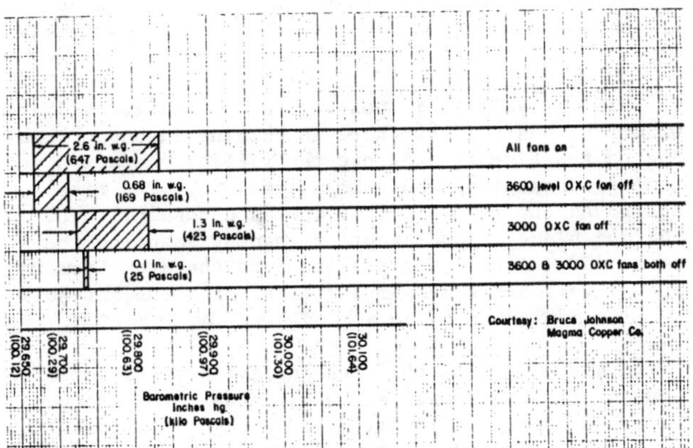

PRESSURE DIFFERENTIAL ACROSS THE 3400 LEVEL EAST BULKHEAD OFF NO. 3 SHAFT
Figure 7

A fan on the 3600 level pulled air from the No. 3 shaft through an equipment room where switchgear was located. The 3000 level fan (Fig. 7) was pushing air from No. 3 shaft across the level so that the secondary escapeway could be maintained on intake air. The 3600 level fan had the most effect on balancing the pressure across the 3200 and 3400 level bulkheads. Shutting down the 3600 fan caused too much heat to build in the equipment room. As a result, the fan was regulated by by-passing some of the air around the fan. This was not an altogether satisfactory solution, since some heat was generated by the fan and the recirculation.

A regulator was established to the west of the No. 6 shaft on the 3600 level, and was set at 13,000 cfm ($6.14m^3$/sec) and 6.5 in. wg (1617 Pa). This regulation broke down the pressure enough to achieve a better balance across the 3200, 3400, and 3000 level bulkheads. The 13,000 cfm ($6.14m^3$/sec) was thought to be sufficient airflow to prevent spalling.

Shutting down the 3000 level fan helped balance the bulkhead pressure, but the airflow direction changed with exhaust air flowing across the level to No. 3 shaft. It was decided that it would be better to keep the fan running so that the secondary escape route could be maintained in fresh air.

Another location causing concern was a badly fractured section of drift in a mined-out area on the 3000 level. The pressure across the bulkheads and cracks to the old workings was about 1.7 in. wg (423 Pa). Installation of two doors on either end of the caved area with a ventilation bag strung between them was tried, but doors could not be maintained tight enough to accomplish a pressure breakdown. A second fan was installed west of No. 6 shaft, and the two fans were balanced by by-passing air at the No. 3 shaft, 3000 level fan. Because of the distance between the two fans, balancing was fairly difficult but the system was effective.

The doors to the No. 8 shaft, 3600 level fan were also used to regulate the pressure on the 3200 and 3400 level bulkheads. The pressure across all bulkheads and connections to old workings was reduced to 0.2 in. wg (50 Pa) as a result of the regulation and fan changes. A graph from a pressure recorder located in a crosscut near the cracked portion of the 3000 level drift is shown in figure 8.

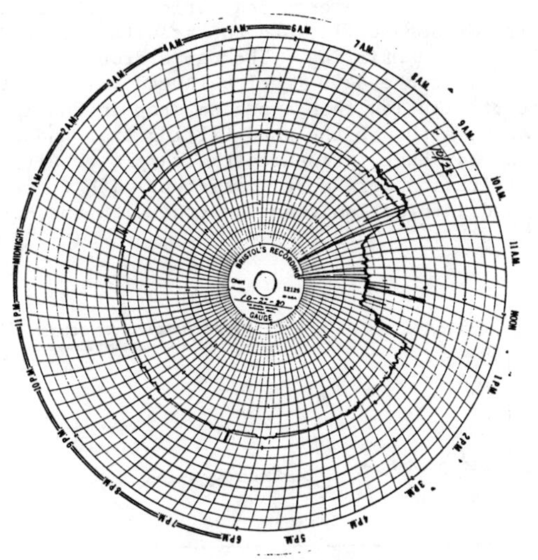

MAGMA MINE
PRESSURE CHART 3000 LEVEL BULKHEAD
OCTOBER 23, 1980
Figure 8

Chromatograph Sampling

The cricket samples for this fire were run on the San Manuel chromatograph. Monitoring and sampling were done at the 3200 level and 3400 level bulkheads, the 3600 level, and at No. 8 shaft. Because it was desirable to keep the slight leakage going into the fire area and not into the intake stream of No. 3 shaft, the samples reflected some dilution; however, as long as the direction and pressure remained constant, we feel that it gave us a fairly good picture of what was happening in the fire area. The samples from No. 8 shaft gave a total amount of CO coming off the fire area, indicating the total amount of wood that probably was burning.

The graphs plotted in figures 9 and 10 show the chromatograph results for the 3400 level bulkhead and No. 8 shaft respectively.

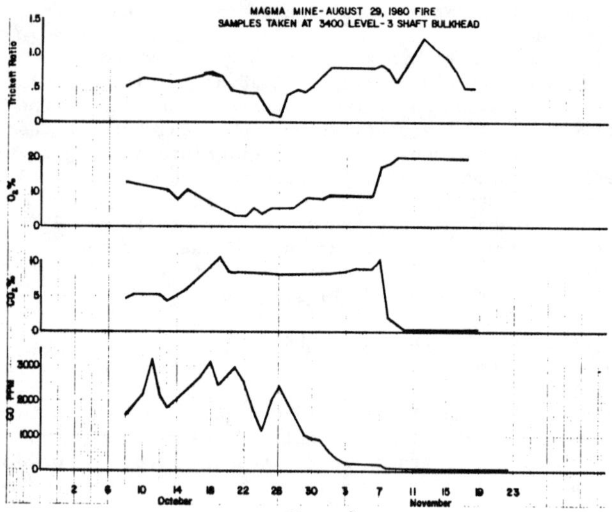

Figure 9

The curves give a good indication of the growth of the fire on October 10, and the results attained by sealing and sand filling. The rapid drop in CO_2 on November 8 at the 3400 level bulkhead was more the result of dilution than a sudden drop in CO_2 in the fire area, although the sand filling that had been in progress may have influenced the sample results. The Trickett ratios were more consistent with those in published information than was true with the May 1980 fire.

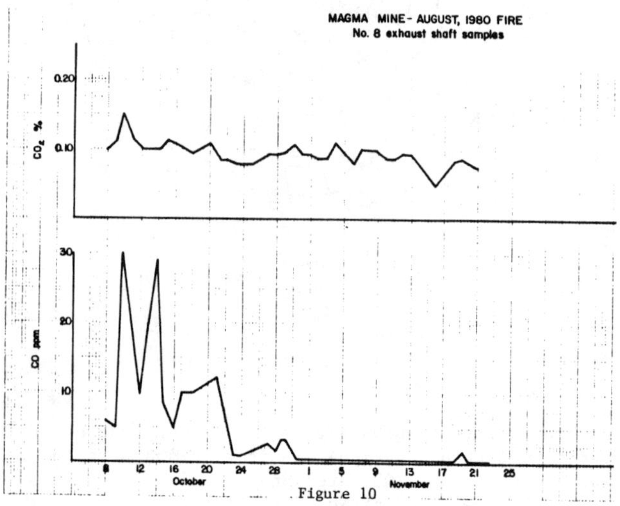

Figure 10

December 1980 Fire

The third fire was found on December 17, 1980, by mine personnel who smelled smoke on the 3200 level. The smoke was emanating from raises in the 6 crosscut area, and was from the old stoping on the 3160 sub-level (Fig. 3). There had reportedly been a sulfur dioxide smell emanating from this area for a number of years which was not uncommon at the Magma Mine.

Once the fire was discovered, steps were immediately taken to isolate the area by bulkheads and sand filling. During the building of two bulkheads to contain the sand fill in the 78 drift, it was decided that sprinklers would be installed on the fire side of the south bulkhead to prevent burning through of the bulkhead. The men were working under apparatus, and a considerable amount of ventilation air was allowed through the drift to keep the work area cool and free of smoke.

When the recovery team was within one hour of completing the sand walls, the fire burned down a raise into the timbered drift. The spread of the fire was very rapid in the timber and the men working on the downstream side of the north bulkhead which caught fire had some difficulty getting out of the area because of dense smoke. After the fire had been extinguished by water quenching, a second bulkhead was constructed farther north.

There was a consensus of opinion after this episode that probably it would have been better to have controlled the air to a minimum amount while building bulkheads. It was felt that the sprinkler inside of the south bulkhead was instrumental in this bulkhead's being saved.

Ironically, in this particular case, a great deal of effort had been expended in putting in copper tubing to the fire area before the filling operation so that good sampling could be maintained. The fire destroyed the sampling line and as a result, there was no good place to sample other than by sampling the total air. A hole was drilled into the area from the 3000 level, and some samples taken from the borehole, but did not prove to be a good sampling location.

An air survey of the area indicated a considerable amount of air leakage to the old stope area around the fan and across the level. The leakage from this location was reduced by two-thirds by putting boards across posts at the fan intake reducing the pressure between 3200 and 3000 levels.

Another ventilation control measure that was effective during this fire was the pressurizing of an airlock to the upcasting 81 winze, which was a secondary escapeway. The normal path for the air was from the 3200 level fire area to the No. 6 shaft, with some leakage through doors to the 81 winze. To prevent leakage of fire gases into the 81 winze, a fan was located near the door on the 81 winze side pressurizing the area between the two airlocks. With this arrangement, air leakage was maintained from the 81 winze into the drift rather than in the normal pattern. The problem with this type of installation is in estimating the leakage through the doors. In this case the fan was capable of delivering about 8,000 cfm ($3.8 m^3/sec$).

Comments

Fan pressure and air volume were accomplished simply and quickly by restricting the fan intake. The start/stop button location is an important consideration, since provision must be made to control the fan.

The installation of a sprinkler at bulkheads being built in a timbered area is well worth the time and effort to prevent a rapid burn-through if the fire suddenly expands.

Particularly where people are working, the amount of air crossing a fire area should be kept to a minimum. With air velocities over 50 ft/min (0.25m/sec), a fire can spread very rapidly.

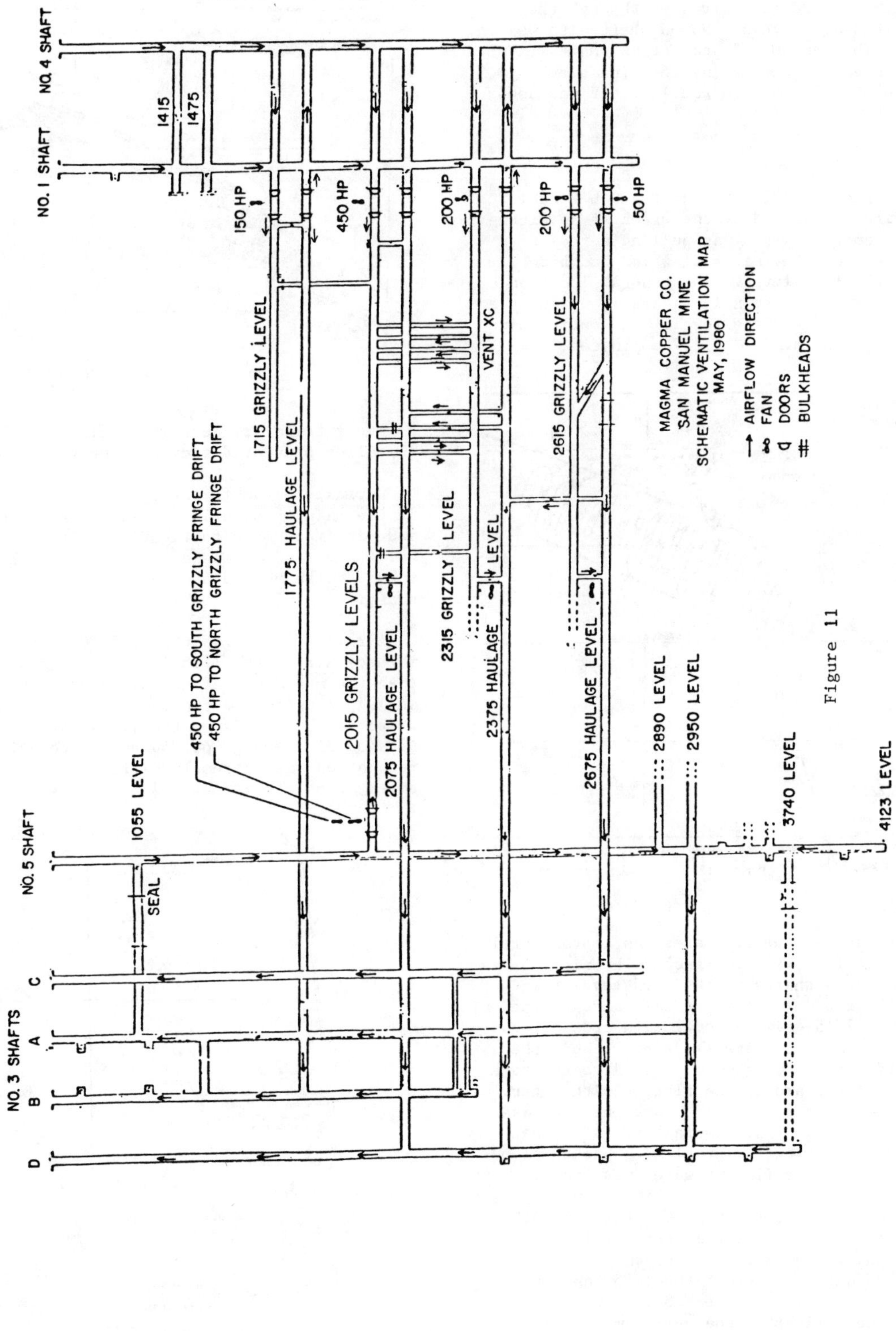
NO. 1 SHAFT
NO. 4 SHAFT
1415
1475
150 HP
450 HP
200 HP
200 HP
50 HP
MAGMA COPPER CO.
SAN MANUEL MINE
SCHEMATIC VENTILATION MAP
MAY, 1980
AIRFLOW DIRECTION
FAN
DOORS
BULKHEADS
450 HP TO SOUTH GRIZZLY FRINGE DRIFT
450 HP TO NORTH GRIZZLY FRINGE DRIFT
1715 GRIZZLY LEVEL
1775 HAULAGE LEVEL
2015 GRIZZLY LEVELS
2075 HAULAGE LEVEL
2315 GRIZZLY LEVEL
VENT XC
2375 HAULAGE LEVEL
2615 GRIZZLY LEVEL
2675 HAULAGE LEVEL
2890 LEVEL
2950 LEVEL
3740 LEVEL
4123 LEVEL
1055 LEVEL
NO. 5 SHAFT
SEAL
NO. 3 SHAFTS
A
B
C
D

Figure 11

SAN MANUEL MINE FIRE

Ventilation

The general pattern of the mine ventilation for the Magma Copper Company's San Manuel Mine is shown in Figure 11. Intake air was routed down No. 1 and No. 4 shafts and across the levels, with the fans located on the grizzly levels directing air down through the raises to the haulage levels. No. 5 shaft also was an intake shaft. The four No. 3 production shafts (A, B, C and D) were used for exhausting the mine air. A total of 1,170,000 cfm (552.2m^3/sec) of air was used to ventilate the mine.

May 1980 Fire

A fire occurred May 23, 1980, at the San Manuel Mine. this fire was caused by the wreck of an eight-ton battery locomotive which was pulling a reel of electrical wire and other material which had been salvaged from the 1775 level. The general location of the locomotive wreck is shown in figure 12 which is a plan view of the 1775 haulage level. The entire mine was evacuated, which involved 1,137 employees.

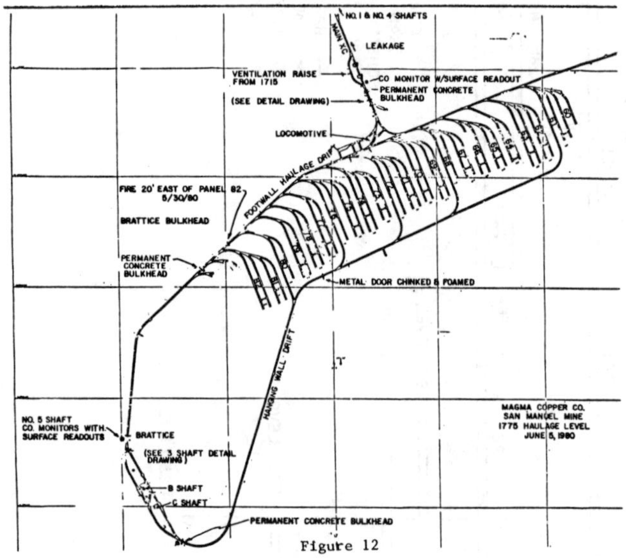

Figure 12

Air velocity across the fire area was approximately 600 to 700 ft/min (3.0- to 3.6m/sec). The airflow was from No. 1 and No. 4 shafts on the 1715 level (Fig. 13), through the fan down an octagonal raise to the 1775 level. The 1715 grizzly level main crosscut had been bulkheaded near the footwall drifts. When the mine rescue crews first went down, they direct-attacked the burning timber around the flatcar with water. They had telephoned to the surface that the fire was nearly out when the water system failed. After many hours of trying to locate the problem, a broken clapper which restricted the flow of water was found.

The rescue team continued to fight the fire with the water available and requested that the airflow be reduced. The opinion was made that the only way to reduce the airflow was to shut off the 1715 fan, which would cause recirculation into the No. 1 and No. 4 intake shafts, contaminating the entire mine. This opinion was substantiated when a fan failure reversed the air and smoke backed toward the intake shafts;

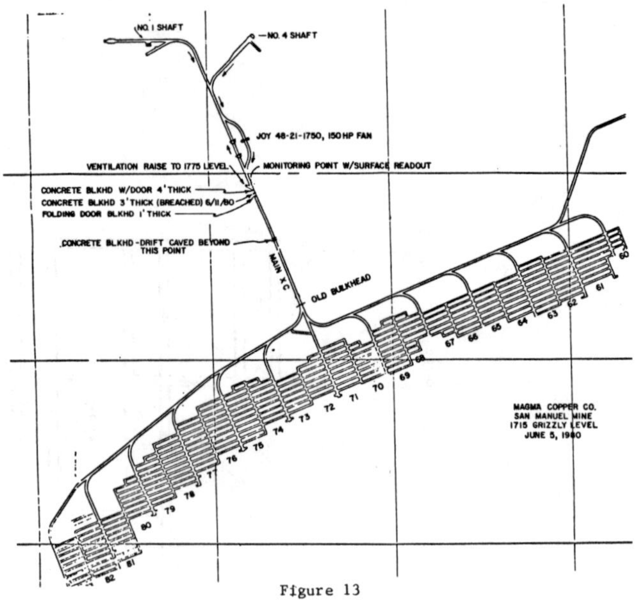

Figure 13

however, temporary doors and bulkheads which had been installed prevented a major problem from developing.

Sprinklers and temporary bulkheads were established in the main crosscut. A fall of ground over the motor made it impossible to fight the fire directly from the No. 4 side, so it was decided to have the team retreat to the bulkhead and approach the fire from the No. 3 shaft.

Air was reduced 45,000 to 22,000 cfm (21.2 to 10.4 m^3/sec) across the 1715 level main crosscut by regulating the man doors at the airlocks. This regulation on 1715 reduced the pressure and leakage across the temporary bulkheads on the 1715 and 1775 levels. Because of the high pressures and velocities, man door regulation was extremely difficult, and easier and safer means of air control should be used.

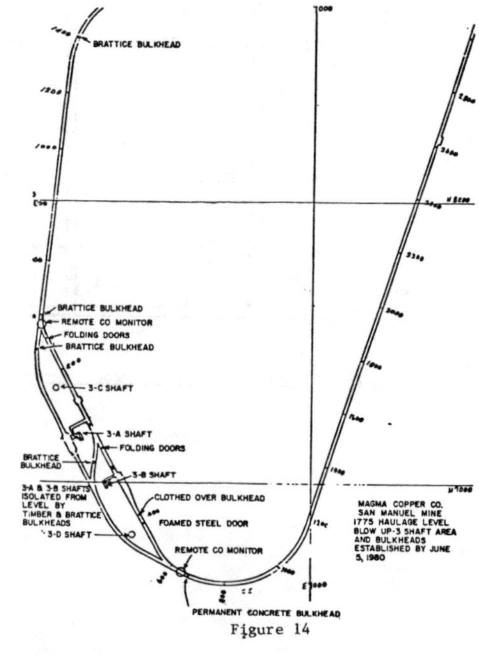

Figure 14

To facilitate exploration of the 1775 level from No. 3 shaft, bulkheads were constructed at the 3A shaft station on the 1775 level. A great deal of time was required to install the bulkheads and temporary doors because of the number of openings connecting the 3A shaft to the level (Fig. 14).

The steel ventilation tubing used for lower level development was disconnected at the 1775 level 3A shaft and reconnected to one of the bulkheads. The surface end of the tubing was connected to three two-stage fans in series. To control airflow, two of the fans were started but tended to go in and out of a stall condition, so the third fan was also used.

Fire was encountered in the north ladder drift just east of panel 82, which ruled out using panel 80 for circulating air while building a bulkhead between panels 78 and 80.

Many problems were experienced in balancing the system during this work on the temporary seal on the 1775 north ladder drift, particularly in the morning at about 3 or 4 a.m. when the air tended to reverse. This was a result of the change in natural ventilation pressure as the air cooled at night.

A turbine-driven fan was installed June 2 in the bulkhead on the No. 4 shaft side of 1775. The air across the fire area could be entirely controlled by regulating the compressed air to the turbine. Only the airflow needed to clear the area where people were working on the bulkhead was used, and the air direction was assured. This ventilation pattern was continued during the pouring of concrete plugs.

Following the completion of the concrete plugs on the 1775 level, which isolated the fire area on that level, the concrete plug on the No. 4 shaft side of the 1715 level, the installation of brattices in nine panels on the 2015 level, and six brattices on the north ladder drift on the 2074 level, the mine was inspected for reopening.

During the reclaiming of the mine, portable Ecolizer 6000 [1/], or Hipster CO monitors, were used extensively and worked out fairly well. Some problems were found in maintaining and learning to use the Hipster, but after operator-training and setting up a maintenance program, the units proved fairly dependable. Hipsters were periodically checked by the operator, using a CO stain tube.

The fire had been effectively sealed by June 6. Before reopening, MSHA requested a 48-hour period after the ventilation had been restored to see if anything developed. The closure was terminated at 4:00 a.m., with the requirement that fire monitoring be continued.

Some CO showed up on the 2075 level; however, the Ecolizer series 3000 had been installed on this level with the monitoring being done on the surface. We felt fairly secure in being able to follow any indications, and the mine continued in production with a close watch of the surface monitor and periodic fire checks on all levels.

The CO monitors were extremely helpful, and prevented an evacuation of the mine when the CO developed on the 2075 level. Mine management felt that the expense of the CO monitor installation was partially offset by avoiding the evacuation.

1/ Reference to specific equipment does not imply endorsement by the Mine Safety and Health Administration.

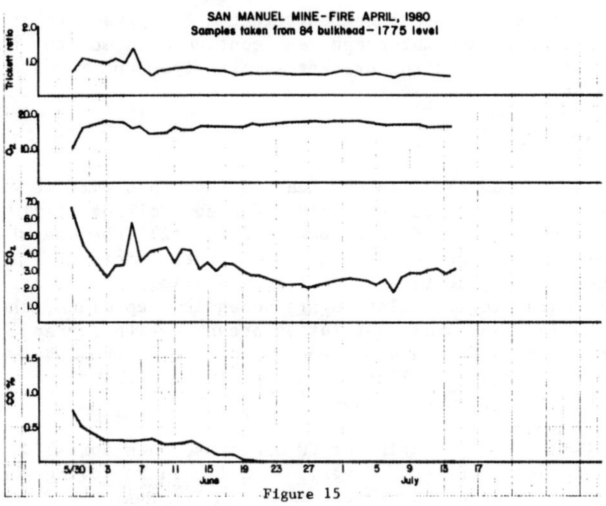

Figure 15

Chromatograph Results

The results of chromatograph analyses of the 1775 level are shown in figure 15. The results of sampling done at the No. 3 A and B shafts are shown in figure 16. Comparatively little CO was detected in the No. 3 C and D shafts.

The CO had dropped to 100 ppm by June 19 following which there was a gradual drop to a trace. The CO_2 dropped until July 9, when a slight increase was noted and a slight decrease in O_2 was seen at the same time. This was watched closely to ensure that heating was not occurring; however, the absence of CO indicated there was no rekindling of fire. The CO_2 was still high in the sealed area when checked by the author a year later.

The Trickett ratio followed the expected pattern while the fire was actively burning, and continued to drop gradually after the CO had disappeared. There was no dramatic drop in Trickett ratio to indicate that there was no fire.

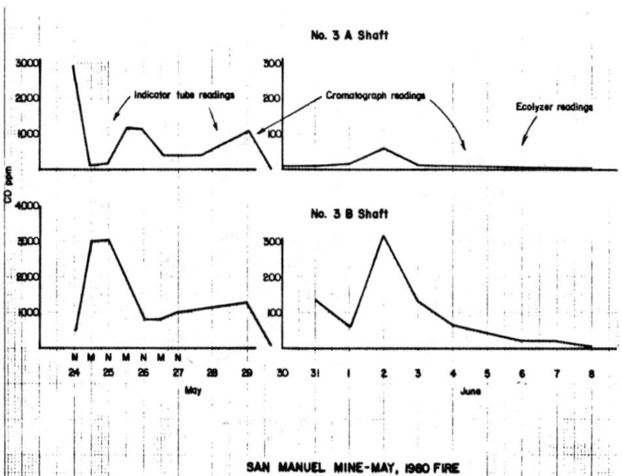

Figure 16

The Magma Copper Company elected to upgrade their laboratory chromatograph and continued to monitor the sealed area and run cricket samples from the Magma Mine.

Comments

Two important lessons learned from this fire were: (1) means should be provided for controlling airflow in areas with a high timber loading, (2) the use of air turbine-driven fans was an effective method of controlling ventilation during a sealing operation. In balancing ventilation pressures in deep mines, the change in natural ventilation pressure with day and night temperature differences can reverse the leakage of air through bulkheads and should be allowed for when the balance is made.

The use of continuous CO monitors, both portable and permanent, proved very effective in relieving the need for using large numbers of stain tubes.

LAKE SHORE MINE FIRE

Ventilation

Noranda Exploration Incorporated's Lake Shore Mine is divided into two sections, the oxide section, which was being mined, and the sulfide section, which was not being mined. The fire occurred in the sulfide section. The escape routes are unusual in that the primary escapeway is in exhaust air and the secondary escapeway is in intake air.

The general ventilation pattern is shown in Figure 17. The air intakes were the Oxide shaft and the No. 2 vent shaft, which had two parallel 96-inch (244cm) Joy axivane surface fans. Air exhaust through the north and south declines, No. 5 and 6 vent shafts and No. 1 vent shaft, which also had duplicate parallel surface fans.

March 1981 Fire

A fire occurred on the 200 level maintenance bay March 19, 1981 (Fig. 18). The bay had been abandoned and lockers were being cut loose for moving (Fig. 19). Upon discovery, the afternoon shift of 72 employees

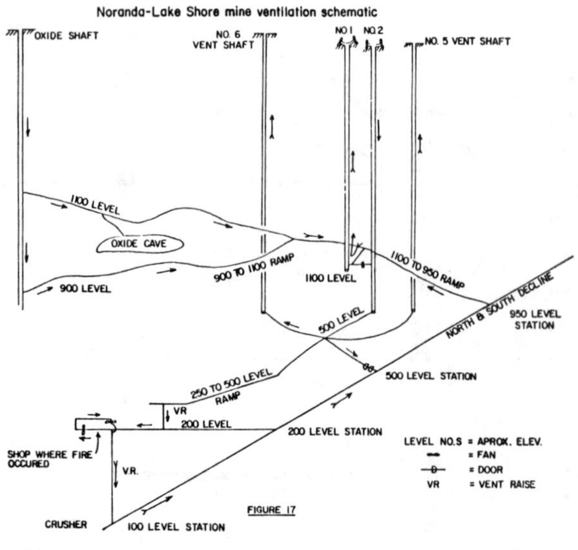

FIGURE 17

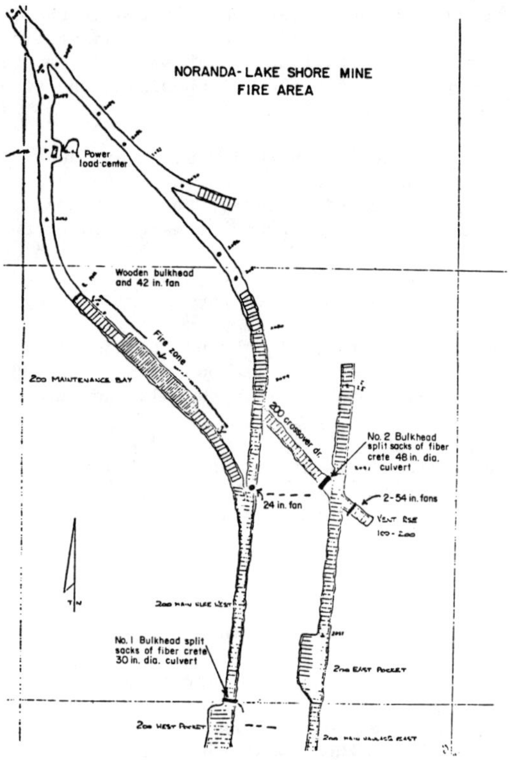

Figure 18

was evacuated up the decline with some self-rescuers used. As evacuation was in the return system, there was some smoke in the escape route, and the highest CO was measured at 200 ppm.

The power to the mine was cut off, which killed the booster fans. The recovery team entered the mine and was able to install brattice bulkheads within 15-feet (4.6m) of the bay entrances by pushing back the smoke with water spray. The immediate concern was that the fire might spread to the 200/100 vent raise (Fig. 17 and 18) and extend to the 100 level, where the fire would have caused extensive damage. The fans at the top of the vent raise had stopped when the mine power was shut off, but the flow remained downcast east to the 100 level, resulting in smoke from the fire going up the two declines (Fig. 17). While the team was building the last brattice bulkhead, an explosion occurred in the bay. Since oxygen, acetylene, and oil drums were stored in the bay, it was decided that rather than risk any danger to the recovery crew, the area should be sealed. The crew left fog nozzles running when they blocked out the bay entrance.

Particularly interesting in this sealing was the use of 50-pound sacks of fibercrete, which were slit and sprayed with water. In addition, the No. 2 bulkhead had a 48-inch (122cm) steel vent tubing installed prior to the placement of the fibercrete, and a 30-inch (76cm)-diameter steel vent tubing was placed in the No. 1 bulkhead. These openings were particularly useful when reentering the area.

With the fire area sealed, and monitoring set up to give any warning of increased problems in the fire area, people were allowed to go back to work in the Oxide area. Cricket samples were taken at the bulkheads using a battery-powered dust sampler and newly designed sampling T discussed under "Sampling from Bulkheads."

A meeting was held with mine management and MSHA personnel March 30 during which the gas analyses and Trickett ratios were reviewed. The evaluation of the gas analyses indicated there was a good chance that the fire might be rekindled if air was allowed into the area. The decision was made to wait until April 7 before reentry. A duct (Fig. 20) would be installed to pipe the fire gases to a return rather than evacuate the mine while the area was being cleared and examined.

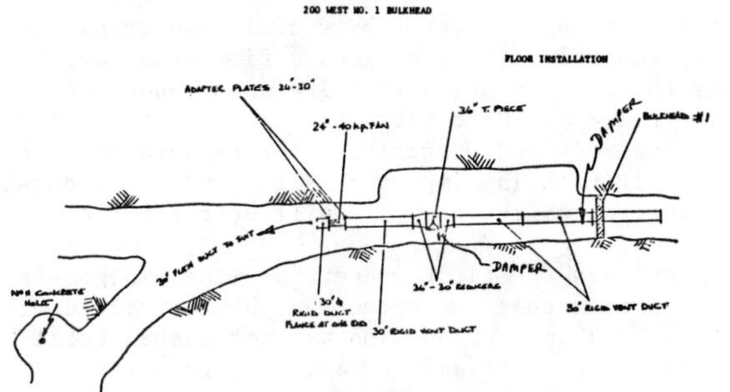

Figure 20
SCHEMATIC ARRANGEMENT OF 200 LEVEL VENT DUCTS, FAN & REGULATOR

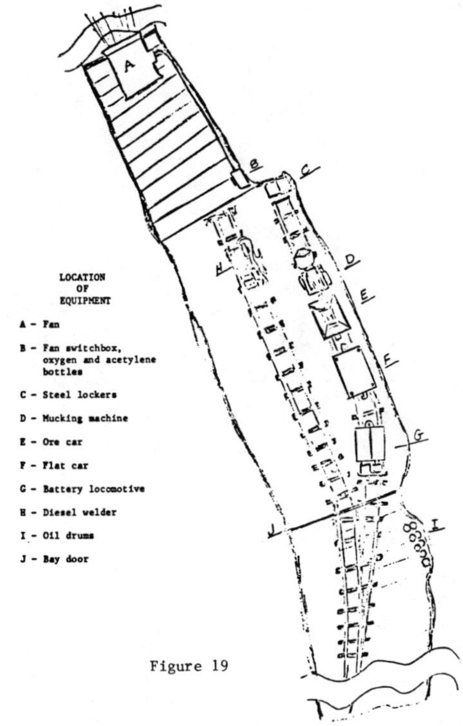

Figure 19

The piping was completed April 8 and recovery crews assembled. The duct was connected to No. 1 bulkhead and the 48-inch (122cm) vent tube in No. 2 bulkhead broached. The air through the bay was controlled by a bypass at the fan, letting 800 cfm through the area, which was later regulated to 1200 cfm (0.57m/sec). The recovery crew entered the fire area through the 48-inch (122cm) vent tube and found no fire.

Chromatograph Results

When sealing the fire area had been decided on, a request was made by the Phoenix Subdistrict for the chromatograph. The chromatograph was dispatched by van to the mine. Results obtained from the cricket samples were, by far, the best achieved at any of the fires. The cricket sample analyses had excellent duplications.

Sample analyses are shown in Figure 21. The author feels that evaluating the trends of all four graphs was important in the decision about the timing for unsealing the fire area. Although the CO concentration had not dropped dramatically, it was decreasing. The O_2 readings indicated that no oxygen was being used up in burning, and the CO_2 was decreasing. This combination of factors indicates no burning.

Comments

The three things about this fire that are particularly interesting are:

1. The use of sacks of fibercrete as a bulkhead and 48-inch vent tubing for reentry worked out exceptionally well. The use of vent tubing should be considered in more cases where there is a need for reentry.

2. Venting the fire area directly to a return gave excellent results.

3. The company is working toward changing the primary excapeway from exhaust air to intake air. With this particular fire location, it would have been better if the miners had gone to the second escapeway and gone out in fresh air. A more important point is that prompt action was taken in evacuating, the miners and the mine personnel showed good reactions as a result of excellent training.

CONCLUSION

Ventilation plays an important part in the safe escape of persons from underground in a fire. During the firefighting, sealing, and mine recovery, ventilation control is equally important. The spread of a

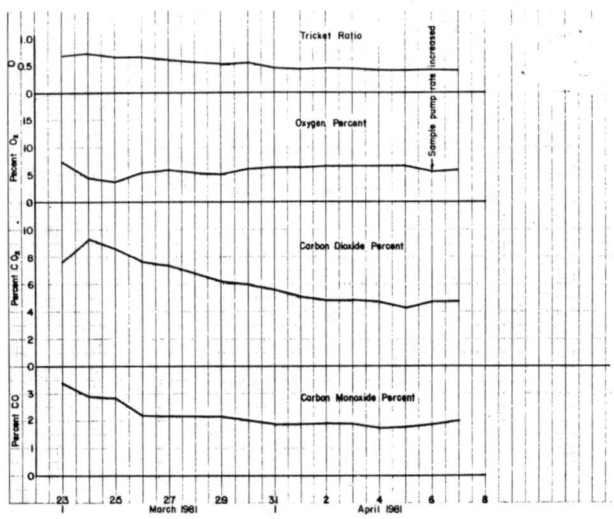

Figure 21 Average chromatograph results Lake Shore Mine fire Mar. 23 to Apr.8, 1981

fire can often be averted by ventilation regulation, and by control of leakage into a fire area, which means the time to bring the fire under under control can be reduced. Mine fires are always potentially very dangerous and expensive. Any reduced exposure during firefighting improves safety and any reduced time in controlling the fire will decrease the cost.

Safety was the first concern of mine management in all the fires that the author has been involved with. Getting back into production was not pushed until the mine was felt to be safe by all involved.

The ventilation engineer should take an important part in the decision-making during all phases of a fire as he is in the best position to recommend ventilation control measures. The chromatograph was an excellent tool for assisting in making decisions. Having been in a position to watch fires with and without the use of chromatographs, the author found it much easier to recommend actions with some scientific basis than by relying solely on experience. A portable chromatograph for instant analyses on the job site makes for safer and smoother running operation than forwarding the sample to Denver, Pittsburgh, or some other place and having a two to three day wait before information is available.

The Trickett ratio is not a magic number and should only be used in conjunction with a comparison of the fire gases, CO, CO_2, and O_2. All the fire gas results need to be considered, not just the CO or Trickett ratio number. Even though the fire gas concentrations detected are extremely low, exhaust air sampling is very important as an indication of the burning rate and as corroboration of samples from the sealed area.

Continuous CO monitors have improved over the past few years and they can aid in safety when correctly used.

5
Methane and Explosive Gas Control

Chapter 23

PREDICTING METHANE EMISSION IN REAL TIME FOR LONGWALL FACES

Richard Dunmore

Head of Firedamp Prediction Group
Mining Research and Development Establishment
National Coal Board
United Kingdom

INTRODUCTION

Tests of the MRDE firedamp prediction method have been carried out in a number of longwall districts with different conditions. The method itself and its development have been described previously.[1,2,3] At many test sites the prediction has been good, i.e. the calculated gas emission levels have agreed with measured values, averaged over a period of many weeks, within a few percent. In other cases, where there have been larger discrepancies between calculation and measurement, investigations have shown that these may be attributed to particular conditions of mining or geology. Recognition of these conditions in advance of mining is clearly desirable to improve the accuracy of the initial prediction.

It has been shown that variations in gas emission rates depend largely on coal production in terms of the weekly advance of the coal face. Although there appears to be a random element present, variations in gas emission levels calculated from changing face advances agree well with measurement. The accuracy of prediction therefore depends on how well face advance can be forecast and needs to be updated as the working proceeds in the light of the face performance actually achieved.

Continuous monitoring of the mine environment provides the information necessary to compare actual gas emission with predicted levels, and computerised systems make it possible to carry this out in real time. Predictions can thus be adjusted to take account of emission levels measured in previous weeks. These adjustments, whether they are made after the first few weeks or subsequently during the district life, would also be used to alert management to particular abnormalities or significant changes in gas emission.

PREDICTION TEST RESULTS

Cases of Good Overall Accuracy

In tests of the prediction method the calculated methane flow rates were compared with measurements made in the district. This was done for a period of many weeks, in some cases for a period of two years. To measure all the district methane, continuous recordings were made at the outbye end of the district of airflow in the roadway, total gas flow in the drainage range and methane concentration in both roadway and range.

The calculation of week-by-week variations in the predicted gas emission level had been made possible by using a computer program[4] developed at MRDE. An example of the resulting comparison is shown in Figure 1. It may be seen that although there are random discrepancies throughout the period, the overall trends of change in total gas flow are reproduced well. In this example the average error over the period of 93 weeks was +2.6 $\pm$ 18.9%, the latter figure representing the deviation from week to week. This random deviation is thought to be due to the irregular progress, as the face advances, of the de-stressed zone in which gas emission takes place.

Cases of Larger Error

Investigations into the mining and geological conditions at sites where the prediction calculation was in error by a significant amount (more than 20%) have provided explanations which fall into a number of identifiable categories.

All methods of prediction contain the related concepts of the zone of gas emission and the degree of emission function.[1] The emission zone consists of the volume of strata de-stressed by the working and containing the gas bearing coal seams. The degree of emission function gives the proportion of the initial gas content of the source coal seam which is considered to enter the working. The value of the emission function decreases with distance from the worked seam; the edge of the emission zone may be taken to be where the degree of emission function falls to a negligible level.

The cases of larger error may be considered therefore in terms of enlargement or reduction in the emission zone or, alternatively, an increase or decrease in the degree of emission function.

<u>Under-estimation</u> The case of gross under-estimation of gas emission has been encountered in three longwall districts in the Wigan Four Foot at Sutton Manor Colliery, NCB Western Area. These faces are within a block of coal with major faulting on three sides. Detailed examination of the firedamp drainage system in one of these districts revealed a large flow of gas in boreholes distant from the face (i.e. outbye), in a region where gas emission would normally have fallen to a low level. Furthermore analysis of the gas composition showed it to be from a different source compared with the other district gas. The prediction calculation accounted for the 'normal' district gas accurately and it was concluded that the zone of emission was enlarged in some way, the effect perhaps being due to the fault zones allowing extraneous gas (up to 70% of the district total) to enter the workings (Figure 2).

The enlargement of the zone of gas emission may occur in other circumstances, e.g. when the working is adjacent to a previously worked area which has produced a de-stressed zone, and this is incompletely degassed. It is clearly important to recognise such cases before the working commences.

<u>Over-estimation</u> Two distinct cases have been recognised where the prediction method has over-estimated the gas emission rates actually encountered. One of these is that of a first longwall district in a virgin area of coal. In this situation the de-stressed zone may not fully develop immediately, so that the emission level remains low. It would be wrong to conclude from this that the seam is 'not gassy', as a second parallel longwall district developed subsequently is likely to experience

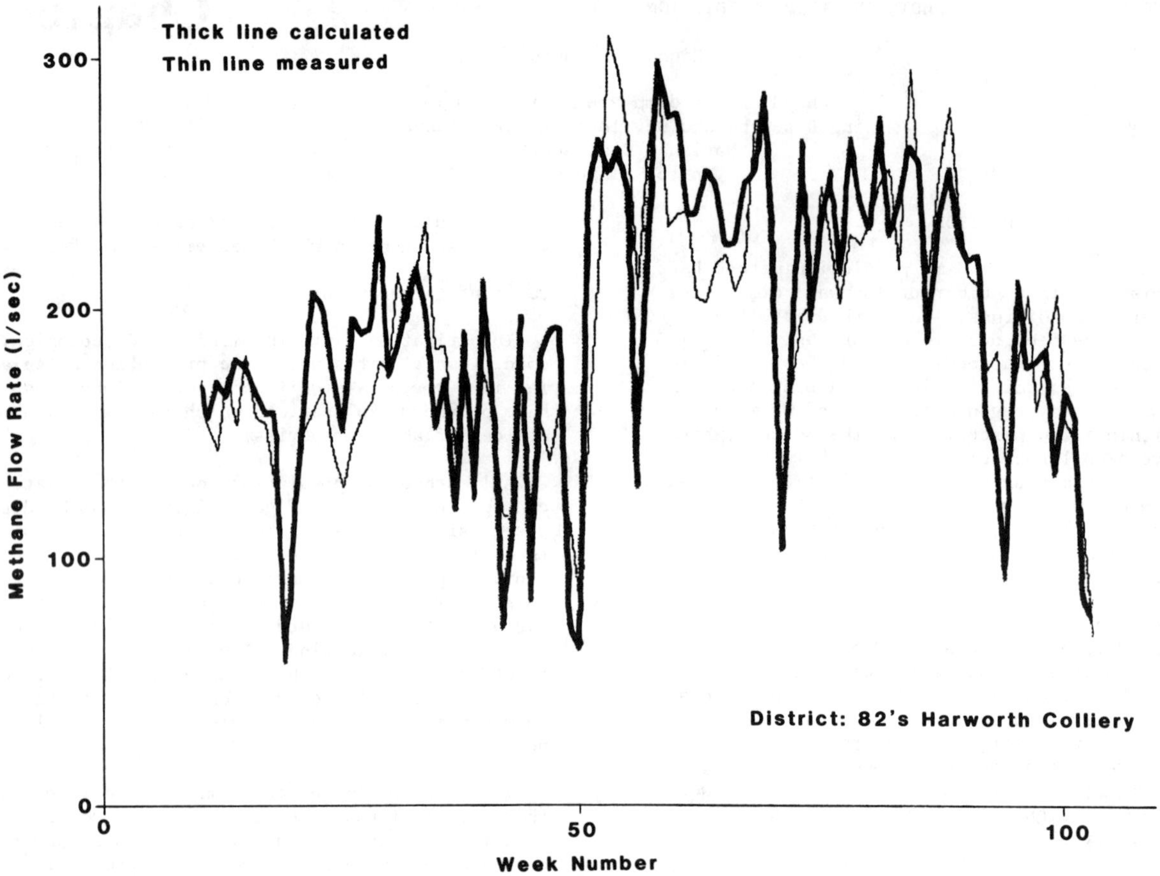

FIGURE 1 Comparison of Prediction with Measurement Over 93 Weeks

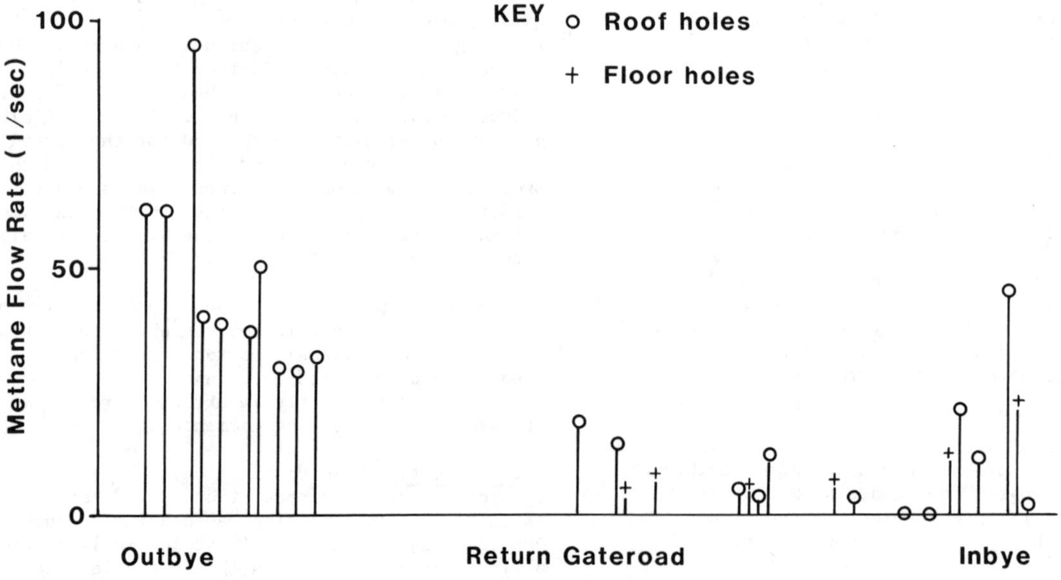

FIGURE 2 Methane Flow in Firedamp Drainage Boreholes in W20's District, Sutton Manor Colliery

larger gas emission rates than predicted, due to the emission zones linking up. In the current MRDE programme of tests the first faces in the Silkstone seam at Kellingley Colliery, NCB North Yorkshire Area, appear to be following this pattern.

The second case is one where the emission zone may be permanently reduced in size because of the presence of some particularly strong bands of rock in the strata which limit the process of de-stressing associated with the working. An example of this may be in seam 5B at Waterschei Colliery in Belgium, where the first 80 m of floor strata contain a total thickness of 27 m of sandstone. Calculations of gas emission fit measured values reasonably well provided that floor emission is assumed to be suppressed (Figure 3).

In some cases of this type, a strong bed of rock may be of critical or varying thickness, so that fracture of the bed leads to a sudden emission of gas. The seriousness of this type of incident is reflected by the detailed nature of the studies which have been made, e.g. by Morris.[5] Knowledge that district emission is less than predicted, would combine with information on the geology of the strata section to alert management of such a potentially dangerous situation.

Variation Patterns

Even when the absolute level of prediction is in error compared with measurement, the pattern of variation with changing weekly face advance is still reproduced well. This may be illustrated by reference to a district in the seam Robert at Gneisenau Colliery in West Germany, which provided another case of under-estimation. (Because of absence of information it was not possible to establish the conditions responsible for the extra gas.) By scaling up the calculated values the changes in level of the prediction can be seen to fit the measured changes very much better (Figure 4). This is important because, in real time calculations, it gives confidence to the current prediction if calculations for past weeks can be made to fit actual observations well.

PREDICTION PROPER

The Initial Prediction

The tests described above have shown that the prediction is often correct within close limits, but that under a number of conditions, some of which have been identified, over- or under-estimation does occur. Clearly it is necessary to make a prediction based on the best information available and to pay particular attention to the possible existence of those conditions which would produce the anomalous emissions described above, and make appropriate allowances for them.

Subsequent Updating

Although the MRDE prediction method is not suitable for calculating the detailed build-up of gas emission in the early weeks of the district life, any gross deviations from predicted levels would begin to be apparent by about 10 weeks. After that time with knowledge of actual emission levels it is possible to compare calculation with measurement and adjust the prediction if this is necessary. An example of this is shown in Figure 5. In this example, calculation is matched to measured emission up to 26 weeks by scaling the calculation down by about 11% compared with the original calculation. The prediction beyond that time is based on projected advance of the face, allowance being made for holiday weeks, which produce the sharp drops in emission shown; a planned increase in production at 51 weeks raises the general level at the end of the predicted emission curve. The actual emission experienced compared with the predicted levels is shown in Figure 6. The main cause of the differences between calculated and measured levels was that the weekly advance was often higher or lower than expected. Nevertheless it can be seen that the general trends can be anticipated reasonably well. This implies that minor variations in face advance rates, which cannot be foreseen in detail, do not appear to have any great consequence in terms of gas emission.

REAL TIME PREDICTION

The updating process thus consists of fitting calculation to measurement for past weeks and predicting subsequent emission on the basis of weekly face advances estimated for future weeks. Given the availability of up-to-date information, updating may be repeated as often as required in the light of changes in basic data, particularly the weekly face advance achieved and that anticipated. Any major adjustments in calculated emission levels either early in the district life or at any later stage would signal the possibility of the type of undesirable emission situations described above. The main objective would be to alert management to the potential situation before an alarm condition could arise.

Computerised comprehensive monitoring systems which include environmental packages are being installed widely in UK coal mines. These provide data on methane emission levels as well as face advance. The 'MINOS' system[6] has a two level computing facility, the primary to deal with immediate on-line analysis for day to day control purposes. The secondary system is intended as a management information computer. Its purpose is to receive data from the primary system, analyse these and produce summary reports required by management. With suitable software, this would be the appropriate unit to carry out weekly firedamp prediction calculations, compare them with measurement and forecast the methane conditions ahead for estimated weekly face advances. Establishing the details of the output, i.e. the type of information to be displayed, will require liaison with the potential user, i.e. management and specialist ventilation personnel.

CONCLUSION

Development and testing of the MRDE firedamp prediction method have identified situations where the calculations are accurate and others where they are in error. The latter have been subject to further investigations revealing particular mining or geological conditions to be responsible. Identification of these conditions makes possible an improvement in the initial prediction for such cases.

As the face advances, measured data provide the basis for correcting the prediction by matching calculation to measurement for past weeks. The prediction then depends on estimating future face advance rates. Computerised environmental monitoring systems equipped with suitable software

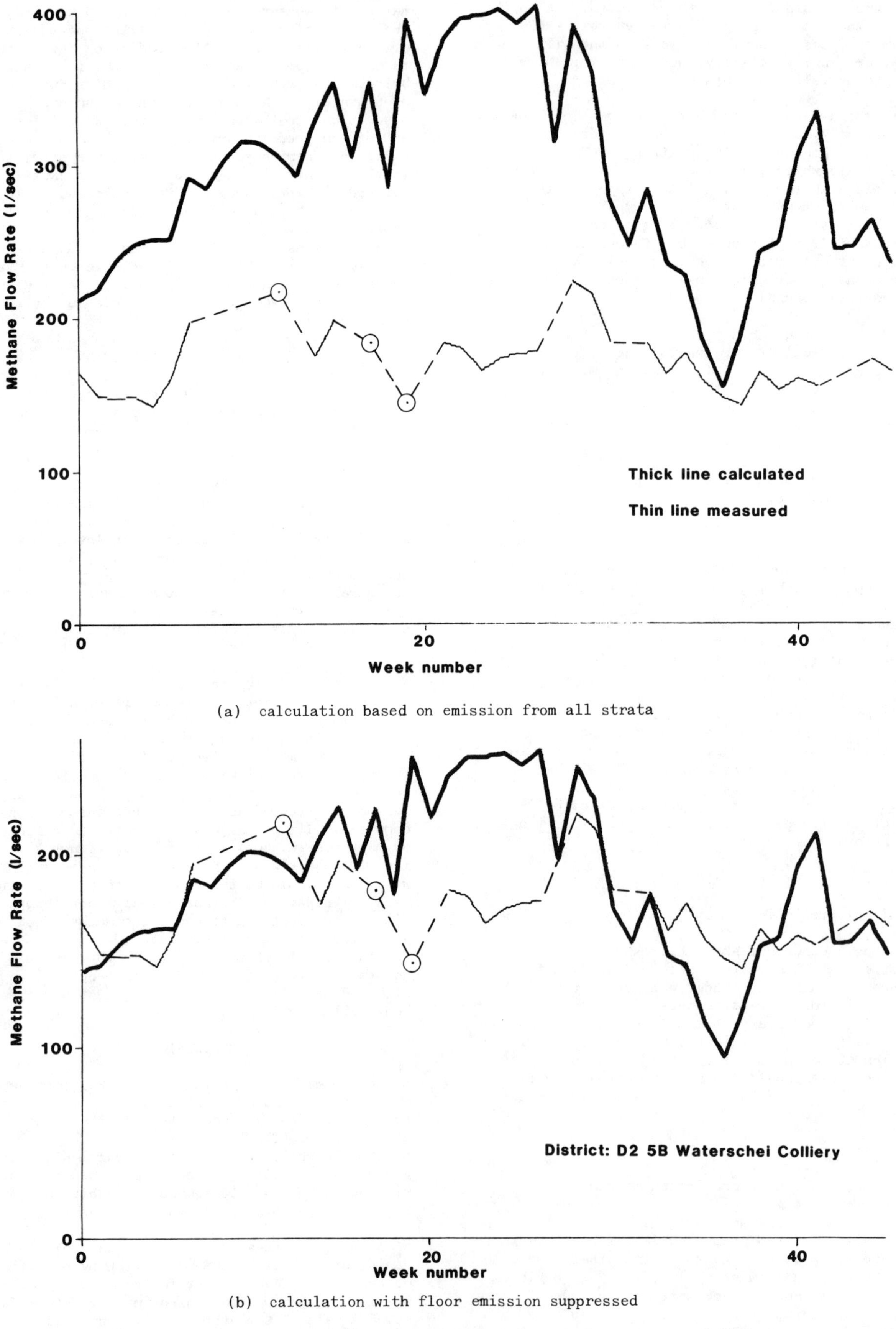

(a) calculation based on emission from all strata

(b) calculation with floor emission suppressed

FIGURE 3 Prediction Comparisons for Site with Sandstone in Floor Strata

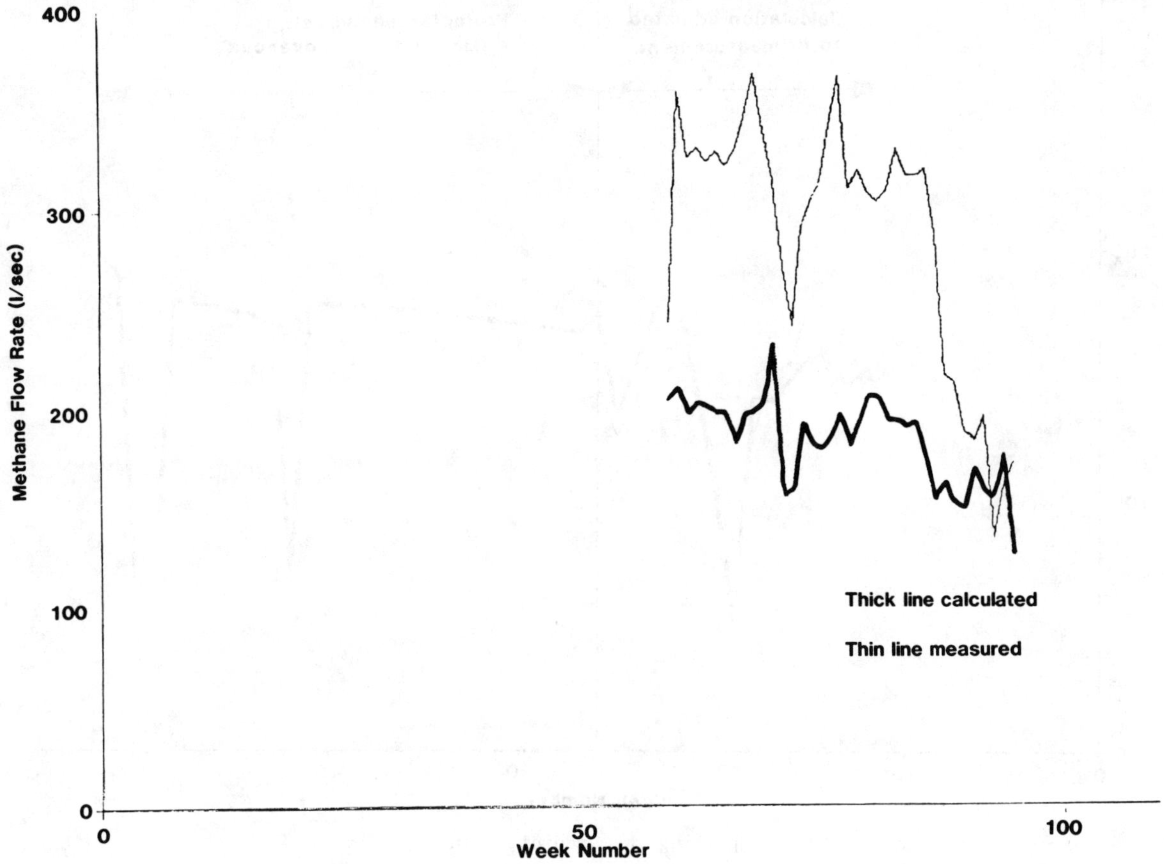

(a) original under-estimation

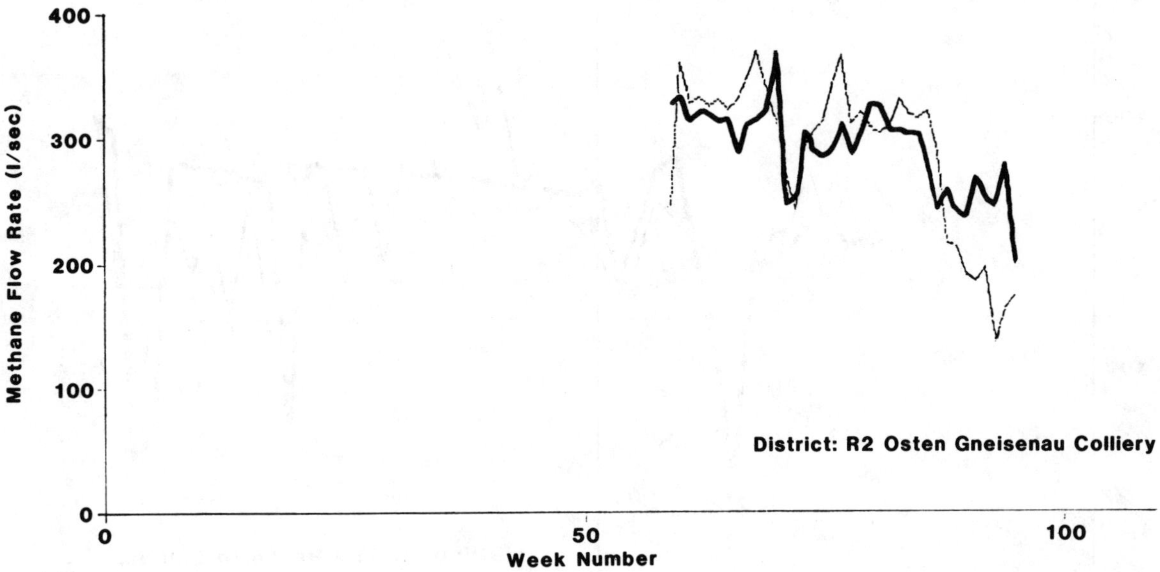

(b) calculation scaled to give best fit

FIGURE 4 Emission Calculation Adjusted to Fit Measurement

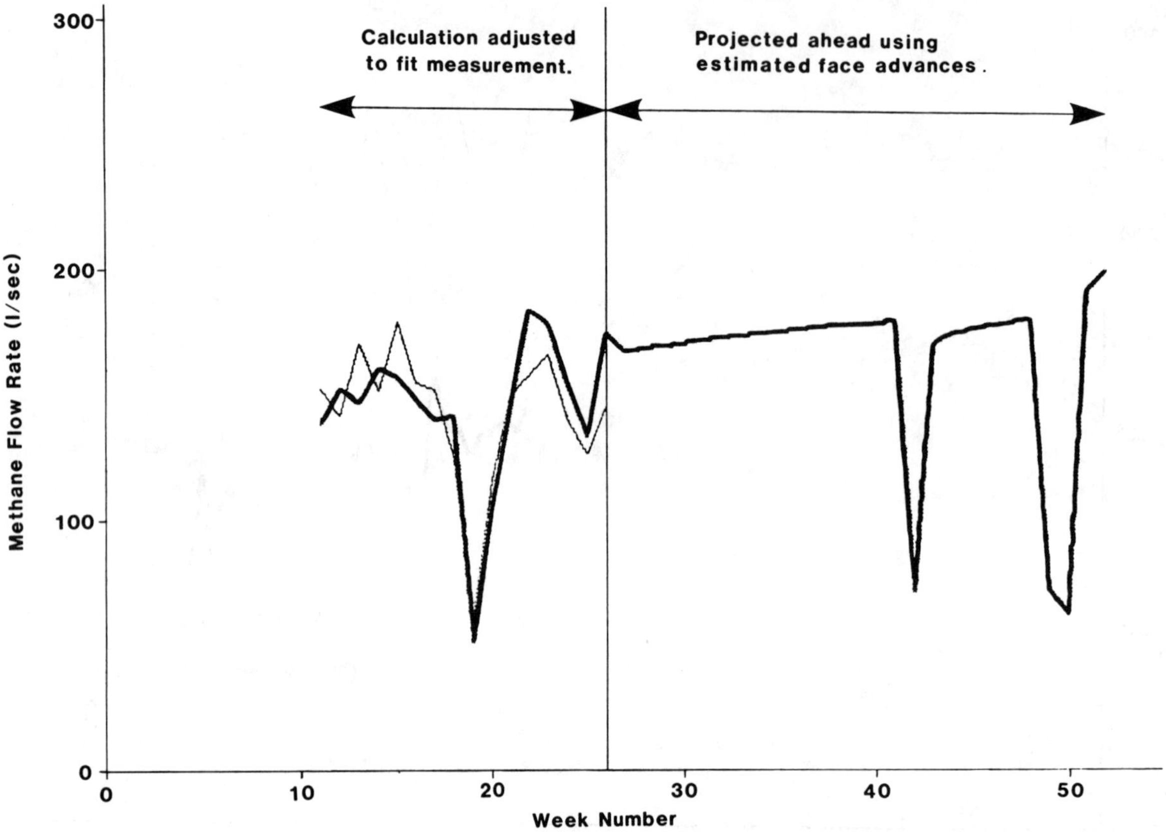

FIGURE 5 Updating the Prediction in Real Time

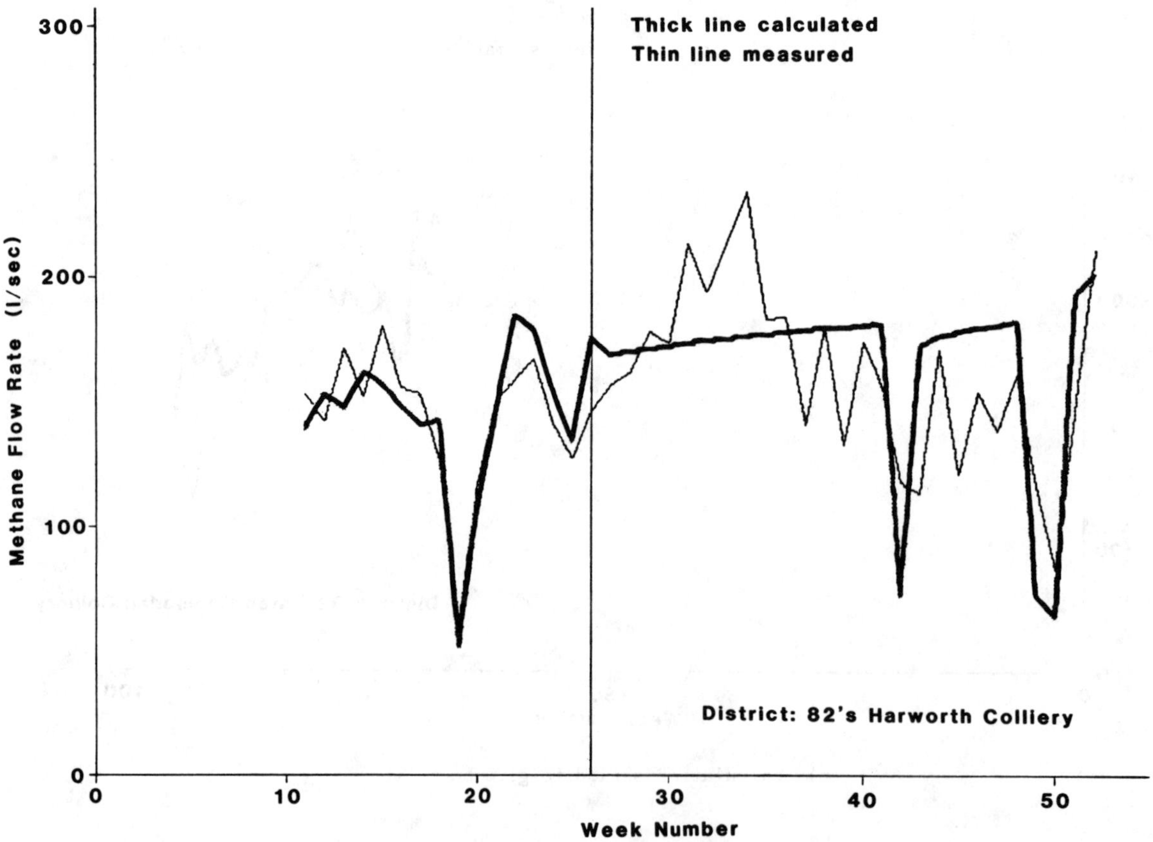

FIGURE 6 Comparison with Actual Emission measured Subsequently

would make these adjustments and forecasts in real time as the district progresses. Any initial abnormality or subsequent change in emission levels compared with prediction would be used to alert mine management to potentially hazardous situations.

REFERENCES

1. Dunmore, R., 1979, "Towards a Method of Prediction of Firedamp Emission for British Coal Mines" 2nd Int. Mine Vent. Congress, Reno, Nevada, USA.

2. Dunmore, R., and Creedy, D. P., 1980, "Measuring Methane in Coal Seams and Predicting its Rate of Emission", Symposium: Methane, Climate and Ventilation in the Coalmines of the European Communities, Luxembourg, Colliery Guardian.

3. Dunmore, R., 1981, "Prediction of Gas Emission from Longwall Faces", Mining Engineer, pp 565-572.

4. Kershaw, S., 1980, Private Communication.

5. Morris, I. H., 1974, "Substantial Spontaneous Firedamp Emissions", Mining Engineer, pp 407-427.

6. Bexon, I., 1977, "Remote Control of Colliery Operations by Computer", Int. Conf. Remote Control and Monitoring in Mining, Birmingham, England.

ACKNOWLEDGEMENTS

The author thanks Mr. P. G. Tregelles, Director of Mining Research and Development of the National Coal Board for permission to publish this paper. He acknowledges the continuing assistance of his colleagues in Environment Branch at MRDE Bretby. The work is supported financially by the European Coal and Steel Community and the prediction examples form part of the MRDE contribution towards a joint study of prediction methods undertaken by the Commission of European Communities' Firedamp Prediction Working Group, of which the author is a member. The views expressed are those of the author and not necessarily of the National Coal Board.

Chapter 24

CENTRIFUGAL FANS AND THEIR USE IN BLEEDER APPLICATIONS

David W. Kennedy

Supervisor - Engineering Department
Buffalo Forge Company
Buffalo, NY

Stephen P. Harrison, P. E.

Project Engineer
Consolidation Coal Company
Pittsburgh, PA

ABSTRACT

In the past few years, a new system has been developed for ventilating longwall panels in some gassy coal mines in the Northern West Virginia area. This system utilizes a five or six foot diameter air shaft to ventilate a group of three or more longwall panels dependent on conditions. This shaft exhausts all the air needed to ventilate the active longwall and adjacent gob areas. By law, this bleeder fan can handle double the maximum methane concentration allowed in a normal return air course. As a result, the methane that must be diluted by the primary ventilation circuits is reduced.

This new ventilation system was first conceived by mine management as a means of coping with a sudden shift in production mode that occurred a few years ago. Formerly, production often consisted of room and pillar mining with continuous miners. The high methane liberation rate was handled by driving entries around a large block of coal prior to mining. This isolated the future mining area from the adjacent coal seam and allowed a major portion of the trapped in situ methane to bleed off. Also, these bleeders served to exhaust return air from the mining area.

Once the bleeders were established, productive mining could start. When the decision was made to shift production mode from continuous miners to longwall mining, some of the bleeders did not encompass large enough unmined areas under the old ventilation system to initiate longwall production. Also, many other established continuous miner section bleeders were either improperly positioned or too restricted for longwall ventilation. Unfortunately, standard bleeder development can take up to two years' time. In order to eliminate the old style bleeder system and maintain or increase the margin of safety for ventilation, the bleeder shaft system was conceived. With this system, the only entries developed are those necessary for the longwall, significantly reducing the production delay. In this area, the Pittsburgh #8 coal seam dips to the northwest. As a result of this dip, the bleeder shaft is constructed in the northernmost set of tailgate entries. In this manner, groundwater accumulations can be pumped to the surface and bleeder entries kept dry. During the initial longwall entry development, the bleeder fan can be used to ventilate entries adjacent to virgin coal where significant methane liberation often occurs.

Once longwall production starts, the bleeder shaft exhausts the entire air quantity necessary to safely ventilate the longwall. As the bleeder entries deteriorate, the air quantity and pressure characteristics must be increased. The extreme flexibility of the centrifugal fan used at the surface allows the air quantity and pressure to be adjusted without stopping the fan. This improves safety by decreasing the possibility of a methane buildup resulting from the fan being out of service.

The bleeder shaft system of ventilation, developed by mine management in Northern West Virginia, can provide a safe means for effectively ventilating longwall panels in excessively gassy mines. This system should be considered where mining conditions are comparable.

FIGURE 1. Coal Mine With Wooden Centrifugal Fan

The specific use of a centrifugal fan for this longwall bleeder configuration is an example of a reevaluation of an historical application (Figure 1). This shows how a more sophisticated examination of an outmoded fan type can lead to an improvement of developed systems and newer applications.

INTRODUCTION

Recently, centrifugal fans have been purchased for mine use by Consolidation Coal Company for a special application. This application involves using these fans in conjunction with 5 and 6 foot diameter air shafts to ventilate active longwall panels. This new system limits the required longwall development in specific cases and enhances the flexibility of the ventilation system. Centrifugal fans appear to be ideally suited for this application.

HISTORY

With this reintroduction of the use of centrifugal fans, the old adage, "History Repeats Itself," seems to apply.

Just as the original natural draft ventilation stacks (Figure 2) and grate furnaces of the late 1800's were replaced by wooden housing paddle-wheel centrifugal fans (Figure 3), these were, in turn, replaced by steel-construction, forward-curved centrifugal fans (Figure 4). Some mines with low pressure requirements installed disc propeller fans (Figure 5).

Until 1950, centrifugal mine fans were still fairly common (Figure 6). The ventilation requirements for most mines at that time were inadequate by today's standards. Axial flow fans in the proper size and rating were not readily available. Fewer entries were driven for development, and lower quantities of air were required for face ventilation. The typical mine fan was a relatively inefficiently rated, forward-curved centrifugal fan (Figure 7) capable of high pressure at relatively low air quantities which did not properly match the mine system.

The forward-curved centrifugal fan was primarily chosen for its small physical size for a given fan rating when compared to other types of fans available at that time. Common rating practice was to physically undersize the fan, operating the fan off peak efficiency towards free delivery. The net result was an inefficiently rated fan with low first cost.

During the past 30 years (Figure 8), centrifugal fans have been replaced by lower-pressure, higher-capacity axial flow fans (Figure 9) that required less power having the mine system resistance match the axial fan curve at an efficient rating point. For most applications, axial flow fans have worked very well. However, when the pressure required to move a specific quantity of air against the mine resistance exceeds the fan pumping limits, an axial flow fan will stall (Figure 10). This is characterized by excessive noise and vibration. Stall can cause severe damage to a mine fan as well as an immediate loss of air quantity. Also, with an axial flow fan, the maximum usable pressure limit drops when the blade pitch is reduced to one of its minimum settings. As a result, reducing the blade pitch to cut down the fan air quantity will not necessarily prevent stall.

A similar stall condition can occur with a forward-curved centrifugal fan which has a similar volume pressure curve shape (Figure 11).

Backward-inclined centrifugal fans work very well in high-pressure applications and, when properly selected, are relatively stall resistant. This can be seen by the lack of a break or "dip" in the fan characteristic curve shape.

Two developments that have occurred since the early 1950's are the advent of the variable inlet vane (VIV) (Figure 12) and airfoil wheel design on centrifugal fans (Figure 13).

These innovations in the design and control of centrifugal fans have improved the efficiency and broadened their operating characteristics so as to compete with adjustable blade axial flow fans in certain applications.

The variable inlet vane is a set of movable guide vanes in the centrifugal fan inlet that control the operating characteristics by creating a vortex in the fan inlet in the direction of rotation. This develops a net reduction in speed between the incoming air and the fan wheel. These guide vanes are positioned simultaneously from a single control lever (Figure 14) on the exterior of the fan. Moving the VIV control is comparable to making blade pitch changes on an axial flow fan. The advantage of the centrifugal fan is that the fan curve, at various VIV settings, shows a constant shutoff pressure available throughout the VIV control range. Another advantage is the fan characteristics can be changed with the fan in operation. The control mechanism is not rotating and remains less troublesome than variable in-flight axial fans. The airfoil wheel design consists of airfoil-shaped blades instead of flat or curved blades. This further improves the efficiency. Typically, the area where centrifugal fans can compete most effectively with axial flow fans is at higher pressure/volume ratios, pressures greater than 8 in. W.G. and air quantities less than 250,000 CFM. It must be emphasized that centrifugal mine fans are for specific applications and it is not intended that they be used for main ventilation unless the mine requires an unusually high-pressure volume ratio.

The aerodynamic and physical size match of a fan to a system may not in all cases present a true picture of the economics of a system. Other related items such as first cost, projected life, and the availability of high or low technology maintenance may dictate the final fan selection.

Backward-inclined centrifugal fans work very well in high-pressure applications and, when properly selected, are relatively stall resistant. This can be seen by the lack of a break or "dip" in the fan characteristic curve shape.

The following chart (Table A) is a general comparison of different types of fans with various system control devices.

FIGURE 2. Natural Draft Ventilation Stack

TABLE A. FAN COMPARISON

FAN TYPE	LOW FIRST COST	HIGH FIRST COST	LOW INSTALLA-TION COST	HIGH INSTALLA-TION COST	HIGH EFFICIENCY RATING	HIGH EFFICIENCY REDUCED RATING	MEDIUM EFFICIENCY REDUCED RATING	LOW EFFICIENCY REDUCED RATING	LOW TECH. MAINTEN-ANCE	HIGH TECH. MAINTEN-ANCE
Forward Curved Centrifugal	X		X					X	X	
Adjustable Pitch Axial	X		X		X		X		X	
Adjustable In-Flight Axial		X	X		X		X			X
Adjustable Speed Axial		X	X		X	X	X			X
Axial With VIVs	X						X		X	
Backward Inclined Centrifugal	X			X	X				X	
Backward Inclined Centrifugal w/VIVs		X		X			X		X	
Adjustable Speed Backward Inclined Centrifugal		X		X	X	X				X

FIGURE 3. Wooden Paddle-Wheel Centrifugal Fan

FIGURE 4. Steel Forward-Curved Centrifugal Fan

FIGURE 5. Disc Propeller Fan

FIGURE 6. Centrifugal Fan (circa 1950)

FIGURE 7. Forward-Curved Centrifugal Fan Wheel

FIGURE 8. Axial Flow Fan (circa 1950)

FIGURE 9. Modern Axial Flow Mine Fan

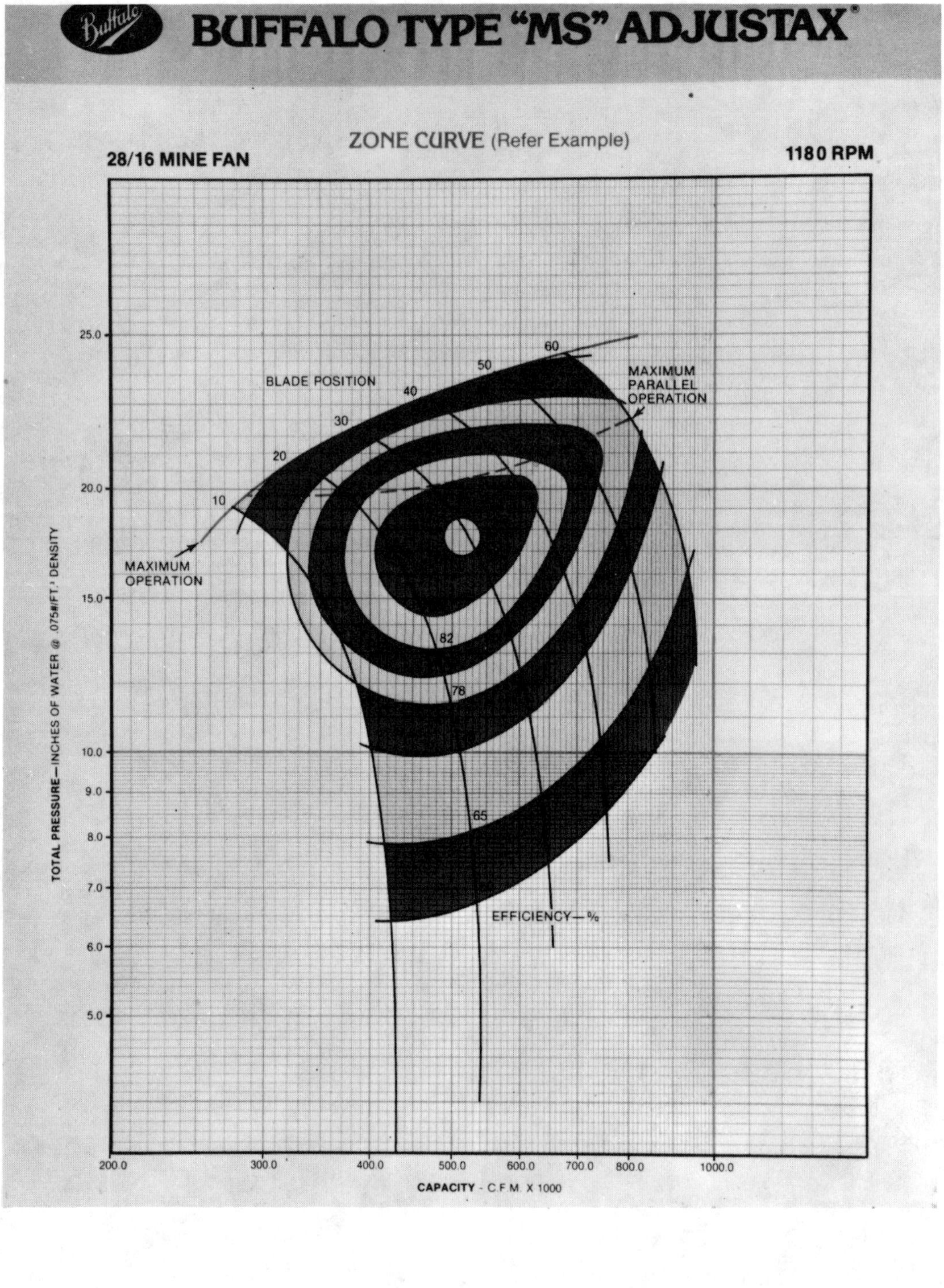

FIGURE 10. Axial Flow Fan Curves
(In. W.G. x 248 = Pa, CFM x .000472 = M^3/S)

FIGURE 11.

CENTRIFUGAL FAN CURVE CHARACTERISTICS

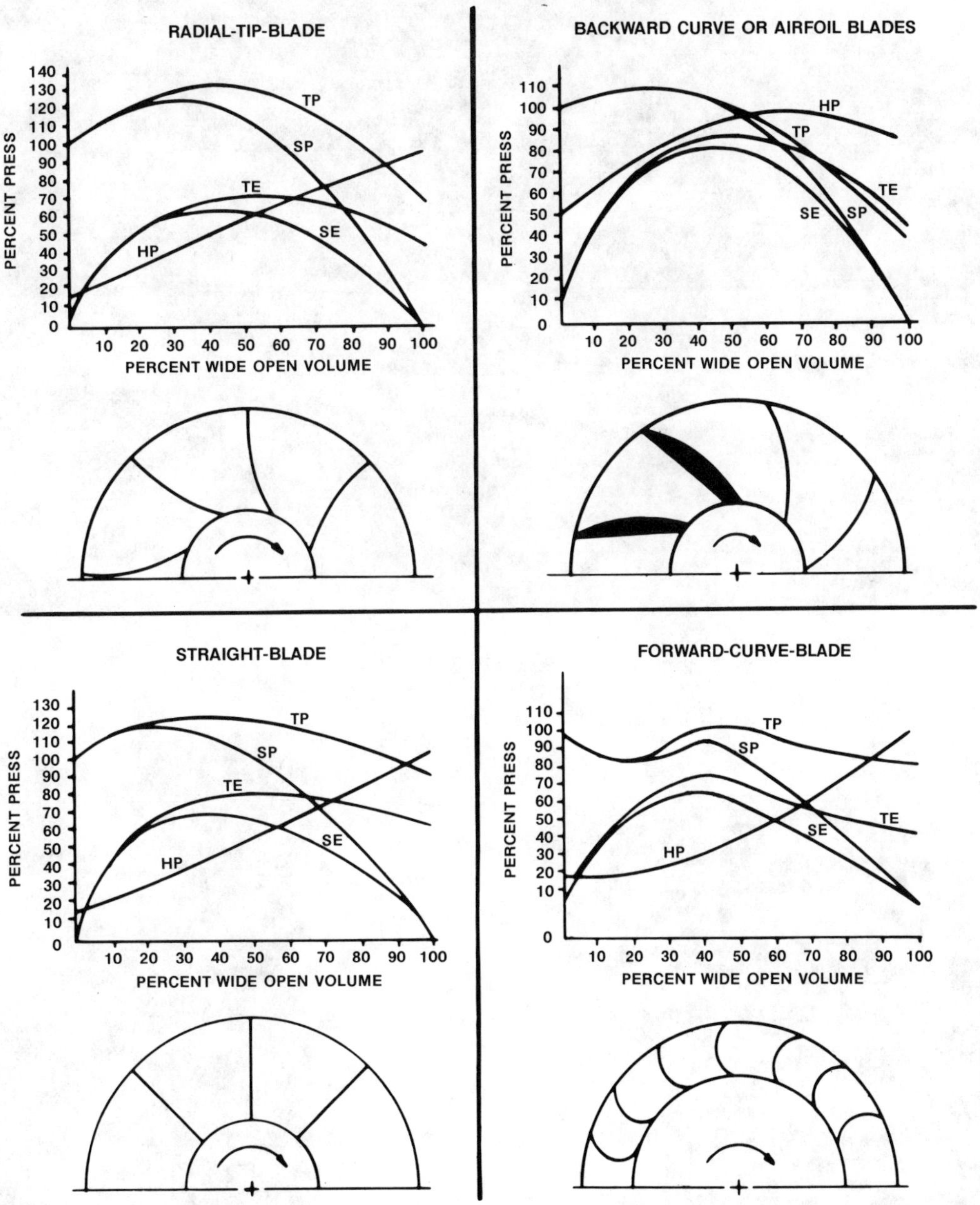

FIGURE 12. Variable Inlet Vanes (VIV)

FIGURE 13. Centrifugal Fan Wheel and Housing

FIGURE 14. VIV Control Lever

BLEEDER SHAFT SYSTEM

Introduction

A few years ago there was an abrupt change in production method at a number of Consol mines in the Northern West Virginia area. Longwalls were scheduled for mines that had been designed for continuous miner production. In order to adapt these mines for longwalls and keep development to a minimum, a new system was conceived by regional management.

Previously it had been standard operating procedure, in gassy areas, to drive bleeder entries around a block of coal prior to mining. This system isolates the mining area from the adjacent coal seam and prevents the further migration of methane. When continuous miner and longwall development sections advanced, they tied into the bleeder prior to retreat mining. This maintains the ventilation focal point at the back end and guarantees a significant pressure differential across the gob. This type of bleeder requires a 2-year development period and must be maintained for fireboss examination.

The new bleeder system (Figure 15) first consisted of locating a 5-foot diameter return air shaft at the back end of the first set of tailgate entries in a projected mining area. As the longwall panels were individually retreated, the extreme flexibility of the centrifugal fan (Figure 16) on the bleeder shaft enabled the system quantity and pressure to be regulated. In this manner, the longwall and bleeder ventilation was adjusted and controlled. With this system, a positive pressure could always be maintained across the gob and independent ventilation assured. This system also reduces the methane and return air quantity that are handled by the primary ventilation fans. The configuration of each longwall mining area, where the bleeder shaft system was to be applied, depends upon several variables. Timing, anticipated methane liberation, water content of overlying strata, coal seam dip, gas well locations, roof control and methane drainage plans all affect the physical design parameters. In each case, a conservative approach was taken in defining mining plans in a given area.

Bleeder Shaft Construction

Currently, there are five bleeder shafts (Figure 17) in operation in Northern West Virginia.

FIGURE 17. Bleeder Shafts

Shaft	Depth	Diameter
Thorn	700' (213m)	5' (1.5m)
Southgate	1000' (305m)	6' (1.8m)
Sloan	520' (158m)	5' (1.5m)
Lowe	800' (244m)	5' (1.5m)
Worley	520' (158m)	5' (1.5m)

All these shafts were blind hole drilled by necessity. The selection of the more expensive (Figure 18) blind hole shafts was justified by development timing.

FIGURE 15.

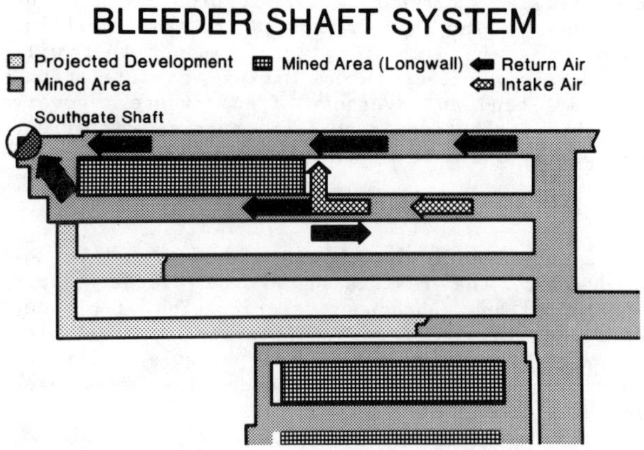

FIGURE 16. Surface Centrifugal Fan Installation

With the sudden emphasis on longwall mining, development plans had to be reviewed and amended. These plans did not allow sufficient time for raise drilling. In each case, bleeder shaft ventilation had to be established when the area was reached underground. The shaft size was selected to meet specific ventilation requirements. The five (1.5m) and six foot (1.8m) diameter shafts were designed to exhaust a maximum flow of 125,000 CFM (59 m^3/s) and 150,000 cfm (71 m^3/s) respectively.

FIGURE 18. Blind Hole Shaft Sinking Cost

520 ft. (158m) Hydrostatic Steel Liner	$130,000
Pre-grouting	90,000
Grouting Liner In Place	36,000
Blind Hole Drilling	393,000
Site Preparation	85,000
Special Reclamation	10,000
	$744,000
	or $1,430/ft.

(NOTE: These cost figures reflect June 1981 prices and are site specific.)

If, in the future, timing allows for shafts to be raise drilled, a separate set of problems must be addressed. The gob generated by the drilling process must be disposed of in a manner that will not adversely affect future mining or ventilation. Gob rooms tend to adversely affect future recovery and gob distribution in existing entries may affect ventilation and must be closely supervised.

Fan Selection

The fan selected for this application must be able to overcome the resistance in the bleeder shaft (Figure 19) and subsequent deterioration of bleeder entries. This fan must be easily controllable, economical, stall resistant and portable. Current plans are to relocate each bleeder fan after area mining is completed.

FIGURE 19. Bleeder Shaft Shock Loss [5 Ft (1.5m) Diameter] 520 Ft (158m Depth)

	Air Quantity	Shaft Pressure Loss
1.	50,000 CFM (24 m^3/s)	1.44 in. W.G. (.36 kPa)
2.	75,000 CFM (35 m^3/s)	3.25 in. W.G. (.81 kPa)
3.	100,000 CFM (47 m^3/s)	5.78 in. W.G. (1.43 kPa)
4.	125,000 CFM (59 m^3/s)	9.02 in. W.G. (2.24 kPa)

For comparison, the operating characteristics of a centrifugal fan with a 73 inch (185 cm) wheel is compared to an axial flow fan with a 60 inch (152 cm) wheel (Figure 20).

It should be noted that the power cost of operating the centrifugal fan below 12 inches of water gage (3 kPa) exceeds the cost of operating the axial flow fan (Figure 21). Also, the fan curves show that the axial flow fan is more efficient in a wider range.

FIGURE 20. Economic Comparison of Axial Flow and Centrifugal Fans

	Operating Point	Buffalo Forge Model 1460 Centrifugal Fan Horsepower Requirements	60 Inch (1.5m) Axial Fan Horsepower Requirements	Horsepower Difference	Relative Power Cost Difference Centrifugal vs. Axial*
1.	190,000 CFM at 3 in. W.G. (89 m^3/s at .7 kPa)	330 (246 kw)	200 (149 kw)	130 (97 kw)	$2,119/mo.
2.	170,000 CFM at at 6.6 in. W.G. (80 m^3/s at 1.6 kPa)	350 (261 kw)	250 (186 kw)	100 (75 kw)	$1,630/mo.
3.	140,000 CFM at 11.5 in. W.G. (66 m^3/s at 2.9 kPa)	355 (265 kw)	350 (261 kw)	5 (4 kw)	$81.5/mo.
4.	136,000 CFM at 12 in. W.G. (64 m^3/s at 3.0 kPa)	350 (261 kw)	350 (261 kw)	0 (0 kw)	0
5.	132,000 CFM at 12.5 in. W.G. (62 m^3/s at 3.1 kPa)	350 (261 kw)	365 (272 kw)	-15 (11 kw)	-$244/mo.
6.	130,000 CFM at at 13 in. W.G. (61 m^3/s at 3.2 kPa)	352 (262 kw)	380 (283 kw)	-28 (21 kw)	-$456/mo.
7.	125,000 CFM at 13.5 in. W.G. (59 m^3/s at 3.4 kPa)	355 (265 kw)	380 (283 kw)	-25 (19 kw)	-$407/mo.
8.	120,000 CFM at at 14.2 in. W.G. (57 m^3/s at 3.5 kPa)	345 (257 kw)	DNA	DNA**	DNA
9.	100,000 CFM at 16.2 in. W.G. (47 m^3/s at 4.0 kPa)	320 (239 kw)	DNA	DNA	DNA
10.	70,000 CFM at 17.4 in W.G. (33 m^3/s at 4.2 kPa)	260 (193 kw)	DNA	DNA	DNA

Based on $.03/kwh
**DNA - does not apply

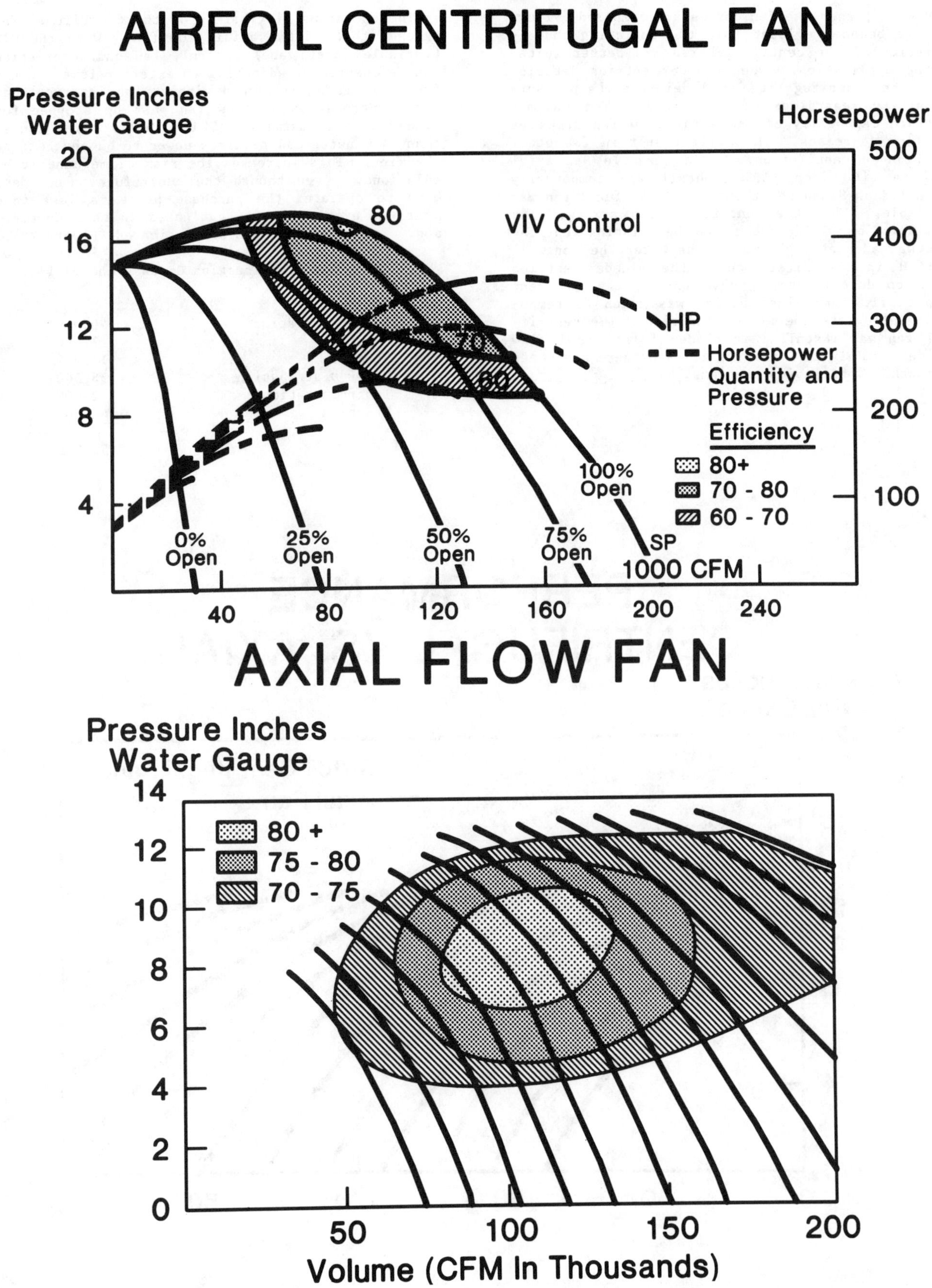

FIGURE 21. Centrifugal and Axial Flow Fan Characteristics
(In. W.G. x 248 = Pa, CFM x .000472 = M^3/S)

However, when the fan curves are compared (Figure 22), it becomes evident that the operating characteristics of the centrifugal fan are better suited to the application. When bleeder entries deteriorate, the centrifugal fan will maintain its pressure capability regardless of flow rate. The maximum pressure capability of the axial flow fan drops as air flow decreases. This will result in fan stall and reduced ventilation. As an example, an axial flow fan [60 inch (1.5m) wheel] was temporarily installed on Sloan Shaft until a centrifugal fan was available. When the centrifugal fan arrived, installation was delayed because the existing ventilation was adequate. A few months after the longwall started in the first panel, the bleeder entries began to deteriorate. Subsequently, the fan went into stall. Immediate action was taken to temporarily alleviate the stall condition. The centrifugal fan was installed in place of the axial flow fan during July 1981 (Figure 23) and there have been no further ventilation problems.

The operating flexibility of the centrifugal fan exceeds that of the axial flow fan. With the VIV (variable inlet vane) control, fan characteristics can be changed by adjusting an exterior lever on the fan. This is accomplished without fan interruption. A conventional axial flow fan must be stopped for blade pitch adjustment. This usually takes longer than 15 minutes and requires power to be shut off at the mine. This increases the risk of methane accumulations. Even though the centrifugal fan costs more to operate, the purchase price is less than half the axial flow fan evaluated in this comparison. Fan erection costs are indicated (Figure 24).

FIGURE 24. Fan Erection Costs (June 1981)

Foundation	$ 5,000
Duct Fabrication	20,000
Motor	17,000
Starter	11,000
Power Distribution	55,000
Instrumentation	2,000
Erection	48,500
	$158,500

FIGURE 22.

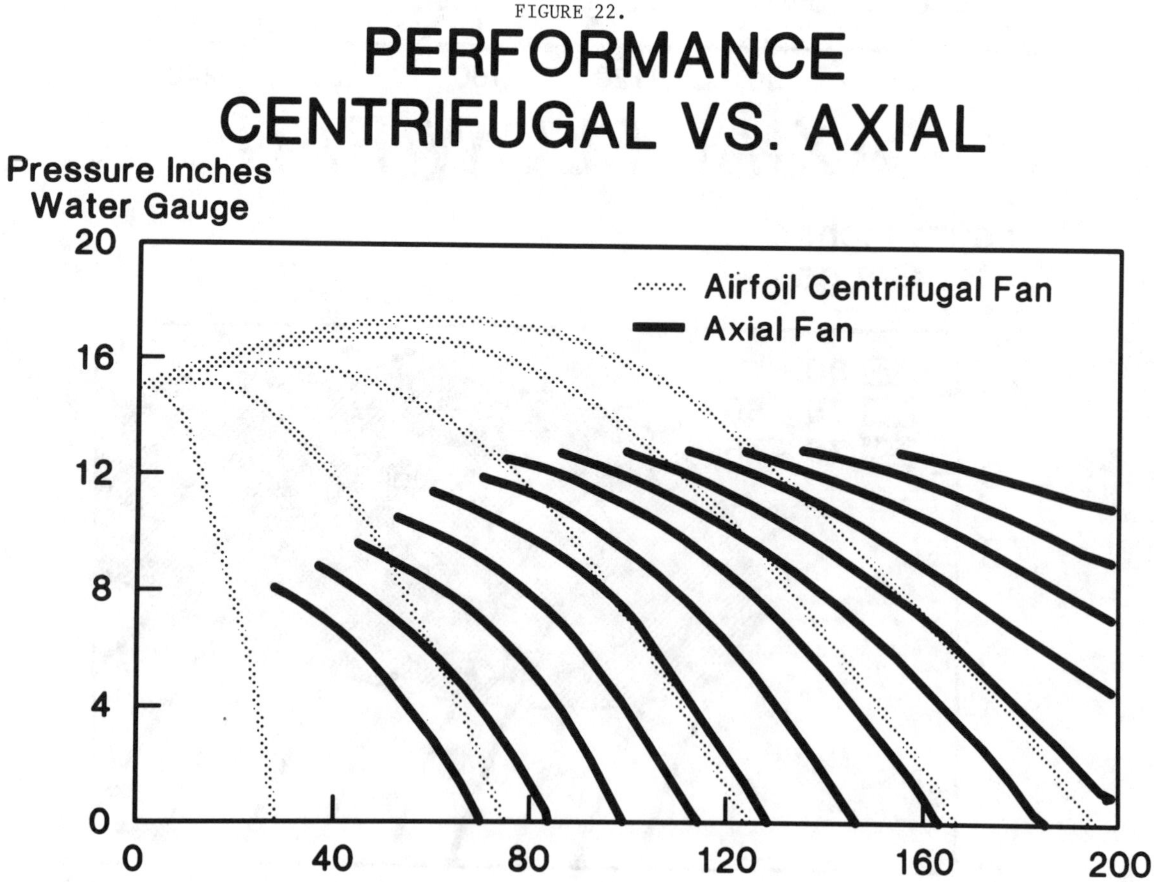

FIGURE 23. Axial Flow Fan Being Replaced By Centrifugal Fan

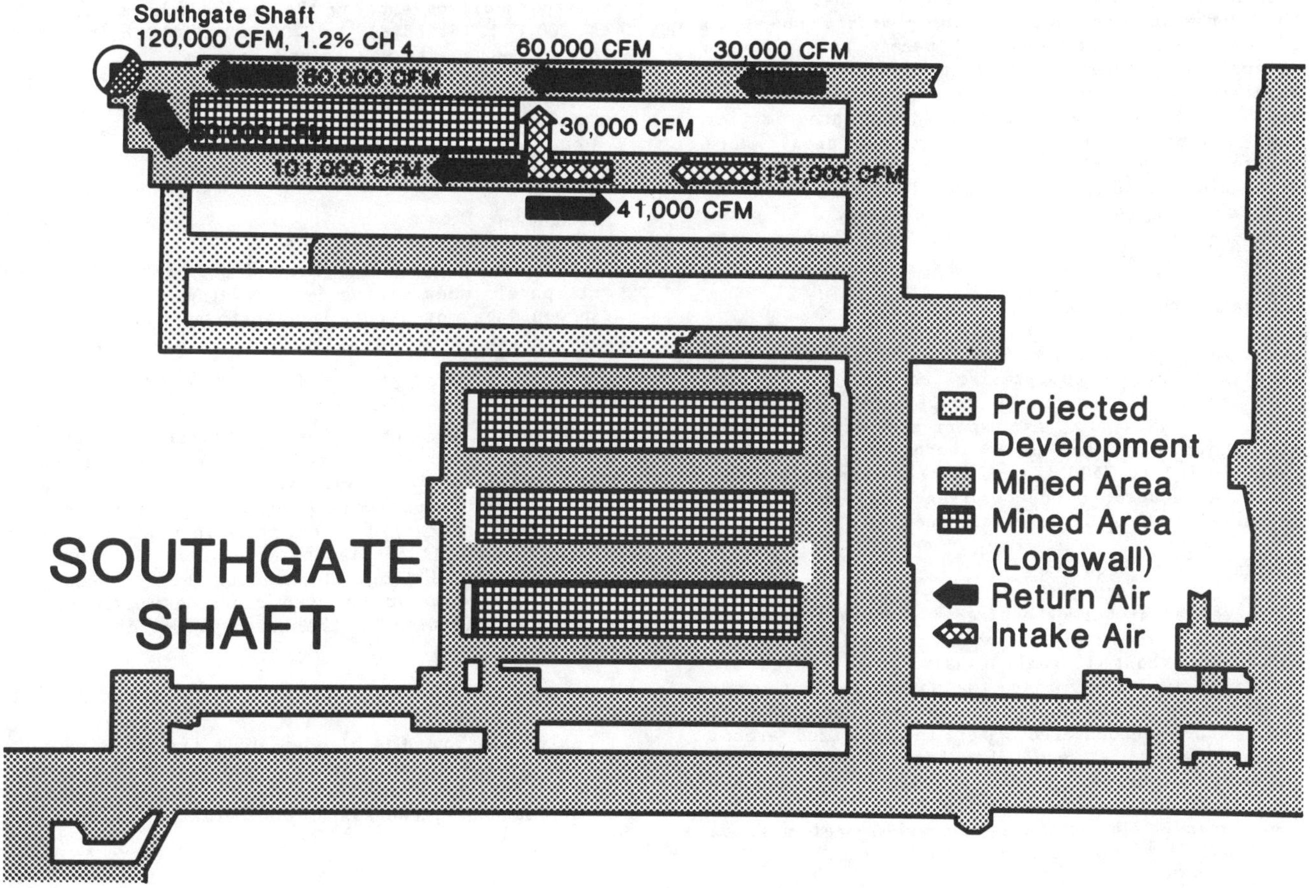

FIGURE 25. Bleeder Shaft System Panel Configuration

Longwall Panel Configuration

Panel configuration (Figure 25) is dependent upon timing, anticipated methane concentration, water inflow and mine roof conditions.

To date, the longwall panels utilizing the bleeder shaft system have been grouped in sets of three or four separated by a barrier pillar. The longwall panel length varies from 6600 to 7500 feet (2011 to 2286m). The face width varies from 550 to 600 feet (168 to 183m) with a mining height between 7 and 8 feet (2.1 and 2.4m).

The longwall development panels are driven with four entries on single split. Large enough pillars are used to protect the tailgate entries. The number of entries driven in the first longwall tailgate development section is primarily dependent on the anticipated methane liberation. In gassy areas, as many as seven headings (minimum four) are used to develop the first set of tailgate entries. The larger number of entries allows for decreased resistance to ventilation and better dilution of methane along the solid rib. Generally, the first set of tailgate entries are driven into virgin coal.

In most areas where the bleeder shaft system is being used, there is a northwest dip in the coal seam (Pittsburgh #8). Longwall panels are generally grouped in sets and independently ventilated by bleeder fans. The bleeder shaft is constructed in the northwest corner of the mining area adjacent to a large sump. The number of panels per group is determined by timing requirements, anticipated roof stability and methane liberation. Typically, a 3-panel group can be developed and retreated earlier than a 4-panel set. Also, in mining areas where methane is prevalent, the additional panel gas liberation could eventually overload the bleeders. Roof control is a third consideration. Existing bleeders would need to remain open longer with larger panel sets.

Longwall Panel Length

Longwall panel length has been adjusted in this area in an effort to optimize recovery and production rate. The earlier longwall panels started at 3000 feet (914m). As experience with longwall mining improved, the panels were lengthened [6600 feet (2011m) to 7500 feet (2286m)]. Longwall moves could be reduced to one occurrence per year. This coincided with required maintenance.

Longwall Panel Sequence

In the area where the bleeder shaft system is in use, the coal seam (Pittsburgh #8) dips to the northwest. Most longwall panels in this area are pulled up dip to eliminate the possibility of water accumulation at the face. In the past, two longwall panels have encountered water in localized areas. In each case, the result has been poor production and extended delays. However, this panel sequence does inherently delay longwall production in northern areas. The furthest longwall panel must be developed first to ensure proper drainage.

Water Control

Each group of longwall panels is isolated from the next group by a barrier pillar. Although this pillar serves to prevent weight transfer to the next panel, its primary purpose is water control. Any water generated during mining is channeled to the sump adjacent to the bleeder shaft. The water will be conducted to the surface through a 16 inch (41 cm) diameter borehole. A submersible deep-well pump (maximum capacity 2000 GPM (.13 m^3/s) with permissible cable will handle any water accumulation. The maximum sump capacity should be 200,000 gallons ($757m^3$).

Methane Control

Longwall panels are currently limited to locations where two or less gas wells require plugging. The approximate cost for plugging a single well is $70,000.

To further control methane, two drainage holes are drilled per panel. These holes [7-1/2 (19 cm) to 8-1/2 (22 cm) inches diameter] are drilled through the overburden to the Redstone coal [30 feet (9m) above the Pittsburgh #8 Seam]. A mechanical packer is set below the Waynesburg sandstone (major water-bearing strata) in competent rock and grouted to the surface. The pipe below the Waynesburg is perforated to facilitate methane movement. After the roof collapses below the drainage hole, as much as 200,000 ($5,663m^3$) to 400,000 ($11,327m^3$) cubic feet of methane is vented per day.

The bleeder fan also provides a means for methane control. As the methane concentration rises, the VIV (variable inlet vane) can be adjusted easily and the methane dilution improved.

Abandonment

Current plans are to seal off each group of longwall panels when mining is completed. The bleeder fan would be moved to a new shaft site and a pressure relief valve installed on the old shaft.

New Developments

Recently, some unusual roof stability characteristics have been discovered in these panel groups. The bleeder entries driven at the back end of these panels (north-south direction) are remaining open. Plans are underway to use this in future mining plans. This will enable one bleeder shaft to ventilate two sets of longwall panels. The only changes in the scheme would be sump construction and controlled ventilation at the northwest corner of the next mining area.

CONCLUSION

In conclusion, the bleeder shaft system of ventilation provides considerable flexibility for longwall mining in gassy areas. As the methane liberation fluctuates, the system can be easily changed to compensate.

The centrifugal fan is ideal for this application because of its ability to generate high pressure at relatively low air quantities. The operational advantages of this system should be considered where longwall mining is to be applied in a relatively gassy area.

It must be emphasized that centrifugal fans are ideal for specific applications, but may not be suitable for primary ventilation. Axial flow fans offer a high efficiency over a wider range than centrifugal fans and are better suited to variable systems. The centrifugal fan is better suited to high pressure applications with relatively fixed systems.

ACKNOWLEDGEMENTS

The authors wish to express their gratitude to Mr. Don Born, P. E. (Senior Mining Engineer, Northern West Virginia Region, Consolidation Coal Company). Without Mr. Born's expertise and direct knowledge of this project, this paper would not have been possible.

COMMERCIAL COAL DEGASIFICATION - A CASE HISTORY

Chapter 25

Kenneth L. Ancell

Enhanced Energy Resources, Inc.
Houston, Texas

ABSTRACT

This paper traces the history and performance of a coal degasification project in the Black Warrior Basin of Central Alabama. The primary coal beds being degasified are the Mary Lee and Blue Creek. The area being developed for gas production is also scheduled for underground mining in two to seven years.

The paper discusses the early data collection and evaluation. The coals are extremely gassy with some measurements in excess of 500 SCF/ton. Five experimental wells were drilled and produced. Geologic heterogeneities were encountered which affect the performance of the project.

Drilling and completion practices together with well design considerations compatible with underground mining are discussed. Finally, the gas gathering and sales system together with the early production history is illustrated.

INTRODUCTION

The Black Warrior Basin is noted for its gassy coal mines. Jim Walter Resources, Inc. is mining the Mary Lee and Blue Creek coal seams in Tuscaloosa County, Alabama. Four mines are in existence and operate at depths of from 1300 ft to 2200 ft. All mines experience high gas emissions and early experience led to the investigation of coal degasification as an attempt to produce a portion of the gas prior to mining.

An experimental program to assess the feasibility of a degasification program was started in 1978. It was expanded in 1979. Performance for the experimental program in late 1980 and early 1981 resulted in commitment to a large scale degasification program of 25 wells for 1981 and 1982. The program is to drill wells and produce gas to be sold in the interstate market.

Sales of gas commenced in February 1982. The 25 well program is in the final stages of completion. The gas sales system is by truck with a pipeline connection planned in the near future.

Experimental Wells

A single well was drilled and completed in 1978. The objectives of this well were (1) to demonstrate the viability of the vertical well degasification method and (2) to provide mining information by placing the well in a position where it would be mined through at an early date. The results of this first well were reported in detail in early 1979.[1] Briefly, the well produced excessive volumes of water with very low volumes of gas. Analysis of this performance led to the conclusion that the single well was connected with a large underground fault or fracture system. In that case, several additional wells were needed to facilitate the production of water which must precede any large gas production.

Four additional wells were drilled in 1979. Figure 1 shows the location of the five experimental wells in relation to the current underground mine and projected mining plan. Well No. 1 was mined through in August 1980, with no adverse effects.

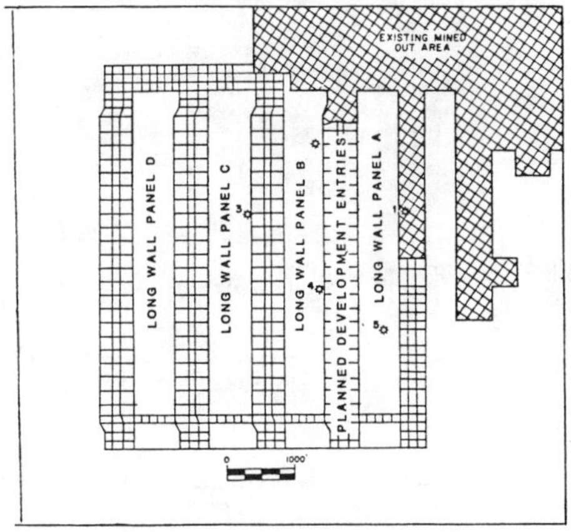

FIGURE 1. LOCATION OF EXPERIMENTAL WELLS

The four wells drilled in 1979 were drilled with a conventional rotary drilling rig. Casing was set through the coal seams and cemented. The wells were all nominally 2000 ft deep. Figure 2 is a type log showing the coal seams in question. The Mary Lee and Blue Creek seams contain about 6.5 ft of coal in an interval of about 11 ft. Casing size was 4½ inch O.D. with a single joint of fiberglass pipe set through the interval to be mined. This feature was instituted to facilitate mining through the wells.

The wells were all cut through the fiberglass casing using a sand water slurry at high pressure

References appear at end of paper.

through a nozzle. Each well was then hydraulically stimulated using about 40,000 gal of water and 40,000 lb of sand. The wells were then placed on production using sucker rod pumps operating on electric power. Figure 3 shows the total gas and water production of the experimental wells for the period January 1980 through May 1981.

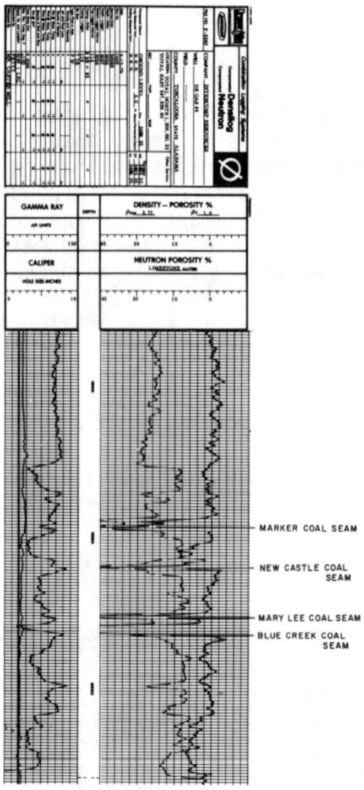

FIGURE 2. DEGASIFICATION WELLS TYPE LOG

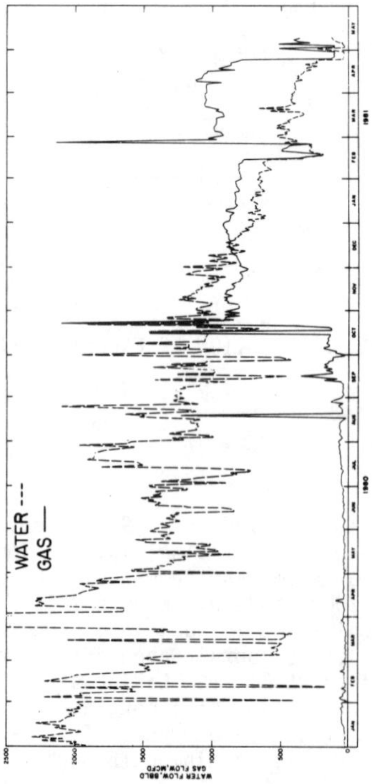

FIGURE 3. GAS AND WATER PRODUCTION HISTORY EXPERIMENTAL WELLS

As can be seen on Figure 3, production through the first nine months of 1980 was high water volumes. Initial design of pumping equipment was for 200 bbl/day per well. With the high water volumes, the pump capacity was rapidly increased by increasing pump size and stroke length. Finally, in March 1980 two casing pumps with 4 inch plunger diameters were installed. This resulted in water rates of about 4000 bbl/day in late March 1980. This is shown as the off scale portion in Figure 3.

In early April 1980, the adjacent mine intercepted a faulted, fractured zone underground. A series of small displacement normal faults and numerous natural fractures were encountered which produced large volumes of water and some gas. Response in the well pattern was dramatic and immediate. The decline in water production starting in April 1980 and continuing through the present is real and not related to mechanical problems.

Starting in June 1980, a noticeable increase in gas production started. This increasing trend continued through mid-October. Several gas production "spikes" started to occur in August and finally in mid-October a continuous, mobile gas saturation was established and the wells all started producing gas in commercial quantities. The total held steady at about 1000 Mcf/d for several months. Mechanical problems with the pumping equipment account for the depressed gas production in April and May 1981.

The composition of the gas was determined to be commercial quality and compatible with natural gas. Table 1 shows the chemical composition of a typical gas sample.

TABLE 1

GAS COMPOSITION

Component	Mole %
Nitrogen	1.13
Carbon Dioxide	0.03
Methane	98.84
Ethane	0
Propanes plus	0
	100.00

ENGINEERING ANALYSIS

The production history described in the previous section validates several conclusions that have been drawn about the coal-methane-water system. These are summarized as follows:

1. The methane is stored in the coal matrix itself by the physical process of adsorbtion.

2. The coal cleat or natural joint system is highly permeable and can transport large fluid volumes with a small pressure gradient.

3. Water is being introduced into the cleat via a fault system from other formations below and above the coal, possibly even from the surface.

4. Large volumes of gas can be produced once sufficient water is produced to cause interference effects between mine and wells.

To assist in the understanding of the physics of the process, we employed a reservoir simulator which rigorously accounts for all the variables that affect the coal degasification process. The description of the simulator is included in the literature and is not repeated here.[2] The simulator rigorously calculates the two phase flow of gas and water and incorporates the adsorbed gas phase as the principal storage mechanism. The simulator describes the wells and mine as pressure sinks and with appropriate values for the pressure boundaries; the output is the flow rates at each well as well as pressure and saturation distributions as a function of time.

The mine was described in the simulator as an area of low pressure which advanced with time according to the mining history. The wells were described as point sources. The coal reservoir data are shown in Table 2.

TABLE 2

RESERVOIR DATA

Gas content	500 cu ft/ton
Diffusion Parameter	$1 \times 10^{-9} sec^{-1}$
Initial Pressure	834 psia
Reservoir Temperature	90°F
Porosity	4%
Permeability	75 md
Coal Height	6.5 ft
Coal Depth	2000 ft

The only remaining unknown was the effect of the fault fracture zone that was suspected from the well performance and later verified by the mine development. This phenomenon was simulated by a zone of very high permeability across the mine-well pattern area connected with a large water source on either side of the area in question. This zone of high permeability is shown on Figure 4.

The 1980 gas production is compared with the simulation of the same period on Figure 5. The result of this work is that the behavior of this pattern of wells is entirely predictable once the system can be adequately described.

This description of the area, developed in early 1981, could finally be used to attack the problem that we set out to solve in 1978. That problem is how can a well pattern adequately degasify coal ahead of mining.

Once the description of the system was set, the simulator could then be used to investigate the degasification effects. Several well spacing configurations and time frame specifications were attempted. The results of one of these simulations is shown on Figure 6. This is a cross section across a well pattern with five rows of wells on 1050 ft centers at the end of a five year period.

The meaning of this figure is that at the end of five years, the gas content of the coal is reduced to approximately 50% of original. This then appeared to be a realistic goal to attempt. It is important to note that at the outer edge of the well pattern, the gas content increases very rapidly with distance. This means the gas content is significantly affected only in the area subtended by the pattern.

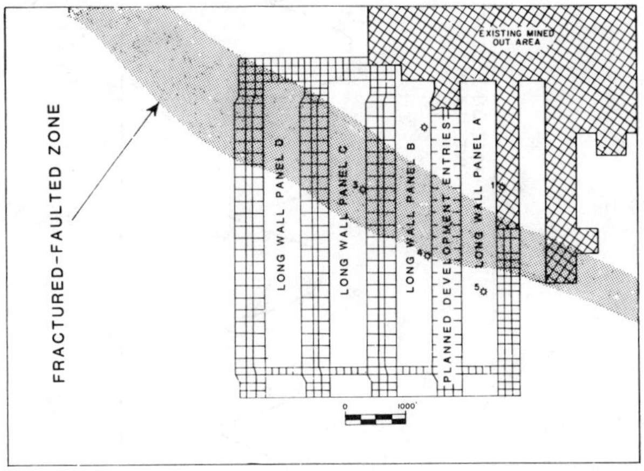

FIGURE 4. LOCATION OF FAULT-FRACTURE SYSTEM

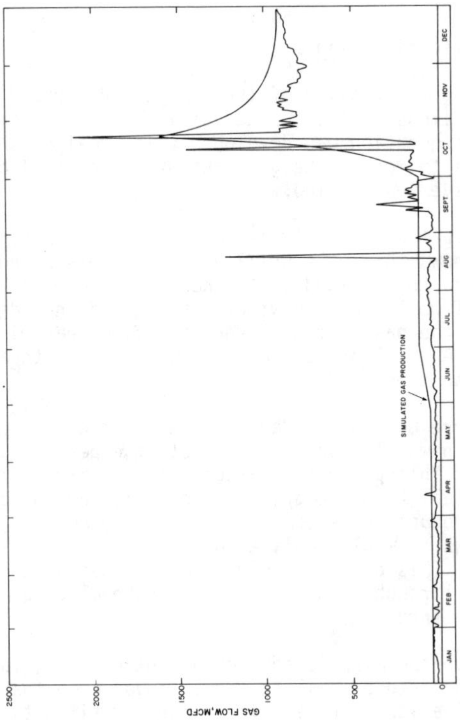

FIGURE 5. COMPARISON OF ACTUAL AND SIMULATED GAS PRODUCTION

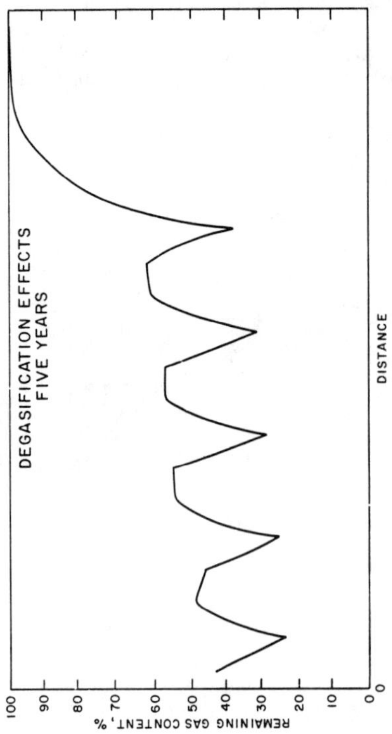

FIGURE 6. DEGASIFICATION EFFECTS OF A WELL PATTERN AT THE END OF FIVE YEARS

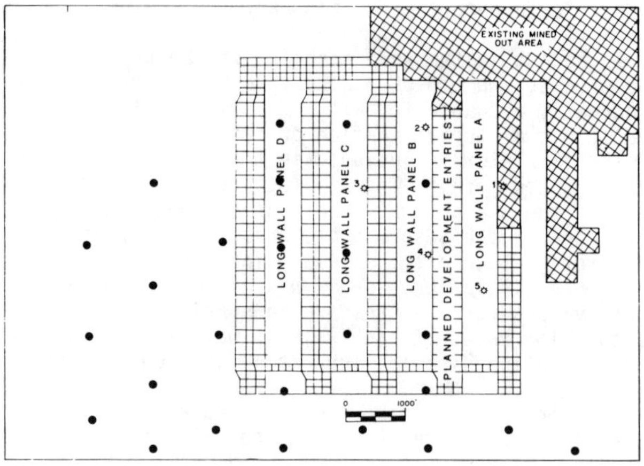

FIGURE 7. LOCATION OF DEGASIFICATION WELLS

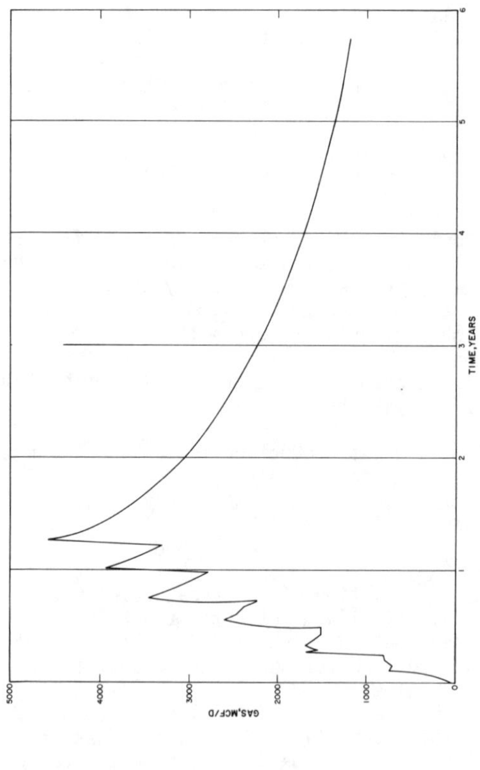

FIGURE 8. PRODUCTION PROJECTION OF INITIAL 25 WELL DEGASIFICATION PLAN

Degasification Plan

Using the results of the analysis described in the previous section, a degasification plan for the area in question was developed. The goal of the plan is to reduce the gas content of the coal by 50% in a five year period.

The pattern finally selected is shown on Figure 7. This shows the wells drilled on 27 acre spacing in the area to be mined in the next five years. Two rows of wells on much wider spacing are included to reduce the gas content beyond the five year mining plan. A total of 25 new wells were needed to accomplish this.

Figure 8 shows the anticipated production from the 25 new wells and the 4 existing experimental wells. This projection assumes a 15 month period to get all 25 wells drilled, completed and producing. Peak production is computed to be in excess of 4000 Mcf/d which would decline to about 1200 Mcf/d in six years. This decline would be aggravated by the mining through of wells which is scheduled to begin in year three.

Since the mine is advancing continuously, more wells need to be drilled far in advance so that more than five years are available for degasification. This was done in the simulator by adding five wells per year on the south and west sides of the pattern and eliminating five wells per year inside the five year mining plan. Figure 9 shows the effects of this program. This plan shows a sustained deliverability of 2,600 Mcf/d.

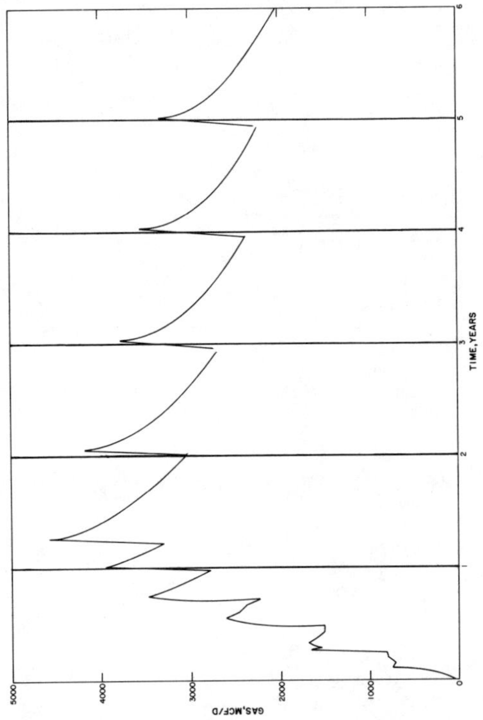

FIGURE 9. DEGASIFICATION PLAN INCLUDING DRILLING FIVE WELL PER YEAR

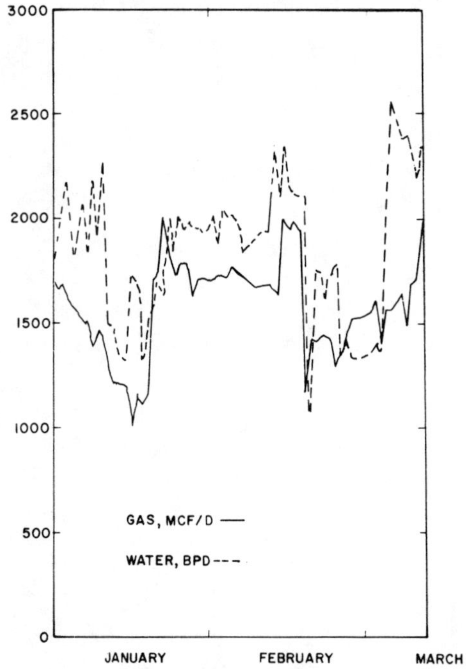

FIGURE 10. 1982 PRODUCTION STATISTICS FOR THE DEGASIFICATION PATTERN

Current Progress

At this writing all 25 wells have been drilled, 19 wells are in some stage of completion, and 10 wells are producing. Production statistics for the year 1982 to date are shown on Figure 10.

Current well design is essentially the same as used for the experimental wells except the fiberglass casing has been eliminated. Production casing is now set above the coal seams with a packer shoe. The packer shoe is then drilled out and the wells are completed open hole. Stimulation procedures are essentially the same as described for the experimental wells.

A gas sales contract has been secured and deliveries commenced in February 1982. Gas sales, well spacing, drilling and completion are under the jurisdiction of the Alabama Oil and Gas Board. The wells are being drilled under the Special Field Rules established by the Oil and Gas Board for the Brookwood Coal Degasification Field. The gas pricing is covered by Section 107 of the Natural Gas Policy Act and is treated as "High Cost Natural Gas". The gas sales system includes wellhead compression operating with a suction pressure of 2-5 psig. The gas is gathered into a network of piping operating at about 40 psig. At a central site the gas is further compressed to 2400 psig and loaded into trucks. These trucks haul the gas to the nearest point on the pipeline and discharge the gas through the sales meter.

CONCLUSIONS

At the present time we can make the following conclusions:

1. Gas can be produced from vertical wells in quantities sufficient to materially affect the gas content of the coal.
2. Gas wells that have been produced can be mined through with no adverse effects.
3. Commercial production of coalbed gas is viable as an economic venture.
4. Existing laws and regulations are sufficient to make a degasification program compatible with gas production practices.

REFERENCES

1. Ancell, K. L. and Stevenson, J. W., 1979, "Degasification Parameters and Well Completion Procedures for the Mary Lee/Blue Creek Coalbeds in Tuscaloosa County, Alabama", Methane Recovery from Coalbeds Symposium, April 18-20, 1979.
2. Ancell, K. L., Lambert, S. and Johnson, F. S., 1980, "Analysis of the Coalbed Degasification Process at a Seventeen Well Pattern in the Warrior Basin of Alabama", SPE/DOE 8971.

6
Mine Air Cooling

Chapter 26

ANALYSING VENTILATION AND COOLING REQUIREMENTS FOR MINES

J. van der Walt, E. M. de Kock, L. K. Smith

Engineering Management Services Limited
P.O. Box 585, Johannesburg 2000, R.S.A.

SYNOPSIS

A unique and systematic approach for analysing alternative methods of ventilating and cooling large hot mines is presented. The important factors that have a bearing on ventilation and cooling requirements and practices, such as air flow through the mine, geographical distribution of heat loads, micro ventilation considerations etc., are dealt with. Analysis of these leads to the definitive ventilation and cooling strategies, air and water process design, detail specifications and design of associated plant and equipment.

INTRODUCTION

The method of analysing ventilation and cooling requirements for large hot underground mines presented is that generally followed by the authors. The method has been developed over a period of several years and is based on experience gained on a number of mines in South Africa and overseas. This development would not have been possible were it not for the research and development work by the Chamber of Mines of South Africa (1 to 5). The authors' exposure to mines in other parts of the world enabled them to modify the ventilation and cooling principles established for the South African gold mining industry to suit totally different requirements of a variety of mining methods and climatic conditions.

There are essentially three types of ventilation and cooling projects viz., those associated with

- existing mines where the object is to audit the ventilation and cooling practices with a view to determining and analysing their inefficiencies and to find ways and means of improving the underground environment without having to increase the ventilation and cooling infrastructure significantly (6). As the mine layout and air flow rates are essentially fixed, improvement of the underground environment depends primarily on the success with which the utilisation of air and existing cooling plant and equipment can be improved. In the text these are referred to as type A projects.
- existing mines where the object is to extend mining operations into new ground and often to greater depth. The aim of this type of project is to make the best possible use of existing facilities and to define the additional infrastructure required. An essential part is to determine the influence of worked out areas and extended airways on the underground environment for future mining operations and to revise mining practices and long term production planning to best suit them (7). These are referred to as type B projects.
- new mines where the aim is to create the necessary infrastructure and to analyse proposed mining methods and layouts to establish whether they are conducive to creating an acceptable environment (8). These are referred to as type C projects.

Problems in achieving an acceptable underground environment are often experienced on mines; they are not only costly to rectify but can result in serious consequential losses particularly when these are associated with low productivity. The reasons for these problems are multiple and can be summarised as follows.

- Historical management and organisational developments with respect to environmental control. The need for a well established ventilation or environmental department has not been recognised by many mines, particularly the older ones and those that started by mining the outcrops of an ore body. Hence, long term mine layout and production planning were conducted without the necessary expertise on ventilation and cooling and consequently the associated infrastructure viz., shafts, airways, major exhaust fans, cooling plant and associated facilities became inadequate.
- Ventilation and cooling problems often escalate due to mining operations extending into new ground. The effect of extending airways and keeping worked out areas open usually results in
  - an increase in the flow of pollutants such as heat and radon from the constantly increasing exposed rock surface area.
  - an escalation of the problems associated with the distribution and control of air and cooling water to the working places. These problems are aggravated because of production activities moving from one area to another (4).
- The thermodynamic and thermal behaviour of mines is complex and difficult to quantify. Hence, the significance of phenomena such as auto-compression, the geothermal gradient of the earth's crust, thermal inertia of the side walls of excavations etc., are often not fully understood, resulting in

the engineering of facilities that are inadequate.

- Thermodynamic and process design requirements associated with ventilation and cooling are often in conflict with practical operational constraints. Too often mines resort to what may appear to be the obvious thermodynamic and process design without analysing the operational and maintenance requirements of the various systems.
- Ventilation and cooling engineering on mines are multi-disciplinary and involve the co-ordination and planning of mining, process, mechanical, electrical and instrumentation engineering. Hence, it can be stated that it is project-engineering orientated, which on many mines is not recognised and consequently the ventilation departments are inadequately staffed (9).

ANALYSING VENTILATION AND COOLING REQUIREMENTS

The most important aspect of auditing and engineering ventilation and cooling systems for underground mines is that of preparing definitive ventilation and cooling strategies. The objectives of these strategies are as follows.

- They serve to define the policy to be followed by the mine as well as the operational responsibilities associated with ventilation and cooling which are to be borne by the various departments on the mine viz., mining and production engineering, engineering services, environmental (ventilation) engineering etc. It is therefore essential that the senior personnel of each department, with extensive experience of the mine, provide the input for preparing the strategies. It is furthermore essential that the strategies devised are acceptable and clearly understood by all who will be concerned with their implementation. To ensure that the measures and decisions finally agreed upon are carried out, it is preferable that a senior manager of at least Mine Manager status personally assumes responsibility for co-ordinating their preparation.
- The associated major capital and operating expenses such as shafts, airways, major fans, refrigeration machines, cooling towers, bulk spray coolers, raise and stope air-coolers, pumping and piping requirements are all determined.
- They serve as the basis for the process design of the ventilation air, cooling water distribution system and refrigeration installation, as well as the process analysis of the entire ventilation and cooling system. By process analysis is meant the evaluation of the performance and utilisation of equipment under the various conditions that may exist on the mine.

A project sequence diagram for preparing definitive strategies is shown in Figure 1 for the three types of projects referred to previously. The diagram depicts the various steps and the iterative nature of the preparation.

Production, Ventilation and Cooling Models

Mines are usually too complex and the future production planning too undefined to permit the preparation of detailed strategies. Care must therefore be taken not to create too much work in their preparation as one soon reaches a point of deminishing returns. It is for these reasons necessary to commence the exercise with simplified ventilation and cooling models that accurately represent the current and anticipated future production scenes on the mine.

These models geographically define the production areas as well as the principal routes for ventilating the various areas. A model has to be prepared for each future production phase commencing with a situation, say 3 to 5 years in the future, and further models for say 7 to 9, 12 to 15, 18 to 22 and 25 to 30 years in the future or whatever may be required. In defining the models it is essential that all concerned, but particularly the production engineering staff, are extensively involved under the leadership of a senior manager. Mistakes or erroneous interpretation of production requirements at this early stage will inevitably result in a considerable amount of abortive work, not to mention the consequences of incorrect assumptions and decisions.

Each geographical production area is defined by

- the production for both ore and waste.
- the centre (depth) of production as well as the geographical boundaries i.e., the shallowest and deepest production levels.
- the virgin rock temperature at the centre of production.
- the total length of haulages, airways, accesses as well as their average cross sectional dimensions.
- the number of stoping areas and the average dimensions of stopes and working excavations. Each stoping (mining) method must be treated separately.

A geographical production area is defined in such a manner that a ventilation and cooling strategy can be devised assuming all production is on one level i.e., at the centre of production, without having to treat the extremities of the area differently from the averages. Figure 2 depicts the principle of a simplified model whilst Figure 3 indicates the model for a typical mine.

Analysis of Air Flow through a Mine

Prior to commencing with the preparation and collection of data for determining the heat loads of the various production areas of a model, it is necessary to briefly analyse the air flows and in particular, the associated air velocities and pressure losses to ensure that the models are in fact practical propositions. The following are guidelines for assessing the practicality of models.

- Flow rates should typically be between 3 and 5 m^3/s per k ton/month at standard density. Values outside this range are not necessarily incorrect or unacceptable.
- Pressure losses should typically be between 1000 and 1500 Pa for each of the downcast and upcast shafts and between 2000 and 4000 Pa through the workings. The total pressure losses at this stage of the investigation should not exceed 7000 Pa. For shallow mines these figures may be too high.
- Velocities should preferably be less than 7 m/s in intake airways that are constantly used. In the return airways where there is little activity they should preferably be less than 10 m/s. Experience has shown that mine layouts that have been based on higher air velocities often result in unacceptable pressure losses.

When auditing existing mines (type A projects) it is usually possible to determine fairly accurately the relationship between air flow rates and pressure losses. This also applies to type B projects provided the extensions depend largely on the existing shafts and airways. On type C projects (new mines) the sizes of airways and shafts are usually dictated

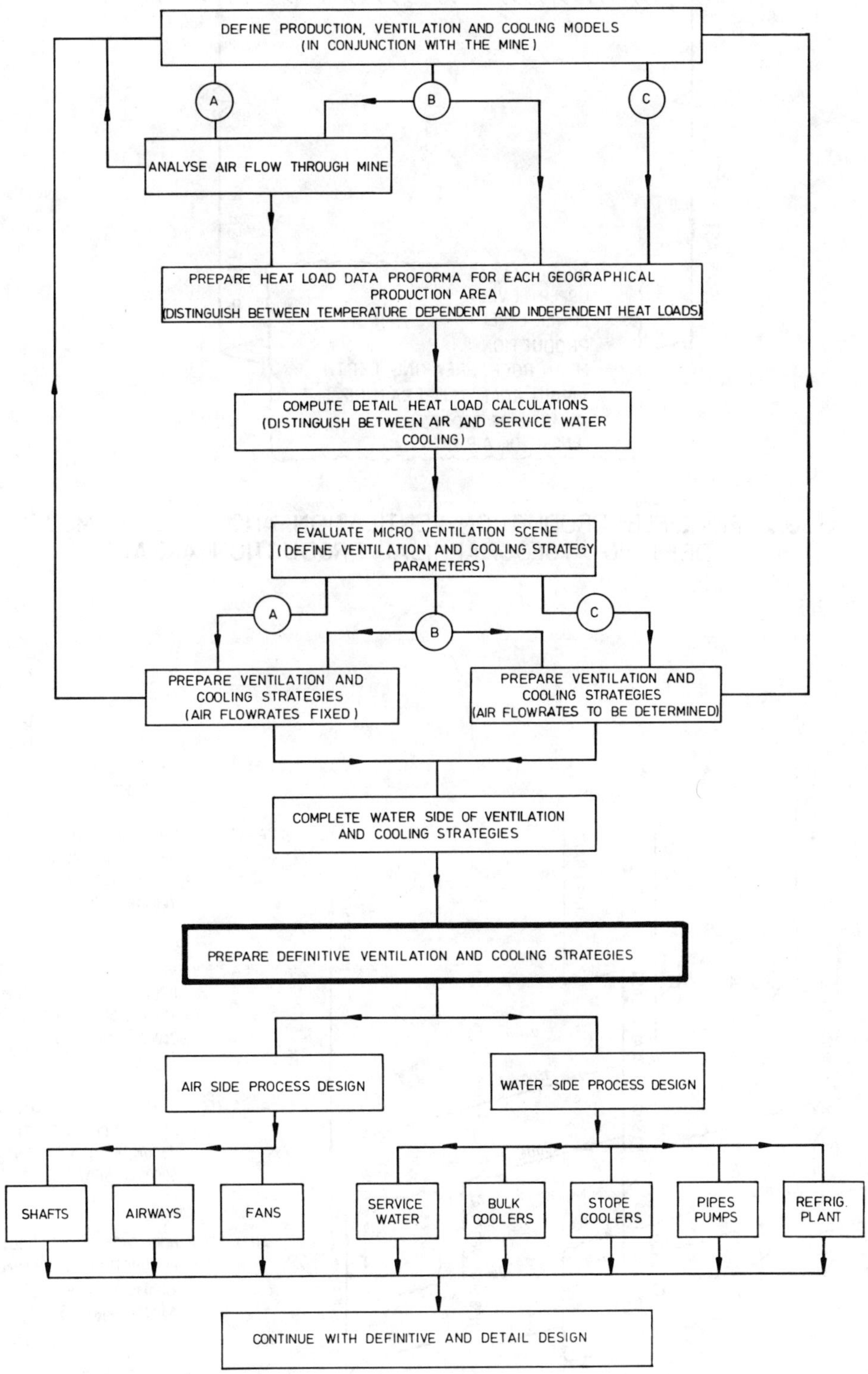

FIG. 1 ANALYSING VENTILATION AND COOLING REQUIREMENTS FOR MINES.

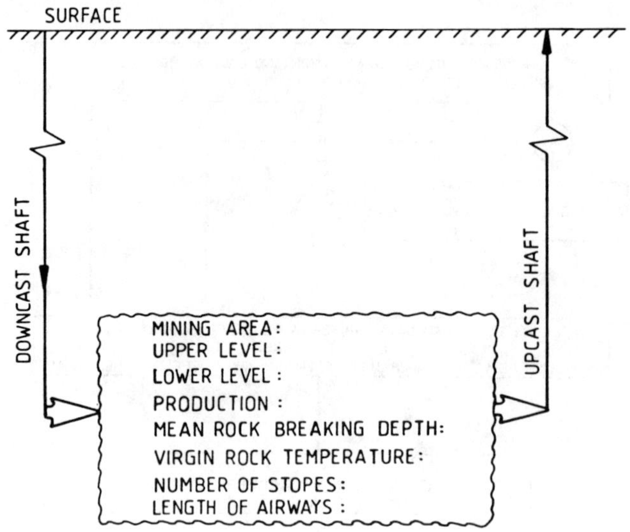

FIG. 2 PRINCIPLE: PRODUCTION, VENTILATION AND COOLING MODEL DEFINING A GEOGRAPHICAL PRODUCTION AREA.

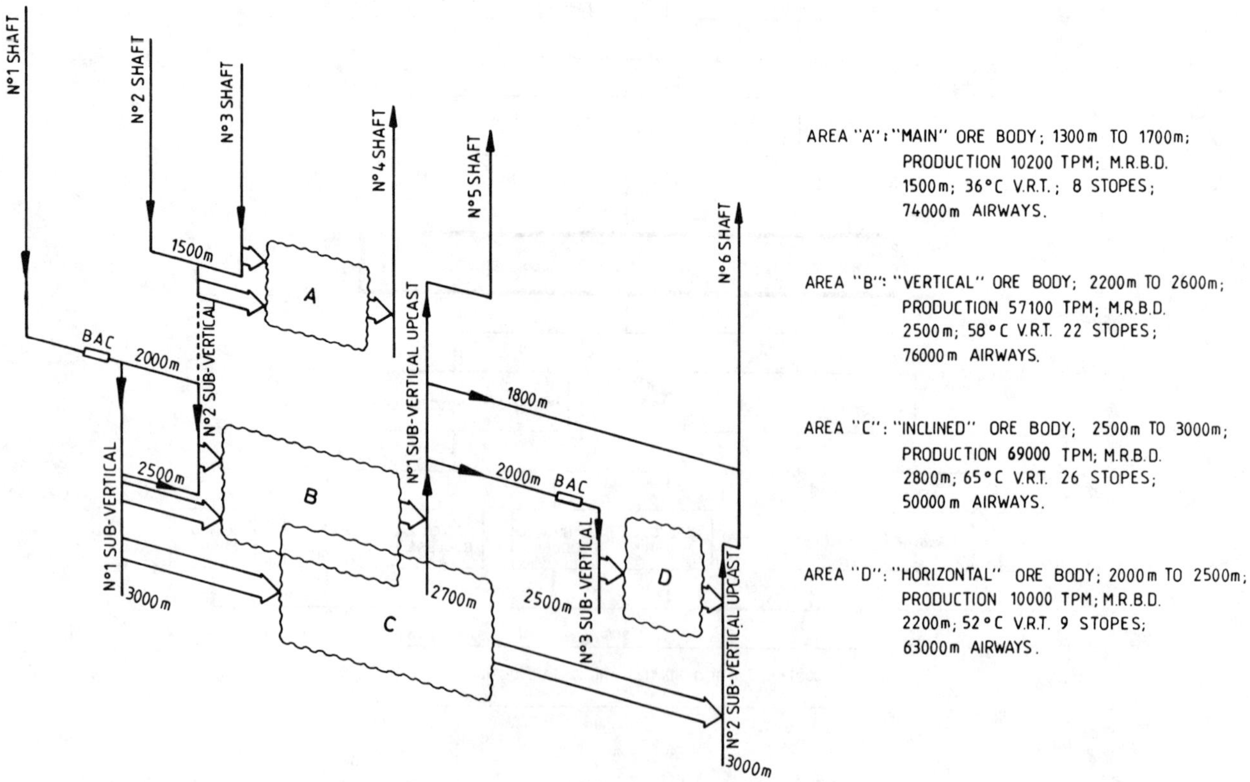

FIG. 3 TYPICAL PRODUCTION, VENTILATION AND COOLING MODEL.

by the air flow requirements and the above guidelines.

Heat Load Data and Data Proforma

Once the models have been prepared, i.e., the mine has been divided into geographical production areas for each of the future production phases, data sheets can be prepared for obtaining the necessary information. The data sheets have to comply with the requirements of the model and must present the results of the heat load calculations in such a way that they can be transferred onto the ventilation and cooling strategy flow diagrams. For this reason the heat load data sheet consists of two parts. The first part is the basic heat load data and the results of the heat load calculations in line with the information as shown on the model. The second part of the tables is a summary and re-arrangement of the results that can be transferred directly onto the strategy flow diagrams.

EXPOSED SURFACE AREA	m^2
THERMAL DIAMETER	m
THERMAL AGE	mths
HEAT FORM SIDE WALLS	kW

TABLE 1. HEAT LOAD DATA FOR THE SIDE WALLS OF SHAFTS.

DEVELOPMENT PRODUCTION (ROCK BREAKING)	tons/mth
HEAT FROM BROKEN ROCK	kW
AIRWAYS AND DRIFTS:	
TOTAL LENGTH	m
AVERAGE WIDTH	m
AVERAGE HEIGHT	m
AGE OF PRODUCTION AREA	mths
HEAT FROM SIDE, HANGING AND FOOT WALL	kW
NUMBER OF DEVELOPMENT HEADINGS	-
FACE ADVANCE FOR ALL HEADINGS	m/mth
FACE ADVANCE PER HEADING	m/mth
DEVELOPMENT CYCLE	mths
HEAT FROM SIDE, HANGING AND FOOT WALL	kW

TABLE 2. HEAT LOAD DATA FOR THE WALLS AND BROKEN ROCK IN AIRWAYS.

STOPE PRODUCTION (ROCK BREAKING)	tons/mth
HEAT FROM BROKEN ROCK	kW
NUMBER OF STOPES	-
AVERAGE LENGTH OF STOPES	m
AVERAGE WIDTH OF STOPES	m
AVERAGE HEIGHT OF STOPES	m
AVERAGE TIME BETWEEN BLASTS	mths
EXPOSED ROCK SURFACE AREA	m^2
HEAT FROM SIDE, HANGING AND FOOT WALL	kW

TABLE 3. HEAT LOAD DATA FOR THE WALLS AND BROKEN ROCK IN STOPES.

	kW RATING	%UTL	% HEAT	kW
SHAFT & SHAFT STATIONS:				
FANS				
PUMPS				
HOISTS				
OTHER				
			TOTAL	
AIRWAYS:				
LOCOMOTIVES				
PUMPS				
TRANSFORMERS (kVA*)				
OTHER				
			TOTAL	
STOPES:				
FANS				
SCRAPERS				
LHD VEHICLES				
OTHER				
			TOTAL	

TABLE 4. HEAT LOAD DATA FOR ALL ELECTRICALLY AND DIESEL POWERED EQUIPMENT.

In the first part, the tables distinguish firstly between temperature dependent and independent heat loads and secondly between heat generated in downcast shafts, main interconnecting airways, airways in the geographical production area and finally stopes (refer to Tables 1 to 4).

In the second part, the figures are rearranged and distinguish firstly between the geographical distribution of heat loads for each of the various areas i.e., shafts and/or interconnecting airways, airways in the production area and finally stopes. For each of these, the tables then distinguish between temperature dependent and independent heat loads with the added complication of defining what portion of each of these should be countered by the incoming ventilation air and/or by the use of chilled service water (refer to Tables 5 to 7).

Although it is possible to standardise on the format for calculating and presenting heat load data, care must nevertheless be taken to ensure that a systematic and logical set of data sheets is prepared for each model and that a clear definition of the geographical boundaries and equipment for each of these is given.

Heat Load Calculations

The various sources of heat in the mine can be placed in either one of two categories, viz., temperature dependent or temperature independent loads. Temperature dependent heat loads are those of which the magnitude depends on the temperature of the ventilation air. There are essentially three sources of heat in this category, viz., from the side walls of excavations and the broken rock, from water on the footwall and lastly from fluids in pipes, e.g., ground water that is pumped out of the mine in an uninsula-

	COLD AIR kW	COLD SERVICE WATER kW
TEMPERATURE DEPENDENT:		
EXPOSED ROCK		
GROUND WATER		
TOTAL		
TEMPERATURE INDEPENDENT:		
FANS		
PUMPS		
HOISTS		
OTHER		
TOTAL		

TABLE 5. HEAT GENERATED IN THE DOWNCAST SHAFT.

	COLD AIR kW	COLD SERVICE WATER kW
TEMPERATURE DEPENDENT:		
EXPOSED AND BROKEN ROCK		
GROUND WATER		
TOTAL		
TEMPERATURE INDEPENDENT:		
LOCOMOTIVES		
PUMPS		
TRANSOFRMERS		
OTHER		
TOTAL		

TABLE 6. HEAT GENERATED IN THE AIRWAYS.

	COLD AIR kW	COLD SERVICE WATER kW
TEMPERATURE DEPENDENT:		
EXPOSED ROCK		
BROKEN ROCK		
GROUND WATER		
TOTAL		
TEMPERATURE INDEPENDENT:		
FANS		
SCRAPERS		
LHD VEHICLES		
OTHER		
TOTAL		

TABLE 7. HEAT GENERATED IN THE STOPES.

ted pipe. Temperature-independent heat loads are those of which the magnitude is independent of the temperature of the air e.g., all heat generated by electrically driven equipment such as pumps, hoists, scrapers, fans etc., as well as heat from diesel driven equipment such as loco's and LHD vehicles.

Experience on a number of mines has shown that typically 70 to 80 percent of all the heat generated comes from 2 or 3 sources. On the gold mines in the R.S.A. heat from the exposed rock surfaces dominates the scene to such an extent that other sources of heat are often neglected. On many coal or base mineral mines, heat from mechanical equipment and in particular from diesel driven LHD vehicles is the major source of heat. On other mines it may be the ground water or, where blind hole mining takes place, it could be the fans for ventilating the blind hole working ends, etc.

Much has been written on the subject of sources of heat in mines (2,3,10 to 13). Reference is therefore only made to the sources of heat that often dominate the heat load scene.

Heat from the rock.

Heat from the exposed rock surface areas, i.e., from the side, hanging and footwalls of airways and stopes is often the major source of heat. Although the Starfield (10) method is the theoretically correct one for calculating the heat load in airways it is often difficult to apply because it depends on the air velocities. These are not always known with an accuracy that justifies the use of this method over others, e.g., the Goch Patterson method (11), which assumes that the rock face temperature is the same as that of the air. Also, the accuracy with which lengths and average cross sectional dimensions of airways can be obtained or anticipated is usually far worse than the improved accuracy obtainable with the Starfield method. When using the Goch Patterson method it is likely that the heat load calculation for airways will be over-estimated by about 10 to 20 percent, and for this reason it is advisable not to allow for a contingency in the heat load. Care should furthermore be taken when using the Goch Patterson tables that the correct value is chosen, as two values are given for each time interval. The first relates to the instantaneous heat load after a period of time has elapsed and should only be used for shafts and connecting airways that are old, i.e., 5 or more years after the excavations have been completed. The other value is the time average heat load and is a measure of the amount of energy stored in the rock walls and should be used in all cases where excavation and development work is a continuous and ongoing activity.

Calculating the heat from side walls in stopes is usually difficult and requires a fair amount of investigation before a reliable and satisfactory method can be found. This is because of the great variety of mining methods and the odd shapes and sizes of stopes associated with them. For narrow reef mining as found in the gold mines in the R.S.A. and for some room and pillar mining methods, the method proposed by Van der Walt and Whillier (2) will give satisfactory results. For other mining methods it is best to relate stopes to cylindrical equivalent shapes and to apply the Goch Patterson (11) method.

Heat from the broken rock is usually small compared with the heat from the side walls of excavations and is calculated assuming that the heat is dissipated at an average rate while the broken rock is cooled from the virgin rock temperature to a reference temperature usually taken as 30°C. The other temperature-dependent heat loads, i.e., ground water and water on the footwall can be, and often are neglected. As these can be detrimental to the environment they should be countered by containing the water in a pipe which is preferably insulated. Seldom can the cost associated with increasing the air flow rate and refrigeration requirements be justified to counter the effect of heat from this source although instances do exist (8).

An essential part of the calculations associated with the preparation of a ventilation and cooling strategy flow diagram is that of calculating the marginal heat loads. These are the portion of the temperature-dependent heat loads due to the actual temperature of the air being above or below the reference temperature (often taken as 30°C).

Heat from underground ventilation fans.

Underground fans could be a major contributor to temperature-independent heat loads as all the energy required for the fans is converted to heat. A psychrometric phenomenon that could easily be overlooked in respect of fans and the associated rise in wet bulb temperature of the air as it passes through the fan, is the amount of cooling that is required to recool the air back to its original wet bulb temperature. Large underground booster fans may thus require as much as twice the fan power in cooling capacity to offset the effect of the fan power on the wet bulb temperature of the air. Auxiliary fans require somewhat less depending on the type of installation and the length of discharge ducts. It is for this as well as for operational reasons that mines should always give preference to employing large exhaust ventilation fans on surface rather than underground fans. In a number of cases where auxiliary fans have been used to force air through coolers and into stopes, the effect of the fan on the wet bulb temperature was greater than the cooling effect of the cooler resulting in a net increase in the temperature of the air passing through the system. In such cases preference should be given to different methods of ventilation and mine layout with a view to eliminating as many of the underground fans as possible.

Heat from mechanical mining equipment.

All the energy that is required to power mining equipment is converted to heat and results in an increase in the wet bulb temperature of the ventilation air. The heat generated from the equipment depends on

- the rated capacity.
- the percentage of time it is utilised which can vary from as little as 5 percent for poorly utilised scrapers to as much as 80 percent for continuous mining equipment. When mechanised equipment is a major contributor to heat it is essential that this aspect is carefully evaluated in conjunction with the mining and operational staff even possibly employing time and motion analysis.
- the extent to which the rated capacity is employed, which once again could be low for tramming equipment and high for continuous mining equipment. This is probably the most difficult element to evaluate for mechanised equipment and time, motion and loading studies are often the only means of establishing the power consumptions.
- whether it is driven by internal combustion engines or electrically. In the latter the rated capacity serves as a good basis for determining the heat load. The heat load associated with internal combustion engines could be much higher than the rated capacity and it is therefore necessary to evaluate, in addition to the abovementioned parameters, the thermodynamic efficiency of the engines under operating conditions. This is once again a difficult problem that may necessitate involving the designers of the equipment, as the heat generated can typically be between 1 and 2 times the rated capacity. In a number of cases where serious problems with the thermal environment have been experienced, a change from equipment driven by internal combustion engines to electrically driven equipment alleviated the problem to such an extent that no further action was necessary.
  In type A projects it may be possible to determine the electrical power and fuel consumption for the various geographical districts on the mine; this is however difficult and must be controlled carefully. It is not possible to establish, in this way, future heat loads or heat loads for new mines.

Pumps and hoists located underground are also contributors to temperature independent heat even though they are not necessarily major contributors. In this regard it should be borne in mind that most of the energy required to raise matter such as sludge, water or rock goes into increasing its potential energy. When lowering matter or transporting it horizontally all the potential energy or the energy required to move it horizontally is converted to heat. For pumping water up shafts it is suggested that an amount equal to about 25 to 35 percent of the pumping energy is considered as heat from the pumps. In the case of hoists an amount equal to between 75 and 100 percent of the mass hoisted or lowered should be considered as heat. It should be noted that these heat loads are not based on the rated capacity of the equipment but on the actual mass and height it is raised or lowered during a given period of time.

Micro Ventilation and Cooling Strategy Parameters

Prior to preparing the strategies it is necessary to evaluate the micro ventilation scene, i.e., in stopes, drawpoints etc., with a view to

- analysing and defining the layouts and design principles in respect of ventilating and cooling the underground environment taking cognisance of all related matters.
- defining air flow requirements.
- defining the desired air entering and leaving wet bulb temperatures in view of the anticipated temperature distribution in these areas.

When analysing the heat dissipated by the exposed rock surfaces in working places it should be borne in mind that it is time dependent and that it is often not possible to provide cooling to offset the heat dissipated in the first few days after a blast has taken place. The reason for this is that a large surface area at a temperature close to the virgin rock temperature is exposed and it usually requires several hours or days to cool the skin of the rock to the "reference" temperature. The heat load calculation for the purpose of determining air flow rates in working places as well as the required wet bulb temperatures for the air entering and leaving them is therefore based on the average heat dissipated in the first 15 to 20 percent of the period between consecutive blasts. This implies that tempera-

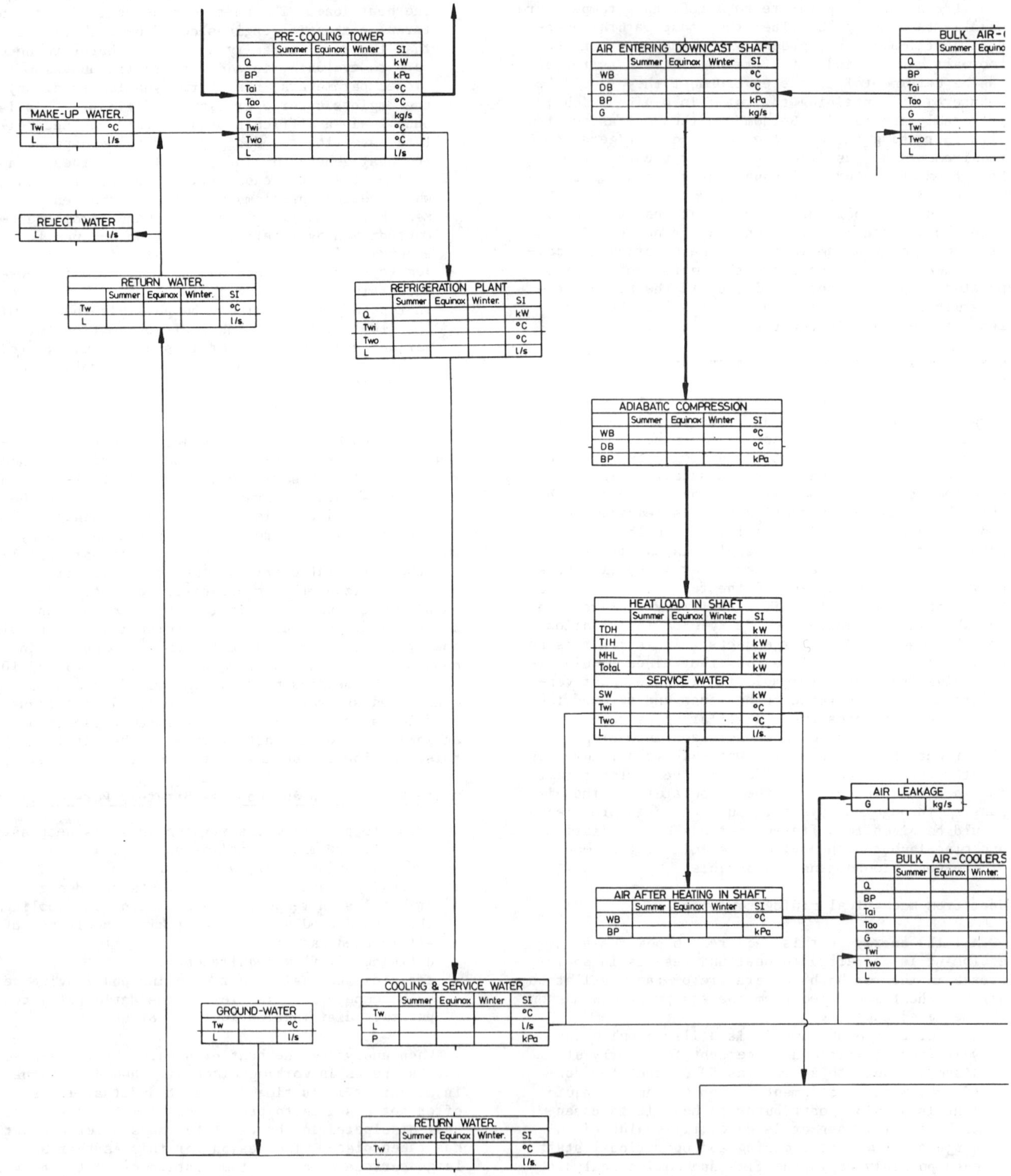

FIG. 4. PRINCIPLE : VENTILATION AND

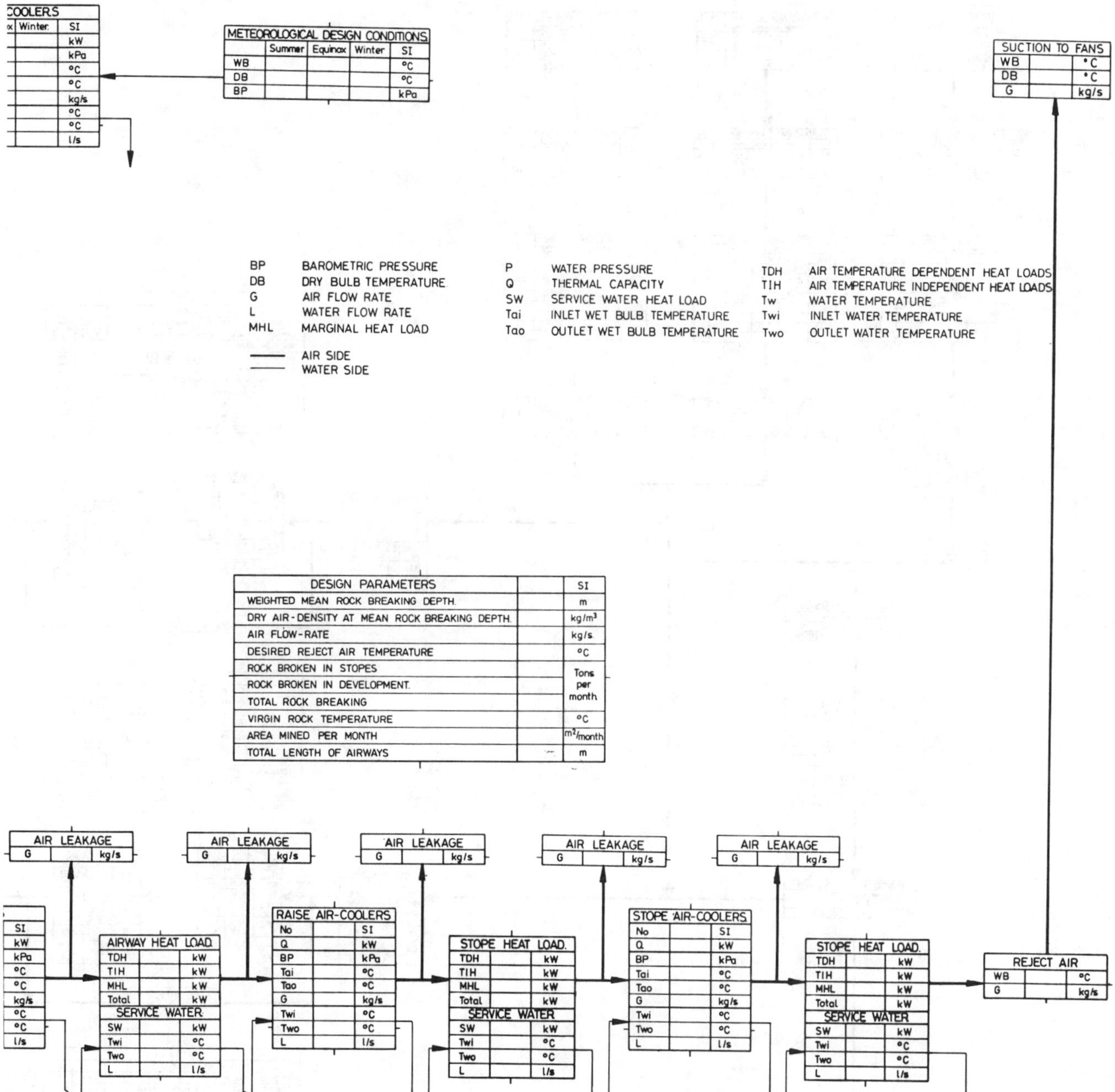

COOLING STRATEGY FLOW DIAGRAM.

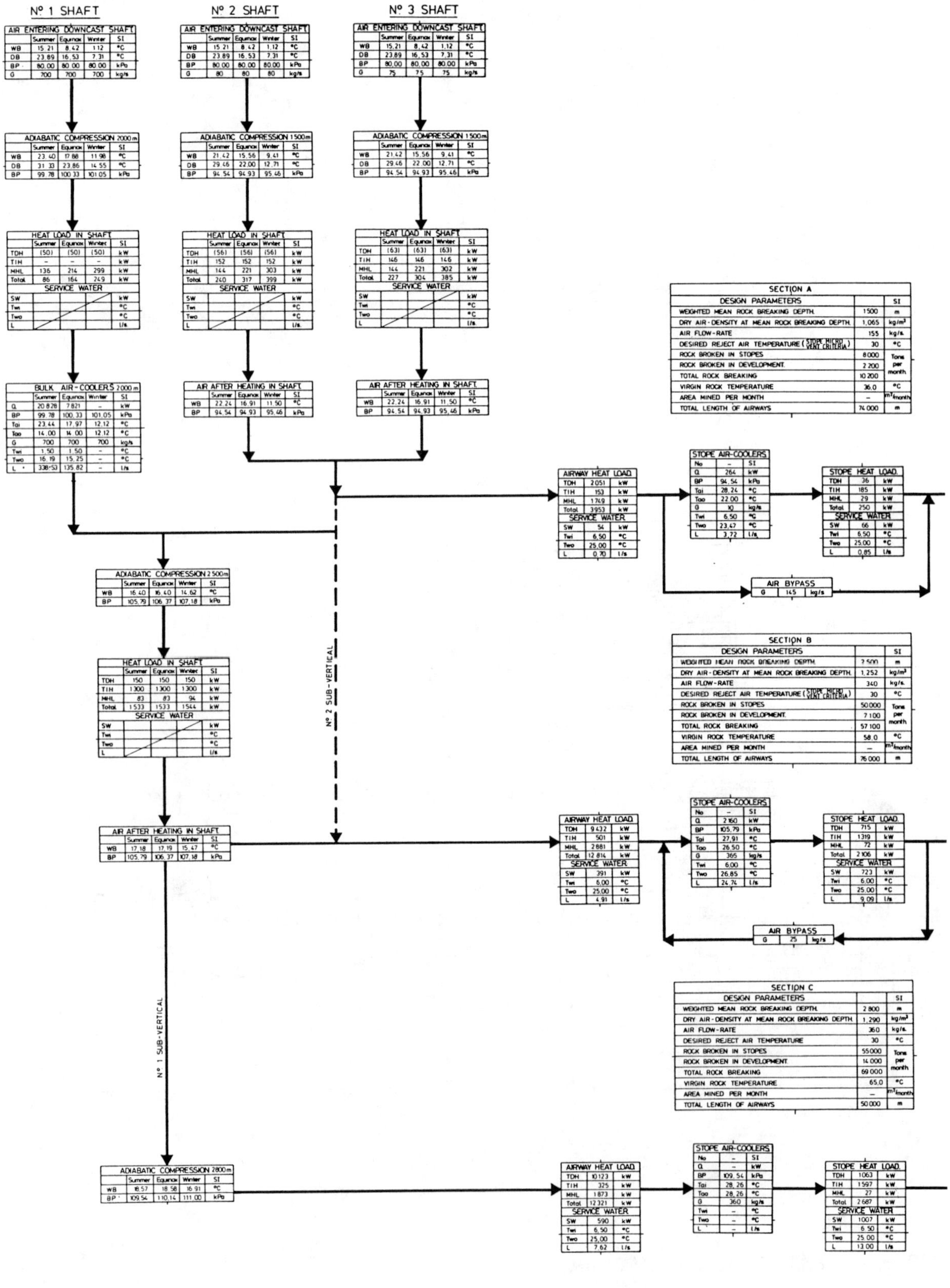

FIG 5. TYPICAL MINE VENTILATION AND

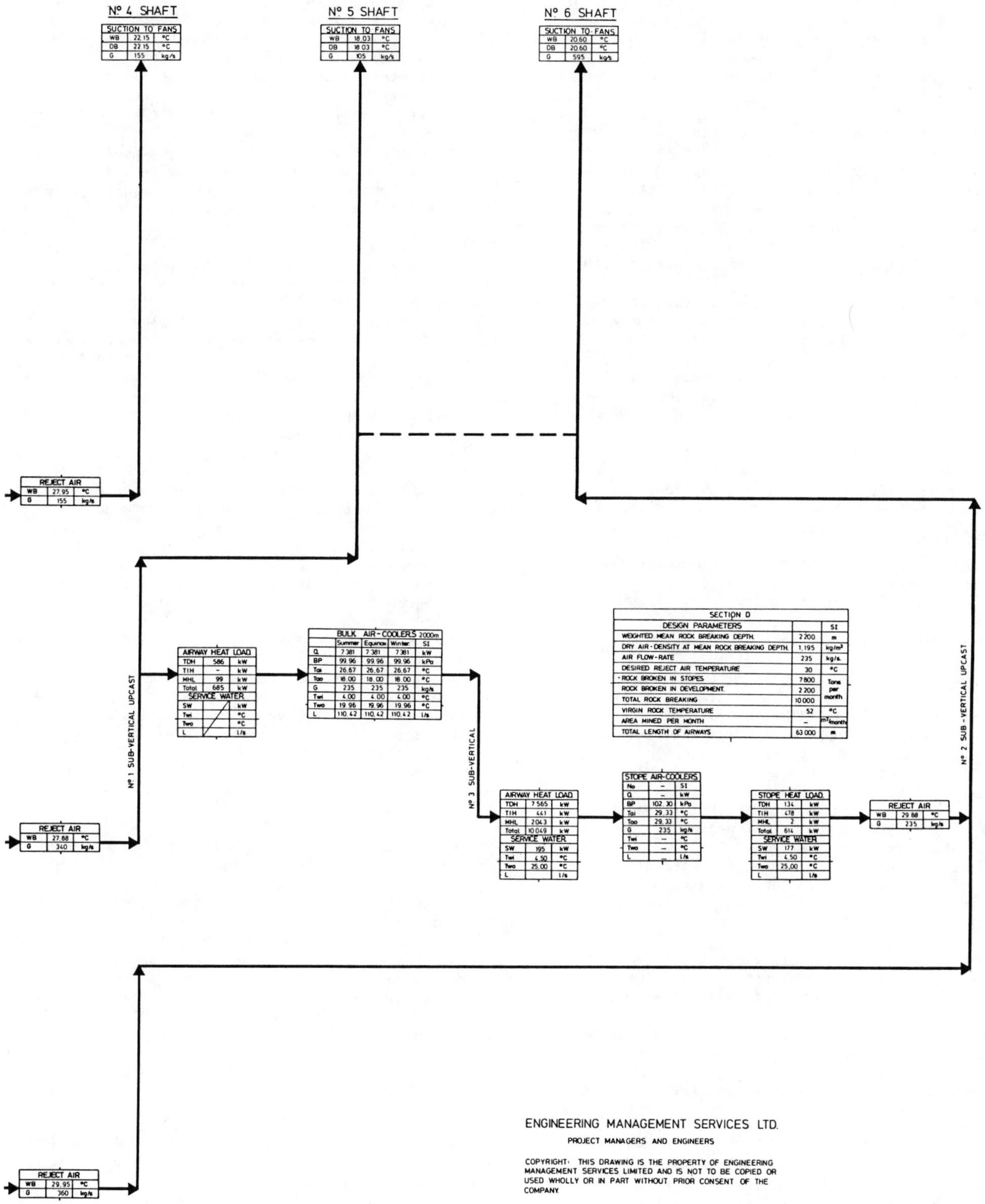

COOLING STRATEGY FLOW DIAGRAM.

tures will generally exceed the desired wet bulb temperature for this period. In the case of continuous mining, this problem does not exist and the heat load calculation is fairly straight forward. When blasting of the working face takes place once a day or once every two days it is treated as continuous mining.

One of the most important aspects to consider when evaluating working place layouts and design practices is the preference for through flow ventilation. This is the principle of passing the majority of the air on a particular level through each of the working places with a minimum of air bypassing them. This generally has the following advantages over the alternative of forcing air to the working places with fans.

- The temperature distribution is generally between the temperature of the air entering and the temperature of the air leaving. In the case of blind hole mining where it is impossible to employ through flow ventilation, the temperature distribution in stopes is largely unpredictable.
- The need for underground auxiliary fans reduces significantly.
- The wet bulb temperature of the air entering the working place can often be significantly higher without jeopordising the thermal environment.
- Less air is required to pass through the mine.
- Less refrigerated cooling is required.

The main disadvantage of through flow ventilation is the requirement for additional airways and ventilation doors which could be costly.

Preparation of Ventilation and Cooling Strategies

The model together with the results of the associated heat load calculations and strategy parameters can be presented on a ventilation and cooling strategy flow diagram as depicted in Figures 4 and 5. Figure 4 demonstrates the principles employed in the strategies, whilst Figure 5 is a typical ventilation and cooling strategy flow diagram. In Figure 4 the basic flow path for the ventilation air is shown along with the various cooling elements, heat loads in the shaft, airways and stopes. The flow diagram may also be referred to as the macro ventilation and cooling scene applicable to the entire mine as opposed to the micro scene which is only applicable to the various working places.

Although there are several types of strategies with varying degrees of complexity and sophistication, they can generally be placed into one of two catagories.

- Those where the air flow rates are determined by factors that are not related to the thermal environment such as
  - the availability and sizes of existing shafts and airways and the ventilation and cooling infrastructure.
  - the dilution of polutants other than heat, i.e., dust, radon, noxious vapours, gases etc.

  This category is largely limited to type A projects, and type B projects where the extension of mining operations depends heavily on the utilisation of the existing infrastructure. Experience has shown it is likely that too little air will be available from the thermal environment viewpoint.
- Those where the air flow rates are to be determined with the aid of the strategies, i.e., heat is the major polutant. The amount of air passing through the mine in this instance is determined by factors such as
  - the micro ventilation requirements in respect of the wet bulb temperature of the air entering and leaving stopes and working places.
  - the preference to bulk cool the ventilation air rather than to cool the air at entrances to stopes and in stopes for practical considerations.

  This category of strategies is applicable to type C projects (new mines) as well as type B projects where extending mining operations requires a major new surface and underground infrastructure.

It can be stated that a strategy is a compromise between practical and environmental considerations on the one hand, thermal and thermodynamic considerations on the other and then of course in the final analysis, economical considerations play an important role. Care must be taken not to select what appears to be an obviously economical strategy without giving due consideration to consequential hidden expenses and problems that may be difficult to quantify particularly in respect of the effect of the thermal environment on production. Other factors that need to be considered are training requirements, norms and customs of the underground work force and also, the factors that could cause the ventilation and cooling system to malfunction. A commitment to a cooling strategy that has little flexibility can be very costly and, in fact, may limit the life of the mine.

The cooling elements involved in a strategy are listed below in the sequence in which they should be considered when preparing such strategies.

- Cooling of all of the mine service water that is used for dust suppression in the rock drills, etc.
- Bulk cooling of the downcast ventilation air on surface.
- Bulk cooling of the ventilation air underground.
- Cooling of the air at the entrances to stopes and in raise air-coolers.
- Cooling of the air in stopes.

Cooling of service water.

Service water is an integral part of the mining operation and will thus provide effective cooling in remote working places where men are using the water and where wet bulb temperatures are often high due to problems with ventilation. Cooling with service water is essentially a direct form of cooling, i.e., it is not used to cool the ventilation air but used to remove heat directly from certain sources thus preventing them from raising the temperature of the air. It is for this reason that the heat load calculations distinguish between heat that can be countered by cooling the air and by providing cold service water (refer to Tables 5 to 7) and it is thus integrated with the block for heat loads on the strategy flow diagram.

Because this form of cooling is so effective in practice it is essential that consideration is given to artificially increasing the quantity of water used. This may not always be possible because of restrictions imposed on its use by mining practices, the infrastructure for pumping and perhaps the type of rock that is mined.

Bulk cooling of air on surface.

This type of cooling should be afforded consideration in all cases where the average daily wet bulb temperature for the hottest month in a year is above about 16°C. The factors that should be evaluated are the magnitude of the temperature dependent heat loads in the shaft and intake airways, air leakages and the year-round utilisation of equipment associated with the cooling of the air on surface. When evaluating a strategy employing this form of cooling, the marginal heat load is calculated and shown on the flow diagram in the same block that gives the details of the temperature dependent and independent heat loads. It is thus possible to assess the amount of cooling that has to be wasted when the air is cooled on surface. Experience has shown that when air flow rates are low or when shafts are wet due to a continuous inflow of ground water it is most unlikely that cooling of the ventilation air on surface can be justified.

Surface air-coolers and the associated refrigeration plant should be designed to cool all the air that decends the shaft from the average wet bulb temperature for the hottest month of the year to about 10°C. The cost of an installation providing more cooling can seldom be justified because of the reduced year round utilisation. In the R.S.A. a practical yard-stick and objective is to create what one may term equinox conditions at the shaft inlet throughout the summer period.

The main advantage of cooling the air on surface is that it reduces the amount of cooling to be done underground and hence the cooling water flow rate which is often a major problem on large installations. The operating and capital cost per unit of cooling is low when compared with other alternatives. The success with which surface bulk air-coolers can be operated and maintained generally offsets many of the inefficiencies associated with them such as the additional heat pickup in shafts and airways as well as the losses associated with the leakage of air directly to return airways.

Bulk cooling of air underground.

In all cases where underground cooling of the air is required it is essential that consideration be given to cooling the air in bulk near to the shaft. This is particularly the case where the effect of auto-compression and shaft heat loads on the temperature of the air is appreciable. The factors that should be evaluated are the effect of the cold air leaving the bulk coolers on the temperature dependent heat loads in the airways to the stopes and the amount of air that leaks through worked out areas and is therefore not beneficially used.

When air flow through a mine is limited to what can pass through the shafts and airways the thermal capacity of the air after it leaves the bulk coolers is often inadequate. resulting in a rapid rise in the temperature of the air due to the increase in temperature dependent heat loads. In such cases there is little merit in cooling the air in bulk as it leaves the shaft unless the air flow rate can be increased. In type B or C projects where the extension to mining operations will require additional infrastructure and shafts, the flow rate through the mine is usually determined by the requirement to cool the air in bulk without any subsequent cooling.

The coolers should be sized to cool the downcast air from about the average wet bulb temperature at which it will arrive at the cooler during the hottest months of the year to roughly 18°C. Although there is often merit in cooling below 18°C this should preferably not be considered when preparing strategies as it would allow no flexibility or safety factor should this become necessary at some future date. In the R.S.A. a practical yard-stick and objective is to create underground winter conditions throughout the year.

The relative ease, reliability and success with which bulk coolers can be operated generally justifies the inefficiencies associated with air leakages and the increase in the temperature-dependent heat loads in airways. They remain an essentially fixed installation over the life of the mine and hence are more accessible and require less manpower for maintenance, reduce the amount of water to be handled in stopes and airways and have a lower capital and operating cost per unit of cooling over the life of the mine than those of the remaining alternatives (4).

Cooling of air at the entrances to stopes and working places.

Although this is not desirable for many practical reasons (4) it is often unavoidable for type A projects and certain type B projects, viz., when the air flow through the mine is inadequate and recirculation of the air is necessary to introduce the required cooling. For practical and economical reasons cooling of air <u>at the entrances</u> is preferred to cooling <u>in</u> the working places. Most strategies involving coolers at the entrances to stopes (raise air-coolers) will require less refrigeration capacity than a bulk air-cooling installation for the same end result. The reason for this is the loss of cooling associated with air leakages and the increase in temperature dependent heat loads in the airways when employing bulk air-coolers. Care should however be taken that this phenomenon does not become a decisive factor when deciding on alternative strategies as there are many practical considerations that are much more important than this apparent saving.

The disadvantage of the raise air-cooler is that it is a movable form of cooling installation which if not fully integrated with the mining operation and properly maintained can be very ineffective. Experience with them on R.S.A. gold mines has shown that they are expensive and difficult to maintain and advance timeously as new stopes are being developed and brought into production. In very few cases in practice can it be stated that their use has achieved the design objective. In all of these cases it was noted that very little cooling was required to achieve acceptable thermal environmental conditions and that the refrigeration requirement was of the order of 50 kW (R) per k ton/month. Furthermore, the life of stopes in all of these cases was of several years duration and the installations were in fixed positions during this period.

Cooling of air in stopes.

This should if possible be avoided as it usually creates serious problems in stopes and is seldom worth the trouble and expenditure. It can often only be overcome by increasing the flow rate or by recirculating some of the air that passes into the stopes.

Both this and the previous method can seldom be afforded if the inlet wet bulb temperature during the summer months at stopes and working places is below about 24°C as air at this temperature is often acceptable. Only when heat loads are exceptionally high and provided enough air is available should consideration be given to cooling the air to below 24°C, in which case the use of spray coolers is preferred to cooling coils. The cooling range for both types of installations varies from 0 to about 5 deg C depending on the temperature of the air, the type of cooler and the air and water flow rates; part of the strategy exercise is to determine these requirements.

The regulation of water to the various cooling installations is one of the main problems that is encountered. This is because the pressure losses through the cooling water pipes are usually much higher than those across the cooling units (particularly cooling coils), and with a multiple of parallel paths and continuous changes to the system it becomes totally impossible to balance (4).

Water Side of Ventilation and Cooling Strategies

Once the air side of the strategy has been defined the water side can be completed and a first estimate of the refrigeration plant requirements and pipe sizes made. Figure 4 depicts cooling water being supplied to the various users from a common source with the return through a common pipe system. The latter is not necessarily correct as the strategy only serves as a basis for the process design of the air and cooling water distribution system. What is however important is to define the water requirements and temperatures of the water entering and leaving the various cooling elements (users).

An important rule with regard to the temperature at which water is supplied to the various elements is that it must at all times be as low as practically possible. The reason for this is to keep water flow rates to an absolute minimum thus permitting the maximum amount of cooling to be distributed for a given quantity of water.

The temperature at which water is delivered to the users depends on the type of refrigeration plant that is considered for the application and the location of the plant whether on surface or underground. The following are rough indications of the temperature at which water will leave the shaft when mining operations are at a depth of about 2 km.

- A conventional plant producing 4°C water on surface will provide water at about 9°C. This is generally too high resulting in large water flow rates. Consequently, there are currently several engineering developments in the R.S.A. to reduce the temperature with a view to reducing the flow rate.
- A conventional surface plant producing 4°C water together with an energy recovery device will provide water at about 5.5°C.
- A plant producing 0°C water on surface will provide it underground at about 5°C.
- A plant producing 0°C water on surface combined with an energy recovery device will provide water at about 1.5°C.
- An ice-making plant on surface will provide water underground at 0°C.

A conventional underground refrigeration plant producing 4°C water will probably provide it at a temperature somewhat higher than this depending on the difference in depths between the plant and the users.

The principle for supplying chilled water to bulk air-coolers on surface is different to that of supplying cold water to other users, as no restrictions to water flow rates exist. In the former the supply water temperature depends on the wet bulb temperature of the ambient air and it is for this reason desirable to have a refrigeration plant separate from the other plants supplying water to the underground users. The engineering of refrigeration plant is the subject of other publications (14 and 15) and will therefore not be dealt with any further.

The following is a rough indication of the temperatures at which water will return from the various users.

- Service water. It is desirable that the temperature of the service water on the footwall should rise to about 25°C. Temperatures above this level may and often do impose a secondary heat load on the air, which is undesirable. On the other hand, temperatures below this level will require an increase in the amount of water and may aggravate the problems on the footwall, increase the possibility of mud-rushes in box holes, etc.
- Surface bulk air-coolers. From a process and equipment utilisation viewpoint the temperature should be about 4 deg C higher than the temperature at which it is supplied, which in turn depends on the ambient wet bulb temperature (15).
- Underground bulk air-coolers. The temperature depends to a large extent on that of the air entering the cooler as well as the design of the cooler and can be calculated reasonably accurately. A return water temperature of between 19 and 21°C is generally considered good.
- Air-coolers at the entrances to and in the stope working places. Here, the return temperatures could vary over a wide range depending to a large extent on
  - the temperature of the air entering the cooler.
  - the type of cooler, i.e., spray type or cooling coil.
  - the water flow rate through the cooler.

  For a spray-type cooler handling inlet air around 30°C, a return water temperature of about 25°C is considered good. For cooling coil installations with an abundance of water, cooling air from about 27°C, a return temperature of roughly 16°C is typical. This demonstrates the variation and uncertainty in return water temperatures. It is for this reason necessary to carefully analyse the cooling of the air and to allow generously on the water side process design as well as the process design of the refrigeration plant to accommodate large variations in water flow rates and return temperatures.

Definitive Ventilation and Cooling Strategies

Preparation of the strategies is an iterative process and often results in a re-investigation of the mine layout and the ventilation and cooling models. This is demonstrated on the project sequence diagram, Figure 1. It should be borne in mind that the definitive strategies for each production phase should logically be developed from the present to the foreseable future of the mine.

Once a set of strategies has been prepared and prior to preparing the definitive set, it is essential that extensive discussion is held on the mine to ensure that the consequences of the strategies are fully understood by the mine management, senior engineering and production staff. To successfully investigate and prepare definitive strategies requires several months and involves many hours of senior staff on the mine. Although the preparation may therefore appear to be a costly exercise the financial savings and benefits achieved by successfully cooling the mine by far offsets the cost and time associated with the preparation.

Process Design of Ventilation and Cooling Water Distribution Systems

After definitive strategies have been prepared the process design and analysis of the ventilation and cooling water distribution systems as well as the process analysis of the various cooling elements and the refrigeration plant can commence.

Air side process design.

This involves the preparation of flowsheets and the analysis of the shafts and airways with respect to

- air flow rates (mass and volume) for each leg.
- pressure losses through each leg.
- air velocities in each airway.
- dimensions of each airway.
- air control facilities.

During the process design, consideration should be given to eliminating areas where the pressure losses are relatively high by providing additional airways. The natural distribution of the air through the mine should be carefully evaluated and may require the use of computer programmes as these calculations are often of an iterative nature that cannot be done manually. Control of air distribution underground should preferably be done by employing air control doors and regulators rather than fans, as the latter introduces heat into the mine, is less flexible and can fail, resulting in a maldistribution of air.

Once flowsheets have been prepared for each of the strategies over the life of the mine, the exhaust fans and associated equipment can be specified to meet with the long term requirements of the mine.

Water side process design.

This involves the preparation of quantitative as well as engineering flowsheets for

- the overall water distribution and reticulation system.
- the bulk air-coolers including the air side.
- stope and raise air-coolers including the air side.
- the refrigeration plant (14, 15).
- water settling and treatment plant.

Quantitative flowsheets essentially depict the design energy and mass balances, i.e., air and water mass flow rates, pressure losses, temperatures and heat transfer rates across heat exchangers such as air-coolers, evaporators, condensers and cooling towers etc. Engineering flowsheets depict pipe and valve sizes, control gear and instrumentation, electrical equipment etc. For the air side, only quantitative flowsheets incorporating the engineering flowsheets are required. Once the process design for each cooling strategy over the life of a mine has been completed, the specifications for the various pieces of equipment can be prepared.

Process analysis of the ventilation and cooling system.

With the aid of the various flowsheets but prior to finalising them, it is necessary to analyse and determine the process characteristic of each of the cooling elements and major pieces of equipment with respect to daily fluctuations, seasonal variations and lifetime changes to water and air flow rates and temperatures, as well as the utilisation of equipment.

It should be borne in mind that shafts, airways and pipes serve a mine throughout its life and failure to meet with this requirement could be costly and, in fact may force the mine to seriously interrupt production. Major capital equipment such as fans, refrigeration plant and bulk air-coolers should be selected accordingly and their installation scheduled to meet with the various production phases throughout the life of the mine and yet offer a neatly engineered installation that meets at all times with the requirements of simplicity and ease of operation and maintenance.

CONCLUSION

A systematic method for analysing and eventually defining the ventilation and cooling requirements for hot underground mines has been presented. The importance of preparing definitive ventilation and cooling strategies in conjunction with senior personnel on a mine prior to any process or detail design work has been demonstrated.

ACKNOWLEDGEMENT

The authors are indebted to Engineering Management Services Limited and their clients who enabled them to develop this approach and trust that all who endeavour to create a better underground working environment will benefit from this paper.

REFERENCES

1. Van der Walt, J., 1978, "Cooling of Gold Mines in The Republic of South Africa", The Professional Engineer, Vol. 7, No. 1, January, pp. 5-11.

2. Van der Walt, J. and Whillier, A., 1979, "Heat Pick-up from the Rock in Gold Mines: The Water-rock Thermal Balance and the Thermal Efficiency of Production", Journal of the Mine Ventilation Society of South Africa, Vol. 32, No. 7, July, pp. 125-144.

3. Van der Walt, J. and Whillier, A., 1977, "Prediction of the Refrigeration Requirements for Cooling the Service Water and the Ventilation Air in South African Gold Mines", Chamber of Mines Research Report No. 28/77, July.

4. Van der Walt, J. and Whillier, A., 1978, "Considerations in the Design of Integrated Systems for Distributing Refrigeration in Deep Mines", Journal of the Mine Ventilation Society of South Africa, Vol. 31, No. 12, December, pp. 217-243.

5. Van der Walt, J. and Whillier, A., 1978, "The Cooling Experiment at the Hartebeestfontein Gold Mine", Journal of the Mine Ventilation Society of South Africa, Vol. 31, No. 8, August, pp. 141 - 147.

6. Smith, L.K. and Van der Walt, J., 1981, "Cooling Strategy and Cooling Requirements for the No. 6 and No. 6 North Shafts of the Hartebeestfontein Gold Mine - Stilfontein", Engineering Management Services, Contract No. 4257, June.

7. Van der Walt, J. et al., 1981, "Report on the Infrastructure Required to Maintain and Extend Mining Operations at the Homestake Gold Mine - South Dakota", Engineering Management Services, Contract No. 4284, June.

8. Van der Walt, J. and Smith, L.K, 1980, "Cooling Strategy and Cooling Requirements for Section 31 of the Nose Rock Mine - New Mexico", Engineering Management Services, Contract No. 4263, December.

9. Van der Walt, J., de Kock, E.M. and Smith, L.K, "Managing Environmental Engineering Requirements for Underground Mining Operations". To be published.

10. Starfield, A.M., 1966, "Tables for the Flow of Heat into a Rock Tunnel with Different Surface Heat Transfer Coefficients", Journal of the South African Institute of Mining and Metallurgy, Vol. 66, No. 12, July, pp. 692-694.

11. Goch, D.C. and Patterson, H.S., 1940, "The Heat Flow into Tunnels", Journal of the Chemical Metallurgical and Mining Society of South Africa, Vol. 41, No. 3, September, pp. 117-128.

12. Whillier, A. and Ramsden, R., 1976, "Sources of Heat in Deep Mines and the Use of Mine Service Water for Cooling", Proceedings, First International Mine Ventilation Congress, The Mine Ventilation Society of South Africa, pp. 339-346.

13. Fenton, J.L., 1972, "Survey of Underground Mine Heat Sources", M.Sc. Thesis, Montana College of Mineral Science and Technology, Butte, Montana.

14. Van der Walt, J., 1979, "Engineering of Refrigeration Installations for Cooling Mines", Journal of the South African Institute of Mechanical Engineers, Vol.19, No. 10, October, pp. 360-372.

15. Van der Walt, J., de Kock, E.M. and Smith L.K., "Recent Developments in the Engineering of Refrigeration Installations for Cooling Mines". To be published.

Chapter 27

DEVELOPMENT OF A HIGH PERFORMANCE, LOW MAINTENANCE, IN-LINE WATER SPRAY COOLER FOR MINES

Kenneth Heller[1]
John McCoy[1]
Edward Thimons[2]

[1]Foster-Miller, Inc.
350 Second Avenue
Waltham, MA 02254

[2]U.S. Bureau of Mines
Bruceton Research Center
Pittsburgh, PA 15236

INTRODUCTION

As the working levels in domestic mines are developed to greater depths and work areas move further from the shafts, maintaining a safe and healthful working environment becomes more difficult. Increased temperature and humidity at greater depths becomes first a factor in productivity and then a threat to the miner's safety. Initially, a miner might develop listlessness leading to reduced productivity and reduced alertness. Continued overexertion can lead to heat stroke and possibly death. Mine ventilation engineers, therefore, must not only provide clean air to work areas, but also must provide cooled air. The following article describes the United States Bureau of Mines (USBM) and Foster-Miller's successful development of a high performance, low maintenance direct water to air contact heat exchanger for mine cooling.

While heat problems frequently do not develop until mining operations descend below sea level, problems will arise any time mining penetrates high temperature rock. Hot fissure water will increase the severity of the situation, causing increases in wet bulb temperature - the temperature to which perspiring miners reject metabolic body heat. All energy expended in mining adds to the heat load which must be absorbed by the ventilation air. Just when the cooling capacity of ventilation air is needed most, adiabatic compression (autocompression) of air descending the shaft raises air temperatures 1.39°C (2.5°F) to 2.94°C (5.3°F)/305m (1000 ft) depending on humidity. Increased temperatures decrease the ability of the air to absorb heat from the mine, and work area temperatures rise.

Once forced ventilation air can no longer maintain safe work area temperatures, refrigeration methods must be employed. Many mines now use cooling coils for this purpose. A cooling coil uses refrigerated mine water to cool the air by passing the air through banks of copper tubing filled with cold water. Heat from the air is transferred to the water through the pipe with the aid of fins. Conventional water to air heat exchangers prove troublesome for many mines because of the high levels of dust that are encountered, causing these exchangers to operate at relatively low efficiencies. A more satisfactory approach to workplace cooling is desired - one that offers reduced maintenance and increased cooling efficiency.

In trying to eliminate maintenance needs while maintaining good performance, a direct water to air contact heat exchanger or Spray Cooler was developed under a USBM contract by Foster-Miller of Waltham, Massachusetts. The prime advantages of the Spray Cooler over available cooling methods are the very low maintenance requirements and high water efficiencies that are obtained. Hot, humid, dusty air is brought into the bottom of the spray cooler and directly contacted with cold water both in droplets and on mesh pads. Heat and moisture are then transferred from the air to the water, while dust and other particulates are removed by the waste water. The warmer water is drained out of the cooler. A mist eliminator pad removes the remaining liquid from the air so that cooler, drier, cleaner air leaves through the top of the unit. Figure 1 shows the Spray Cooler.

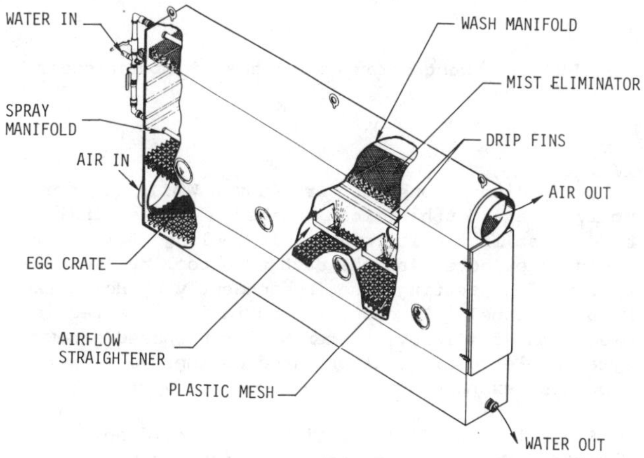

FIGURE 1. - New design spray cooler (3050 × 406 × 1680mm)

Thirty Spray Coolers are currently operating in the Homestake Gold Mine at Lead, South Dakota. Twelve additional units are on order. Many of these coolers have replaced cooling coils. Ten of the coolers, supplied by the USBM, have been monitored by Foster-Miller for one year. Observations have shown the coolers can cool effectively for as long as nine months without any maintenance. Both supervisory personnel and contract miners at Homestake are very happy with the Spray Cooler.

The following discussion traces the cooler's development from an original duct type cooler to the present Spray Cooler. Also, closer examination is given to the cooler's operation and performance at various conditions.

SPRAY COOLER DEVELOPMENT

Development of the Spray Cooler began with testing by the USBM of a spray system that mounted in line in the ventilation tubing. This system consisted of a series of axially mounted spray nozzles in a 406 mm (16-in.) diameter duct as shown in Figure 2. Preliminary testing at the Homestake Mine demonstrated both good cooling performance and the need for some mechanical redesign. As a result, Foster-Miller was contracted by the Bureau of Mines to optimize the Spray Cooler concept. Factors to be considered included direction of spray (counterflow, crossflow, or parallel flow), nozzle spacing, type of nozzle, and number of nozzles. Water carryover into the downstream ducting also had to be eliminated so that puddling in the vent tubing did not occur. Puddling had caused the vent tubing to sag and tear under the weight of the water.

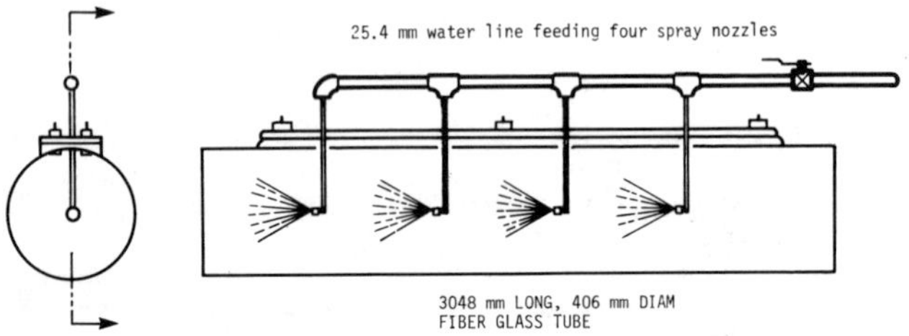

FIGURE 2. - Duct spray cooler tested at Homestake

Early discussions with Dr. Austin Whillier, formerly of the South African Chamber of Mines, indicated that little improvement in cooling efficiency could be expected for the duct type cooler. This was confirmed by testing done independently by Homestake Mine personnel. In order to provide the desired increase in efficiency, Foster-Miller proposed a new spray cooler configuration based on improved counterflow heat exchange principles.

Since initial field testing was to be done at the Homestake Mine, the new configuration was based on Homestake's physical and environmental requirements. The coolers were to be located against the ribs in the drifts without interfering with mine traffic. The air inlet and outlet connections were to accept 406 mm (16-in.) mine vent tubing. This 406 mm (16-in.) dimension also determined the maximum cooler width. Other cooler dimensions were determined by practical constraints of moving coolers into the mine with minimum effort. Design water and air flows were based on typical inlet conditions that existed for the cooling coils that were operating. The efficiency of the spray cooler had to be equal to or exceed that of the cooling coils. Table 1 is a summary of the desired cooler design specifications.

TABLE 1. - Initial specifications of spray cooler

Maximum dimensions	406 mm (16 in.) wide 1,680 mm (66 in.) high 3,050 mm (120 in.) long
Inlet air flow	2,36 m^3/s (5000 fpm)
Inlet air temperature	26,67°C (80,0°F) wet bulb 29,44°C (85,0°F) dry bulb
Inlet water flow	0,76 ℓ/s (12 US gpm)
Inlet water pressure	689 kPa (100 psi)
Inlet water temperature	10°C (50°F)
Minimum water efficiency	80% (this implies a cooling duty of 42,2 kW)
Other	No significant mist shall be present in the downstream air flow since the weight of water deposited in ventilation ducting can lead to tubing support failure.

The design of the first prototype cooler is presented in Figure 3. Air enters horizontally into the bottom of the cooler where its velocity slows and turns upward. Spray nozzles located on a manifold at the top of the chamber spray downward into the airflow. Above the manifold, a mist elimination pad removes excess water from the existing air. After the air passes through the pad it turns horizontally and accelerates before entering the outlet duct. The velocity of the air entering and leaving the cooler is ten times that of the vertical velocity in the cooler. Minimizing pressure losses during these velocity changes and ensuring uniform distribution of airflow throughout the cooler cross section were primary considerations in the design process.

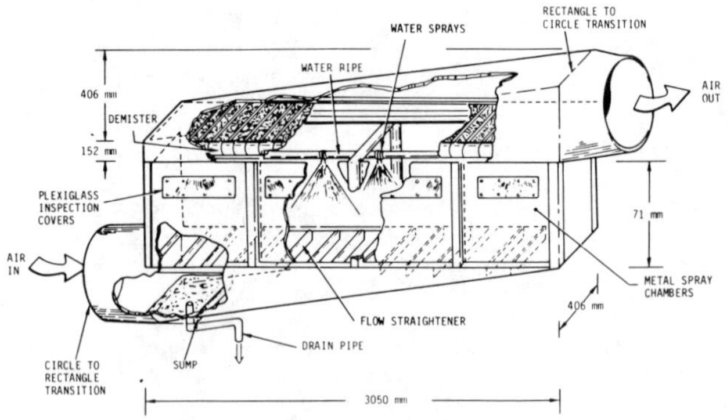

FIGURE 3. - Original spray cooler design

For initial testing, a prototype cooler was built from 1.2 mm (18 gauge) galvanized sheet metal. This unit was installed in a specially constructed test facility that provided operating conditions which closely simulated those expected underground. A series of tests were run to evaluate and to optimize the new concept. Factors evaluated during the testing included:

- Techniques for turning the air upward while simultaneously distributing it evenly in the spray chamber
- Positioning of the manifold to evaluate spraying with or against the airflow
- Various types and numbers of nozzles
- Drip fins on the cooler's walls to re-entrain water running down the walls (reentrainment of the water into the airstream improves heat exchange).

After the various elements had been tried, a modified cooler configuration resulted as shown in Figure 4. The changes made to the original prototype were:

- The spray manifold was now located near the bottom of the cooler spraying upwards. Testing had shown slightly higher efficiencies spraying upward versus downward. This was suspected to be due to both a more uniformly wetted mist eliminating pad and a mixture of cross-flow and counter-flow heat exchange. The mist eliminator contributes significantly to the overall heat exchange.
- The nozzle chosen for optimum cooling performance was a solid cone, 46° angle spray. Its good coverage and minimum contact with the sidewalls of the cooler provided the best heat exchange.
- A triple layer of egg-crate material, added to the bottom of the cooler, provided a simple solution to slowing, turning, and evenly distributing the air.
- Air flow straighteners were added between each set of nozzles to direct the air in a vertical pattern after being turned upward.
- Between the air flow straighteners, plastic mesh pads were installed to enhance heat transfer from the air to the water. The mesh, as shown in the sketch, was actually added during in-mine testing, not during laboratory testing.

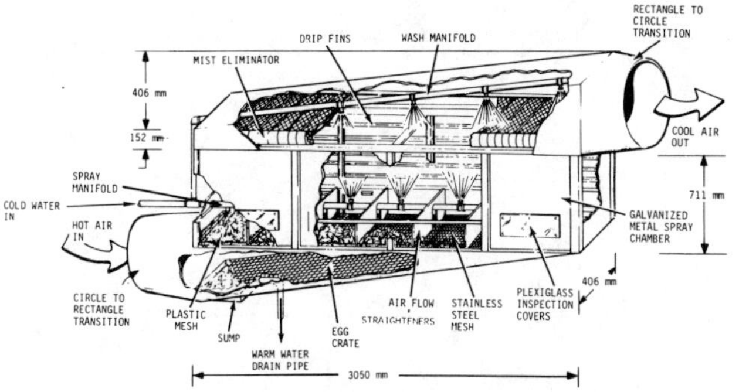

FIGURE 4. - Final configuration of prototype spray cooler

After laboratory testing, the above features were incorporated into the cooler. A few additional operating features were added for the extended in-mine testing. A wash manifold of four wide-angled sprays was installed over the mist eliminator for occasional washing of the pad to prevent excessive dust accumulation. A Griswold constant-flow valve was added to the inlet water pipe to prevent high water consumption in the event of nozzle wear. Upstream of the Griswold valve a "y" strainer was installed to prevent nozzle plugging due to water contamination.

Before shipment to the mine, the cooler was tested in the laboratory under conditions specified in Table 1. A measured 81.3% water efficiency and 46 kW (13 ton) cooling capacity exceeded initial design requirements. The cooler was now ready for in-mine testing.

IN-MINE TEST

In-mine installation of the prototype cooler was at the 6,200 ft level of the Homestake Gold Mine in Lead, South Dakota. A scram located in a crosscut about 61m (200 ft) from the cooler was to be the target for the cooled air. No problems were encountered bringing the Spray Cooler to its designated location.

In-mine inlet conditions were found to be different from laboratory conditions. This resulted in the amount of cooling also differing, since for a given cooler, water cooling efficiency and cooling duty depend on inlet conditions. To evaluate performance of a cooler, a parameter independent of inlet conditions is necessary. Such a parameter known as the F-factor has been developed although not commonly known in the United States. This parameter allows evaluation of changes in performance due to modification and wear. The following section on performance describes the F-factor.

Further modification to increase cooler performance was made during underground testing. A plastic mesh material was added below the spray manifold which increased water-air contact, noticeably improving cooler performance.

Immediately after installation and again after one month of operation, performance of the spray cooler was evaluated. Overall acceptance of the spray cooler was also evaluated by Homestake, USBM, and Foster-Miller personnel. The cooling performance was excellent and is discussed in detail later in this paper. Underground acceptance of the cooler was also excellent. During the first month of operation, the only maintenance necessary on the cooler was a daily flush of the "y" strainer located in the inlet water piping. This procedure took only a few seconds. Homestake personnel and miners who had been working near the cooler all voiced approval of its cooling ability and minimal maintenance. The excellent results of the one month operation and positive response from Homestake personnel prompted the fabrication and extended underground evaluation of nine additional coolers, in accordance with the terms of the USBM contract.

Cooler design was modified prior to fabrication of the additional coolers. Redesign incorporated:

- Lower cost fabrication features
- Simpler access to internal parts
- Better external shape for the Homestake Mine geometry
- Addition of a lower mesh for additional heat exchange.

Figure 1 shows the resulting redesigned cooler.

Following underground installation of the nine additional coolers, tests were conducted on all units. Results of these tests, plus the favorable reaction of Homestake personnel led to an order for twenty additional Spray Coolers.

At the time of this writing, the 30 coolers have been in continuous operation underground at Homestake Gold Mine for up to 18 months. Reports on the operation and maintenance of the coolers continue to be favorable.

SPRAY COOLER PERFORMANCE

Performance of a spray cooler must be carefully defined. For a given cooler, water effectiveness or efficiency is defined as the change in cooling water temperature across the cooler divided by the temperature difference between inlet wet bulb air temperature and the inlet cooling water temperature, that is:

$$E_w = \frac{\text{temp. water out - temp. water in}}{\text{air wet bulb temp. in - temp. water in}} \quad (1)$$

Cooling power or cooling duty is the kilowatts or refrigeration tons provided by the cooler. Both these quantities are unsatisfactory for describing cooler performance since, even for a given cooler, both water effectiveness and cooling duty are strong functions of cooler inlet temperature and flow rates. For example, Table 2 compares the cooling duty of the same cooler for two inlet water conditions with the air condition held constant.

TABLE 2. Comparison of cooling duty for a cooler at two locations

Location A	Location B
Inlet:	Inlet:
Air; 26.7°C wb at 2.4 m^3/sec	Air; 26.7°C wb at 2.4 m^3/sec
Water; 4.4°C at 0.76 l/sec	Water; 10°C at 0.76 l/sec
Cooling duty:	Cooling duty:
58.7 kW	44.3 kW
Water effectiveness:	Water effectiveness:
E_w = 0.825	E_w = 0.833

Note that the colder water in Location A contributed to a much greater cooling duty than achieved in Location B while water effectiveness or efficiency was higher in B than A. Clearly a measure of performance intrinsic to the cooler independent of operating conditions is necessary. A method is needed to account for conditions different from laboratory conditions such as occurs from place to place in the mine.

A rating called the "Factor of Merit" or "F-factor" developed by Whillier (1) and improved by Bluhm (2) was used to characterize Spray Cooler performance. For airflows within the normal operating range, the F-factor is independent of inlet fluid temperatures, flow rates, and barometric pressures. It is equivalent to knowing UA of a heat exchanger. For extreme airflows, F-factor ratings may become less meaningful. Very low air flows can change air distribution within the unit, while very high flow can distort the unit. Both these limiting factors will create erroneous F-factors.

Figure 5 shows the parametric relationship of F-factors. The capacitance ratio, R, of a direct air-water contact cooler is the ratio of the thermal capacity of the inlet water stream to that of the air stream. When R values are plotted against water efficiency of a cooler, test points are found to lie on a single curve which is described by the equation developed by Bluhm (2). This is nicely illustrated in Figure 6. The graph shows various data points of R vs E_w for laboratory tests at different inlet conditions. As can be seen, the points tend to form a single curve within experimental error. The actual curve shown is predicted by Bluhm's equations for F=0.626. This also predicts how the cooler will perform at various inlet fluid temperatures, flows, and barometric pressures. Table 3 summarizes the laboratory data and shows the condition variations. Bluhm's equation was used to calculate F-factor values from the experimental test data.

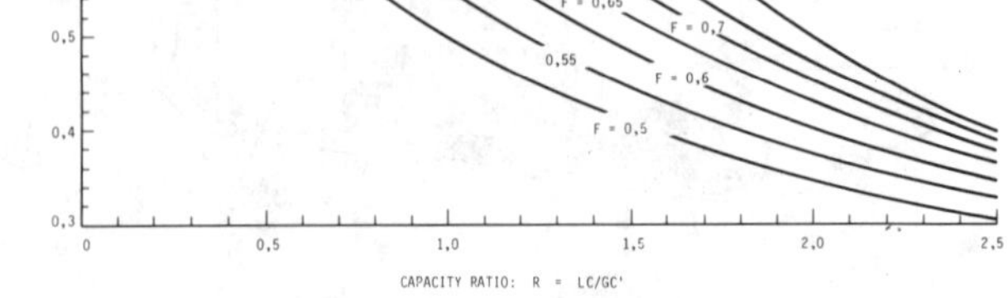

FIGURE 5. Performance curves of counterflow direct-contact heat exchanger.

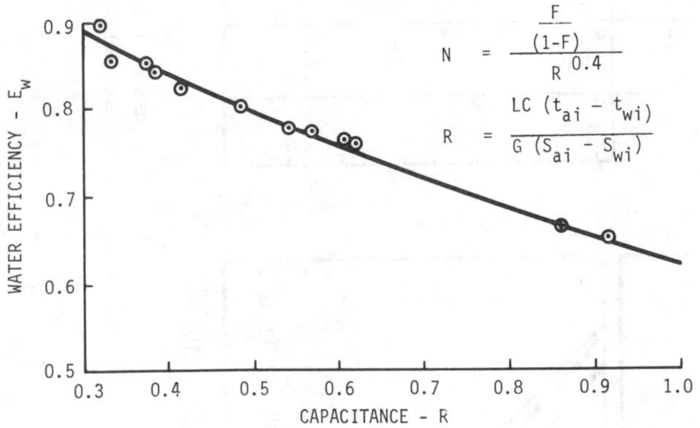

$$E_w = \frac{1 - e^{-N(1-R)}}{1 - Re^{-N(1-R)}}$$

$$N = \frac{\frac{F}{(1-F)}}{R^{0.4}}$$

$$R = \frac{LC\,(t_{ai} - t_{wi})}{G\,(S_{ai} - S_{wi})}$$

FIGURE 6. - Factor of merit curve for laboratory test data-tests B1-B12

Using inlet conditions specified in Table 1, the resultant cooler performance showed water efficiencies in the range of 81%, average cooling duty of 46 kW (13 tons) and an F-factor of 0.626. This data is based on a few selected coolers of the ten units that were initially shipped to the Homestake Mine. Laboratory inlet conditions, as shown in Table 3, produced widely varied water efficiencies and cooling duty. Results show that by increasing water flow through the cooler with the use of higher flow nozzles, total cooling duty can be substantially increased to 70 kW (20 tons), but at the same time water efficiency is decreased to 66%. Higher inlet air flows up to 2.83 m^3/sec (6000 cfm) produced slightly higher water efficiencies and cooling duty. Overall spray cooler performance in the laboratory exceeded the initial specifications of a 0.58 F-factor established in conjunction with the USBM and the Homestake Mine.

When laboratory testing was concluded a set of graphs based on an F-factor of 0.626, Figure 7, was drawn to assist in predicting how the cooler would perform in various environments. Prediction curves such as those shown would not be possible without the Factor of Merit unless testing was done at all given conditions.

TABLE 3. Operating parameters of spray cooler laboratory test data*

Test No.	Air Flow (m^3/sec)	Water Flow (kg/sec)	Temperature Air In-Wet Bulb (°C)	Temperature Air Out-Wet Bulb (°C)	Temperature Water In (°C)	Temperature Water Out (°C)	Capacitance (R)	Water Efficiency E_W (%)	F-Factor (F)	Cooling Duty (kW)
B1	2.12	0.801	27.06	22.83	12.78	24.56	0.424	82.9	0.622	38.7
B3	2.34	0.808	26.33	23.00	14.00	24.44	0.382	84.5	0.619	34.8
B4	2.35	0.808	27.89	24.06	13.22	25.78	0.377	85.5	0.626	41.9
B2	2.46	0.808	33.56	29.39	14.11	31.61	0.320	89.9	0.646	58.4
B5	2.66	0.808	27.89	24.50	13.11	25.78	0.333	85.6	0.607	42.2
B7	2.05	1.150	27.72	21.44	13.72	23.67	0.619	76.2	0.633	47.2
B6	2.09	1.150	26.61	21.44	13.78	23.67	0.609	76.8	0.635	46.8
B8	2.26	1.150	27.06	27.72	12.67	23.83	0.569	77.4	0.627	52.8
B9	2.35	1.140	27.44	22.22	12.61	24.22	0.541	78.0	0.623	54.6
B10	2.64	1.140	27.28	22.56	12.61	24.44	0.484	80.5	0.623	55.6
B11	2.31	1.880	27.39	19.67	12.56	22.28	0.915	65.2	0.626	75.3
B12	2.49	1.880	26.61	19.56	12.56	22.00	0.860	66.8	0.625	72.9

Average $\overline{F}$ = 0.626

Standard Deviation, σ = 0.009

*All data calculated by using an adjusted airflow to force a heat balance.

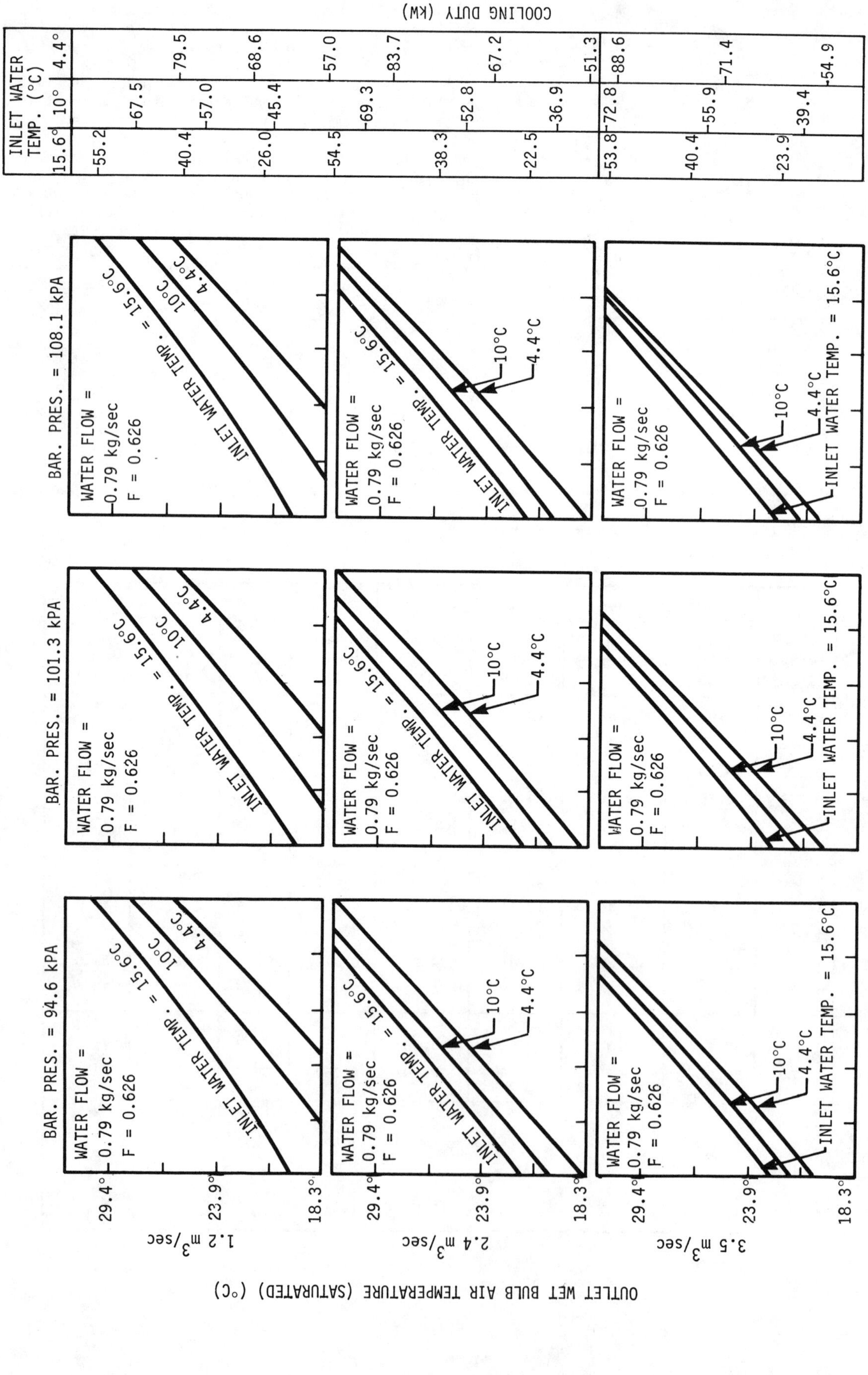

FIGURE 7. - Predicting spray cooler performance with F = 0.626*

*For prediction curves with English units contact principal author.

Data points of R vs E_w for testing done at the Homestake Gold Mine are shown in Figure 8. Table 4 shows the conditions for these data points. The Homestake Gold Mine data are more scattered than laboratory data due to a larger margin of field measurement error. Calculated average F-factor for the various points is 0.593.

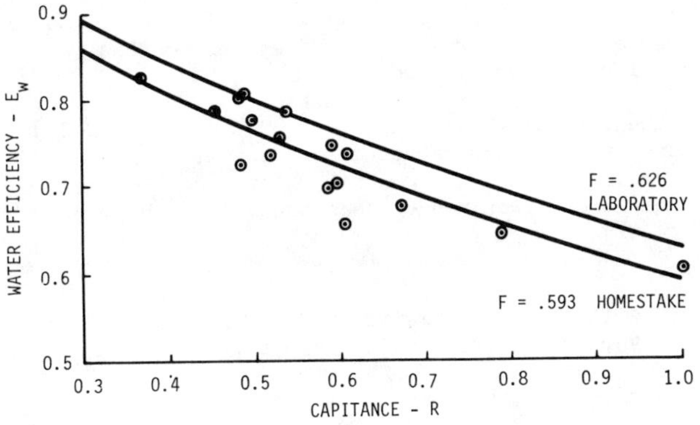

FIGURE 8. - Factor of merit curve for Homestake test data-tests H25-H43

Cooler performance at the Homestake Gold Mine showed water efficiencies in the range of 75% and average cooling duties of 33 kW (9.5 tons). The lower values of the water efficiencies and cooling duty as compared to laboratory tests can be attributed to the much lower air flows and water temperatures encountered at Homestake. Since the field testing did not take place until four months after the coolers' installation the following occurred which degraded cooler performances, lowering the F-factor from 0.626 to 0.593 on average:

- Water pressure dropped due to nozzle wear (brass nozzles were used) creating less uniform mixing of the air and water.
- Lime deposits on the bottom mesh due to the water hardness produced areas of uneven air flow.
- During some in-mine installations flow diverters had been jostled out of place creating uneven air flow.
- Air flows were very low (sometimes less than half laboratory tested conditions) contributing to uneven air distribution.

The performance decrease was really minimal considering no maintenance was performed on any cooler.

TABLE 4. - Operating parameters of spray cooler Homestake test data*

Test No.	Air Flow (m^3/sec)	Water Flow (kg/sec)	Temperature Air In-Wet Bulb (°C)	Temperature Air Out-Wet Bulb (°C)	Temperature Water In (°C)	Temperature Water Out (°C)	Capacitance (R)	Water Efficiency E_w (%)	F-Factor (F)	Cooling Duty (kW)
H26	1.27	0.820	28.78	23.67	15.56	24.56	0.671	67.7	0.576	30.3
H27	1.72	0.719	27.00	21.72	10.44	23.78	0.489	80.4	0.624	39.6
H28	1.51	0.782	26.44	20.67	11.06	22.44	0.609	73.7	0.608	36.8
H30	1.62	0.782	30.44	23.44	9.11	25.89	0.539	78.5	0.626	54.4
H31	1.83	0.745	23.44	18.78	9.11	20.28	0.499	77.7	0.604	33.4
H32	1.66	0.883	29.78	22.72	9.67	24.78	0.591	74.9	0.612	55.3
H33	1.37	0.808	26.39	23.39	17.94	23.56	0.608	65.8	0.540	18.4
H34	1.65	0.808	29.11	25.22	17.44	26.83	0.486	80.2	0.621	31.2
H37	1.61	0.808	24.77	19.89	11.06	20.78	0.596	70.5	0.576	32.3
H38	1.48	0.751	25.44	23.22	19.00	23.78	0.520	73.5	0.576	14.5
H39	1.63	0.795	27.33	22.44	12.72	23.78	0.531	75.4	0.596	36.2
H40	2.39	0.827	29.78	25.61	12.33	26.78	0.370	82.6	0.596	49.4
H41	1.77	0.726	24.89	21.00	12.33	22.22	0.465	78.5	0.598	29.5
H42	1.09	0.833	28.67	22.61	15.06	23.89	0.799	64.5	0.593	30.2
H43	1.51	0.770	25.78	24.89	23.00	25.06	0.485	72.4	0.554	6.1

Average $\overline{F}$ = 0.593

Standard Deviation, σ = 0.0250

*All data calculated by using an adjusted airflow to force a heat balance.

Other inspections done on the cooler have shown they can operate satisfactorily up to nine months without maintenance but periodic attention (every few months) to ensure their performance meets initial specifications is recommended. Maintenance may include a bottom mesh change or a nozzle change, both of which are easily accomplished due to the cooler's design.

CONCLUSIONS

The spray cooler concept began as a research and development project that resulted in a manufactured item now used extensively at the Homestake Gold Mine and is available for use by other hot mines. The concept has advantages over other mine cooling devices including:

- Simple maintenance procedures; required only infrequently
- High water efficiencies to maximize the cooling obtained from the mines chilled water system
- Simple installation and operation procedures
- A configuration well-suited to underground passage shapes.

The present spray cooler has produced cooling in excess of 70 kW (20 tons) with appropriate air and water inlet conditions. Many mines need devices with cooling capacity up to 350 kW (100 tons) to cool larger areas. The present design can be modified to facilitate these requirements. Additional research is currently underway to determine the efficiency which may be expected with scaled-up spray coolers.

ACKNOWLEDGEMENTS

This work was performed under U.S. Bureau of Mines Contract No. H0395058 to Foster-Miller, Inc. The Homestake Mining Co. and Homestake Gold Mine personnel especially Mr. Jerry Pontius have been most helpful in all phases of this work.

The late Dr. Austin Whillier of the Chamber of Mines of South Africa served as consultant to this program. Without his contributions and tutoring, we doubt success would have been so easily achieved.

REFERENCES

1. Whillier, A., "Predicting the Performance of Forced-Draught Cooling Towers", Vol. 30, Journal of the Mine Ventilation Society of South Africa, January 1977.
2. Bluhm, S.J., "Predicting Performance of Spray Chambers for Cooling Air", Heating Refrigeration and Air Conditioning (South Africa), November 1980, pp. 27-39. (Paper presented at "Frigair '80" conference, Pretoria, June 1980.)

IN-LINE WATER SPRAY COOLER FOR HOT MINES

K.S. Heller and J.F. McCoy

Contribution by

J. van der Walt and E.M. de Kock
Engineering Management Services Limited

Although the practice of cooling air at the entrances to and in stopes has been successful for many years, experience has shown that problems are often experienced in creating and maintaining an acceptable thermal working environment when the refrigeration requirements on a mine increase to above about 50 kW per k ton/month. Cognisance should therefore be taken of these problems and the field of application of the type of cooler described in the paper.

Problems

Distribution of Cooling Water.

When the number of small air-cooling units becomes large, it is often difficult to control the distribution, i.e., supply and return of cooling water to each of the units, and to balance the supply so that each unit receives the correct amount of water. The main reason for the problem is the friction loss associated with the long lengths of pipes and the difference in the length of pipes serving each air cooling unit. The problem is aggravated by

- the often inadequate pipe diameters.
- changes to the distribution system because of production requirements.
- temporary changes to the system during maintenance periods (1).

It is often argued that this type of cooler is open ended and that pipe friction should not impose water balancing problems. This is not necessarily true as the total amount of water still has to be balanced with the amount that is supplied by the refrigeration machines. Furthermore, the water needs to be returned to the cooling plant or to settlers by pumps and pipes, which means that the advantage of an open ended system is largely defeated. The situation could in fact be aggravated due to the return water pumps having their own pressure-flow characteristic that complicates the process design of the reticulation system. On some mines the return water flows on the footwall (floor) back to the shaft settlers, which in turn can impose problems with mud in haulage ways and particularly in box holes, etc. The water on the footwall may also impose a secondary heat load as it usually comes into contact with the cool intake air.

Advancement of Cooling Units

Advancement of cooling units is usually a problem when there are a number of them, and they have to be advanced frequently as a considerable effort by a large workforce is required to dismantle, move and re-erect the units and associated water pipes, ducts, fans, pumps, switchgear, etc.

In practice, they are often not advanced timeously or installed, and hence thermal environmental problems develop in working places. This can often not be rectified because of the thermal load and inertia building up to a point where it is well in excess of the capacity of such a unit.

Cost of Cooling Installations

One of the major problems of cooling distribution systems involving a number of small units is that of reliably determining the cost of the system (1) as there are many indirect and consequential costs that are difficult to anticipate and quantify such as those associated with maintenance, breakdowns, monitoring and performance.

A system consisting of a number of these small units usually requires less refrigeration capacity than the alternatives (2), which have fairly obvious inefficiencies associated with air leakage and an increase in heat load from air temperature dependent sources. Care should be taken not to be misled by these rather obvious inefficiencies.

Process Considerations

When considering spray cooling units of this kind, it is necessary that the entire unit and associated equipment be evaluated as both the fan and duct have an influence on the net cooling effect. A psychrometric phenomenon that could easily be overlooked in respect of the fan is the amount of cooling that is required to re-cool the air to the wet bulb temperature before the fan, and typically 1.4 to 1.8 times the fan energy is required as cooling. A 10 kW fan connected to a 30 kW air cooler would result in a net cooling effect of about 15 kW as about 5 kW is associated with the increase in air pressure across the fan. A 20 kW fan would result in a negligible change in wet bulb temperature across the system, whereas a fan absorbing more energy could result in a net increase in the wet bulb temperature. This phenomenon can be observed on many small mine air cooling installations.

Applications

Although there are many problems associated with employing this type of cooler and its use may not be desirable, there are nevertheless circumstances where the alternatives cannot be employed. This is normally the case when

- the air flow through a mine is inadequate and the mine is forced to cool air at the entrances to and inside working places.
- the air temperature dependent heat loads and air leakages are high and the use of bulk air-coolers cannot be justified.
- the principle of through flow ventilation cannot be applied to working places and they have to be ventilated with auxiliary fans and ducts regardless of whether cooling is required or not.

Conclusion

Although there are many reservations about the use of in-line spray coolers and cooling coils, the U.S. Bureau of Mines, Foster-Miller Associates and the late Dr. Whillier of the Chamber of Mines of South Africa should be congratulated with the development of this spray cooler as it appears without doubt an improved piece of equipment that will find many applications in the mining industry.

References

(1) Van der Walt, J. and Whillier, A., Considerations in the Design of Integrated Systems for Distributing Refrigeration in Deep Mines, Journal of the Mine Ventilation Society of South Africa, Vol. 31, No. 12, pp. 217-243 (December 1978).

(2) Van der Walt, J., de Kock, E.M. and Smith, L.K., Analysing Ventilation and Cooling Requirements for Mines, Paper presented at the 1st Mine Ventilation Symposium at the University of Alabama, Tuscaloosa (March 1982).

IN-LINE WATER SPRAY COOLER FOR HOT MINES

K.S. Heller and J.F. McCoy

Contribution by

Bruce Johnson
Magma Copper Co.

One thing that might be added to the problems with using such a cooler is the problem of water supply. This might become a large problem in arid areas where extensive use of coolers is contemplated.

Reuse of water is of course possible in such situations, but here the problem of concentration of dissolved solids and dust on the water may become significant.

Chapter 28

THERMAL RECOVERY SYSTEM FOR REDUCING MINE REFRIGERATION AND ENERGY NEEDS

Richard J. Kline - Senior Engineer[1]
Edward D. Thimons - Supervisory Physical Scientist[2]

[1]Foster-Miller, Inc.
350 Second Avenue
Waltham, MA 02254

[2]U.S. Bureau of Mines
Pittsburgh Research Center
Pittsburgh, PA 15236

BACKGROUND

Converting the energy of falling water into useful work is a well developed engineering concept. Much of the electric power consumed in the United States is generated in this manner. It is easy to conceive that the water which flows down long vertical pipelines into deep mines could be utilized in a similar manner by allowing it to operate a turbine at the bottom of the pipelines.

The mines in the United States are continually descending to greater depths in search of mineable deposits. Large quantities of water are carried down long vertical pipelines in order to be used for dust suppression, motor cooling, and refrigeration. Pumping the used water back to the surface is also required. The electrical costs associated with mining are also constantly escalating due to both increased demand and rising kilowatt hour costs. To help deal with this problem, the United States Bureau of Mines (USBM) has awarded a contract to Foster-Miller, Inc. of Waltham, MA, to investigate, evaluate, and demonstrate commercially available means of energy recovery for the deep mining industry. In other words, it demonstrates an efficient system for converting the energy of water flowing down vertical pipelines into useful mechanical or electrical energy.

PROGRAM REVIEW

The program sponsored by the USBM is comprised of four phases. Phase I is a feasibility study which is currently nearing completion. During this phase, existing water turbine technology will be reviewed. Various types of turbines will be evaluated and their operating advantages and disadvantages recorded. The minimum limits of economic feasibility for thermal energy recovery systems will be established in terms of water heads and flows and electrical costs. Finally, alternative systems for using the high pressure water, such as shell and tube heat exchangers, high pressure closed circuit water lines, and hydrolift systems, for example, hydrotransformers, will be evaluated and compared to the turbine system concept.

In Phase II, various turbine-pump and turbine-generator system designs will be developed. These will be complete system designs and will take into account the sizing of required water reservoirs and emergency backup pressure reducing systems in the event of turbine failure. Cost/benefit analysis will be done for each of the designs under a variety of flow, head, and energy cost conditions.

Phase III involves the procurement, installation, operation, and testing of a complete thermal energy recovery system in a deep United States metal/nonmetal mine. This system will be designed to meet the specific requirements of this mine.

The Phase IV efforts will run concurrently with the other phases and will consist of technology transfer of the findings of this program to the United States mining industry. This will be done by means of written reports, industry seminars, and a handbook which will act as a guideline for mines in selecting and designing energy recovery systems to meet their specific needs. This handbook will outline the system options and technical and economic tradeoffs which each mine operator should consider before selecting a particular system.

To obtain information on the first industry seminar or to contribute information on energy recovery systems, contact:

Mr. Edward D. Thimons
U.S. Bureau of Mines
Pittsburgh Research Center
P.O. Box 18070
Pittsburgh, PA 15236
(412) 675-6683

TECHNICAL DISCUSSION

Using high pressure water to operate a turbine can result in three major benefits to mines:

1. The turbine can generate a significant amount of power. Modern turbines can convert almost 90 percent of the water's potential energy into useful mechanical or electrical energy. The turbine can be directly connected to a pump which lifts used water out of the mine. It can be connected to a generator and the electric power generated can be phased into the mine's grid. Figure 1 shows a schematic of one thermal recovery concept. In this case, water from the surface drives a turbine, which in turn operates a generator set. A Pelton wheel (impulse turbine) offers the ability to maintain high efficiency over a wide range of flows (30 to 110 percent of design flow). This is ideal for any site where large daily or weekly fluctuations in flow due to changes in daily mine operations are expected.

2. The turbine also yields thermal energy savings equal to the recovered mechanical energy.

FIGURE 1. Schematic of one thermal recovery concept.

Without a turbine, the potential energy of the water is converted to heat. The water temperature will automatically rise approximately 1.28°F for every 1000 ft of drop. This rise is independent of any heat transfer in the shaft. Conversion of this potential energy to heat can be minimized by converting most of the potential energy into the form of work. For example, a turbine may extract 85 percent of the energy from water flow as work. This leaves only 15 percent of the potential energy available to generate heat. In this case, the water temperature would gain only 15 percent of 1.28°F or 0.19°F for each 1000 ft of drop. Lower water temperatures resulting from turbine usage can be expressed directly in additional cooling capacity.

3. A third benefit results from the turbine performing the necessary water pressure reduction. Mining people are familiar with the headaches associated with pressure regulators and cascading dams. These methods of pressure reduction are eliminated by the turbine.

The Phase I feasibility analysis will include a review of all existing literature and current technology related to mine energy recovery. This includes evaluation of available types of turbines and a comparison of turbines and reversed pumps. Various electrical and mechanical output options will be considered as power forms. The overall economic feasibility of the several turbine/pump-driven system options will be determined in absolute and relative terms. Although only Pelton wheels and reversed centrifugal pumps seem appropriate for high heads, Francis turbines will also be considered for those underground applications where heads on the order of 1000 ft are available.

A comparative evaluation of power conversion techniques will be performed to determine the "pros and cons" of direct mechanical transmission and conversion to electrical power. This task requires careful analysis of typical mine layouts, mine power systems, pumping locations and pumping requirements. Both capital and operating costs, efficiencies, and maintenance requirements will be considered.

Furthermore, based on the most attractive systems from the above tasks, the economic feasibility limits will be determined. The economics will include cooling and power savings as well as capital costs considerations with operating and maintenance costs. Due to the "nonlinearity" of some costs, it is believed that economic feasibility will be a strong function of the specific mine application. The analysis will indicate "breakeven" points for the optimum system or systems in terms of flows, heads, and electrical power costs.

Although the initial capital investment in an energy recovery system is significant, once this investment is made, the three above-mentioned benefits are realized at little additional operating cost. Figure 2 graphically illustrates the potential electrical savings to a mine when a Pelton wheel is installed under the following assumed basic operating conditions: electrical costs of \$0.045/kWhr, and average water flow into the mine of 500 gal/min at a head of 2000 ft. For the electrical savings only, the total annual savings, assuming continuous operation at 85 percent efficiency, would be over \$63,000/year. The total savings for any specific mine must be determined from the total system savings which depends on whether the mine uses the water for cooling of motors or other energy saving systems.

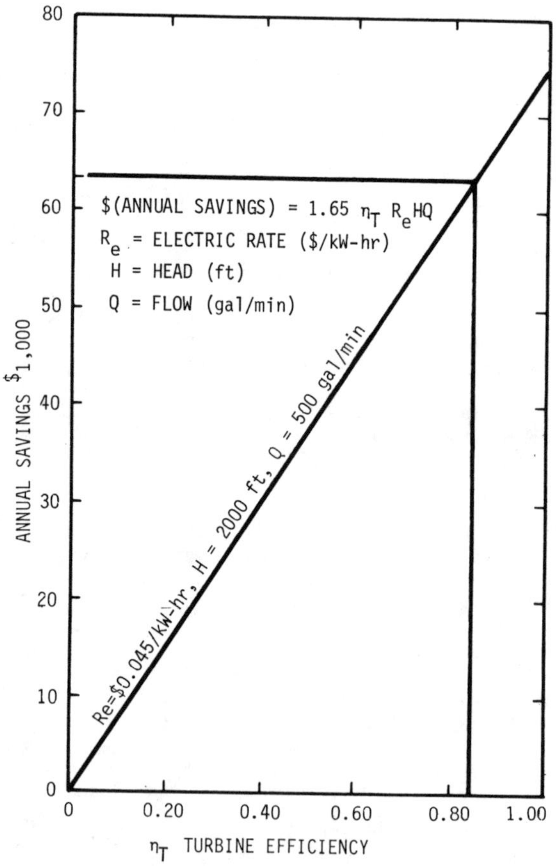

FIGURE 2. Effect of turbine efficiency on annual savings.

The pump or turbine-driven energy recovery systems developed in the previous task will be compared with alternate applications of the available incoming water. To this end, models of the alternate systems will be developed in the detail required to adequately identify critical performance and cost parameters. Each system model will then be evaluated to yield the data necessary to compare the alternate systems with energy recovery systems developed earlier. These comparisons will be made for a wide range of pipe lengths, water flow rates and electrical costs.

As was mentioned earlier, the Phase I effort is nearing completion now. Although the data were not available for presentation at this meeting, an oral review is scheduled for 5 and 6 April 1982 in Reno, NV where results of Phase I will be presented as part of a 2-day technical transfer seminar sponsored by the USBM.

An item worth noting as part of the Phase II effort is the production of a handbook for the mining industry. This thermal energy recovery handbook will include criteria for selecting an appropriate system for a given set of mine conditions, a cost benefit analysis of the various alternative systems, a turbine selection guide, and the advantages and disadvantages of each type of system.

SUMMARY

In these days of rising costs, alternate methods of making the mining operation more efficient are constantly being sought. The utilization of energy, especially falling water, should be evaluated. This USBM program will develop the information needed for a mine to make an evaluation about the usefulness of energy recovery systems.

THERMAL ENERGY RECOVERY FOR DEEP MINES

R.J. Kline and E.D. Thimons

Contribution by

J. van der Walt and E.M. de Kock
Engineering Management Services Limited

The approach the U.S. Bureau of Mines has adopted to investigate the feasibility of energy recovery systems will no doubt lead to results that would be of great value and importance to the entire mining industry, and they should be congratulated with their efforts.

As stated by the authors, the need for employing systems for recovering energy from water descending mines has been prompted by several developments in recent years, inter alia,

- to counter the effect of auto-compression (friction) on the temperature of water in an open circuit that would increase at a rate of 2.34 deg C per kilometer without an energy recovery system.
- to convert the recovered energy into some useful work such as pumping water to surface or generating electrical power.
- to purposely reduce the maximum power demand and the associated cost penalties by synchronising the maximum supply of water with the maximum demand for electrical power.

The design and type of energy recovery device depends on the application and the objective of the system. When large volumes of water are required for mining operations, the objective would be to recover energy and to apply it to do some useful work. On the other hand, when water is required for cooling, the prime objective is to counter the effect of auto-compression with a view to reducing the amount of water to be circulated underground. Of secondary importance is the saving associated with the recovered energy as this is typically about half that associated with the entire ventilation and cooling system which, in turn, is only a portion of the overall production cost on a mine. Hence care must be taken not to over emphasize energy recovery systems by neglecting considerations that could be of greater importance, such as the consequences they may have on production and maintenance operations. There are also alternatives that could be of greater importance such as cooling of air on surface, an underground refrigeration plant or perhaps an ice plant on surface. All of the latter three will generally reduce the amount of water to be sent and distributed underground and to be pumped to surface by significantly more than an energy recovery system would do. The findings of an investigation with respect to these have recently been recorded (1).

Mines employing energy recovery systems are experiencing some of the following problems with them:

- Water requirements, particularly the run of mine water (service water), fluctuate widely and it is found that the fluctuations are often outside the permissible range of a turbine. It is also difficult for the process designers of a water supply system to accurately predict the future water requirements.
- When a turbine is connected to a pump, the synchronization of the water supply and the pumping of return water is virtually impossible, particularly when the water requirements are dominated by the service water. It should be borne in mind that the service water normally returns to the shaft after several hours and probably at a time when little is required.
- The above two points are aggravated on deep mines where it is necessary to have two or more energy recovery systems operating in series.
- These systems appear to be of little importance when it comes to production and maintenance of mining and hoisting equipment. Hence operational and maintenance problems are being experienced due to a lack of attention and attendance. It cannot be expected of mine managers not to bypass the system if it interferes with production as the slightest loss can easily offset all the benefits.

Experience gained thusfar on mines operating a Pelton Wheel coupled to an electrical power generator (squirrel cage induction generator) indicates that if the system is well engineered from the outset, it stands a better chance of success than the alternatives. This is due to

- the simplicity of the installation and the operation thereof.
- the wide range of water flow rates that can be accommodated without sacrificing the efficiency which can be around 80%.

Reference

(1) Van der Walt, J., de Kock, E.M. and Smith, L.K., Recent Developments in the Engineering of Refrigeration Installations for Cooling Mines, to be published (1982).

THERMAL ENERGY RECOVERY FOR DEEP MINES

R.J. Kline and E.D. Thimons

Contribution by

Bruce Johnson
Magma Copper Co.

To the various alternative methods presented in the paper of handling the high-pressure water may be added the possibility of installing refrigeration machinery underground and using the high-pressure water as condenser water, then using the chiller of the machine to cool water used in the actual air cooling at or near the faces.

While this might seem to be similar to using water - water heat exchangers, and indeed the ultimate heat flow is water - water, heat exchangers have the double disadvantage of (1) a poor temperature approach between the entering high-pressure entering water and the leaving low-pressure water on the shell side (45°F high-pressure water will often result in 55°F leaving low-pressure water to the cooling coils), and (2) a high pumping head loss across the shells of the heat exchangers. Whereas the use of the high-pressure water as condenser water can result in virtually any desired leaving low-pressure water from the chiller, if the machine is properly designed, the pumping head loss through the tubes of a chiller is much less than the loss through the shell of a heat exchanger.

However, as all three of the excellent papers presented in the Wednesday morning session emphasize, a number of alternatives exist, and each alternative must be carefully examined for each individual case.

7
General Ventilation

Chapter 29

CONTROL OF RADON DAUGHTERS IN UNDERGROUND URANIUM MINES

Wade E. Cooper and Pete Turcic

Mining Engineer
Physical Agents Division
Denver Safety and Health Technology Center
Mine Safety and Health Administration
Denver, Colorado

Mining Engineer
Technical Support
Mine Safety and Health Administration
Arlington, Virginia

Many factors influence radon-222 influx and radon-daughter concentrations in underground uranium mines. A brief review of some of these factors is presented. Current radon-222 and radon-daughter control principles and practices for U.S. underground uranium mine atmospheres are discussed and a brief review of the applicability of other possible control methods such as rock sealants, radon-222 removal, and radon-daughter removal is given.

INTRODUCTION

The uranium-238 decay series (Table 1) contains several elements which are believed to be the primary cause of the excess lung cancer rate among early uranium miners. Radon-222 (radon) in this series, is a noble gas which diffuses into the mine air. Once this gas enters the mine air it decays progressively into its four short-lived daughters: RaA, RaB, RaC, and RaC'. These daughters, commonly referred to as radon daughters, are particulates which can deposit in the respiratory tract when inhaled. They present a potential health hazard because of their release of high-energy alpha particles in a relatively short period of time. Experimental studies have indicated that alpha particles of this energy can deposit more than enough energy to destroy a cell's reproductive capacity (National Academy of Sciences, 1976).

Table 1. - Uranium disintegration series *(Rock and Walker, 1970).

Isotope	Half life	Principal radiations
Uranium 238	4.49 x10^9 years	alpha
Thorium 234	24.1 days	beta
Proactinium 234	1.17 minutes	beta
Uranium 234	2.48 x 10^5 years	alpha
Thorium 230	8 x 10^4 years	alpha
Radium 226	1622 years	alpha
Radon 222	3.825 days	alpha
Polonium 218, RaA	3.05 minutes	alpha
Lead 214, RaB	26.8 minutes	beta, gamma
Bismuth 214, RaC	19.7 minutes	beta, gamma
Polonium 214, RaC'	2.73 x 10^{-6} minutes	alpha
Lead 210	22 years	beta, gamma
Bismuth 210	5.02 days	beta
Polonium 210	138.3 days	alpha
Lead 206	stable	

*Minor percentage isotopes excluded.

The ventilation engineer's task of controlling radon-daughter concentrations is highly complex. As shown in this paper, many factors in mines affect radon influx and the resulting radon-daughter concentrations. This paper also provides the ventilation engineer with the fundamentals of radon-daughter control in underground uranium mine atmospheres. These control measures have been directed principally at uranium mines but are also applicable to other mines with radon daughters.

RADON SOURCES

One of the first considerations for control of radon daughters is the identification of radon sources. Once the radon sources have been identified, measures can be taken to reduce the radon-daughter concentrations resulting from them. As the quantity of radon allowed to enter air in or upstream from active work areas increases, so do the countermeasures required to maintain the same low radon-daughter concentration.

Two sources of radon release into mine atmospheres are of primary concern: (1) radon emanation from water courses and (2) radon emanating from exposed surfaces of radium-226 bearing rock. The fraction of total radon contributed by each source has been found to vary among mines. For instance, in some non-uranium mines the radon was derived almost entirely from water sources (de Villiers and Windish, 1964; Snihs, 1974), and some studies of some uranium mines indicated that the fraction from water courses was insignificant (Misaqi, 1975; Rock and Beckman, 1977).

Water

The radon or radium-226 in rocks can be dissolved or leached into the water prior to entrance of the water into the mine. The efficiency of these processes is dependent on many factors such as chemical composition, temperature, porosity, and pressure. Highly acidic waters and waters containing large amounts of calcium and barium are usually highly efficient in leaching the radium-226 (Misaqi, 1975).

Upon release of the water into the mine, a major portion of the radon in the water is usually emanated into the mine air. That portion of radon not emanated upon release of the water into the mine opening is usually released within the first 300 meters of the drainage ditch (Bykhovskii, et al., 1969). This radon emanation, however, is dependent on such factors as the water and air temperature, degree of water agitation with air, radium-226 content of the water, and radon concentration of both the air and water. Perhaps the most important of these factors is the degree of water agitation with air. The potential radon emanation due to radium-226 decay in the water while in the mine can usually be ignored, because the half-life of radon (3.8 days) is usually large in comparison to the length of time that the water is in the mine. Table 2 presents the growth of radon from radium-226 as a function of time. Using this table, 1 pCi of radium-226 in the water produces approximately 0.166 pCi of radon in one day. Schroeder and Evans, 1969, indicate that at normal mine temperatures, the water loses radon until the radon concentration of the air is approximately three times the radon concentration of the water. Procedures for determining the radon-222 and radium-226 content of mine waters have been presented by Misaqi, 1975.

Table 2. - Growth of radon from radium-226 based on 3.825 days as the half-life of radon (Misaqi, 1975).

Days	Growth	Days	Growth
0	0.000	16	0.945
1	0.166	17	0.954
2	0.304	18	0.962
3	0.419	19	0.968
4	0.516	20	0.973
5	0.596	21	0.978
6	0.663	22	0.981
7	0.719	23	0.985
8	0.765	24	0.987
9	0.804	25	0.989
10	0.837	26	0.991
11	0.864	27	0.993
12	0.886	28	0.994
13	0.905	29	0.995
14	0.921	30	0.996
15	0.934		

Rock Surfaces

Radon diffusion rates from rock surfaces have been researched extensively. Many rock properties such as radium-226 content, porosity, permeability, mineralogy, moisture content, and temperature affect the radon diffusion rate from rock surfaces. Several authors have indicated (Austin and Droullard, 1978; Bykhovskii, et al., 1969) that emanation coefficients (ratio of radon diffused and radon produced) from rocks can vary from nearly 0 percent to nearly 100 percent. Average emanation coefficients of many domestic uranium ores have been reported by Austin and Droullard, 1978. Estimation of the average radon diffusion rate from rock surfaces of large areas prior to mining appears to be extremely difficult at the present time because of the large number of parameters affecting radon diffusion rates and the variability of these parameters within a mine.

Influences of barometric pressure changes on radon influx from rock surfaces have been noted by several authors (Kraner, et al., 1964; Pohl-Ruling and Pohl, 1969; Franklin, et al.,1979). In general, their conclusions were that radon influx from rock surfaces follows an inverse correlation with barometric pressure changes. Franklin, 1981, also notes that previous studies have shown that a steadily dropping barometric pressure for several days does not necessarily cause a steadily rising radon concentration.

Some mining activities have also been shown to significantly increase radon influx (Franklin, 1981; Ertle, et al., 1981). Franklin, 1981, has reported radon concentration increases in mine areas of up to 400 percent due to blasting and up to 40 percent due to slushing or mucking. The increase is, of course, dependent on many factors such as size of blast, degree of fragmentation, and radon diffusion characteristics of the rock. The ventilation engineer should be especially aware of these increases in radon influx to properly plan control measures.

RADON-DAUGHTER ESTIMATION

The fundamental formulas behind growth and decay of radon daughters provide the ventilation engineer with a tool for assessing radon-daughter concentrations and the best radon-daughter control method. Schroeder and Evans, 1969, present perhaps the most practical method for understanding the radon-daughter concentration change between two serially ventilated points. Their method involves dividing the radioactive constituents of mine air into three components: (1) the working level (WL)[1/] growth due to the radon concentration (pCi/l) of the mine air at the starting point (starting point is upstream from ending point); (2) the WL growth due to radon diffusion from rock surfaces between the starting point and ending point; and (3) the WL decay or decrease of the WL concentration at the starting point. The resultant WL concentrations for each component at the ending point are estimated and summed to obtain the total WL at the ending point.

The WL estimate for each component is based on the curves shown in Figure 1 (Evans, 1969). These curves show the growth of each of the daughters and their sum in WL per 100 pCi of radon per liter of air as a function of time. The following formula can be used to approximate ($\pm$ 5 percent) the WL growth curve presented in Figure 1 for periods up to 40 minutes (Evans, 1969):

$$WLc = 0.023t^{0.85} \qquad (1)$$

Where t = time in minutes
WLc = working level concentration at t per 100 pCi radon/l.

The calculations of the curves in Figure 1 assume that the activity of radon does not decrease during the periods presented. Use of this assumption tends to overestimate the WL because some radon activity is lost through decay. The amount of overestimation, however, should be less than 2 percent because the half-life of radon (3.8 days) is at least 78 times larger than the periods given in Figure 1. The curves in Figure 1 are made with the additional assumption that radon and the resulting daughters remain suspended in the air. This assumption can contribute significant error (overestimation) in WL estimates because some radon daughters attach (plate out) on surfaces. The degree of plate out is highly

1/ Working level (WL) is defined as that concentration of short-lived radon-daughter products in a liter of air that will yield 1.3 x 10^5 million electron volts of alpha energy in decaying through RaC' (Holaday, et al., 1957).

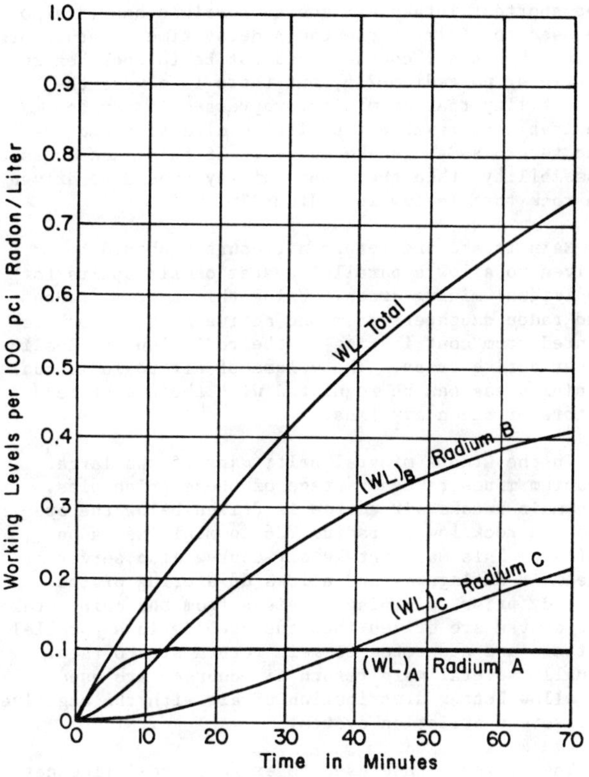

Figure I. Growth of working levels in initially pure Radon. (Evans, 1969)

variable, depending on such factors as shape, size, and velocity of the air course, and the type and amount of condensation nuclei in the air. The following formulas do, however, provide an understanding of the potential growth and decay of radon daughters.

The WL concentration at the ending point due to radon at the starting point is estimated using Figure 1 and the following formula:

$$WL_1 = \frac{R}{100} WL_t \quad (2)$$

Where R = radon concentration (pCi/l) at the starting point.

WL_t = the WL from Figure 1 corresponding to the residence time (time for air to travel from starting point to ending point in minutes).

WL_1 = the WL concentration at the ending point due to the radon concenration at the starting point.

For example, if the radon concentration at the starting point is 180 pCi/l, and the residence time is 5 minutes, then the WL concentration at the ending point due to radon at the starting point is estimated as shown below:

$$WL_1 = \frac{180}{100} (0.10 \text{ WL}) = 0.18 \text{ WL}$$

The WL concentration at the ending point due to radon diffusion from the surface of rocks between the starting point and ending point for residence times up to 40 minutes is estimated using the following formula:

$$WL_2 = \frac{JP(t)^{1.85}}{134A} \quad (3)$$

Where J = average radon diffusion rate between the starting point and ending point.

P = average perimeter distance (m) of the mine opening between the starting point and ending point.

A = average cross sectional area (m^2) of the mine opening between the starting point and ending point.

t = residence time (min).

WL_2 = the WL concentration at the ending point due to radon diffusion from rock surfaces between the starting point and ending point.

For example, if the average radon diffusion rate (J) is 2.0 pCi/m^2sec, the average perimeter distance (P) is 12 m, the average cross sectional area is 9 m^2, and the residence time is the same as before (5 minutes), then the WL concentration at the ending location due to radon diffusion from rock surfaces between the starting point and ending point is estimated as:

$$WL_2 = \frac{(2)\ (12)\ (5)^{1.85}}{134\ (9)} = 0.39 \text{ WL}$$

The WL concentration at the starting point decreases with residence time (provided no radon is present). This decrease is dependent on the age of air (WL concentration of each of the daughters at the starting point). The WL concentration at the ending point due to the WL concentration at the starting point is estimated using 1 and the following equation:

$$WL_3 = WL_s \left(\frac{WL_{(a+t)} - WL_t}{WL_a} \right) \quad (4)$$

Where WL_s = the WL concentration at the starting point.

WL_a = the WL from Figure 1 corresponding to the age of the air (minutes) at the starting point.

WL_t = the WL from Figure 1 corresponding to the residence time.

$WL_{(a+t)}$ = the WL from Figure 1 corresponding to the age of air (minutes) at the starting point plus the residence time.

WL_3 = the WL concentration at the ending point due to the WL concentration at the starting point.

For example, suppose the WL concentration at the starting point is 0.33 WL, the age of air at the starting point is 10 minutes, and the residence time is the same as before (5 minutes). In this case the WL concentration (WL_3) at the end location due to the WL concentration at the starting point is:

$$WL_3 = (0.33) \left(\frac{0.24 - 0.10}{0.17} \right) = 0.27 \text{ WL}$$

In other words, the radon daughter concentration at the starting point (0.33 WL) will have decayed to 0.27 WL by the time the air reaches the ending point.

The total WL at the ending point is then the sum of the three WL contributions (provided no other radon or radon-daughter sources are introduced between the starting point and ending point) as follows:

$$WL_f = WL_1 + WL_2 + WL_3 \qquad (5)$$

Where WL_1, WL_2, and WL_3 are estimated with formulas (2), (3), and (4).

WL_f = total WL concentration at ending point.

For our example, the WL_f at the ending point is:

$$WL_f = 0.18 + 0.39 + 0.27 = 0.84 \text{ WL}$$

Some very important aspects of radon-daughter growth and decay can be derived from these formulas. Firstly, Figure 1 depicts the WL growth per 100 pCi of radon per liter of air as a function of time. Radon-daughter concentrations from constant radon concentrations develop approximately in proportion to the 0.85 power of residence time for residence times up to 40 minutes. The working level develops very rapidly during the first few minutes and levels off to a maximum as residence time reaches about 3 hours. Secondly, radon-daughters concentrations resulting by radon diffusion from the surface of rocks increase as almost the square of residence time. This in effect means that decreasing the flowrate in an area by one-half almost quadruples the WL caused by radon diffusion in the area. This is because there has been twice the time for radon to diffuse (hence double the radon input) and this radon has had twice the time to grow into daughter products. The effects are approximately multiplicative and so an almost fourfold WL growth occurs. Because the largest percentage of radon influx in most mines is by diffusion from rock surfaces, it is important to ventilate radium-bearing rock surface areas rapidly. Thirdly, the WL decrease rate of radon-daughters is dependent on the age of air. Provided no radon is present, the working level decreases but the rate of decrease is dependent on the age of air. Older air has a faster rate of decrease than younger air.

Difficulties in using these formulas to accurately estimate WL concentrations in mines occur because of difficulties in obtaining accurate estimates of the average radon diffusion rates and radon-daughter plate out effects. The large number of parameters influencing each of these estimates and the variability of these parameters within a mine make their accurate estimation extremely difficult.

CONTROL

Proper mine planning and effective ventilation are currently the primary radon-daughter control methods in underground uranium mines. To minimize employee radon-daughter exposures, the mine should be designed to both minimize radon influx and reduce radon-daughter growth time. Ventilation helps control radon daughters by both diluting the radon daughters and reducing the radon decay time.

Mine Planning and Ventilation

One of the most effective methods of minimizing radon influx is to drive as much of the intake air course as possible in rock that is low in radium-226. The shortest intake air course possible should also be used to minimize the radon decay time. Return or main exhaust air courses need not be in rock bearing little or no radium-226, but there is always the possibility that as mining progresses it may be desirable to reverse the airflow direction and use the return as an intake airway. If this is seen as a possibility, then the return airway should be driven in rock that is low in radium-226.

Main intake and return air courses should be driven to allow a parallel system of air splits to individual mining areas. Using this system, radon and radon daughters from one active area can be prevented from contributing to the radiation problem in other active areas. The volume of air to individual mining areas can be regulated with the use of regulators or secondary fans.

In the Grants mineral belt, many of the larger uranium mines take advantage of these principles. The main intake air course is driven below the orebody in rock low in radium-226 to minimize radon influx. This main intake air course also serves as the main haulageway and allows dewatering of the orebody prior to mining. Raises from the main intake air course are driven into the orebody in a parallel pattern and connected to main return air courses. Usually several main return air courses are used to allow better distribution of air with the negative pressure ventilation system.

The mining method used to extract ore influences the amount of radon influx. The amount of surface area of radium-226 bearing rock exposed has an effect on the radon influx. Large surface areas are usually more conducive to radon influx into the mine than small surface areas. Ideally, the mining method should expose no more ore surfaces in or upstream from the employee than necessary. The amount of broken ore upstream from employees should also be minimized. Mining on the retreat (toward intake air) helps minimize both exposed surface areas upstream from active areas and radon decay time.

The location of radon sources with respect to the location of employees and the direction of airflow between the two locations dictates the radon-daughter exposure of the employees. To minimize exposures, the ventilation system and the mine should be designed to minimize placement of employees downstream from areas where mining activities such as slushing, ore transfer and blasting (known to increase radon influx) are occuring. Sometimes the exposures can be avoided or reduced through proper timing of the activities.

Fan shutdowns normally result in high radon-daughter concentrations in uranium mines. Many factors such as air distribution efficiency, volume of workings, fresh air flowrate, and time span of fan shutdown determine how long the fan must be left on before the radon-daughter concentrations return to normal. Determining this time is usually performed by a trial and error process. This process normally involves turning the fans on and measuring the radon-daughter concentrations in poorly ventilated areas or exhaust air at various times. Even in small mines it usually requires at least 2 hours of fan on-time to return concentrations to normal once the fans have been turned off overnight.

Old mined-out areas upstream from active work areas can contribute to the radon-daughter hazards if they are not ventilated properly. Radon and radon-daughters can build up to high concentrations in these nonventilated mined-out areas. This buildup is largely dependent on the average radon diffusion rate, volume of air space, and the exposed rock surface area in the mined-out area. Bulkheads are used to prevent leakage of the contaminated air into the intake air but some radon leakage almost always occurs unless a negative pressure is maintained in the sealed area. The leakage usually occurs during periods of decreasing barometric pressure and may not be detected during periods of increasing barometric pressure. A negative pressure in the sealed area helps assure that the airflow direction of the bulkhead leakage is into the sealed area rather than into the intake air. A small bleeder pipe and fan connected to a return airway will usually provide sufficient negative pressure. Many types of materials have been used to construct these bulkheads. Coating materials should be applied on the higher pressure side of the bulkhead and on the rock surface within several meters of the bulkhead to prevent leakage.

Recirculation of air in mines should be avoided if possible. Recirculation increases the radon decay time, thus effectively increasing the radon-daughter concentration in the area. Recirculation in mines is usually the result of improper fan placement or improper fan or vent tubing selection. Secondary ventilation systems used to ventilate dead-end drifts in uranium mines are usually of the push-type (fresh air blown into dead-end heading). For these systems, the fan should normally be placed at least 3 m (10 ft) into the intake side of the primary air supply to preclude recirculation. To effectively sweep contaminants at the face, the discharge end of the vent tubing should, as a general rule of thumb, be kept within 9 m (30 ft) of the face. Normally, for this type of system, the flowrate of the primary air supply must be at least twice the flowrate of the secondary system to prevent recirculation (Rock and Walker, 1970).

Radon from water leakage into the mine is often neglected as a source of contamination, yet it is sometimes the easiest to control. To minimize radon-daughter exposures, emphasis should be placed on keeping the water away from intake air. Sumps should be placed in return airways if possible. Drill holes or pipes can usually be used to effectively direct water to return airways. Unnecessary agitation of the water with intake air should be avoided as much as possible.

A mathematical model and some tests (Bates and Edwards, 1981; Franklin, 1979) have indicated that a positive pressure ventilation system is more conducive to the control of radon influx into a mine than a negative pressure ventilation system. Yet, only one large uranium mine (Thebeau, 1978) is known to have adopted a positive pressure ventilation system.

Backfilling of worked-out areas in uranium mines has been suggested for environmental, ground support, and ventilation reasons. Reductions in radon concentrations due to backfilling with classified mill tailing in a uranium mine have been described by Franklin, 1981. Whether or not backfilling will increase or decrease the radon-daughter hazard must be evaluated on an individual mine basis.

Other Possible Control Methods

One possible control method which shows some promise is the application of sealants on rock surfaces to reduce radon diffusion rates. Many rock sealants have been tested for their applicability as a method to reduce radon diffusion in uranium mines (Franklin, et al., 1975; Franklin, et al., 1977; Lindsay, et al., 1981). The two most effective materials found by these authors were a water-base epoxy and a water-base acrylic latex. Field tests of these products showed that 50 to 75 percent of the radon can be retained in the rock at material costs varying from \$0.30 to \$1.10 per square foot. Sealants are currently not used extensively because of the relatively high costs.

Arthur D. Little, Inc., 1975, assessed the probable efficiency, reliability, safety, and cost of a number of possible methods for removing radon from mine air. Those radon gas removal methods assessed were activated charcoal, organic fluids, semipermeable membranes, molecular sieves, cryogenics, chemical reaction, centrifuge, and several combinations of the methods listed. Each method considered had to be able to handle 2.4 m^3/sec (5000 cfm) at a radon removal efficiency of 90 percent. The findings indicated that the charcoal method and cryogenic distillation method were the least costly. The authors concluded that "the removal of radon from mine air appears to be technically feasible, but generally rather expensive, and it will probably not be widely used in the near future."

Filtering of radon daughters from mine air has been tested using several types of units (Chakravatti, 1979; Lawter, et al., 1981). These studies indicate that the filtration units can be cost-effective in special situations such as remote areas where ordinary ventilation practices would be impractical. Because the radon is not removed, radon daughters quickly regenerate. Lawter, et al., 1981, has reported 1977 operating costs of filtration units, including annual capital charges at 20 percent of installed costs at from \$0.002 to \$0.009 per 28 m^3 (1000 ft^3) of air cleaned (over 90 percent efficiency). U.S. underground uranium mines have used filtration in the past, but are not currently using radon-daughter filtration units.

The use of electrostatic precipitators to remove radon daughters from mine air has also been tested. Lawter, et al., 1981, reports, "Between filtration and electrostatic precipitation, there was little to recommend one over the other."

CONCLUSIONS

Many factors affect radon influx and the resulting radon-daughter concentrations in underground uranium mines. Formulas for estimating radon-daughter concentrations have been presented by Schroeder and Evans, 1969, but the accuracy of these formulas appears limited because of the difficulty in obtaining accurate estimates of the average radon diffusion rate and radon-daughter plate out effects. The primary radon-daughter control methods used in U.S. underground uranium mines center around proper mine planning and effective ventilation. Principles of these methods are aimed at minimizing radon influx and growth time in air.

REFERENCES

Arthur D. Little, Inc., 1975, "Advanced Techniques for Radon Gas Removal," U.S. Bureau of Mines Contract Report H0230022, ADL Reference No. 75946, 170 pp.

Austin, R. L., and R. F. Droullard, 1978, "Radon Emanation from Domestic Uranium Ores Determined by Modifications of the Closed-Can Gamma Only Assay Method, "U.S. Bureau of Mines, RI 8264, 74pp.

Bates, R. C., and J. C. Edwards, 1981, "The Effectiveness of Overpressure Ventilation: A Mathematical Study, "presented at "Radiation Hazards in Mining: Control, Measurement, and Medical Aspects," Oct. 4-9, 1981. Published by Society of Mining Engineers of AIME, 1981, pp. 149-154.

Bykhovskii, A. V., et al., 1969, "Experience with Radon Control in Mining," Atomizdat Press (Translated from Russian). Published by Amerind Publishing Co. Pvt. Ltd., New Delhi, 1974, 192 pp. Available from National Technical Information Service, Springfield, Virginia 22151.

Chakravatti, J. L., 1979, "Air Filtration as a Ventilation Strategy for the control of Radiation in Underground Uranium Mines," presented at "Second International Mine Ventilation Congress," Nov. 4-8, 1979. Published by Society of Mining Engineers of AIME, 1980, pp. 420-424.

deVilliers, A. J., and J. P. Windish, 1964, "Lung Cancer in a Flourspar Mining Community-Radiation, Dust, and Mortality Experience," British Journal of Medicine, pp. 94-108.

Ertle, H. J., et al., 1981, "Investigations into Methods for the Protection of Underground Workers Against Radon and Its Daughter Products," presented at "Radiation Hazards in Mining: Control, Measurement, and Medical Aspects," Oct. 4-9, 1981. Published by Society of Mining Engineers of AIME, 1981, pp. 275-282.

Evans, R. D., 1969, "Engineers Guide to the Elementary Behavior of Radon Daughters," Health Physics, Vol. 17, pp. 229-252.

Franklin, J. C., L. T. Nuzum, and A. L. Hill, 1975, "Polymeric Materials for Sealing Radon Gas Into the Walls of Uranium Mines," U.S. Bureau of Mines, RI 8036, 26 pp.

Franklin, J. C., T. O. Meyer, and R. C. Bates, 1977, "Barriers for Radon in Uranium Mines," U.S. Bureau of Mines, RI 8259, 24pp.

Franklin, J. C., et al., 1979, "Monitoring and Control of Radon Hazards," presented at "Second International Mine Ventilation Congress," Nov 4-8, 1979. Published by Society of Mining Engineers of AIME, 1980, pp 405-411.

Franklin, J. C., 1981, "Control of Radiation Hazards in Underground Uranium Mines," presented at "Radiation Hazards in Mining: Control, Measurement and Medical Aspects," Oct. 4-9, 1981. Published by Society of Mining Engineers of AIME, 1981, pp. 441-445.

Holaday, D. A., et al., 1957, "Control of Radon and Daughters in Uranium Mines and Calculations on Biological Effects," U.S. Public Health Service publication 494, 81pp.

Kraner, H. W., et. al., 1964, " Measurements of the Effects of Atmospheric Variables on Radon-222 Flux and Soil Gas Concentrations," Chap. in Natural Radiation Environment, Univ. of Chicago Press, pp. 191-215.

Lawter, J. R., D. B. Lindsay, and I. Jashnani, 1981, "Air Cleaning to Remove Radon Daughters: A State-of-the-Art Study," presented at "Radiation Hazards in Mining: Control Measurement, and Medical Aspects," Oct. 4-9, 1981. Published by Society of Mining Engineers of AIME, 1981,pp. 787-789.

Lindsay, D. B., G. L. Schroeder, and C. H. Summers, 1981, "Polymeric Wall Sealant Test for Radon Control in a Uranium Mine," presented at "Radiation Hazards in Mining: Control, Measurement, and Medical Aspects," Oct. 4-9, 1981. Published by Society of Mining Engineers of AIME, 1981, pp. 790-793.

Misaqi, F. L., 1975, "Monitoring Radon 222 Content of Mine Waters," Mining Enforcement and Safety Administration, IR 1026, 18pp.

National Academy of Sciences-National Research Council, EPA 520/476-013, 1976, "Health Effects of Alpha Emitting Particles in the Respiratory Tract," National Academy of Sciences, Washington, pp. 10-11.

Pohl-Ruling, J., and E. Pohl, 1969, "The Radon-222 Concentration in the Atmospheres of Mines as a Function of the Barometric Pressure," Health Physics, Vol. 16, May, pp. 579-584.

Rock, R. L., and D. K. Walker, 1970, "Controlling Employee Exposure to Alpha Radiation in Underground Uranium Mines," Vol. 1 of two volumes, U. S. Bureau of Mines, 72 pp.

Rock, R. L., and R. T. Beckman, 1977, "Measurements of Some of the Factors Which Influence the Radon Daughter Health Hazard, "Mining Enforcement and Safety Administration, IR 1067, pp. 9-11.

Schroeder, G. L., and R. D. Evans, 1969, "Some Basic Concepts in Uranium Mine Ventilation," Society of Mining Engineers of AIME, Transactions-Vol. 244, pp. 301-307.

Snihs, J. O., 1974, "The Approach to Radon Problems in Non-Uranium Mines in Sweden, "Third International Congress of International Radiation Protection Assoc. CONF -- 73097 -- P2, pp, 900. Available from NTIS, U.S. Dept. of Commerce, Springfield, Va. 22151.

Thebeau, D., 1978, Industrial Control and Procedures to Reduce Radon Daughter Radiation Injury," presented at "Conference/Workshop on Lung Cancer Epidemiology and Industrial Applications of Sputum Cytology, Nov. 14-16, 1978. Colorado School of Mines Press, Jan, 1979, pp. 208-307.

IMPROVEMENT OF VENTILATION SYSTEM EFFICIENCY THROUGH THE ANALYSIS OF AIR LEAKAGE

Chapter 30

E. Topuz, S. S. Bhamidipati, M. Bartkoski

Department of Mining and Minerals Engineering
Virginia Polytechnic Institute and State University

ABSTRACT

Air leakage in underground mines is the most common cause of inefficient distribution of ventilating air. Depending upon various factors, more than one half of the fresh air entering a mine short-circuits to the return airways before reaching the working faces. Consequently, large leakages create an additional demand for increased intake air, which substantially affects the efficiency of a mine system.

This paper deals with the identification of the physical variables of air leakages and offers an overview of the methods of measuring leakage.

INTRODUCTION

The provision of adequate fresh air to working sections is the primary objective of mine ventilation. It is essential to dilute and carry away any contaminant gases and particulate matter, and to maintain the temperature and humidity of work places within acceptable limits. The efficient fulfillment of these requirements is one of the main concerns of a mine ventilation engineer. The need for efficient ventilation has been increased as mines get deeper and utilize larger and more powerful mining machinery. These machinery have permitted both production concentration at faces and larger production from fewer mines. But, they have also resulted in increased gas and dust liberation. Moreover, ventilation standards have risen steadily and stringent regulations have been enacted. All these factors call for increased sophistication in planning and implementing mine ventilation systems for higher efficiencies along with the need for increased air quantities.

Air leakage can be defined as the short-circuiting of intake air into return airways before reaching the working faces. It was believed that a moderate amount of air leakage helps to make the return airways safer to travel and hence considered beneficial (Graham, 1931). It was suggested by Roberts (1960) that a reasonable allowance should be made for air leakage in ventilation planning. Leakage through caved ground can help prevent the accumulation of strata gases. Bleeder ventilation applications which serve the above purpose can be seen as controlled leakage practices. In addition, the leakage paths can be viewed as high resistance splits which are parallel to the airways, thus reducing the mine resistance. Therefore, a completely leak-proof ventilation network, besides being an impossible and costly alternative, is technically undesirable. Air leakages, however, are frequently found in places where neither required nor desired, and are of such dimensions as to render the ventilation system inefficient. In operating underground coal mines, the air leakage usually exceeds 50 percent of the intake air quantities. An opinion survey, conducted by the authors among the engineers of several underground coal mines in Southern Appalachia, indicates similar figures as shown in Table I. It is not uncommon to observe even higher figures reaching 70 to 80 percent (Herbert, 1953; Hartman, 1961; Kawenski et al, 1965). The expression of leakage as a percentage of the intake air neglects the extension of a ventilation network. For example, a 50 percent leakage may not be considered excessive for a large mine having numerous sections and a large ventilation network. Thus, the percentage figures alone do not permit us to quantify the leakage conditions. The large percentage figures, however, may help to illustrate the importance of reducing air leakage.

Air leakage directly influences the efficiency of the ventilation system because the system's energy consumption is approximately proportional to the cube of the intake air quantity. For example, in a mine with 50 percent leakage, if a modest improvement to a 40 percent leakage level can be attained, this will result in an energy savings of about 40 percent in the system. Therefore, leakage has a very pronounced effect on the efficiency of a mine ventilation system.

PHYSICAL VARIABLES OF LEAKAGE

Air tends to flow from a high pressure zone to a low pressure zone in an attempt to create equilibrium. Thus, whenever the intake is at a higher pressure level, the air leaks into the nearby return airways through the separating permeable medium.

Several studies conducted in the past contribute to the knowledge that the leakage through fine openings in a separating medium is dependent only on the pressure differential across the medium and the permeability of the medium. It may be understood that the permeability in turn is affected by several factors: the age of the mine, the method of mining and ventilation, and the materials and methods used in the construction of ventilation control devices.

The Differential Pressure

Studies done by the U.S. Bureau of Mines on stoppings indicate that the amount of air leaking across a stopping, Q, can be expressed as:

$$Q = aP^{n}$$

where (P) is the pressure differential across the stopping in inches of water gauge ; (a) and (n) are coefficients for a given stopping (Kawenski and Mitchell, 1963). The coefficient (a) is the

Table I - Air Leakage Survey Among Selected Underground Coal Mines

Mine No.	Age of Mine (yr)	Annual Production Tonne x 10^6 (Tons x 10^6)	Number of Fans	Total Mine Air m^3/s (cfm x 10^5)	Average Mine Head kPa (in. WG)	Estimated Leakage (%)	Suspected Source of Leakage
1	75	2.4 (2.7)	5	708 (15.0)	0.92 (3.7)	50	Sp; OW
2	2	0.5 (0.5)	1	85 (1.8)	0.30 (1.2)	n.a.**	Sp; CG
3	3	0.4 (0.4)	1	189 (4.0)	1.24 (5.0)	55	Sp.
4	5	4.3 (4.7)	2	566 (12.0)	2.01 (8.1)	50	Sp; OC
5	34	3.6 (4.0)	3	354 (7.5)	1.37 (5.5)	n.a.	OW
6	7	2.5 (2.8)	4	708 (15.0)	1.00 (4.0)	50	Sp.
7	5	0.2 (0.2)	4	71 (1.5)	1.00 (4.0)	n.a.	Sp; CG
8	13	3.3 (3.6)	3	189 (4.0)	1.42 (5.7)	50	Sp.
9	6	0.2 (0.2)	2	326 (6.9)	0.82 (3.3)	45	Sp; OW
10	16	3.6 (4.0)	2	283 (6.0)	1.74 (7.0)	34	Sp; OC

* Sp = Stoppings; OW = Old Works; CG = Caved Ground; OC = Overcasts
** n.a. = not available.

leakage quantity in cubic feet per minute per 100 square feet (0.000051 cubic meters per square meter) area of the stopping. Values for (a) and (n) vary depending upon the construction methods and materials of the stopping (Table II). Leakage quantities for wet and dry constructed stoppings are illustrated in Figure 1.

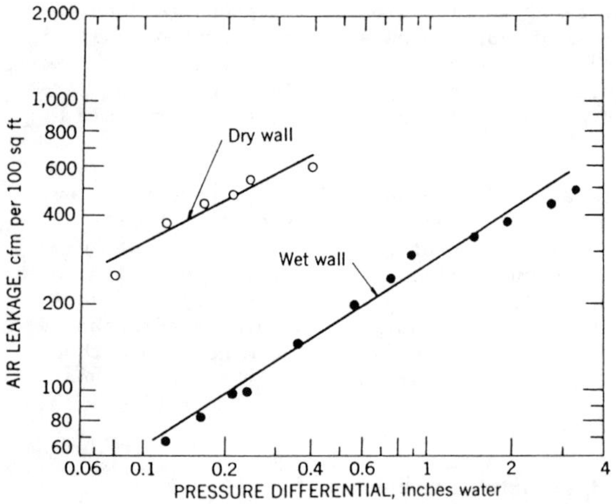

Figure 1 - Air Leakage Through Cinder-Block Stoppings (After Kawenski et al, 1965)

The relationship proposed by Briggs to express leakage across a stopping was linear and it can be stated as the following (Ramani, 1977):

Table II - Air Leakage Coefficient for Mine Stoppings (Condensed from Kawenski and Mitchell, 1963)

Stopping Construction	Coefficient a	Exponent n
Concrete Block:		
Hollow core, dry-wall	900	0.3
Solid core, dry-wall	1,200	0.9
Hollow core, dry-wall mortar-coated	5	0.9
Hollow core, dry-wall foam-coated	10	0.8
Hollow core, wet-wall	50	0.8
Solid core, wet-wall	20	0.9
Gravel Block:		
Hollow core, dry-wall	900	0.3
Hollow core, wet-wall	55	0.8
Cinder Block:		
Hollow core, dry-wall	1,060	0.5
Hollow core, wet-wall	300	0.7
Hollow core, wet-wall, mortar-coated	5	0.8
Slag Block:		
Hollow core, dry-wall	1,400	0.3
Hollow core, wet-wall	340	0.7
Sheet Stopping:		
Brattice cloth	35,500	0.7
Plastic brattice	5	1.2
Brattice cloth, latex-coated	10	0.8
Metal lath, foam-coated	15	0.2
Metal lath, mortar-coated	5	1.2

$$f = b(P_1 - P_2) = b P$$

where f is the quantity flowing through the stopping and b is a constant dependent on the construction of the stopping.

A linear relationship between the pressure differential and the leakage quantity may hold true, although a theoretical basis for this relationship has yet to be obtained. The argument may be supported because leakage is typically a laminar flow of Newtonian fluid in a permeable medium.

The Permeability

The permeability of a material may be defined as the state of having pores or small openings that permit liquids or gases to seep through it. In mine ventilation, the amount of leakages are closely related to the permeability of materials. There are several well-known studies in the area of air leakages through mine stoppings (Holland and Sargeant, 1958; Kawenski and Mitchell, 1963; Kawenski et al, 1965; Timko and Kissell, 1982). These studies indicate that permeability of stoppings shows large variations depending upon the materials and methods used in their constructions. The two major types of stoppings are block and sheet stopping.

Block stoppings are constructed of aggregate blocks of various materials. The four most common block materials listed in their decreasing order of permeability characteristics are slag, cinder, gravel, and concrete. Besides the material, the block formation, whether hollow or solid, has a very significant effect on the leakage rates as shown in Table II.

Another major factor that influences permeability of a stopping is the method of construction. A dry stopping is formed by stacking blocks on top of each other without mortar between the joints. A wet stopping is constructed with the aid of mortar between the stacked blocks' joints. The mortar will decrease the leakage through the joints and give added strength to the stopping. The leakage difference between the two types of construction is very large. Using a hollow concrete block, the leakage is three times more in a dry stopping as compared to a wet. With a solid concrete block, the leakage difference is sixty-time more for dry compared to that of wet construction (Kawenski and Mitchell, 1963). While the majority of leakage in a dry stopping is through the cracks between the blocks, the leakage through a wet stopping must permeate through the block. Since the block is relatively impermeable, leakage in wet stoppings is minimal. Coated stoppings, sealed on one or both sides with a mortar, foam or latex paint, have considerably different leakage characteristics. Test results show that a correctly coated stopping seals very effectively against leakage. The results also revealed that when the stopping was coated, the block type or stopping construction had no effect on the leakage characteristics.

Sheet stoppings are normally constructed of brattice cloth or reinforced plastic. When constructed of brattice cloth, the amount of air leakage will be influenced by type and density of the weave, dust concentration, and moisture. Even if the conditions are ideal for the cloth installation, excessive leakage must be expected. One method of decreasing leakage in a brattice stopping is by coating the cloth with a sealent. The reduction of leakage would be extensive but the flexibility would be greatly impaired.

Reinforced plastic sheeting has an excellent leakage rate and is considered practically impermeable. Tests have shown that the majority of leakage in plastic stoppings occurs around the perimeter (Kawenski and Mitchell, 1963). Thus, if greater care is exercised by securely fastening the plastic to the ribs, roof, and floor, a large amount of leakage could be prevented.

SOURCES OF LEAKAGE

The occurrence of air leaks in a mine can happen in a number of locations including the ventilating shafts, stoppings, doors, overcasts, caved ground and abandoned workings.

Leakage at the top of the ventilating shaft can sometimes reach disturbing proportions. The highest pressure differential occurs at the top of the ventilating shaft. Excessive use of this shaft for the purpose of hoisting and inadequate lining and air locks create a leakage situation in a blower system and recirculation in an exhaust system.

The primary ventilation system which extends from surface to the last open cross-cut has a relatively stable configuration. It is normally several thousand feet long and the pressure differential encountered in this region is high. The ventilation control devices used in this system are relatively permanent installations.

Stoppings and overcasts are used to separate intake and return airways. The number of stoppings in a primary ventilation is often a few hundred. In comparison to a few overcasts, the stoppings play a significant role in contributing to air leakage. Although leakage from a single well-constructed stopping is negligible, the cumulative effect of these small leakages can reach significant proportions. This amount further increases as cracks form as a result of strata pressures and ground movements. Therefore, older stoppings are more susceptible to leakage because of increased permeability.

Doors are installed to provide an access between intake and return airways. They are commonly installed in pairs to prevent high quantities of leakage. The proper construction and provision of sills for doors at the bottom helps to reduce leakage.

The secondary ventilation system extends from the last open cross-cut to the face. In this system, which rarely exceeds a few hundred feet, air is coursed making use of line brattice, or ventilation tubing with auxiliary fans. The air loss per linear foot in the secondary system ranges from 100 to 200 times that of the primary system (Dalzell, 1966). This system is very dynamic in configuration and the control devices for separating intake from return keep extending forward in cyclic fashion.

The absence of safety hazards partly makes the system of line brattice a common practice. The U. S.

Bureau's survey indicates that a loss of 95 percent of air in 100 feet (30 meters) of line brattice is not uncommon. Proper installation and selection of suitable brattice line are essential to ensure adequate air circulation. Stone dusting and moisture are seen to improve the effectiveness of line brattice (Graham, 1931).

Problems in recirculation in gassy mining conditions have dampened the practice of using ventilation tubing; although, increased usage of continous miners favors its application. Leakage at the joints become crucial with the usage of small diameter pipes.

With an ideal system, normally leakage may not exceed 25 percent (Hartman, 1961). The possible locations for excessive leakage in their decreasing order of importance are the fan installations, caved ground, stoppings, overcasts, and doors. Leakage from line brattice is very pronounced, and is the biggest source of inefficiency among the face ventilation methods.

LEAKAGE MEASUREMENTS

The magnitude of leakage varies substantially depending on the location. Large leakage quantities may easily be identified by visual inspections and by air quantity surveys. Often as with the case of a single stopping, the quantity of leakage can be so small that special measurement techniques are required.

The U. S. Bureau of Mines developed two methods to aid in determing the leakage quantities. The Brattice Window Method (BWM) was developed by Vinson et al (1977) for identifying excessively leaky stoppings in a ventilation system. The old stoppings that are nearest to the intake and return portals leak excessively as compared to others. The BWM consists of erecting a temporary fabric stopping on the return side of the leaky permanent stopping. Two windows are cut into the temporary test stopping. From the volume of air passing through these windows, the leakage quantity is measured. This effective, practical method for underground use is fairly inexpensive. It is a very time-consuming method, however, and requires two people and is applicable to only one stopping at a time.

The Tracer Gas Method (TGM) developed by Thimons and Kissell (1974) is a more versatile method and helps in detecting large-scale leakages. Leakage through caved ground, leakage in a section or panel, leakage into nearby old workings, are some of the possibilities that could be analyzed by TGM. This method consists of releasing a known amount of sulfur hexafluoride (SF_6), a tracer gas, into the mine atmosphere and collecting the air samples in glass syringes at certain locations in the down stream. These collected samples are then analyzed by gas chromatography, for the purpose of determining leakage quantities. The TGM is a simple, accurate and more versatile method, but it could become costly and time-consuming unless gas testing facilities are nearby.

Recently, a number of sensitive and accurate pressure measuring devices have become available. Despite these devices, leakage measurements are not amenable through pressure surveys because of the inherent pressure fluctuations in a ventilation system. Air quantity surveys, however, can be used to identify the total leakage in a section of a mine. These surveys may provide a general idea about the leakage levels and the likely places where further investigations are needed.

CONCLUSIONS

Undesirably large leakage levels have been observed in many ventilation networks. These large quantities render the entire ventilation system inefficient. The need for higher efficiencies is likely to increase as the mine workings become deeper, larger and extend farther away from the ventilating shafts. The increase in power costs is another factor emphasizing efficiency.

Individual leakage quantities are often small and therefore difficult to detect. The cumulative effect of these small quantities, however, can reach large proportions. In underground coal mines, it is not uncommon to observe air leakages exceeding 50 percent of the intake air quantities. Even a small reduction in leakage may result in comparatively large savings in the cost of energy.

Air leakages can be reduced in several ways. During the planning stages an efficient mine layout should be chosen with a proper selection of ventilation methods and materials. Avoidance of pressure buildup through the maintenance of obstruction-free airways has a direct implication on leakage control. Finally, the periodic ventilation surveys conducted in a mine must include inspections to identify places of high leakages. With these suggestions in design and operations, along with an increased awareness for the importance of air leakages, the efficiency of mine ventilation systems can be substantially improved.

REFERENCES

Dalzell, R. W., 1966, "Face Ventilation in Underground Bituminous Coal Mines," Report of Investigations, 6725, U. S. Bureau of Mines, 30 pp.

Graham, J. Ivon, 1931, "Air-Leakage Through Brattice and the Combustibility of this Material," Transactions, Institution of Mining Engineers, 1930-31, Vol. 80, pp. 452-467.

Hartman, H. L., 1961, Mine Ventilation and Air Conditioning, Ronald Press Co., New York, New York.

Herbert, C. A., 1953, "Some Factors Affecting and Suggested Ways for Improving Coal Mine Ventilation with Particular Reference to Mines in Illinois, Indiana, and Western Kentucky," Information Circular, 7656, U. S. Bureau of Mines, 15 pp.

Holland, C. T. and L. Sargeant, 1958, "An Investigation of Material Used for Stopping Construction in Coal Mine Ventilation Systems, " Proc. Coal Min. Inst. America, Pittsburgh, PA, pp. 60-80.

Kawenski, E. D. and D. W. Mitchell, 1963, "An Evaluation of Stopping Construction," Mining

Congress Journal, Sept., pp. 57-61.

Kawenski, Edward M., Donald W. Mitchell, George R. Bercik, Angelo Frances, 1965, "Stoppings for Ventilating Coal Mines," Report of Investigations, 6710, U. S. Bureau of Mines, 20 pp.

Ramani, R. V., R. Stefanko, and G. W. Luxbacher, 1977, "Sensitivity of Leakage and Friction Factors," Volume IV in Advancement of Mine Ventilation Network Analysis From Art to Science, Final Report, Prepared for the U. S. Buruea of Mines, June 30.

Roberts, A., 1960, Mine Ventilation, Cleaver-Hume Press, Ltd., London.

Timko, Robert J. and Fred N. Kissell, 1982, "New Techniques for Reducing Stopping Leakage," Pre-print, No. 82-89, SME-AIME Annual Meeting, Dallas, Texas, February 14-18.

Thimons, E. D., F. N. Kissell, 1974, "Tracer Gas as an Aid in Mine Ventilation Analysis," Report of Investigations, 7917, U. S. Bureau of Mines, 17 pp.

Vinson, Robert P., Edward D. Thimons, Fred N. Kissell, 1977, " Brattice Windown Method for Measuring Leakage through Mine Stoppings," Report of Investigations, 8240, U. S. Bureau of Mines, 17 pp.

Chapter 31

THE USE OF THE S.I. UNIT SYSTEM
IN MINE VENTILATION CALCULATIONS AND PRACTICE

Jan Wolski
Associate Professor of Mining Engineering
New Mexico Institute of Mining & Technology
Socorro, New Mexico

INTRODUCTION

Today's rapidly shrinking world requires a common language of measurements, a consistent system which could be used by everyone. The S.I. system of units is generally accepted as such and is already in use by a great majority of nations in the world. The change to S.I. units is also slowly taking place in U.S. engineering practice. The transition period is difficult and long because it takes time to get familiar with the new system, to start "feeling" values in new units. Instruments, formulas, computer programs, survey grids, etc. have to be converted or exchanged. The transition period should be used not only for substitution of units, but also for rationalization of presently used calculation methods.

This paper examines the major parameters of mine ventilation from the above point of view and discusses a system which is much simpler and more consistent than the American Unit System still in use in the United States.

In October 1979, the Society of Mining Engineers of AIME decided to introduce the S.I. unit system in all its publications including the periodical "Mining Engineering", and since February 1980 the system has been used consistently, not only in SME publications but in all conferences and meetings organized by this Society.

The authors have no intention of explaining the details of the S.I. system in general. Any reader can find this information in many available publications and conversion tables, and in particular in the ASTM Standard for Metric Practice, E380-79.

A comparison is made between various units used in mine ventilation from the past and those presently in use. This will allow an easy translation from one system to another, important when reading publications from abroad.

S.I. and customary U.S. formulae are shown side by side for easier visualization of similarities and differences.

Recommendations for the necessary steps in order to switch from American to S.I. units conclude the paper.

MINE VENTILATION S.I. UNITS AND MEASUREMENTS

The main parameters used in mine ventilation are air quantity and density, pressure, airway resistance and fan power. Each of these parameters will be discussed below. Formulae, symbols, subscripts and constants are shown in Tables 1 to 4.

Quantity of Air Flow.

To determine air quantity flowing through an airway per unit time (q), area (A) and velocity (v) need to be measured.

Then:

$$q = A\,v \qquad (1)$$

where q = quantity of flow in m^3/s, written here on purpose as a lower case 'q' in order to distinguish it from the capital "Q" in 10^5 ft^3/min,
A = cross-section area of airway, m^2,
v = air velocity, m/s.

To visualize better the range of values in the new unit system, it is good to remember that:
- air velocity 1 m/s is equal roughly 200 ft/min (involved error is 1.6%),
- air flow 1 m^3/s is equal roughly 2 000 ft^3/m (involved error is 5.6%).

In deep mines and where significant changes in air temperature occur, volumetric air flow can vary to such an extent that it is better to use mass flow instead:

$$m = \rho\, q = \rho\, A\, v \quad \text{kg/s} \qquad (2)$$

where ρ is air density, kg/m^3.

Obviously mass flow rate obeys the law of continuity and is constant along its path throughout the mine, unless additional components mix with air (water vapor, compressed air and mine gases).

Air Density,

In the S.I. unit system, air mass density can be calculated from formulae:
- without taking in consideration air humidity:

TABLE 1

MAIN FORMULAE USED IN THIS PAPER
(See Table 2 for Glossary of Symbols)

DESCRIPTION	U.S. System Formula	U.S. System Unit	S.I. System Formula	S.I. System Unit
Quantity of Air Flow	$Q = A\ V\ 10^{-5}$	10^5 ft³/min	$q = A\ v$	m³/s
Mass (or by Weight) Air Flow	$W = \omega\ Q$	10^5 lbf/min	$m = \rho\ q$	kg/s
Air Density	$\omega = \dfrac{1.325\ (H_b - 0.378\ H_{vv})}{459.7 + t_d}$	lbf/ft³	$\rho = \dfrac{3.484\ (P_b - 0.378\ P_{vv})}{273.15 + t_d}$	kg/m³
Pressure From Barometer Reading in Pressure Head	$p = \omega_{fl}\ H_{fl}$	lbf/ft²	$p = g\ \rho_{fl}\ H_{fl}$	Pa
Velocity Pressure	$H_v = \omega(\frac{V}{1\ 098})^2$	in WG	$Pv = \frac{\rho v^2}{2}$	Pa
Shock Pressure Loss	$\Delta H_x = X\ H_v$	in WG	$\Delta P_x = X\ P_v$	Pa
Airway Resistance Factor	$R = \frac{K\ 10^{10}}{5.2}\ \frac{C\ L}{A^3}$	$\frac{\text{in WG} \cdot \text{min}^2}{\text{ft}^6\ 10^{10}}$	$r = \frac{f}{8}\ \frac{C\ L}{A^3}$	m^{-4}
Airway Friction Pressure Loss	$\Delta H = R\ Q^2$	in WG	$\Delta p = \rho\ r\ q^2$	Pa
Airway Resistance Per Unit Length	$R_u = 100\ \frac{K\ 10^{10}}{5.2}\ \frac{C}{A^3}$	Per 100 ft	$r_u = 100 \frac{f}{8}\ \frac{C}{A^3}$	Per 100 m
Shock-Loss Equivalent Length	$L_e = \frac{5.2\ \omega\ X}{(1\ 098)^2 K}\ \frac{A}{C}$	ft	$L_e = \frac{4X}{f}\ \frac{A}{C}$	m
Airway Resistance Expressed as $f(r_u, L_e)$	$R = R_u\ (L + L_e) 10^{-2}$	$\frac{\text{in WG} \cdot \text{min}^2}{\text{ft}^6}$	$r = r_u\ (L + Le) 10^{-2}$	m^{-4}
Fan Power (on Motor Shaft)	$P = \frac{\Delta H_f\ Q\ C_f}{6\ 350\ \eta}\ 10^5$	hp	$P = \frac{\Delta P_f\ q\ C_f}{10^3\ \eta}$	kW

$$\rho = 1.20\ \frac{p_b}{101.325}\ \frac{294.15}{273.15 + t_d} =$$

$$= \frac{3.484\ p_b}{273.15 + t_d} \quad \text{kg/m}^3 \qquad (3)$$

- with influence of air humidity:

$$\rho = \frac{3.484\ (p_b - 0.378\ p_{vv})}{273.15 + t_d} \quad \text{kg/m}^3 \qquad (4)$$

where:

ρ = air mass density, kg/m³
p_b = atmospheric pressure, kPa
p_{vv} = water vapor pressure, kPa
t_d = temperature in degrees Celsius

Under the standard atmospheric conditions, where p_b = 101.325 kPa and t_d = 21 °C, dry air density becomes 1.20 kg/m³ which in turn is adopted as standard in mine ventilation practice in many countries (examples: Poland, USA, USSR, South Africa). (1)

- actual water vapor pressure (Carrier 1951):

$$p_{vv} = p_{vsw} - \frac{(p_b - p_{vsw})(t_d - t_w)}{1523.4 - 1.3\ t_w} \quad \text{kPa} \qquad (5)$$

or (McPherson 1979, formula (31)):

$$p_{vv} = p_{vsw} - 6.44 \times 10^{-4}\ p_b\ (t_d - t_w) \quad \text{kPa} \qquad (5a)$$

The latter gives a maximum error of about 6% which is small considering that the value p_{vv} is small also in relation to total atmospheric pressure.

Symbols used above are:

p_b = barometric pressure, kPa
t_d = temperature on dry thermometer, °C
t_w = temperature on wet thermometer, °C
p_{vsw} = saturation vapor pressure at dew point at wet bulb temperature which can be calculated from (Whillier 1971, McPherson 1979):

$$p_{vs} = 0.6106 \exp\left(\frac{17.27\ t}{237.3 + t}\right) \qquad (6)$$

for $t = t_w$; the above formula is accurate to within 0.1% for the temperature 0 to 60 °C.

The density of moist air can be easily calculated using (4), (5) or (5a) and (6).

(1) A great number of "standard" temperatures can be found in the U.S. and abroad, the most common being 0, 4, 15, 20 °C and 50 or 70 °F. Even in the 1981 edition of the ASHRAE Handbook of Fundamentals, two standard temperatures are indicated: 15 °C on page 5.1 and 20 °C on page 35.1. The values of 21 °C and 1.20 kg/m³ adopted in this paper as standard can be treated only as an alternative for discussion.

TABLE 2

GLOSSARY OF SYMBOLS

DESCRIPTION	U.S. SYSTEM Symbol	U.S. SYSTEM Unit	U.S. to S.I.(1) Multiplier	S.I. SYSTEM Symbol	S.I. SYSTEM Unit
Cross-Section Area of Airway	A	ft^2	0.092 903 04*	A	m^2
Airway Perimeter	C	ft	0.304 800 6	$\mathcal{C}$	m
Energy	E(2)	Btu	1.054 68	E	kJ
Pressure Head	H	in.	25.4*	H	mm
Air Friction Factor	K	$\frac{lbf \cdot min^2}{ft^4}$	$1.235\ 5 \times 10^7$ (3)	f	—
Airway Length	L	ft	0.304 800 6	L	m
Mass Air Flow	M	lb/min	$7.559\ 873 \times 10^{-3}$	m	kg/s
Power	P	hp	0.746*	P	kW
Pressure	p	lbf/ft^2	47.880 26	p	kN/m^2
Pressure Head	H	in. WG	248.59 (3)	p	kN/m^2
Quantity of Flow	Q	$10^5\ ft^3/min$	$4.719\ 474 \times 10$	q	m^3/s
Airway Resistance Factor	R	$\frac{in.\ WG \cdot min^2}{ft^6 \cdot 10^{10}}$	$9.3\ 007 \times 10^{-2}$ (3)	r	m^{-4}
Absolute Temperature	T	^0R	---------	T	^0K
Temperature	t_F	^0F	$t_C = (t_F - 32)/1.8$	t_C	^0C
Air Volume		ft^3	$28.316\ 85 \times 10^{-3}$		m^3
Air Velocity	V	ft/min	5.08×10^{-3}*	v	m/s
Air Flow by Weight	W	10^5 lbf/min		Not Used	
Shock Loss Factor	X	—	1	X	—
Pressure Loss or Gain	ΔH	in. WG	248.59(3)	Δp	Pa
Fan Efficiency Factor	η	—	1	η	—
Air Mass Density	ρ	lb/ft^3	16.018 46	ρ	kg/m^3
Air Specific Weight	ω	lbf/ft^3		Not Used	

Notes:
* Exact Value
(1)ASTM Standard for Metric Practice, 1980, or as noted.
(2)At 60 ^0F
(3)At Standard Atmospheric Conditions

Pressures.

Air pressure is one particular area of mine ventilation practice that the changeover from conventional U.S. to S.I. units really simplifies and makes much easier to comprehend. An inch of water (in. WG) is not well defined and varies with temperature. This variation is also true of any other liquid column used to measure pressure such as mercury, alcohol or oil. The influence of liquid density is small in measurements of differential pressures but can be significant in measurements of absolute pressures where column of liquid is high.

For the above reason and in order to be consistent throughout the S.I. system, the use of pressure head is not recommended and instruments for absolute pressure measurements should be calibrated directly in pressure units (Pa).

If instruments are used with pressure head reading, the actual pressure is:

$$p = g\ \rho_{fl}\ H_{fl} \quad \text{Pa} \qquad (7)$$

TABLE 3
GLOSSARY OF SUBSCRIPTS

a	Dry Air
b	Barometric (Static Condition)
d	Temperature on Dry Thermometer
f	Fan
fl	Fluid used in Barometer Scale
ref	Reference Value
v	Velocity
vs	Saturated Water Vapor
vsw	Saturated Water Vapor at Wet Bulb Temperature
vv	Water Vapor
w	Temperature on Wet Thermometer
1	Inlet
2	Outlet

TABLE 4

LIST OF CONSTANTS

	SYMBOL	U.S. System		S.I. SYSTEM	
		Value	Unit	Value	Unit
Thermodynamic Constant for Dry Air	c	0.286	---	0.286	---
Gravitational Acceleration	g	32.174 05	ft/s^2	9.806 65	m/s^2
Standard Atmospheric Pressure	p_{bst}	2 116.22	lbf/ft^2	101.325*	kPa
Standard Air Temperature	t_{st}	70	°F	21	°C
Standard Air Density at Standard Pressure and Temperature (dry condition)	ρ_{st}	0.075	lb/ft^3	1.20	kg/m^3
	ω_{st}	0.075	lbf/ft^3	Not Used	

*exact value

where:

ρ_{fl} = density of liquid used in a barometer, kg/l
H_{fl} = pressure head, mm of liquid
g = 9.806 65 = gravitational constant, m/s^2

If an altimeter is still used for ventilation pressure measurements, conversion has to be done according to well known methods used in survey works. Instead of using conversion from feet of altitude to inches of water, conversion will take place from meters of altitude to pascals in S.I. system.

The density of mercury, the most common liquid used in barometer scales, can be calculated from formula (Kirk-Othmer, 1967, p. 220):

$$\rho_{Hg} = \frac{13.5955}{1+0.181\ 82\times10^{-3}t+0.0078\times10^{-6}t^2}\ kg/l \qquad (8)$$

where t is the temperature in degrees Celsius.

Appropriate density tables can be used for other fluids such as alcohol, water or oil.

The basic unit of pressure in the S.I. system is the N/m^2 or Pa (pascal). This is a small pressure unit equal to about 1/10 of 1 mm on a water gauge, but it is perfect for measurements of pressure drop along airways due to friction, which is usually small as well. This pressure drop will be called Δp in the next chapters.

A different situation exists in absolute pressure measurements. Atmospheric pressure at sea level is about 100 000 Pa and to avoid large numbers, the use of kilopascals (kPa) is more appropriate. Nevertheless high precision instruments are required for measurements in absolute pressure, with resolution at least between 1 and 5 pascals.

To better "feel" the order of magnitude of the new pressure unit, it is worthwhile remembering that 1 inch of water gauge (1 in. WG) is equal to about 250 Pa, and the error involved is very small:

at the temperature of 21 °C and standard atmospheric pressure, water density is 997.99 kg/m^3, then

1 in. WG = 9.806 65 x 997.99 x 0.0254 = 248.59 Pa

and the error involved is only 0.6%.

The total pressure in an airway is composed of static and velocity pressures, the latter being equal to:

$$p_v = \frac{\rho v^2}{2}\ Pa \qquad (9)$$

ρ = air density, kg/m^3
v = air velocity, m/s.

Then air velocity can be found if velocity pressure is measured:

$$v = \sqrt{\frac{2\ p_v}{\rho}}\ m/s \qquad (9a)$$

Shock pressure losses caused by local obstructions in airways are usually expressed in the form (Hartman, 1961, p. 98):

$$\Delta p_x = X\ p_v \qquad (10)$$

X - dimensionless coefficient or shock-loss factor which can be calculated from appropriate formulae for particular obstruction conditions (Hartman 1961, Table A-3, p. 373-375).

Shock-loss factors are the same in any unit system.

Friction Factor and Airway Resistance.

In mine ventilation practice the square law can be used in most cases to determine the relationship between pressure friction loss and air quantity in the condition of the fully developed turbulent flow. The last condition is valid for most mining airways.

The basic formulae used in the U.S. are:
- the square law:

$$\Delta H = RQ^2 \quad (11)$$

- resistance factor (Atkins formula):

$$R = \frac{K\ 10^{10}}{5.2} \frac{C\ L}{A^3} \quad \text{in. WG} \cdot \text{min}^2 \cdot 10^{-10}/\text{ft}^6 \quad (12)$$

where:

ΔH = pressure head loss due to friction, inch WG
Q = air flow quantity, $10^5 ft^3/min$
K = friction factor, $\frac{lbf \cdot min^2}{ft^4}$
C = airway perimeter, ft
L = airway length, ft
A = airway area, ft^2
5.2 = factor (5.2 lbf/ft^2 per one in. WG).

The relationship between the airway friction factor K and the unitless coefficient of friction f is:

$$K = \frac{\omega\ f}{8 \times 60^2\ g} = 1.079\ 20 \times 10^{-6}\ \omega\ f \quad (13)$$

where

ω = specific weight of air, lbf/ft^3
g = 32.174 - gravitational accleration ft/s^2
60^2 = factor, $(s/min)^2$.

For standard atmospheric conditions(2), specific gravity of dry air equal to ω_{st} = 0.075 lbf/ft^3. Then, at this standard specific gravity of air, the relation K_{st} and f will take the following form:

$$K_{st} = 8.0940 \times 10^{-8}\ f \quad (14)$$

Relation between actual and standard friction factors is:

$$K = K_{st} \frac{\omega}{\omega_{st}} = K_{st} \frac{\omega}{0.075} \quad (15)$$

This way standard values K_{st}, given in tables and graphs, can be used in calculations where the specific gravity (ω) differs from the standard value 0.075 lbf/ft^3.

Some obvious inconveniences exist in using the above formulae:

1st - both airway resistance (R) and friction factor (K) depend on air density. Adjustments must be made every time conditions vary from standard values.
2nd - friction factor K, defined above, has been used only in mine ventilation, and is not compatible with other branches of science and engineering where the general friction coefficient f is being used.
3rd - friction factor K is not unitless, and therefore, its value will be different in every unit system.

It is felt that during the switch to the S.I. system would be a good period to improve and rationalize ventilation calculation methods. It seems that one of the major improvements is to go back to the original Chezy-Darcy equation and the dimensionless friction coefficient. This way the mine ventilation engineer would overcome many of the difficulties which arise out of the variations in air density in a mine network. Mine ventilation would also come into agreement with fluid flow calculations involved in other engineering disciplines and thus would facilitate mutual understanding and communication.

Using the same symbols as in equations (11) (12), the Chezy-Darcy formula can be written in the following form (Burrows 1975, McPherson 1971, Hall 1981)

$$\Delta p = \rho\ r\ q^2 \quad (16)$$

$$r = \frac{f}{8} \frac{C\ L}{A^3} \quad (17)$$

Δp = pressure loss due to friction, Pa
ρ = air density, kg/m^3
r = airway resistance factor, m^{-4}
f = friction factor in fully developed (*) turbulent flow, dimensionless
C = airway perimeter, m
L = airway length, m
A = airway cross-section area, m^2
q = air flow, volumetric, m^3/s

As can be seen, resistance factor r has only a geometrical dimension (m^{-4}) and therefore it is independent from air density. Symbol q for air flow is used instead of Q in formula (11) because flow is now calculated per second and not per minute as before. Friction pressure loss is expressed in pascals in the formula above and, as a consequence, the simplest way would be to take measurements in pascals as well. It is recommended that instrument manufacturers put on the market a new generation of precision barometers for ventilation surveys calibrated in kPa and Pa.

For reasons of consistency and precision, pressure heads expressed in mm of mercury or in cm WG should be avoided. Precision of pressure head measurements suffers since the density of liquids varies with the temperature, unless correction factors are applied. Of course, the best way to determine the friction and resistance factors for a given airway is to calculate them from equations, after measuring the actual pressure drop underground. If the above method cannot be used, some estimate can be done with the experimental values published in mining literature.

Airway Resistance per Standard Length

For the purpose of comparison it is convenient to use airway resistance calculated for a standard length, 100 ft, 100 yd or 100 m in S.I. units. This unit resistance is very useful, especially in the design of ventilation networks where particular types of airways can be standardized by shapes (C), areas (A) and friction factors (f).

(2) Standard atmospheric conditions:
pressure: 101 325 Pa. (2 116.22 lbf/ft^2)
temperature: 21 °C (70 °F)

* In some countries symbol λ (lambda) is used instead of f : both symbols are equivalent and should not be confused with symbol f used in England and some other countries. To avoid confusion, the last friction factor will be given a symbol "f(Brit)" and f=4 f(Brit).

The resistance factor defined above can be expressed in the following form (using formula (17)):

$$r_u = 100 \frac{r}{L} = 100 \frac{f}{8} \frac{C}{A^3} \ m^{-5} \quad (18)$$

Values r_u can be found also from direct measurements of airway resistance in mines. Standardized unit resistances can be found in many publications (Roberts, Skochinski and Komarov, Simode, Bertard and others).

Calculation of Shock Pressure Losses by Equivalent - Length Method.

It is very useful to present local pressure losses in form of the equivalent airway length (Hartman 1961, p. 99). In S. I. this can be found from equations (9) (10) (16) (17):

$$\Delta p_x = X \ p_v = X \ \frac{\rho \ v^2}{2} = \frac{X \ \rho}{2 \ A^2} \ q^2$$

$$\Delta p_x = \rho \ r \ q^2 = \frac{\rho \ f \ C \ L_e}{8 \ A^3} \ q^2$$

Then the length of an airway equivalent to the shock pressure loss Δp_x is:

$$L_e = \frac{4 \ X \ A}{f \ C} \ m \quad (19)$$

where

X = dimensionless shock-loss factor
A = airway area, m^2
C = airway perimeter, m
f = dimensionless friction factor.

Finally the total airway resistance can be calculated using resistance per 100 m (r_u) and equivalent length (L_e):

$$r = r_u \ (L + L_e) \ 10^{-2} \ m^{-4} \quad (20)$$

The obvious advantage of using standardized values of r_u and L_e can be seen here.

Fan Power.

Required fan power can be calculated from formulae (McPherson 1971, p. 735):

$$P = \frac{\Delta p_f \ q_1 C_f}{10^3 \ \eta} \ kW \quad (20a)$$

where:

P = power developed by the motor shaft, kW
Δp_f = pressure difference developed by the fan, Pa
q_1 = inlet air flow rate, m^3/s
C_f = compressibility coefficient
η = efficiency coefficient

$$C_f = \frac{p_1}{c \ (p_2 - p_1)} \left[\left(\frac{p_2}{p_1} \right)^c - 1 \right] \quad (20b)$$

where:

c = 0.286 (exact for dry air)
p_1, p_2 = absolute static pressures at fan inlet (1) and outlet (2), Pa.

The total fan power is found if total differential pressure (Δp_f) is used. Similarly static power can be calculated with the use of static differential pressure.

COMPARISON BETWEEN VARIOUS FRICTION FACTORS.

Several friction factors were, and some still are, in use in the world literature. It is irritating to find some values which are difficult to compare with those currently used by the reader. For this reason it seems useful to supply the means for simple comparison of the friction factors used in different unit systems. In the following discussion, various systems will be described using the same symbols given in the previous formulae.

Only fully developed turbulent flows in enclosed ducts will be considered.

1. Dimensionless S.I. friction factor derived from the Chezy-Darcy formula (16):

$$\Delta p = \rho \ r \ q^2 \text{ and } r = k \ \frac{C \ L}{A^3} \quad (21)$$

where $k = \frac{f}{8}$ = dimensionless friction factor.

This form of the friction factor was proposed by McPherson 1971, Swilski 1978, and others. The only advantage of introducing the k factor above would be the slight simplification of the Chezy-Darcy formula. On the other hand, the friction coefficient f is widely used in engineering and everybody knows what it is. Therefore, it is recommended that the f factor rather than k be used. Note that French mining engineers have always been using coefficient f in form of the λ (Bertard 1962, Simode 1967).

2. Dimensioned S.I. friction factor was proposed by Burrows and McPherson, both in 1975, as another attempt to rationalize the friction factor.

Dimensioned friction factor:

$$K(SA) = \frac{f(Brit)}{2} \ 1.2 \ kg/m^3 \text{ or } N \cdot s^2/m^4 \quad (22)$$

where:

$f(Brit) = \frac{f}{4}$ = British friction coefficient,
1.2 kg/m^3 is the air density at standard atmospheric condition

Resistance factor (same as in (17)):

$$r = \frac{K(SA)}{1.2} \ \frac{C \ L}{A^3} \quad (23)$$

A comparison of formulae (17) (22) and (23) gives the relation:

$$f = \frac{8 \ K(SA)}{1.2} \quad (23a)$$

As an explanation McPherson (1975, p. 152) stated that "Despite the rationale of using the dimensionless coefficient, the fact of the matter is that generations of ventilation engineers have become very familiar with the Atkinson type of friction factor."

The above form of friction factor will be made obvious by looking at the pressure loss formula which becomes:

$$\Delta p = \rho \, r \, q^2 = \frac{\rho}{1.2} \; K(SA) \, \frac{C \, L}{A^3} \, q^2 \qquad (24)$$

K(SA) is then the friction factor at the standard atmospheric condition. The only problem is that these chosen standard conditions vary from one country to another.

Even if the barometric pressure is usually adopted as 101.325 kPa, the "standard" temperature can be 0, 4, 15, or 20 °C, or 50 or 70 °F. Consequently "standard" air density is sometimes 1.2927, 1.2255 or 1.20 kg/m³. In the condition above, formulation of the friction factor does not seem to be the right way to go. Besides, it is dimensioned and therefore its value is specific to the S.I. system only.

3. U.S. versus S.I. system. Both systems were already described. The following relations can be derived from formulae (11) (12) (16) (17):
   - Relation between pressure friction losses:

$$\Delta p = 248.59 \, \Delta H \quad * \; \text{Pa} \qquad (25)$$

ΔH = pressure head loss in in WG

   - Relation between resistance factors:

$$\rho \, r = 11.160\,83 \times 10^{-2} \, R \qquad (26)$$

where:
R = U.S. airway resistance
ρ = air density, kg/m³

   - Relation between friction factors:

$$\rho \, f = 1.482\,58 \times 10^{7} \; K \qquad (27)$$

   - As above, but at standard air:

$$f = 1.2355 \times 10^{7} \; K_{st} \qquad (27a)$$

The standard friction factors K_{st} were tabulated by McElroy 1935. Equivalent f values can be found in Hall 1981, p. 64.

It should be noted that in some publications Q was expressed in thousands of ft³/min in formula (11). As a consequence the related friction factor K' is 10^6 times greater than K value used in U.S. customary system (Roberts 1960, p. 263-268).

4. The Old Metric System used kgf as a unit of force and the unit of air resistance, called a "murgue"**, was defined as the air resistance in an airway where air flows at the rate of 1 m³/s and causes pressure drop of 10^{-3} kgf/m². Therefore:

$$\Delta p_m = 10^{-3} \, R_m \; q^2 \qquad (28)$$

where:
Δp_m = friction pressure loss, kgf/m²
q = air flow rate, m³/s
R_m = resistance factor expressed in murgues.

   - Relation between friction pressure loss in S.I. and old metric system is:

$$\Delta p = 9.806 \, \Delta p_m \;\; \text{Pa} \qquad (29)$$

because 1 kgf = 9.806 N and 1 kgf/m² = 9.806 Pa.

   - Relation between air resistance in S.I. and old metric systems:

$$\rho \, r = 9.806 \times 10^{-3} \, R_m \qquad (30)$$

For example French standard resistance factors are calculated at a barometric pressure of 760 mm Hg and a temperature of 15 °C which result in air density ρ = 1.226 kg/m³ (Simode 1976, p. 232). Then to calculate air resistance in S.I. units, the following relation has to be used:

$$r = \frac{9.806 \times 10^{-3}}{1.226} \; R_m = 8.00 \times 10^{-3} \, R_m$$

The important role of French mining engineers in the measuring and tabulating of resistance factors for a wide range of airway types (Simode 1976, Bertard 1962 and others) should be emphasized here.

There are two types of friction factors in the old system:

Type 1 (Simode 1975):
The resistance is calculated from the following formula:

$$R_m = \frac{10^3}{8g} \; \frac{\lambda \, \omega \, C \, L}{A^3} \;\; \text{murgues} \qquad (31)$$

λ = dimensionless friction factor
ω = specific weight of air, kgf/m³

By comparing (16), (17), (28), (29) and (31) we can find that $\lambda = f$, therefore both λ and f are the same dimensionless friction factors.

As we can see, we have three dimensionless factors in world literature:

- dimensionless friction factor called "f", used in U.S. and Canada,
- dimensionless friction factor called "f", used in England, South Africa and some other countries. To avoid misunderstanding the last factor will be called here f(Brit) and its value is f(Brit) = 0.25 f,
- the same as above two, but called "λ", used in continental Europe.

Extreme caution is needed when reading publications from abroad, since it is easy to confound the above friction factors which are written in the same way but represent different values. For this reason, it is worthwhile to show the Chezy-Darcy formula using both factors:

$$\Delta p = \frac{\rho}{8} \; f \, \frac{C \, L}{A^3} \, q^2$$

$$\Delta p = \frac{\rho}{2} \; f \, (\text{Brit}) \, \frac{C \, L}{A^3} \, q^2$$

Type 2 (Skochinsky - Komarov 1968, p. 249):
Resistance is calculated from the formula:

$$R_m = 10^3 \, \alpha \; \frac{C \, L}{A^3} \quad \text{kgf} \cdot \text{s}^2/\text{m}^8 \qquad (32)$$

where: α = friction factor, kgf · s²/m⁴, and both R_m and α refer to standard air density 1.2 kg/m³; using (17), (30), (32):

* This relation is exact only at the temperature 21 °C, at water density 997.99 kg/m³.

** 1 "weisbach", used in German publications is equal to 1 kilo murgue or 1 000 murgues.

$$f = \frac{8 \times 9.806}{1.2} \alpha = 65.38\,\alpha \qquad (33)$$

Relationships between f, r from one side and α, R_m from the other, reflect the main difference between the old metric and the new S.I. systems. The basic unit of force is one kilogram-force (kgf) in the first system, and one newton (N) in the latter (one kgf is equal 9.806 N).

Relations between the friction and resistance factors are shown on Tables 5 and 6.

Friction factor K and airway resistances R, R_m depend on air density. Therefore, their "standard" values (K_{st}, R_{st}) are taken for comparison with other equivalent factors. As for R_m, its standard values vary from country to country and therefore the air density must be known (ρ_{ref}).

Example:

U.S. friction factor K_{st} for a slightly obstructed straight airway in sedimentary rock has an average value of 60×10^{-10} (McElroy 1935). What is the equivalent friction factor f? Using Table 5, we find:

$$f = 1.2355 \times 10^7\,K_{st} = 1.2355 \times 10^7 \times 60 \times 10^{-10} = 0.074\,13$$

This value can be verified by using Table 1, Hall: "Mine ventilation engineering", p. 64. The equivalent value is 75×10^{-3}, obviously rounded for good appearance.

Using Tables 5 and 6 any friction or resistance factors can be easily compared. This way the large quantity of existing data from abroad could be more easily studied and used. Lack of U.S. research in this field for many years, if not decades, can be observed.

TABLE 5

RELATIONS BETWEEN DIMENSIONLESS f (or λ) AND OTHER FRICTION FACTORS

Equation Number	Relation	Unit
21	$f = 8k$	--
22	$f = 4f(Brit)$	--
23a	$f = 6.67\,K(SA)$	$N \cdot s^2/m^4$
27a	$f = 1.2355 \times 10^7\,K_{st}$*	$lbf \cdot min^2/ft^4$
33	$f = 65.38\,\alpha$	$kgf \cdot s^2/m^4$

* at air density 1.20 kg/m³

TABLE 6

RELATIONS BETWEEN VARIOUS AIR RESISTANCE FACTORS

Equation Number	Relation	Unit
26	$r = 9.3007 \times 10^{-2}\,R_{st}$	in WG $\cdot$ min² $\cdot\ 10^{-10}$/ft⁶
30	$r = \frac{9.806 \times 10^{-3}}{\rho ref} R_m$	murgue

NOTES:
1. Unit of r resistance is m^{-4}
2. R_{st} is the U.S. resistance factor at air density 1.2 kg/m³ (0.075 lb/ft³)
3. Unit murgue = 10^{-3} kgf $\cdot$ s² $\cdot$ m^{-8}
4. Value of the reference air density ρref (kg/m³) depends on "standard" density used in particular country

CONCLUSIONS

The S.I. system is much more than just the use of meters, kilograms, newtons, etc. Once fully adopted, the world will speak and understand the same language of measurements, and all international industry and commerce will work with the same standards.

As was shown in this paper, the S.I. system is much more uniform and logical than the customary U.S. unit system. A lot of effort could be saved in measurements, calculations and analyses of results if the S.I. system were adopted in U.S. mine ventilation practice.

Basically there are three prerequisites for the changeover to the S.I. system of units:
- equipment, instruments and tools,
- theory and formulae for computation,
- training in usage of the above.

All of them are of equal importance.

The equipment is available already, at least partially, on the U. S. market. The preliminary investigation gave the following results:

- Length. Tapes in meters and centimeters are off-the-shelf items from Lufkin, K & E and other manufacturers, in varying lengths, most common 3 and 20 m.
- Air Velocity. Vane type anemometers, most common and most reliable in mining environment, are already available with metric scale (Davis, for example). There are a number of sensor velocity meters, but they are not released or too delicate for underground usage. Some of them are very fine instruments like TSI model 1650-2 with direct reading of velocity in three zones, up to 3, 12 or 30 m/s and the temperature -10 to + 90 °C.
- Temperature. Large selection of fluid or digital thermometers are available with reading in °C. As for psychrometers, the most popular is the "sling" psychrometer readily available with Celsius scale. Others are spring driven Assmann (unfortunately, large and heavy), electric driven Kahlsico (still encumbrant) or digital like Atkins Model 44000-C equipped with special probe for wet bulb temperature measurements.
- Air Pressure. Instruments for absolute pressure measurements presently in use are calibrated in inches Hg, mm Hg or in feet of altitude (Taylor, Paulin, Wallace and Tiernan). The ideal barometer for mine pressure survey would be one calibrated in kilopascals (kPa) and pascals (Pa) with resolution to between 1 and 5 Pa. Unfortunately some salesmen of

large instrument manufacturing companies still do not know what pascal is (authors' investigation). The most appropriate units in differential pressure measurements would be centimeters and millimeters of water gauge, because influence of fluid density in this case is small.

- Electrical, Mechanical. Measurement instruments such as volt meters, ammeters, wattmeters are based on S.I. system. As for rotational speed, number of revolutions per second or hertz (Hz) is rather used in S.I. instead of revolutions per minute (RPM) in present tachometers.

The theory of air flow through mine workings should supply practical methods and formulae for everyday engineering practice. These methods depend on degree of precision required and type of the mine environment. In most cases simplified methods still can be used as described in this paper, with the exception of deep and hot mines where thermodynamic analysis of flow would be necessary. Advanced theories of flow through mine workings can be found in recently published "Mine Ventilation Engineering" by C.J. Hall, in works of Bystron, Hall, McPherson, Simode and others. The introduction of more precise but much more elaborate methods is possible nowadays with use of advanced computer programs. New programs should be written in metric system for simplicity and to avoid later conversions, very difficult and time consuming.

Education of S.I. unit system and training in usage of new instruments and methods has already started in some schools and through symposia and short courses. Teaching practical usage of S.I. system in schools of all levels is absolutely essential, but lack of suitable handbooks creates a serious obstacle in introducing S.I. into Mining Engineering curriculum. The important role of AIME, which fully introduced and consequently promotes S.I. unit systems in all its publications and conferences, is in reality pushing students and other readers as well as authors themselves into the S.I. unit system by necessity. The MSHA Mine Ventilation Training Course taught periodically in Denver has started to familiarize students with S.I. units and their field usage.

One of the main obstacles in switching to this new system of units is the "feeling" of engineering values; it is not enough to understand the new units and to be able to manipulate them, but it is absolutely necessary for an engineer to feel whether a particular value is high or low in comparison with some actual values. For example, the ventilation engineer knows that air velocity 1 600 ft/min in mining workings is high and can be unacceptable for the comfort of miners or for dust control, but what is 8 m/s, which in this case is equivalent to 1 600 ft/min ?

It is strongly felt that this "feeling" of engineering values is the most difficult barrier in changing to a new, even much better, unit system.

To facilitate the transition period from U.S. to S.I. units, it is good practice to remember a few approximate conversion factors which were already mentioned in previous chapters. These are the following (exact values given in brackets):

Air velocity:
To find air velocity in meters per second (m/s), divide number of feet per minute (ft/min) per 200 (196.85) error: 1.6%.

Volumetric air flow:
To find flow in cubic meters per second (m/s), divide number of cubic feet per minute (ft^3/min) per 2 000 (2118.9) error: 5.6%.

Air density:
To find air density in kilograms per cubic meter (kg/m^3), multiply number of pounds per cubic foot (lb/ft^3) by 16 (16.0185) error: 0.1%.

Air pressure:
To find pressure in pascals (Pa), multiply number of inches of water gauge (in. WG) by 250 (248.59) error: 0.6%.

Temperature:
To find temperature in degrees Celsius, substract 32 from number of degrees Fahrenheit, multiply result by 5 and divide per 9 (exact value).

Power:
To find power in kilowatts (kW). multiply number of horsepowers (hp) by 3/4 (0.746) error: 0.5%.

Energy:
Energy (work) in kilojoules (kJ) is about equal to the number of British thermal units (Btu). Exact value: 1 kJ = 0.948 Btu. Error: 5.5%.

Of course, the above approximative values should not be used in calculations. Their role is only to assist in the mental conversion of U.S. units to S.I. system. If, for example, somebody is telling you that a particular fan is giving 100 m^3/s air at 2 500 Pa, the approximative values allow you to make a fast, rough estimation as to whether this fan is small, average or large, or low or high pressure, without using a calculator or sheet of paper and pencil.

It takes time, but after a while no such conversion will be needed. As in learning a new language, after some time it is no longer necessary to translate mentally foreign words and sentences into the mother tongue to be able to understand and to speak.

RECOMMENDATIONS

1. New instruments for measurements of air velocity, temperature and pressure should conform to S.I. system of units. In particular, precision barometers to have the direct reading in kilopascals (kPa) and pascals (Pa), with resolution between 1 and 5 pascals.
2. The dimensionless air friction fractor f to be generally adopted instead of any other friction factors presently in use in mine ventilation calculations.
3. Standardization of air friction factor and unit resistance for typical mining conditions should be initiated. Field measurements are needed for various types of airways which are different from those at the time of the McElroy study. Large numbers of existing data from abroad could be used as well.
4. Standard atmospheric conditions to be established. The barometric pressure was defined as 101.325 kPa

(ASTM E380-79, p. 24) and is internationally accepted. As for the standard temperature, advantage of adopting 21.0 °C is that at that temperature the density of dry air is 1.20 kg/m³ (0.075 lb/ft³) which again is generally accepted in many countries as the standard air density. Of course, other values could be adopted like 15 °C which is internationally accepted as standard in meteorology.

5. Teaching mine ventilation in S.I. system shall be introduced in all mining schools.

6. New computer programs for ventilation network simulation will be written in S.I. units; possibility of conversion of old programs should be analyzed.

REFERENCES

Berret, P.O., 1977, "Metrication: More Than Just New Numbers", Engineering Education, April, Vol. 67, No. 7, pp. 672-673.

Bertard, C., 1962, "Valeurs Pratiques des Resistances", Document S. I. M., No. 2.

Bunt, H.J. et al., 1960, "Mine Ventilation", Roberts, A. ed., Cleaver-Hume, London, 363 pp.

Burrows, J., 1975, "K - the Friction Factor of an Airway", Journal of the Mine Ventilation Society of South Africa, Vol. 28, No. 4, p. 60.

Carrier, W.H., 1951, "Principles of Air Conditioning", Carrier Corp., Syracuse. pp. 10-11.

Hall, C.J., 1981, "Mine Ventilation Engineering", Society of Mining Engineers of AIME, New York, 344 pp.

Hartman, H.L., 1961, "Mine Ventilation and Air Conditioning", J. Wiley & Sons, New York, 393 pp.

"Heating, Ventilating, Air Conditioning Guide", 1959, American Society of Heating, Refrigeration, Air Conditioning Engineers, New York, 768 pp.

Kirk-Othmer ed., 1967, "Encyclopedia of Chemical Technology", John Wiley & Sons, New York, second edition, Vol. 13, p. 220.

McElroy, G.E., 1935, "Engineering Factors in the Ventilation of Metal Mines", U.S. Bureau of Mines' Bulletin No. 385, 196 pp.

McPherson, M.J., 1971, "The Metrication and Rationalization of Mine Ventilation Calculations", The Mining Engineer, No. 131, Aug.

McPherson, M.J., 1975, "The Resistance of Mine Airways", Journal of the Mine Ventilation Society of South Africa, Vol. 28, No. 10, Oct., pp. 149-157.

McPherson, M.J., 1979, "Psychometry: The Measurement and Study of Moisture in Air", Mining Department Magazine, Vol. XXXI, University of Nottingham, 11 pp.

"Mining Engineering Moves to an SI Format", 1980, Mining Engineering, Feb., p. 155.

Simode, E., 1976, "Valeurs Pratiques des Résistances", Industrie Minerale, Mine 2-76, Aerage, Doc. S.I.M. N3, Saint-Etienne, Avril; pp. 231-256.

Skochinsky, A. and Komarov, V., 1968, "Mine Ventilation", Mir, Moscow, 580 pp.

"Standard for Metric Practice" ASTM Designation: E 380-79, 1980, American Society for Testing and Materials, 42 pp.

Szwilski, A.B., 1978, "Underground Mine Ventilation: A Review". Western Miner, Dec., pp. 21-26.

Whillier, A., 1971, "Psychometric Charts for all Barometric Pressures", Journal of the Mine Ventilation Society of South Africa, Vol. 24, No. 9, pp. 138-143.

LIST OF AUTHORS

Kenneth L. Ancell
Vice President
Enhanced Energy Resources, Inc.
1160 Dairy Ashford, Suite 400
Houston, TX 77079

James L. Banfield, Jr.
Chief, Ventilation
Mine Safety & Health Administration
4800 Forbes Avenue
Pittsburgh, PA 15213

Mark Bartkowski
Graduate Student
Department of Mining & Minerals Engineering
Virginia Polytechnic Institute & State University
Blacksburg, VA 24061

Sastry S. Bhamidipati
Graduate Student
Department of Mining & Minerals Engineering
Virginia Polytechnic Institute & State University
Blacksburg, VA 24061

William E. Bruce
Chief, Ventilation Division
Denver Safety & Health Technology Center
Mine Safety & Health Administration
Box 25367
Denver, CO 80225

Wade E. Cooper
Mining Engineer
Mine Safety & Health Administration
Box 25367
Denver, CO 80225

E. M. de Kock
Engineering Management Services, Inc.
Box 585
Bedfordview 2008, South Africa

Edward F. Divers
Mining Engineer
Pittsburgh Research Center
U.S. Bureau of Mines
Cochrans Mill Road-Box 18070
Pittsburgh, PA 15236

Richard Dunmore
Head of Firedamp Prediction Group
Mining Research & Development Establishment
National Coal Board
Ashby Road, Stanhope Bretby
Burton-on-Trent, Staffs
United Kingdom DE15 0QD

Michael F. Dunn
Mining Engineer
Engineers International, Inc.
5107 Chase Avenue
Downers Grove, IL 60515

Cheri Eby
Eby Constructors
22005 84th Ave. S.
Kent, WA 98031

William Russell Eby
Managing Partner
Eby Constructors
22005 84th Ave. S.
Kent, WA 98031

Ralph K. Foster
Mining Engineer-Ventilation Division
Mine Safety & Health Administration
Box 25367
Denver, CO 80225

John C. Franklin
Supervisory Physical Scientist
Spokane Research Center
U.S. Bureau of Mines
East 315 Mongtomery Avenue
Spokane, WA 99027

Stephen J. Gigliotti
Mining Engineer-Ventilation Branch
Mine Safety & Health Administration
4800 Forbes Avenue
Pittsburgh, PA 15213

A.D.S. Gillies
Assistant Professor of Mining Engineering
University of Missouri-Rolla
110 Mining Building
Rolla, MO 65041

Robert Grayson
Group Chief Mining Engineer
Vesta/Nemacolin Group
Jones & Laughlin Steel Corporation
420 S. Washington Road
McMurray, PA 15317

Robert A. Haney
Mining Engineer-Ventilation Branch
Mine Safety & Health Administration
4800 Forbes Avenue
Pittsburgh, PA 15213

Klaus-Peter M. Hanke
Project Engineer
Thyssen Schachtbau GMBH
P.O. Box 149
Brookwood, AL 35444

Stephen P. Harrison
Project Engineer
Consolidation Coal Co.
Consol Plaza
Pittsburgh, PA 15214

Howard L. Hartman
Professor and Holder of Endowed Chair of Mining Engineering
The University of Alabama
Box 1468
University, AL 35486

Kenneth S. Heller
Project Engineer
Foster-Miller Associates
350 Second Avenue
Waltham, MA 02154

M. E. Herrera
Vice President, Research & Development
American Energy Products Corporation
8210 Industry Avenue
Pico Rivera, CA 90660

Michael Hood
Department of Materials Science & Mineral Engineering
Hearst Mining Building
University of California, Berkeley
Berkeley, CA 94720

Robert Jankowski
Physical Scientist
Dust Control & Ventilation Group
Pittsburgh Research Center
U.S. Bureau of Mines
Box 18070
Pittsburgh, PA 15231

Guy A. Johnson
Group Supervisor
Twin Cities Mining Research Center
Box 1660
Minneapolis, MN 55417

Francis S. Kendorski
Project Manager
Engineers International, Inc.
5107 Chase Avenue
Downers Grove, IL 60515

David Kennedy
Product Engineer
Buffalo Forge Company
490 Broadway
Buffalo, NY 14240

Earl Kennedy
Supervisor, Noise & Dust
Consolidation Coal Co.
Northern W. Va. Region
P.O. Box 1314
Morgantown, WV 26505

Richard J. Kline
Senior Engineer
Foster-Miller Associates
350 Second Avenue
Waltham, MA 02154

Anthony W. Laurito
Mining Engineer
Engineers International, Inc.
5107 Chase Avenue
Downers Grove, IL 60515

George W. Luxbacher
Research Engineer
Occidental Research Corp.
Island Creek Coal Company
444 S. Main Street
Madisonville, KY 42431

John F. McCoy
Senior Engineer
Foster-Miller Associates
350 Second Avenue
Waltham, MA 02154

Malcolm J. McPherson
Professor of Mineral Engineering
Department of Materials Science & Mineral Engineering
Hearst Mining Building
University of California, Berkeley
Berkeley, CA 94720

R. Mike Mishra
Chief Mining Engineer
Emway Resources, Inc.
90 W. Chestnut Street
Washington, PA 15301

Sandip K. Mukherjee
Senior Mining Engineer
Engineers International, Inc.
5107 Chase Avenue
Downers Grove, IL 60515

Jan M. Mutmansky
Associate Professor of Mining Engineering
The Pennsylvania State University
Mineral Sciences Bldg., Room 104
University Park, PA 16802

Duk-Won Park
Associate Professor of Mineral Engineering
The University of Alabama
Box 1468
University, AL 35486

R. V. Ramani
Professor of Mining Engineering
The Pennsylvania State University
Mineral Sciences Bldg., Room 104
University Park, PA 16802

Robert J. Rawicki
President, Underground Surface Pollution Control, Inc.
2040 Regency Road, Suite F
Lexington, KY 40503

Raphael D. Schilz
Joy Manufacturing Co.
338 S. Broadway
New Philadelphia, OH 44663

C. T. Sheeran
Mining Engineer
Spokane Research Center
U.S. Bureau of Mines
East 315 Montgomery Avenue
Spokane, WA 99207

Madan M. Singh
President, Engineers International, Inc.
5107 Chase Avenue
Downers Grove, IL 60515

L. K. Smith
Engineering Management Services, Ltd.
P.O. Box 585
Bedfordview 2008, South Africa

Tribhawan N. Srivastava
Engineer, Ventilation Section
Climax Molybdenum Co.
Henderson Mine
Box 68
Empire, CO 80438

Ralph B. Stevens
Project Engineer
ESD Corporation
328 Brokow Road
Santa Clara, CA 95052

Charles D. Taylor
Industrial Hygienist
U.S. Bureau of Mines
Box 18070
Pittsburgh, PA 15231

Edward D. Thimons
Supervisory Physicist
U.S. Bureau of Mines
Box 18070
Pittsburgh, PA 15231

Entugrul Topuz
Assistant Professor of Mining Engineering
Department of Mining & Minerals Engineering
Virginia Polytechnic Institute & State University
Blacksburg, VA 24061

Peter M. Turcic
Mining Engineer
Mine Safety & Health Administration
Ballston Tower N. 3
4015 Wilson Blvd.
Arlington, VA 22203

J. van der Walt
Engineering Management Services, Ltd.
P.O. Box 585
Bedfordview 2008, South Africa

Jon C. Volkwein
Physical Scientist
U.S. Bureau of Mines
Box 18070
Pittsburgh, PA 15231

Y. J. Wang
Associate Professor of Mining Engineering
College of Mineral & Energy Resources
West Virginia University
213 White Hall
Morgantown, WV 26506

Jan Wolski
Associate Professor of Mining Engineering
New Mexico Institute of Mining & Technology
Socorro, NM 87801

LIST OF REGISTRANTS

AMAN, Joseph P.
Consolidation Coal Company
P.O. Box 598
Pocahontas, VA 24635

ANCELL, Kenneth L.
Enhanced Energy Resources, Inc.
1160 Dairy Ashford, Suite 400
Houston, TX 77079

ANDERSON, Andy
Western Contractors Service Corp.
1420 East Edinger, Suite 221
Santa Ana, CA 92705

ARGENT, Robert
Jim Walter Resources, Inc.
P.O. Box C-79
Birmingham, AL 35283

ASHCRAFT, Michael
Hartzell Fan Company
910 South Downing St.
Piqua, OH 45356

BANFIELD, James L.
MSHA
4800 Forbes Ave.
Pittsburgh, PA 15213

BARRY, Elizabeth A.
Bonanza Fans, Inc.
1622 Browning Ave.
Irvine, CA 92714

BARRY, Jack
Bonanza Fans, Inc.
1622 Browning Ave.
Irvine, CA 92714

BARTKOWSKI, Mark
Virginia Polytechnic Institute & State University
Blacksburg, VA 24061

BECK, Craig
Pickands Mather & Company
P.O. Box 3041
Pikeville, KY 41501

BEECH, John J.
Mineral Engineering Dept.
P.O. Box 1468
University, AL 35486

BELDEN, Ron
Cotter Corp.
9305 West Almeda
Lakewood, CA 80226

BENNETT, Jerry
Mining & Marketing Associates
Route 4, Box A176
Rolla, MO 65401

BENNETT, Mike
Colorado Westmoreland, Inc.
P.O. Box E
Paonia, CO 81428

BIBB, Thomas C.
Westmoreland Coal Company
117 E. 1st St.
Big Stone Gap, VA 24219

BLACK, Robert
USAF
Norton AFB, CA 92409

BLACKWELL, Johnny C.
R. J. Stern Company
4800 1st Ave. South
Birmingham, AL 35222

BOULTON, James
Carbon County Coal Company
P.O. Box 330
Hanna, WY 82327

BRADFORD, Phil
North River Energy Company
234 Vestavia Center
Northport, AL 35476

BRAFFORD, David R.
Rawl Sales & Processing Company
General Delivery
Lobata, WV 25677

BROOKS, Robert
Rockwell International
P.O. Box 800
Richland, WA 99352

BRUCE, William E.
MSHA, Denver Safety & Health Technology Center
P.O. Box 25367
Denver, CO 80225

BUCKS, Scot
Mineral Engineering Dept.
P.O. Box 1468
University, AL 35486

BURTON, Brent
Monterey Coal Company
P.O. Box 496
Carlinville, IL 62626

CAVANAUGH, John E.
Exxon Corp.
P.O. Box 14342
Aurora, CO 80014

CHAPMAN, Donald
TLT-Babcock, Inc.
3340 West Market St.
Akron, OH 44313

CHRISTIANS, William
Jim Walter Resources, Inc.
P.O. Box C-79
Birmingham, AL 35283

CHUKOBASIH, Collins
Mineral Engineering Dept.
P.O. Box 1468
University, AL 35486

CONYERS, Ed
Continuing Engineering Education
P.O. Box 2967
University, AL 35486

COOPER, Wade
MSHA
P.O. Box 25367
Denver, CO 80225

CURRY, Donald
Chicago Blower Corp.
1675 Glen Cllyn Road
Glendale Heights, IL 60137

de KOCK, E. M.
Engineering Management Services, Inc.
P.O. Box 585
Bedfordview, 2008 South Africa

DICK, Alan
James Howden & Co., LTD.
195 Scotland St.
Glasgow, Scotland G5 8PJ

DIVERS, Edward F.
USBM, Pittsburgh Research Center
Cochrans Mill Road, P.O. Box 18070
Pittsburgh, PA 15236

DOTTERBOCK, Eric
Buffalo Forge Company
490 Broadway
Buffalo, NY 14240

DUNMORE, Richard
Firedamp Prediction Group, National Coal Board
Ashby Road, Stanhope Bretby, Burton-on-Trent
Staffs DE15 0QD United Kingdom

DUNN, Michael F.
Engineers International, Inc.
5107 Chase Ave.
Downers Grove, IL 60515

EARLY, Jerry
MSHA
228 West Valley Ave.
Birmingham, AL 35209

EBY, Cheri
Eby Constructors
25047 214th Pl. S.E.
Maple Valley, WA 98038

EBY, Russell
Eby Constructors
22005 84th Ave. South
Kent, WA 98031

ELLISON, James
Peabody ABC Corp.
P.O. Box 77
Warsaw, IN 46580

ENGLISH, Lloyd
Conoco, Inc.
Route 1, Box 119
Morgantown, WY 26505

ERNST, Charles R.
North American Coal Corp.
Powhatan Point, OH 43942

EVANS, George J.
Peabody Coal Company
P.O. Box 328
Morganfield, KY 42437

FLEMING, Robin
James Howden America
P.O. Box 146
East Hartford, CT 06108

FOSTER, Ralph K.
MSHA
P.O. Box 25367
Denver, CO 80225

FRANKLIN, John C.
USBM
East 315 Montgomery Ave.
Spokane, WA 99207

GAMBRELL, Samuel
College of Engineering
P.O. Box 1968
University, AL 35486

GANEY, Ronnie Joe
Mineral Engineering Dept.
P.O. Box 1468
University, AL 35486

GILLIES, A. D. Stewart
University of Missouri
Dept. of Mining Engineering
Rolla, MO 65041

GRAVES, William
Climax Molybdenum Company
Climax, CO 80429

GRAYSON, Robert L.
Vesta/Nemacolin Group
420 South Washington Road
McMurray, PA 15317

GREEN, Peter
Coal Age Magazine
1221 Avenue of the Americas
New York, NY 10020

GREY, Tony
North River Energy Company
234 Vestavia Center
Northport, AL 35476

GRUBBS, David M.
Mineral Engineering Dept.
P.O. Box 1468
University, AL 35486

GRYGIEL, Edward W.
Jim Walter Resources, Inc.
P.O. Box 133
Brookwood, AL 35444

HAGER, Charles J., Jr.
Consultant
1807 Seneca Road
Birmingham, AL 35216

HAGOOD, David
Jim Walter Resources, Inc.
P.O. Box C-79
Birmingham, AL 35283

HANEY, Robert A.
MSHA
4800 Forbes Ave.
Pittsburgh, PA 15213

HANKE, Peter
TMCI Construction, Inc.
P.O. Box 149
Brookwood, AL 35444

HANNA, H. Shafick
Mineral Resources Institute
P.O. Drawer AY
University, AL 35486

HARRIS, Stan M.
Joy Manufacturing Company
338 South Broadway
New Philadelphia, OH 44633

HARRISON, Stephen
Consolidaton Coal Company
Consol Plaza
Pittsburgh, PA 15241

HART, Don
Cementation Company
P.O. Box 9
Brampton, Ontario, Canada L6V 2K7

HARTMAN, Howard L.
Mineral Engineering Dept.
P.O. Box 1468
University, AL 35486

HAYS, Charles
Student Chapter AIME
P.O. Box 1468
University, AL 35486

HEALD, David
4825 South Jackson #1
Tulsa, OK 74107

HEINICKE, Paul M.
Sverdrup & Parcel Associates
801 North 11th St.
St. Louis, MO 63101

HELLER, Kenneth
Foster-Miller, Inc.
350 2nd Ave.
Waltham, MA 02254

HERRERA, Manual E.
American Energy Products Corp.
8210 Industry Ave.
Pico Rivera, CA 90660

HOLLINGSHEAD, John
Dept. of Mineral Engineering
P.O. Box 1468
University, AL 35486

HOLLUB, Ray
Continuing Engineering Education
P.O. Box 2967
University, AL 35486

HONEYCUTT, Don
North River Energy Company
234 Vestavia Center
Northport, AL 35476

HORTON, Keith D.
Penn VA Resources Corp.
P.O. Box 43
Duffield, VA 24244

HUBBARD, Ricky
Jim Walter Resources, Inc.
P.O. Box 133
Brookwood, AL 35444

HUFFER, William
Peabody ABC Corp.
P.O. Box 77
Warsaw, IN 46580

INGLE, Steven R.
Mineral Engineering Dept.
P.O. Box 1468
University, AL 35486

JACKSON, Tennis
6 Bedford Plaza
Tuscaloosa, AL 35406

JANKOWSKI, Robert A.
USBM
P.O. Box 18070
Pittsburgh, PA 15231

JOHNSON, Bruce K.
Magma Copper Company
P.O. Box 37
Superior, AZ 85273

JOHNSTON, David L.
Joy Manufacturing Company
338 South Broadway,
New Philadelphia, OH 44663

JOINER, Thomas J.
Tom Joiner and Associates
1549 37th East
Tuscaloosa, AL 35405

JONES, James E.
Jim Walter Resources, Inc.
P.O. Box 133
Brookwood, AL 35444

JONES, Randy
Mineral Engineering Dept.
P.O. Box 1468
University, AL 35486

JONES, Thomas
Consolidation Coal Company
101 Plaza E. Blvd.
Evansville, IN 47715

KALASKY, John
Island Creek Coal Company
2366 Harrodsburg Road
Lexington, KY 40575

KALUSKI, Chris
Underground Surface Pollution Control
2040 Regency Road, Suite F
Lexington, KY 40503

KAZAR, Robert
TLT-Babcock, Inc.
3340 West Market St.
Akron, OH 44313

KEMP, Alan B.
Stewart Manufacturing Company
2252 Freeport Road
Pittsburgh, PA 15238

KENDRICK, Jeffrey K.
Mineral Engineering Dept.
P.O. Box 1468
University, AL 35486

KENNEDY, David
Buffalo Forge Company
490 Broadway
Buffalo, NY 14240

KENNEDY, Earl
Consolidation Coal Company, N.W. Virginia Region
P.O. Box 1314
Morgantown, WV 26505

KERNS, Edward A.
Giangarlo Scientific Co., Inc.
2500 Baldwick Road
Pittsburgh, PA 15205

KING, Robert
Colorado School of Mines
Golden, CO 80401

KING, Roger
USBM
P.O. Box 18070
Pittsburgh, PA 15236

KLINE, Richard
Foster-Miller, Inc.
350 2nd Ave.
Waltham, MA 02254

KOPAY, Edward M.
Joy Manufacturing Company
3382 South Broadway
New Philadelphia, OH 44663

KOZEL, Fred
Jim Walter Resources, Inc.
P.O. Box C-79
Birmingham, AL 35283

KOZEL, Merialice
Jim Walter Resources, Inc.
P.O. Box 133
Brookwood, AL 35444

LAGUS, Peter
S-Cubed
P.O. Box 1620
La Jolla, CA 92038

LaMOREAUX, Jim
P.E. LaMoreaux and Associates
2610 University Blvd.
Tuscaloosa, AL 35401

LANGSTAFF, Alvin
Cliffs Engineering
P.O. Box 1211
Rifle, CO 81650

LAWSON, John
North River Energy Company
234 Vestavia Center
Northport, AL 35476

LIU, Ching-Hua
Bureau of Mines
2 Chenkiang St.
Taipei, Taiwan, Republic of China

LINDSTROM, C. J.
Pickands Mather & Company
1100 Superior Ave.
Cleveland, OH 44114

LIONEL, Groome
208 Lorraine Dr.
Baire D'urfe
Quebec, Canada H9X 2P7

LUMSDEN, Alastair M.
Golder Associates
1950 Northup Way
Bellevue, WA 98004

LUXBACHER, George W.
Occidental Research Corp., Island Creek Coal Co.
444 South Main St.
Madisonville, KY 42431

McARTHUR, John F., Jr.
Bethlehem Mines Corp.
P.O. Box 143
Eighty Four, PA 15330

McCRORIE, Jim
Jim Walter Resources, Inc.
P.O. Box 133
Brookwood, AL 35444

McGEHEE, Wayne
Mineral Engineering Dept.
P.O. Box 1468
University, AL 35486

McKIE, Ross
Brown Boveri Howden
1519 Birchmount Road
Scarborough, Ontario, Canada M1P 2G6

McNIDER, Thomas
Jim Walter Resources, Inc.
P.O. Box C-79
Birmingham, AL 35283

McDANIEL, Randall E.
U.S. Steel Mining Co., Inc.
Route 2, Box 753
Calera, AL 35040

McPHERSON, Malcolm J.
Hearst Mining Building
University of California
Berkeley, CA 94720

MANCINI, Ernest
Alabama Geological Survey
P.O. Drawer O
University, AL 35486

MARSH, Steve
Cotter Corp.
9305 West Almeda
Lakewood, CO 80226

MATEER, Richard S.
Dept. of Mining Engineering
University of Kentucky, 212 Anderson Hall
Lexington, KY 40506-0046

MATHEWS, Edward M.
Jeffrey Mining Company
274 East 1st Ave.
Columbus, OH 43229

MEADOWS, Bill
MSHA
228 West Valley Ave.
Birmingham, AL 35209

MELCHER, Richard
Old Ben Coal Company
500 N. DuQuoin St.
Benton, IL 62812

METCALF, Charles
Westmoreland Coal Company
P.O. Drawer A
Big Stone Gap, VA 24219

MICHAELS, Tom
Buffalo Forge Company
405 Golf Drive
Birmingham, AL 35226

MILLER, Paul
Amvest Corp.
P.O. Box 1468
Wise, VA 24293

MORROW, Danny
Perry Supply, Inc.
P.O. Box 1237
Birmingham, AL 35201

MOUSSET-JONES, Pierre
Department of Mining Engineering
University of Nevada
Reno, NV 89557

MUKHERJEE, Sandip K.
Engineers International, Inc.
5107 Chase Ave.
Downers Grove, IL 60615

MUTMANSKY, Jan M.
The Pennsylvania State University
210 Mineral Sciences Bldg.
University Park, PA 16802

NARAMORE, Melanie
Mineral Engineering Dept.
P.O. Box 1468
University, AL 35486

NEUDECKER, Joseph W., Jr.
Los Alamos National Laboratory, G-777
P.O. Box 1663
Los Alamos, NM 87545

NOTLEY, Keith
Queen's University
Mining Engineering Dept., Goodwin Hall
Kingston, Ontario, Canada K7

NUGENT, J. W.
TOSCO
11100 East Bethany Drive
Aurora, CO 80014-1439

O'NEAL, J. Y.
U.S. Steel Mining Company
P.O. Box 599
Fairfield, AL 35064

PARK, Duk-Won
Mineral Engineering Dept.
P.O. Box 1468
University, AL 35486

PAROBECK, Paul S.
MSHA
105 Berkeley Road
Pittsburgh, PA 15237

PARROTT, Michael R.
Mineral Engineering Dept.
P.O. Box 1468
University, AL 35486

PATTS, Larry D.
Consolidation Coal Company
200 Hidden Valley Road
McMurray, PA 15317

PETERSON, Randall D.
Plateau Mining Company
P.O. Drawer PMC
Price, UT 84501

PICHERT, Tony
H.A. Simmons
1015 Fourth St. SW, #203
Calgary, Alberta, Canada R2R 1J4

POMROY, William H.
USBM
P.O. Box 1660
Minneapolis, MN 55417

PRESLEY, Charles
Tennessee Consolidated Coal Company
P.O. Box 878
Jasper, TN 38347

PRITCHARD, Christopher
Texasgulf
P.O. Box 100
Granger, WY 82934

PROCARIONE, James
University of Utah
320 WBB
Salt Lake City, UT 84112

PUGH, Denny
Mineral Engineering Dept.
P.O. Box 1468
University, AL 35486

QUE, John
Bechtel Company
45 Fremont St.
San Francisco, CA 94119

RAMPACEK, Carl
Director, Mineral Resources Institute
P.O. Drawer AY
University, AL 35486

RAWICKI, Robert J.
Underground Surface Pollution Control, Inc.
2040 Regency Road, Suite F
Lexington, KY 40503

RAY, R.
Jim Walter Resources, Inc.
P.O. Box C-79
Birmingham, AL 35283

REEL, Nick
Tennessee Colsolidated Coal Company
P.O. Box 878
Jasper, TN 37347

RENDELL, Bruce
Brunswick Mining
P.O. Box 3000
Bathurst, NB, Canada E2A 3Z8

REPCHAK, George J.
Outokumpu Engineering, Inc.
10530 Holland St.
Broomfield, CO 80020

ROZE, V.
Alpha Nuclear Division
6380 B Viscount Road
Mississauga, Ontario, Canada L4V 1H3

RYAN, David B.
Jim Walter Resources, Inc.
P.O. Box 133
Brookwood, AL 35444

RYBICKI, Ted
Whiteriver Oil Shale Co.
P.O. Box 1211
Rifle, CO 81650

SADIK, Wayne
Exxon Minerals Company
P.O. Box 4508
Houston, TX 77210

SANDSTEDT, Jon
Donaldson Company, Inc.
P.O. Box 1299
Minneapolis, MN 55406

SANFORD, Richard
Mineral Engineering Dept.
P.O. Box 1468
University, AL 35486

SANFORD, Wendell S.
Mineral Engineering Dept.
P.O. Box 1468
University, AL 35486

SCHLICK, D. P.
Jim Walter Resources, Inc.
P.O. Box C-79
Birmingham, AL 35283

SHEEHAN, Edward
Perry Supply, Inc.
P.O. Box 1237
Birmingham, AL 35201

SHELLEDY, Ross E.
Peabody Coal Company
50 Jerome Lane
Fairview Heights, IL 62208

SHOTTS, Reynold Q.
Consultant
8 Forest Lake Drive
Tuscaloosa, AL 35401

SHUPE, William D.
Penn VA Resources Corp.
P.O. Box 43
Duffield, VA 24244

SILVER, John
Hartzell Fan Company
910 South Downing Street
Pizua, OH 45386

SIMPSON, Thomas A.
Mineral Engineering Dept.
P.O. Box 1468
University, AL 35486

SINGH, Madan M.
Engineers International, Inc.
5107 Chase Ave.
Downers Grove, IL 60515

SKAGGS, Gary
Y & O Coal
P.O. Box 1000
St. Clairsville, OH 43950

SMITH, Colin
Engineering Management Services
7061 South University Blvd. #304
Littleton, CO 80122

SMITH, George E.
Eastover Mining
P.O. Box E
Brookside, KY 40801

SMITH, Robert Z.
Joy Manufacturing Company
11707 East 51st Ave.
Denver, CO 80239

SOMERS, John
MSHA
228 West Valley Ave.
Birmingham, AL 35209

SPENDRUP, Quentin
Spendrup Fan Company
746 Duray Ave.
Grand Junction, CO 81501

SPENDRUP, Trent
Spendrup Fan Company
746 Duray Ave.
Grand Junction, CO 81501

SRIVASTAVA, Tribhawan
Climax Molybdenum Company, Henderson Mine
P.O. Box 68
Empire, CO 80438

STANCZYK, Martin
USBM
P.O. Box L
University, AL 35486

STEVENS, Ralph B.
ESD Corporation
328 Brokow Road
Santa Clara, CA 95052

STEVENSON, J. W.
Jim Walter Resources, Inc.
P.O. Box C-79
Birmingham, AL 35283

STEVENSON, Robert, Jr.
Peabody ABC Corp.
P.O. Box 77
Warsaw, IN 46580

TALMAN, Woods
Consultant
2624 Thorntree Drive
Pittsburgh, PA 15241

TEASLEY, Frank
Jim Walter Resources, Inc.
P.O. Box 133
Brookwood, AL 35444

THOMAS, Joab
President, The University of Alabama
P.O. Box B
University, AL 35486

TIEN, Jerry
Peabody Coal Company
P.O. Box 1981
Henderson, KY 42420

TINKLEPAUGH, John
North River Energy Company
234 Vestavia Center
Northport, AL 35476

TOLLEY, William J.
Wen-Don Chemical Company
P.O. Box 13905
Roanoke, VA 24038

TOPUZ, Entugrul
Virginia Polytechnic Institute & State University
Blacksburg, VA 24061

TRAVIS, Dan
U.S. Steel Mining Company
P.O. Box 599
Fairfield, AL 35064

TSAY, Frank
Mineral Engineering Department
P.O. Box 1468
University, AL 35486

ULM, R. F.
U.S. Steel Mining Company
P.O. Box 599
Fairfield, AL 35064

VAN DER WALT, Cornelius
Prieska Copper Mines
Private Bag 1
Copperton, 8944 South Africa

VAN DER WALT, J.
Engineering Management Services
P.O. Box 585
Bedfordview, 2008 South Africa

VIOLA, Gary T.
Zeni Drilling Company
324 8th St.
Morgantown, WV 26505

VOLKWEIN, Jon C.
USBM
P.O. Box 18070
Pittsburgh, PA 15236

WALA, Andrezej
University of Kentucky
214 Anderson Hall
Lexington, KY 40506

WANG, George
Mineral Engineering Dept.
P.O. Box 1468
University, AL 35486

WANG, Y. J.
West Virginia University
230 White Hall
Morgantown, WV 26506

WARNOCK, Robert B.
Buffalo Forge Company
490 Broadway
Buffalo, NY 14240

WEAVER, Al
Colorado Westmoreland, Inc.
P.O. Box E
Paonia, CO 81248

WILLIAMS, William
Y & O Coal
P.O. Box 1000
St. Clairsville, OH 43950

WILSON, Jeffrey A.
Rawl Sales & Processing Company
General Delivery
Lobata, WV 25677

WOLSKI, Jan
Dept. of Mining & Geological Engineering
New Mexico Tech
Socorro, NM 87801

WRAY, Mike
North River Energy Company
234 Vestavia Center
Northport, AL 35476

YORK, Don
Jeffrey-Dresser
274 East First Ave.
Columbus, OH 43216

YOUNG, Kenneth D.
Sandia National Labs
P.O. Box 5800
Albuquerque, NM 87185